Reviews and Perspectives on Smart and Sustainable Metropolitan and Regional Cities

Reviews and Perspectives on Smart and Sustainable Metropolitan and Regional Cities

Editor

Tan Yigitcanlar

MDPI • Basel • Beijing • Wuhan • Barcelona • Belgrade • Manchester • Tokyo • Cluj • Tianjin

Editor
Tan Yigitcanlar
Queensland University of Technology
Australia

Editorial Office
MDPI
St. Alban-Anlage 66
4052 Basel, Switzerland

For citation purposes, cite each article independently as indicated on the article page online and as indicated below:

LastName, A.A.; LastName, B.B.; LastName, C.C. Article Title. *Journal Name* **Year**, *Volume Number*, Page Range.

ISBN 978-3-0365-2756-7 (Hbk)
ISBN 978-3-0365-2757-4 (PDF)

Contents

About the Editor

Tan Yigitcanlar is an eminent Australian researcher with international recognition and impact in the field of urban studies and planning. He is a professor of urban studies and planning at the School of Architecture and Built Environment, Queensland University of Technology, Brisbane, Australia. Along with this post he carries out an honorary professor role at the School of Technology, Federal University of Santa Catarina, Florianopolis, Brazil, and the founding director position of the Australia–Brazil Smart City Research and Practice Network. He has been responsible for research, teaching, training, and capacity-building programs in the fields of urban studies and planning in esteemed Australian, Brazilian, Finnish, Japanese, and Turkish universities. He is an eminent Australian scholar with international recognition, reputation and impact on policy, practice, and society. His research aims to address contemporary urban planning and development challenges that are economic-, societal-, spatial-, governance-, or technology-related in nature. The main foci of his research interests, within the broad field of urban studies and planning, are clustered around the following three interdisciplinary themes: 'Smart Technologies, Communities, Cities and Urbanism', 'Knowledge-Based Development of Cities and Innovation Districts', and 'Sustainable and Resilient Cities, Communities and Urban Ecosystems'. His research outputs have been cited over 13,000 times, resulting in an h-index of 67 (Google Scholar). According to the science-wide author databases of standardised citation indicators, he is ranked #1 in Australia and #7 worldwide in 2020. For this achievement he was recognised as one of the 'Australian Research Superstars' in social sciences in The Australian's 2020 Research Special Report.

 *sustainability*

Editorial

Greening the Artificial Intelligence for a Sustainable Planet: An Editorial Commentary

Tan Yigitcanlar

School of Architecture and Built Environment, Queensland University of Technology, 2 George Street, Brisbane 4000, Australia; tan.yigitcanlar@qut.edu.au; Tel.: +61-7-3138-2418

Citation: Yigitcanlar, T. Greening the Artificial Intelligence for a Sustainable Planet: An Editorial Commentary. *Sustainability* **2021**, *13*, 13508. https://doi.org/10.3390/su132413508

Received: 23 November 2021
Accepted: 3 December 2021
Published: 7 December 2021

Artificial intelligence (AI) is one of the most popular and promising technologies of our time. While there is a clearer understanding on the role of AI in boosting the efficiencies at private companies, government agencies and urban management, there is ambiguity on the specific contributions of AI to environmental sustainability. In this editorial commentary: (a) the important role that AI could play in addressing global environmental sustainability challenges is discussed; (b) the need for a consolidated AI approach to support the efforts in addressing global environmental sustainability problems—e.g., meeting the global sustainable development goals, developing smart and sustainable cities and regions, and tackling the climate and biodiversity crises—is identified; (c) the emerging Green AI concept that offers a consolidated AI perspective that is an essential step towards global environmental sustainability is introduced; (d) the adoption of the Green AI approach by industry, government, and not-for-profit organizations for addressing environmental sustainability challenges of the planet and for improving the quality of lives of our societies in cities is advocated. The editorial commentary also introduces the contributions to the Special Issue on reviews and perspectives on smart and sustainable metropolitan and regional cities.

Our planet reached its maximum human carrying capacity with 3.5 billion people in 1970 [1]; where human carrying capacity is defined as "the maximum rates of resource harvesting and waste generation, the maximum load, that can be sustained indefinitely without progressively impairing the productivity and functional integrity of relevant ecosystems wherever the latter may be located" [2] (p. 203). Regrettably, over the last five decades, we continued the practice of rapid population and also urban and industrial growth beyond the limits of our planet's carrying capacity [3].

This carrying capacity overshoot, also coupled with the excessive human activity that is causing environmental degradation, resulted in a degraded carrying capacity state. These overshoot and degraded carrying capacities have been changing the climate and causing biodiversity losses at scale [4]. Biodiversity loss is inextricably linked to anthropogenic climate change that jeopardizes global ecosystems services, including agricultural, forest, marine, reef, costal, polar, mountain, island, and microbial ecosystems [5].

The unavoidable risk of facing catastrophic consequences, such as biodiversity collapse, has triggered some global initiatives with an aim of to slow down or reverse the existing growth processes. As the successor of the 2020 Millennium Development Goals, in 2015, United Nations (UN) adopted 2030 Sustainable Development Goals (SDGs), as the blueprint to achieve a more sustainable future for all [6]. UN's SDGs contain 17 goals that consist of 169 targets [7]. The same year, UN delivered a universal and legally binding climate change deal—i.e., The Paris Agreement on climate change [8]. In 2018, the Intergovernmental Panel on Climate Change (IPCC)—the world's most authoritative body on climate science—released its special report (i.e., Global Warming of 1.5 °C) revealing the impacts of 1.5 °C global warming on natural and human systems [9].

Additionally, 2019 European Green Deal provided action plans to boost the efficient use of resources by moving to a clean and circular economy, restore biodiversity, and cut pollution—2020 US Green Deal and 2021 The Biden Climate Plan also have similar goals [10]. The latest IPCC report, released in 2021, was a code red declaration for humanity, and a warning for "world leaders to drastically scale up their plans to curb carbon dioxide emissions if humanity is to avoid the worst consequences of a warming world" [11] (p. 8).

In an era that is desperately seeking environmental sustainability, besides these global UN-led initiatives, the last few years have also witnessed global level grassroots movements for climate and environmental action. The 'Fridays for Future' global climate strike movement started in 2018, when Greta Thunberg initiated school strikes for climate [12]. In the same year, the Extinction Rebellion global environmental movement began with about 100 academics signing a call to climate action, and continued with civil disobedient events [13]. Following these, some nations (Scotland being the first, 23 national governments have declared a climate emergency) and local governments (Darebin City Council, VIC, Australia being the first, 1990 local councils have declared a climate emergency) declared climate emergencies [14]. Most recently, during the 2021 United Nations climate change conference (COP26), 200 countries have been asked for their plans to cut emissions by 2030 [15].

Whilst the promising efforts at both fronts—i.e., UN and grassroot level initiatives—are putting increasing pressure on political leaders and fossil fuel companies, and rising awareness among the societies, the immediate gains have been so far limited [16]. Nonetheless, there are also other efforts to contribute to addressing global environmental challenges. Benefiting from digital data, technology and innovation is one of them [17–19].

Among the innovative technologies, artificial intelligence (AI) is considered as the most promising and prominent one [20]. AI is being applied in a wide spectrum of areas to address their central challenges in both public and private sectors [21,22]. These areas range from human resources to customer relationship, from law to security, from decision automation to education, from agriculture to mining, transportation to gaming, and many more [23–30].

AI is also being utilized in the environmental monitoring and protection and natural resource management areas [31,32]. Just to give some examples, AI is becoming an integral part of autonomous marine environmental monitoring [33], wildlife monitoring [34], environmental surveillance and flood protection [35], water quality modeling [36], biodiversity assessment [37], detecting natural disasters, damage, and incidents in the wild [38,39], and climate change modeling and preparedness [40].

This viewpoint advocates the need for appropriate technocentric instruments, in parallel to the ongoing policy and awareness efforts, for combatting climate change and its extreme weather events, and delivering SDGs. Against this backdrop, the remainder of this viewpoint focuses on elaborating the important role the Green AI approach could play in tackling global environmental sustainability challenges.

In recent years, the exponentially increasing technical capabilities and rapidly growing application areas have turned AI into one of the most popular technologies of our time [41]. AI, in a broad sense, is seen as "computational agents that act intelligently and perceive their environments in order to take actions that maximize chances of success" [42] (p. 148). AI, more specifically, is an umbrella term used for rules-based systems encompassing, but not limited to, machine and deep learning systems, neural networks, natural language processing, predictive analytics, computer vision, and robotics [43].

At present, there is a relatively clear understanding on AI's role in boosting the efficiencies at private companies and government agencies [44], as it has been experimented or adopted in pretty much in all industry sectors and in a wide spectrum of government services at varying levels [45]. When it gets to the role of AI in addressing global environmental sustainability problems (e.g., degradation of the natural environment and the climate crisis), there is much ambiguity. One of the leading reasons for this ambiguity has been the lack of or limited public (due to political short sightedness) and private (due to no or low profitably) sector interest in developing, adopting, and deploying AI for environmental sustainability—besides some small-scale demonstration projects and government initiatives [46]. However, in recent years we observe an increasing willingness and interest in both public and private sectors. The main reasons for this involve adoption of new national policies, such as Green New Deal, and associated business growth potentials in AI for environmental sustainability [47].

In other words, while AI is at our service for improving efficiencies in businesses and government services, AI utilization for environmental sustainability seems to be, at least for now, at a degree of neglect [48]. Nonetheless, there exist, and some on the horizon, promising research and practical solutions of AI for environmental sustainability. Just to give some examples, AI is currently being utilized for assessing ecosystem services, detection and conservation of species, modeling climate change, natural disaster forecasting, waste and wastewater treatment, and so on so forth [49–51]. While the benefits of AI for environmental sustainability are evident from small-scale project initiatives and hypothetical or qualitative tests [52], these solutions have not been applied at scale in the real-world [53].

On the abovementioned point of the benefits of AI, for instance, the study by [54] (p. 2) scrutinized the role of AI in achieving SDGs and disclosed that "AI may act as an enabler on 134 targets (79%) across all SDGs, generally through a technological improvement, which may allow to overcome certain present limitations. However, 59 targets (35%, also across all SDGs) may experience a negative impact from the development of AI". Likewise, a study by [55] (p. 2), on whether AI enables environmental sustainability, revealed that "when SDGs related to society, economy, and environment were analyzed, it was observed that the environment category has the highest potential, with 93% of the targets being positively affected, whereas society has the largest negative effect with 38% of the targets exhibiting a negative interaction with AI".

The growing literature in the field also supports the abovementioned findings. For instance, a study by [56] (p. 283) disclosed that "AI can represent the vehicle to meet the SDGs allowing for the identification of the cultural change required by enterprises to achieve sustainability goals. Thus, business companies, academic research practitioners, and state policy should focus on the further development of the use of AI for sustainable development". Moreover, the authors of [57] underlined that the integration of AI into the SDGs initially happened through experimentation, and in most recent years through sustainable management and leadership programs. These programs concentrated on: (a) AI and the water crisis; (b) AI and the agriculture; (c) AI and sanitation and health.

Moreover, in [58] the authors warned us about the risks AI poses to the achievement of SDGs, with particular vulnerability for developing countries. Additionally, the authors of [59] revealed insights into public perception of linkages, synergies, and trade-offs between AI and SDGs. Likewise, according to [60] (p. 98), in support of SDG 14, "seas and oceans can be explored using submarine AI robots, such as Stanford's OceanOne. Marine resources can be monitored through AI-driven smart stationary and mobile sensors. Illegal fishing activities as well as marine life migration can be tracked through pattern recognition", and as for SDG 15, "AI can monitor different aspects of life on land, such as species health, land use changes, food security and nutrition, noise levels, weather-related stresses, and disease vectors and outbreaks. Predictive analytics can generate insights about population and desertification trends and the spread of epidemics".

In sum, numerous studies emphasized the critical importance of achieving foundational SDGs in reducing global risks [61], and many others discussed the role AI is playing and also could play in assisting the delivery of these goals by unlocking enormous sensing, data collection, analysis, prediction, and intervention capabilities [62]. Nevertheless, "given the fact that AI's internal decision-making process is non-transparent, some experts consider it to be a significant existential risk to humanity, while other scholars argue for maximizing the technology's exploitation" [60] (p. 95). This brings the importance of AI to be utilized for SDGs based on the principles of the technology being responsible, ethical, trustworthy, explainable, and also sustainable [63]. As elaborated by [64] "laissez-faire AI is a dangerous political choice".

The next section introduces the Green AI approach as an essential instrument for achieving global environmental sustainability.

Addressing the colossal environmental challenges call for a sustainable approach [65]. This approach also requires a new AI conceptualization and practice—i.e., Green AI—that involves a green-based technological perspective in the AI industry [66,67]. The green perspective includes "switching to an environmentally sustainable AI infrastructure, employing green sensing, watching for AI rebound effects, mandating AI transparency, accounting for the entire AI ecosystem, making non-energy policy a standard practice, integrating AI and climate policy, curbing the use of AI to extract fossil fuels, and addressing AI's impact on climate refugees" [68] (p. 1). In other words, as stated by [69] (p. 3) "green AI accommodates green sensing and moves away from short-term efficiency solutions to focus on a long-term ethical, responsible and sustainable practice that will help build sustainable urban futures for all".

This is to say, the Green AI approach moves away from short-term efficiency solutions to focus on a long-term ethical, responsible, and sustainable AI practice that will help build environmentally sustainable futures for all [70,71]. Making AI green, hence, requires adopting bias free, inclusive, trustworthy, explainable, ethical, and responsible principles to the AI technology that aims to alleviate the developmental challenges of the planet in a sustainable way [72,73]. This green approach, using AI to solve environmental sustainability challenges and using AI in a more sustainable way, can also serve as an enabler of smart and sustainable urban transformation or smart city development/transformation—as urbanization accounts for the majority of human activities that generate negative environmental externalities [74].

Growing technical capabilities and increasing application areas have turned AI into one of the most popular technologies of our time. Particularly, AI, when wisely harnessed for sustainability-inducing projects and applications, has the capacity to support SDGs [75]. Fortunately, at a time that we are desperately seeking environmental sustainability to not risk our existence on the planet [76], Green AI comes as a promising concept to adopt. In developing successful Green AI practices, the following AI for environmental sustainability principles are of significant importance [77] (pp. 8–9):

"(a) AI for environmental sustainability should be viewed from a multilevel view, as a multilevel view will help to build better models by capturing the complexity inherent in the real-world; (b) AI for environmental sustainability should be viewed from a system dynamics perspective, as a system dynamics perspective will capture interactions and feedback loops among the technology, users, and other stakeholders; (c) AI for environmental sustainability should be approached from a design thinking approach, as a design thinking approach will help to minimize potential unintended consequences and improve the effectiveness of AI solutions; (d) AI for environmental sustainability should incorporate environmental psychology and sociology perspectives, as understanding the psychological and sociological underpinnings of human response is necessary for effective long-term solutions, and; (e) AI for environmental sustainability should examine the economic value of AI for sustainability to develop our understanding of how AI differs from conventional information systems."

The successful Green AI practice also depends on effective government regulation and administration. Unfortunately, up until now there has been limited efforts to regulate AI [78,79]. The most notable effort is the European Commission's legal framework on AI that addresses the risks of AI [80]. However, there is no clear reference in the European Commission's regulation to the AI for environmental sustainability. Still, substantial work is needed to answer the following questions: what needs be done to develop Green AI at the policy level, how Green AI development can be supported by the administrations, what kind of mistakes should be avoided on the road to Green AI, how we can learn from best or good Green AI practices, and what types of legal liabilities should be considered. These questions form the basis of a new research agenda for scholars to investigate, and also practitioners and government officials get involved in the debates and discussions in establishing a sound policy—in support of Green AI as an essential instrument for achieving global environmental sustainability.

In their paper [81] (p. 13), on the past, present, and future of AI, the authors stated that "nobody knows whether AI will allow us to enhance our own intelligence, as Raymond Kurzweil from Google thinks, or whether it will eventually lead us into World War III, a concern raised by Elon Musk. However, everyone agrees that it will result in unique ethical, legal, and philosophical challenges that will need to be addressed".

Developing and implementing the Green AI approach is one of these challenges that needs urgent attention; as the IPCC report [82] highlights, climate change is widespread, rapid, and intensifying, and our planet will reach temperature rise of about 1.5 °C in only around a decade.

Furthermore, most recently, as an outcome of COP26, the role that advanced digital technologies, such as AI, could play in addressing planetary challenges has been better comprehended by a number of countries; for example, Australian Government's new 'technology-led approach' to emissions reduction is one of the most prominent national climate change strategies in Australia [83]. While for politicians greenwashing [84] and technowashing [85] are, at times, applied methods to divert attention from the core issues, we hope for the sincerity of the current Australian Government administration, and others, for the use of technology for achieving sustainable outcomes and in a sustainable way [86]— as in the Green AI approach.

Against the above editorial commentary, the Special Issue on reviews and perspectives on smart and sustainable metropolitan and regional cities contributes to the efforts in improving research and practice in smart and sustainable metropolitan and regional cities and urbanism. The Special Issue brings together the key literature review and scholarly perspective pieces and forms an open access knowledge warehouse. It offers insights into research and practice in smart and sustainable metropolitan and regional cities by producing in-depth conceptual debates and perspectives, insights from the literature and best practice, and thoroughly identified research themes and development trends. The Special Issue, hence, serves as a repository of relevant information, material, and knowledge to support research, policymaking, practice, and transferability of experiences to address the challenges in establishing smart and sustainable metropolitan and regional cities and urbanism in the era of climate change, biodiversity collapse, natural disasters, pandemics, and socioeconomic inequalities.

The Special Issue includes the following 16 commentary, perspective, review, and research papers with the input of 58 urban scholars from Australia, Bangladesh, Brazil, China, Korea, Saudi Arabia, Spain, Spain, Sweden, Thailand, Turkey, the UK, and the US:

1. Greening the artificial intelligence for a sustainable planet: an editorial commentary.
2. The lived experience of residents in an emerging master-planned community [87].
3. Making the Gold Coast a smart city: an analysis [88].
4. Leveraging smart and sustainable development via international events: insights from Bento Gonçalves Knowledge Cities World Summit [89].
5. Sustainable smart cities and industrial ecosystem: structural and relational changes of the smart city industries in Korea [90].
6. Redesigning the municipal solid waste supply chain considering the classified collection and disposal: a case study of incinerable waste in Beijing [91].
7. Empowering a sustainable city using self-assessment of environmental performance on Ecocitopia platform [92].
8. Sustainability understanding and behaviors across urban areas: a case study on Istanbul city [93].
9. Overview and exploitation of haptic tele-weight device in virtual shopping stores [94].
10. Framing corporate social responsibility to achieve sustainability in urban industrialization: case of Bangladesh ready-made garments [95].
11. Data-driven analysis on inter-city commuting decisions in Germany [96].

12. Exploring the role of digital infrastructure asset management tools for resilient linear infrastructure outcomes in cities and towns: a systematic literature review [97].
13. Blockchain and building information management (BIM) for sustainable building development within the context of smart cities [98].
14. Green artificial intelligence: towards an efficient, sustainable and equitable technology for smart cities and futures [99] (Yigitcanlar et al., 2021).
15. Towards Australian regional turnaround: insights into sustainably accommodating post-pandemic urban growth in regional towns and cities [100].
16. Social capital and sustainable social development: how are changes in neighborhood social capital associated with neighborhood sociodemographic and socioeconomic characteristics? [101].

This collection of papers focused on answering the overall questions of this Special Issue—namely, what the critical aspects of smart and sustainable metropolitan and regional cities are, and how we can construct such cities that are resilient to the increasing severity and frequency of climate change effects, biodiversity loss, natural disasters, and pandemics, and are the generators of socioeconomic equalities, and are the vehicles of delivering SDGs for a sustainable planet.

Conflicts of Interest: The author declares no conflict of interest.

References

1. Cohen, J.E. Population growth and earth's human carrying capacity. *Science* **1995**, *269*, 341–346. [CrossRef] [PubMed]
2. Rees, W.E. Revisiting carrying capacity: Area-based indicators of sustainability. *Popul. Environ.* **1996**, *17*, 195–215. [CrossRef]
3. Dizdaroglu, D.; Yigitcanlar, T. A parcel-scale assessment tool to measure sustainability through urban ecosystem components: The MUSIX model. *Ecol. Indic.* **2014**, *41*, 115–130. [CrossRef]
4. Rees, W.E. Ecological footprints and appropriated carrying capacity: What urban economics leaves out. *Environ. Urban.* **2017**, *2*, 66–77. [CrossRef]
5. Mooney, H.; Larigauderie, A.; Cesario, M.; Elmquist, T.; Hoegh-Guldberg, O.; Lavorel, S.; Mace, G.M.; Palmer, M.; Scholes, R.; Yahara, T. Biodiversity, climate change, and ecosystem services. *Curr. Opin. Environ. Sustain.* **2009**, *1*, 46–54. [CrossRef]
6. Nilsson, M.; Griggs, D.; Visbeck, M. Policy: Map the interactions between Sustainable Development Goals. *Nat. Cell Biol.* **2016**, *534*, 320–322. [CrossRef] [PubMed]
7. United Nations. The 17 Goals, Sustainable Development. Available online: https://sdgs.un.org/goals (accessed on 24 May 2021).
8. Dimitrov, R.S. The Paris Agreement on Climate Change: Behind Closed Doors. *Glob. Environ. Politics* **2016**, *16*, 1–11. [CrossRef]
9. Tollefson, J. IPCC says limiting global warming to 1.5 °C will require drastic action. *Nature* **2018**, *562*, 172–173. [CrossRef]
10. Fernández-Aller, C.; de Velasco, A.F.; Manjarrés, Á.; Pastor-Escuredo, D.; Pickin, S.; Criado, J.S.; Ausín, T. An Inclusive and Sustainable Artificial Intelligence Strategy for Europe Based on Human Rights. *IEEE Technol. Soc. Mag.* **2021**, *40*, 46–54. [CrossRef]
11. Vaughan, A. Will the IPCC report matter? *New Sci.* **2021**, *251*, 8–9. [CrossRef]
12. Bergmann, Z.; Ossewaarde, R. Youth climate activists meet environmental governance: Ageist depictions of the FFF movement and Greta Thunberg in German newspaper coverage. *J. Multicult. Discourses* **2020**, *15*, 267–290. [CrossRef]
13. Gunningham, N. Averting Climate Catastrophe: Environmental Activism, Extinction Rebellion and coalitions of Influence. *King's Law J.* **2019**, *30*, 194–202. [CrossRef]
14. Ruiz-Campillo, X.; Broto, V.C.; Westman, L. Motivations and Intended Outcomes in Local Governments' Declarations of Climate Emergency. *Politics Gov.* **2021**, *9*, 17–28. [CrossRef]
15. Beiser-McGrath, L. COP26: Why Politicians Have Little Incentive to Prepare for Future Climate Change Disasters. Available online: https://theconversation.com/cop26-why-politicians-have-little-incentive-to-prepare-for-future-climate-change-disasters-171539 (accessed on 11 November 2021).
16. Fisher, D.R.; Nasrin, S. Climate activism and its effects. *Wiley Interdiscip. Rev. Clim. Chang.* **2021**, *12*, e683. [CrossRef]
17. Aldieri, L.; Carlucci, F.; Vinci, C.P.; Yigitcanlar, T. Environmental innovation, knowledge spillovers and policy implications: A systematic review of the economic effects literature. *J. Clean. Prod.* **2019**, *239*, 118051. [CrossRef]
18. Lauterbach, A. Artificial intelligence and policy: Quo vadis? *Digit. Policy Regul. Gov.* **2019**, *21*, 238–263. [CrossRef]
19. Nwankwo, W.; Ukhurebor, K.E.; Aigbe, U.O. Climate change and innovation technology: A review. *Renew. Energy* **2020**, *20*, 23.
20. Fujii, H.; Managi, S. Trends and priority shifts in artificial intelligence technology invention: A global patent analysis. *Econ. Anal. Policy* **2018**, *58*, 60–69. [CrossRef]
21. Wirtz, B.W.; Weyerer, J.C.; Geyer, C. Artificial Intelligence and the Public Sector—Applications and Challenges. *Int. J. Public Adm.* **2018**, *42*, 596–615. [CrossRef]
22. Dignam, A. Artificial intelligence, tech corporate governance and the public interest regulatory response. *Camb. J. Reg. Econ. Soc.* **2020**, *13*, 37–54. [CrossRef]

23. Chatterjee, S.; Sreenivasulu, N.S. Personal data sharing and legal issues of human rights in the era of artificial intelligence: Moderating effect of government regulation. *Int. J. Electron. Gov. Res. (IJEGR)* **2019**, *15*, 21–36. [CrossRef]
24. Chatterjee, S.; Bhattacharjee, K.K. Adoption of artificial intelligence in higher education: A quantitative analysis using structural equation modelling. *Educ. Inf. Technol.* **2020**, *25*, 3443–3463. [CrossRef]
25. Simon, J.P. Artificial intelligence: Scope, players, markets and geography. *Digit. Policy Regul. Gov.* **2019**, *21*, 208–237. [CrossRef]
26. Vadlamudi, S. How Artificial Intelligence Improves Agricultural Productivity and Sustainability: A Global Thematic Analysis. *Asia Pac. J. Energy Environ.* **2019**, *6*, 91–100.
27. Yigitcanlar, T. Kamruzzaman Smart Cities and Mobility: Does the Smartness of Australian Cities Lead to Sustainable Commuting Patterns? *J. Urban Technol.* **2019**, *26*, 21–46. [CrossRef]
28. Masakowski, Y.R. Artificial Intelligence and the Future Global Security Environment. In *Artificial Intelligence and Global Security*; Emerald Publishing Limited: London, UK, 2020; pp. 1–34.
29. Nayal, K.; Raut, R.; Priyadarshinee, P.; Narkhede, B.E.; Kazancoglu, Y.; Narwane, V. Exploring the role of artificial intelligence in managing agricultural supply chain risk to counter the impacts of the COVID-19 pandemic. *Int. J. Logist. Manag.* **2021**. [CrossRef]
30. Wheeler, A.R.; Buckley, M.R. *The Current State of HRM with Automation, Artificial Intelligence, and Machine Learning. InHR without People?* Emerald Publishing Limited: London, UK, 2021; pp. 29–44.
31. Cortés, U.; Sànchez-Marrè, M.; Ceccaroni, L.; R-Roda, I.; Poch, M. Artificial Intelligence and Environmental Decision Support Systems. *Appl. Intell.* **2000**, *13*, 77–91. [CrossRef]
32. Coulson, R.N.; Folse, L.J.; Loh, D.K. Artificial Intelligence and Natural Resource Management. *Science* **1987**, *237*, 262–267. [CrossRef]
33. Jones, D.O.; Gates, A.; Huvenne, V.A.; Phillips, A.B.; Bett, B.J. Autonomous marine environmental monitoring: Application in decommissioned oil fields. *Sci. Total Environ.* **2019**, *668*, 835–853. [CrossRef]
34. Corcoran, E.; Denman, S.; Hanger, J.; Wilson, B.; Hamilton, G. Automated detection of koalas using low-level aerial surveillance and machine learning. *Sci. Rep.* **2019**, *9*, 3208. [CrossRef] [PubMed]
35. Pyayt, A.L.; Mokhov, I.I.; Lang, B.; Krzhizhanovskaya, V.V.; Meijer, R.J. Machine learning methods for environmental monitoring and flood protection. *World Acad. Sci. Eng. Technol.* **2011**, *78*, 118–123.
36. Tung, T.M.; Yaseen, Z.M. A survey on river water quality modelling using artificial intelligence models: 2000–2020. *J. Hydrol.* **2020**, *585*, 124670.
37. Li, C. Biodiversity assessment based on artificial intelligence and neural network algorithms. *Microprocess. Microsyst.* **2020**, *79*, 103321. [CrossRef]
38. Weber, E.; Marzo, N.; Papadopoulos, D.P.; Biswas, A.; Lapedriza, A.; Ofli, F.; Imran, M.; Torralba, A. Detecting natural disasters, damage, and incidents in the wild. In *European Conference on Computer Vision*; Springer: Cham, Switzerland, 2020; pp. 331–350.
39. Kankanamge, N.; Yigitcanlar, T.; Goonetilleke, A. Public perceptions on artificial intelligence driven disaster management: Evidence from Sydney, Melbourne and Brisbane. *Telemat. Inform.* **2021**, *65*, 101729. [CrossRef]
40. Huntingford, C.; Jeffers, E.S.; Bonsall, M.B.; Christensen, H.M.; Lees, T.; Yang, H. Machine learning and artificial intelligence to aid climate change research and preparedness. *Environ. Res. Lett.* **2019**, *14*, 124007. [CrossRef]
41. Yigitcanlar, T.; DeSouza, K.C.; Butler, L.; Roozkhosh, F. Contributions and Risks of Artificial Intelligence (AI) in Building Smarter Cities: Insights from a Systematic Review of the Literature. *Energies* **2020**, *13*, 1473. [CrossRef]
42. Paschen, U.; Pitt, C.; Kietzmann, J. Artificial intelligence: Building blocks and an innovation typology. *Bus. Horiz.* **2019**, *63*, 147–155. [CrossRef]
43. Barns, S. Out of the loop? On the radical and the routine in urban big data. *Urban Stud.* **2021**, *58*, 3203–3210. [CrossRef]
44. Kaplan, A.; Haenlein, M. Rulers of the world, unite! The challenges and opportunities of artificial intelligence. *Bus. Horiz.* **2019**, *63*, 37–50. [CrossRef]
45. Watson, G.J.; Desouza, K.C.; Ribiere, V.M.; Lindič, J. Will AI ever sit at the C-suite table? The future of senior leadership. *Bus. Horiz.* **2021**, *64*, 465–474. [CrossRef]
46. Yigitcanlar, T.; Cugurullo, F. The Sustainability of Artificial Intelligence: An Urbanistic Viewpoint from the Lens of Smart and Sustainable Cities. *Sustainability* **2020**, *12*, 8548. [CrossRef]
47. Lee, J.-H.; Woo, J. Green New Deal Policy of South Korea: Policy Innovation for a Sustainability Transition. *Sustainability* **2020**, *12*, 10191. [CrossRef]
48. Frank, B. Artificial intelligence-enabled environmental sustainability of products: Marketing benefits and their variation by consumer, location, and product types. *J. Clean. Prod.* **2020**, *285*, 125242. [CrossRef]
49. Zhou, L.; Wu, X.; Xu, Z.; Fujita, H. Emergency decision making for natural disasters: An overview. *Int. J. Disaster Risk Reduct.* **2018**, *27*, 567–576. [CrossRef]
50. Zhao, L.; Dai, T.; Qiao, Z.; Sun, P.; Hao, J.; Yang, Y. Application of artificial intelligence to wastewater treatment: A bibliometric analysis and systematic review of technology, economy, management, and wastewater reuse. *Process Saf. Environ. Prot.* **2019**, *133*, 169–182. [CrossRef]
51. Nañez Alonso, S.L.; Reier Forradellas, R.F.; Pi Morell, O.; Jorge-Vázquez, J. Digitalization, circular economy and environmental sustainability: The application of Artificial Intelligence in the efficient self-management of waste. *Sustainability* **2021**, *13*, 2092. [CrossRef]

52. Khakurel, J.; Penzenstadler, B.; Porras, J.; Knutas, A.; Zhang, W. The rise of artificial intelligence under the lens of sustainability. *Technologies* **2018**, *6*, 100. [CrossRef]

53. Tanveer, M.; Hassan, S.; Bhaumik, A. Academic Policy Regarding Sustainability and Artificial Intelligence (AI). *Sustainability* **2020**, *12*, 9435. [CrossRef]

54. Vinuesa, R.; Azizpour, H.; Leite, I.; Balaam, M.; Dignum, V.; Domisch, S.; Felländer, A.; Langhans, S.D.; Tegmark, M.; Nerini, F.F. The role of artificial intelligence in achieving the Sustainable Development Goals. *Nat. Commun.* **2020**, *11*, 233. [CrossRef] [PubMed]

55. Gupta, S.; Langhans, S.D.; Domisch, S.; Fuso-Nerini, F.; Felländer, A.; Battaglini, M.; Tegmark, M.; Vinuesa, R. Assessing whether artificial intelligence is an enabler or an inhibitor of sustainability at indicator level. *Transp. Eng.* **2021**, *4*, 100064. [CrossRef]

56. Di Vaio, A.; Palladino, R.; Hassan, R.; Escobar, O. Artificial intelligence and business models in the sustainable development goals perspective: A systematic literature review. *J. Bus. Res.* **2020**, *121*, 283–314. [CrossRef]

57. Goralski, M.A.; Tan, T.K. Artificial intelligence and sustainable development. *Int. J. Manag. Educ.* **2019**, *18*, 100330. [CrossRef]

58. Truby, J. Governing Artificial Intelligence to benefit the UN Sustainable Development Goals. *Sustain. Dev.* **2020**, *28*, 946–959. [CrossRef]

59. Yeh, S.-C.; Wu, A.-W.; Yu, H.-C.; Wu, H.C.; Kuo, Y.-P.; Chen, P.-X. Public Perception of Artificial Intelligence and Its Connections to the Sustainable Development Goals. *Sustainability* **2021**, *13*, 9165. [CrossRef]

60. Khamis, A.; Li, H.; Prestes, E.; Haidegger, T. AI: A Key Enabler of Sustainable Development Goals, Part 1 [Industry Activities]. *IEEE Robot. Autom. Mag.* **2019**, *26*, 95–102. [CrossRef]

61. Cernev, T.; Fenner, R. The importance of achieving foundational Sustainable Development Goals in reducing global risk. *Futures* **2019**, *115*, 102492. [CrossRef]

62. Miailhe, N.; Hodes, C.; Jain, A.; Iliadis, N.; Alanoca, S.; Png, J. AI for sustainable development goals. *Delphi* **2019**, *2*, 207. [CrossRef]

63. Yigitcanlar, T.; Corchado, J.M.; Mehmood, R.; Li, R.Y.; Mossberger, K.; Desouza, K. Responsible urban innovation with local government artificial intelligence (AI): A conceptual framework and research agenda. *J. Open Innov. Technol. Mark. Complex.* **2021**, *7*, 71. [CrossRef]

64. Lahsen, M. Should AI be Designed to Save Us from Ourselves? Artificial Intelligence for Sustainability. *IEEE Technol. Soc. Mag.* **2020**, *39*, 60–67. [CrossRef]

65. Yigitcanlar, T.; Han, H.; Kamruzzaman, M.; Ioppolo, G.; Sabatini-Marques, J. The making of smart cities: Are Songdo, Masdar, Amsterdam, San Francisco and Brisbane the best we could build? *Land Use Policy* **2019**, *88*, 104187. [CrossRef]

66. Dhar, P. The carbon impact of artificial intelligence. *Nat. Mach. Intell.* **2020**, *2*, 423–425. [CrossRef]

67. Schwartz, R.; Dodge, J.; Smith, N.A.; Etzioni, O. Green AI. *Commun. ACM* **2020**, *63*, 54–63. [CrossRef]

68. Dobbe, R.; Whittaker, M. AI and Climate Change: How They're Connected, and What We Can Do about It. Available online: https://medium.com/@AINowInstitute/ai-and-climate-change-how-theyre-connected-and-what-we-can-do-about-it-6aa8d0f5b32c (accessed on 11 November 2021).

69. Yigitcanlar, T.; Butler, L.; Windle, E.; Desouza, K.C.; Mehmood, R.; Corchado, J.M. Can building "artificially intelligent cities" safeguard humanity from natural disasters, pandemics, and other catastrophes? An urban scholar's perspective. *Sensors* **2020**, *20*, 2988. [CrossRef] [PubMed]

70. Yang, X.; Hua, S.; Shi, Y.; Wang, H.; Zhang, J.; Letaief, K.B. Sparse optimization for green edge AI inference. *J. Commun. Inf. Netw.* **2020**, *5*, 1–5.

71. Candelieri, A.; Perego, R.; Archetti, F. Green machine learning via augmented Gaussian processes and multi-information source optimization. *Soft Comput.* **2021**, *25*, 12591–12603. [CrossRef]

72. Yun, J.J.; Lee, D.; Ahn, H.; Park, K.; Yigitcanlar, T. Not Deep Learning but Autonomous Learning of Open Innovation for Sustainable Artificial Intelligence. *Sustainability* **2016**, *8*, 797. [CrossRef]

73. Gould, S. Green AI: How Can AI Solve Sustainability Challenges? Available online: https://www2.deloitte.com/uk/en/blog/experience-analytics/2020/green-ai-how-can-ai-solve-sustainability-challenges.html (accessed on 23 June 2021).

74. Yigitcanlar, T.; Kankanamge, N.; Regona, M.; Ruiz Maldonado, A.; Rowan, B.; Ryu, A.; Desouza, K.C.; Corchado, J.M.; Mehmood, R.; Li, R.Y. Artificial intelligence technologies and related urban planning and development concepts: How are they perceived and utilized in Australia? *J. Open Innov. Technol. Mark. Complex.* **2020**, *6*, 187. [CrossRef]

75. Palomares, I.; Martínez-Cámara, E.; Montes, R.; García-Moral, P.; Chiachio, M.; Chiachio, J.; Alonso, S.; Melero, F.J.; Molina, D.; Fernández, B.; et al. A panoramic view and swot analysis of artificial intelligence for achieving the sustainable development goals by 2030: Progress and prospects. *Appl. Intell.* **2021**, *51*, 6497–6527. [CrossRef] [PubMed]

76. Yigitcanlar, T.; Foth, M.; Kamruzzaman, M. Towards post-anthropocentric cities: Reconceptualizing smart cities to evade urban ecocide. *J. Urban Technol.* **2019**, *26*, 147–152. [CrossRef]

77. Nishant, R.; Kennedy, M.; Corbett, J. Artificial intelligence for sustainability: Challenges, opportunities, and a research agenda. *Int. J. Inf. Manag.* **2020**, *53*, 102104. [CrossRef]

78. Reed, C. How should we regulate artificial intelligence? *Philos. Trans. R. Soc. A Math. Phys. Eng. Sci.* **2018**, *376*, 20170360. [CrossRef]

79. Buiten, M.C. Towards Intelligent Regulation of Artificial Intelligence. *Eur. J. Risk Regul.* **2019**, *10*, 41–59. [CrossRef]

80. Glauner, P. An Assessment of the AI Regulation Proposed by the European Commission. Available online: https://arxiv.org/pdf/2105.15133.pdf (accessed on 23 June 2021).

81. Haenlein, M.; Kaplan, A. A Brief History of Artificial Intelligence: On the Past, Present, and Future of Artificial Intelligence. *Calif. Manag. Rev.* **2019**, *61*, 5–14. [CrossRef]

82. IPCC. Sixth Assessment Report of the Intergovernmental Panel on Climate Change. Available online: https://www.ipcc.ch/assessment-report/ar6/ (accessed on 11 November 2021).

83. Australian Government. Australia's Climate Change Strategies. Available online: https://www.industry.gov.au/policies-and-initiatives/australias-climate-change-strategies (accessed on 11 November 2021).

84. Cislak, A.; Cichocka, A.; Wojcik, A.D.; Milfont, T.L. Words not deeds: National narcissism, national identification, and support for greenwashing versus genuine proenvironmental campaigns. *J. Environ. Psychol.* **2021**, *74*, 101576. [CrossRef]

85. Perakslis, C. Exposing Technowashing: To Mitigate Technosocial Inequalities [Last Word]. *IEEE Technol. Soc. Mag.* **2020**, *39*, 88. [CrossRef]

86. Rasmussen, K.; Thornber, B. How Can Australia Get Cracking on Emissions? The Know—How We Need Is in Our Universities. Available online: https://theconversation.com/how-can-australia-get-cracking-on-emissions-the-know-how-we-need-is-in-our-universities-170374 (accessed on 11 November 2021).

87. Buys, L.; Newton, C.; Walker, N. The Lived Experience of Residents in an Emerging Master-Planned Community. *Sustainability* **2021**, *13*, 12158. [CrossRef]

88. Khanjanasthiti, I.; Chandrasekar, K.S.; Bajracharya, B. Making the Gold Coast a Smart City—An Analysis. *Sustainability* **2021**, *13*, 10624. [CrossRef]

89. Michelam, L.D.; Cortese, T.T.P.; Yigitcanlar, T.; Fachinelli, A.C.; Vils, L.; Levy, W. Leveraging Smart and Sustainable Development via International Events: Insights from Bento Gonçalves Knowledge Cities World Summit. *Sustainability* **2021**, *13*, 9937. [CrossRef]

90. Jo, S.-S.; Han, H.; Leem, Y.; Lee, S.-H. Sustainable Smart Cities and Industrial Ecosystem: Structural and Relational Changes of the Smart City Industries in Korea. *Sustainability* **2021**, *13*, 9917. [CrossRef]

91. Yang, X.; Guo, X.; Yang, K. Redesigning the Municipal Solid Waste Supply Chain Considering the Classified Collection and Disposal: A Case Study of Incinerable Waste in Beijing. *Sustainability* **2021**, *13*, 9855. [CrossRef]

92. Kongboon, R.; Gheewala, S.H.; Sampattagul, S. Empowering a Sustainable City Using Self-Assessment of Environmental Performance on EcoCitOpia Platform. *Sustainability* **2021**, *13*, 7743. [CrossRef]

93. Topal, H.; Hunt, D.; Rogers, C. Sustainability Understanding and Behaviors across Urban Areas: A Case Study on Istanbul City. *Sustainability* **2021**, *13*, 7711. [CrossRef]

94. Farooq, A.; Seyedmahmoudian, M.; Horan, B.; Mekhilef, S.; Stojcevski, A. Overview and exploitation of haptic tele-weight device in virtual shopping stores. *Sustainability* **2021**, *13*, 7253. [CrossRef]

95. Saha, P.K.; Akhter, S.; Hassan, A. Framing Corporate Social Responsibility to Achieve Sustainability in Urban Industrialization: Case of Bangladesh Ready-Made Garments (RMG). *Sustainability* **2021**, *13*, 6988. [CrossRef]

96. Chen, H.; Voigt, S.; Fu, X. Data-Driven Analysis on Inter-City Commuting Decisions in Germany. *Sustainability* **2021**, *13*, 6320. [CrossRef]

97. Caldera, S.; Mostafa, S.; Desha, C.; Mohamed, S. Exploring the Role of Digital Infrastructure Asset Management Tools for Resilient Linear Infrastructure Outcomes in Cities and Towns: A Systematic Literature Review. *Sustainability* **2021**, *13*, 11965. [CrossRef]

98. Liu, Z.; Chi, Z.; Osmani, M.; Demian, P. Blockchain and Building Information Management (BIM) for Sustainable Building Development within the Context of Smart Cities. *Sustainability* **2021**, *13*, 2090. [CrossRef]

99. Yigitcanlar, T.; Mehmood, R.; Corchado, J.M. Green Artificial Intelligence: Towards an Efficient, Sustainable and Equitable Technology for Smart Cities and Futures. *Sustainability* **2021**, *13*, 8952. [CrossRef]

100. Guaralda, M.; Hearn, G.; Foth, M.; Yigitcanlar, T.; Mayere, S.; Law, L. Towards Australian Regional Turnaround: Insights into Sustainably Accommodating Post-Pandemic Urban Growth in Regional Towns and Cities. *Sustainability* **2020**, *12*, 10492. [CrossRef]

101. Eriksson, M.; Santosa, A.; Zetterberg, L.; Kawachi, I.; Ng, N. Social Capital and Sustainable Social Development—How Are Changes in Neighbourhood Social Capital Associated with Neighbourhood Sociodemographic and Socioeconomic Characteristics? *Sustainability* **2021**, *13*, 13161. [CrossRef]

 *sustainability*

Article

Social Capital and Sustainable Social Development—How Are Changes in Neighbourhood Social Capital Associated with Neighbourhood Sociodemographic and Socioeconomic Characteristics?

Malin Eriksson [1,*], Ailiana Santosa [2], Liv Zetterberg [1], Ichiro Kawachi [3] and Nawi Ng [2]

[1] Department of Social Work, Umeå University, Universitetstorget 4, 907 36 Umeå, Sweden; liv.zetterberg@umu.se
[2] School of Public Health and Community Medicine, University of Gothenburg, Medicinaregatan 18A, 413 90 Gothenburg, Sweden; ailiana.santosa@gu.se (A.S.); nawi.ng@gu.se (N.N.)
[3] Harvard T.H. Chan School of Public Health, 677 Huntington Avenue, Boston, MA 02115, USA; ikawachi@hsph.harvard.edu
* Correspondence: malin.eriksson@umu.se; Tel.: +46-907867782

Citation: Eriksson, M.; Santosa, A.; Zetterberg, L.; Kawachi, I.; Ng, N. Social Capital and Sustainable Social Development—How Are Changes in Neighbourhood Social Capital Associated with Neighbourhood Sociodemographic and Socioeconomic Characteristics? *Sustainability* **2021**, *13*, 13161. https://doi.org/10.3390/su132313161

Academic Editor: Tan Yigitcanlar

Received: 19 October 2021
Accepted: 23 November 2021
Published: 27 November 2021

Publisher's Note: MDPI stays neutral with regard to jurisdictional claims in published maps and institutional affiliations.

Abstract: The development of social capital is acknowledged as key for sustainable social development. Little is known about how social capital changes over time and how it correlates with sociodemographic and socioeconomic factors. This study was conducted in 46 neighbourhoods in Umeå Municipality, northern Sweden. The aim was to examine neighbourhood-level characteristics associated with changes in neighbourhood social capital and to discuss implications for local policies for sustainable social development. We designed an ecological study linking survey data to registry data in 2006 and 2020. Over 14 years, social capital increased in 9 and decreased in 15 neighbourhoods. Higher levels of social capital were associated with specific sociodemographic factors, but these differed in urban and rural areas. Urban neighbourhoods with a higher proportion of older pensioners (OR = 1.49, CI: 1.16–1.92), children under 12 (OR= 2.13, CI: 1.31–3.47), or a lower proportion of foreign-born members (OR= 0.32, CI: 0.19–0.55) had higher odds for higher social capital levels. In rural neighbourhoods, a higher proportion of single-parent households was associated with higher levels of social capital (OR = 1.44, 95% CI = 1.04–1.98). Neighbourhood socioeconomic factors such as income or educational level did not influence neighbourhood social capital. Using repeated measures of social capital, this study gives insights into how social capital changes over time in local areas and the factors influencing its development. Local policies to promote social capital for sustainable social development should strive to integrate diverse demographic groups within neighbourhoods and should increase opportunities for inter-ethnic interactions.

Keywords: social capital; neighbourhoods; sustainable social development; ecological study; ordinal logistic regression; northern Sweden

1. Introduction

Following the UN Sustainable Development Goals, set in Agenda 2030, social sustainability has increasingly become a goal for urban policy and planning and for local and regional developmental strategies. Social sustainability is the least-developed dimension of the sustainable development discourse [1], compared to environmental and economic sustainability. Still, there is an agreement that social sustainability implies values such as e.g., social inclusion, social interaction and participation, safety, and a sense of cohesion in local areas [2,3]. Sustainable social development thus indicates a process of change towards specific social values. However, due to "conceptualizations and definitional concerns," assessment and measurement of social sustainability remains a challenge [1]. Social capital, defined as "social networks, the reciprocities that arise from them and the value of these for achieving mutual goals" [4] is viewed as a crucial part of social sustainability [5,6]. Therefore, developing or strengthening social capital could be seen as a means for how the process of sustainable social development could be shaped and assessed [6,7]. The concept of social capital is rather well defined and operationalized [8], with significant contributions from scholars such as Robert Putnam [9,10] and Alejandro Portes [11]; hence, it could be a useful indicator for measuring social sustainability. Social capital is also increasingly acknowledged as key in urban and local development for ensuring and assessing social sustainability. Accordingly, a neighbourhood high in social capital could be viewed as being socially sustainable. However, little is known about the processes through which local social capital is generated and how it changes over time [12]. Most research has focused on the definition and function of social capital. In contrast, less emphasis has been put on how social capital could be developed, what institutional and political conditions facilitate its development, and how cultural changes influence the development of social capital [13]. This study tries to fill this gap by using repeated measures of neighbourhood social capital to investigate how it changes over time and what factors exist that might influence the development of social capital, and thus social sustainability, in local areas.

There are some examples of studies reporting from interventions designed to strengthen social capital in local areas. These studies underline the importance of investments in the physical environment, such as setting up attractive meeting places [14,15], arranging requested and inclusive community activities [16,17], and ensuring access to local meeting places such as libraries and cafés [18,19]. These interventions focus on meso-level interactions for social capital to be strengthened in local communities.

A related question asks if and how neighbourhood social capital is influenced by the sociodemographic and socioeconomic composition of people living in an area. The economic literature has emphasised that community heterogeneity (racial and ethnic diversity and income inequality) tends to erode trust and reduce civic engagement (and hence social capital). Drawing on data from the World Values Survey and the European Values Survey on civic engagement over time in Western European countries, Costa and Khan [20] found that high levels of ethnic heterogeneity were associated with lower levels of participation in various organisations in almost all countries. Interestingly, they found Sweden to be an exception with relatively high levels of civic engagement despite comparable high-income inequality and ethnic heterogeneity [20]. Whether this pattern holds at local and neighbourhood levels in the Swedish setting remains unknown. Further, a study from Philadelphia, USA [21] found that neighbourhood social capital was associated with racial composition (higher levels of social capital in neighbourhoods with less than 50% black residents) and socioeconomic disadvantage (lower levels of social capital in neighbourhoods with a high proportion of people with low income and low educational levels). However, since social capital is context-specific, these patterns may differ in various community contexts and at different spatial levels (national vs regional, municipal, and neighbourhood). Thus, further explorations on how the composition of people influences the development of social capital and social sustainability in different settings are needed, not least in local areas where urban planning and policy for social sustainability take place. In addition, most research on social capital tends to be cross-sectional, making it hard to rule out causal inference and changes over time. This current study contributes to further understanding the factors that may influence social capital development in local areas, by measuring social capital changes over time in 46 neighbourhoods in northern Sweden.

Given its multidimensional and context-specific feature, the measurement of social capital and comparability between studies pose a challenge. Social capital consists of different dimensions (structural and cognitive) and forms (bonding, bridging, and linking) and can be measured at the individual, family, organisational, as well as area levels [8,22]. In addition, area-specific social capital can be simultaneously conceptualized at different levels such as national, regional, municipal, and neighbourhood levels, which is why the level of analysis (spatial scale) needs to be carefully considered in any study. An association that holds true at the country level might not be valid at the neighbourhood level, and vice versa. The effect of social capital on, e.g., well-being, might also differ between various contexts. Based on national data from the Netherlands, Mohnen et al. [23] found a stronger positive health effect of area-specific social capital in urban areas than in rural areas, even though the overall level of social capital was lower in the city. The authors [23] claimed that these results support the hypothesis that access to social capital does not necessarily imply actual benefits from it.

Many studies so far have used aggregated individual data on trust and social participation to measure area-specific social capital, but the need for indicators that relate more clearly to the local area have been raised [22,24,25]. In the present study, we used a previously developed instrument to measure neighbourhood social capital based on questions related to people's perceptions about social values in their neighbourhoods [26]. This instrument was previously used in a baseline survey from 2006 with the same neighbourhood division. In this follow-up study, we analysed the development of social capital in the same neighbourhoods over 14 years.

In sum, in this study we used repeated measures of neighbourhood social capital in urban and rural neighbourhoods in a municipality in northern Sweden, as an indicator for social sustainable development. The overall aims were to examine changes in neighbourhood social capital in relation to neighbourhood-level sociodemographic and socioeconomic characteristics over time and to draw out implications of these findings for local policies aiming to strengthen social capital for socially sustainable development.

In the next section, we present the study context and the linkages of data from the social capital surveys in 2006 and 2020 and the Swedish register data, which provide the ecological, neighbourhood-level social capital, sociodemographic, and socioeconomic data employed in this study. In the results section, we map the distribution and changes in neighbourhood social capital in choropleth maps and analyse sociodemographic and sociodemographic factors associated with neighbourhood social capital. Next, we discuss and contextualise the positive and negative changes in neighbourhood social capital and its associated factors observed in this study. Finally, we reflect upon the strengths and the weaknesses of the study and its implications for local policies aiming to strengthen social capital for socially sustainable development.

2. Material and Methods

2.1. Research Setting

We conducted this study in Umeå Municipality, one of northern Sweden's fastest-growing and most-populous cities. Its population has increased from approximately 111,235 inhabitants in 2006 to 130,224 residents at the beginning of 2020, with an annual growth rate of approximately 1.5% [27]. Umeå Municipality expects to host 200,000 inhabitants by 2050. This continuous and rapid growth rate (at least from a Swedish perspective) poses challenges regarding housing and social sustainability. The municipality has acted to ensure that the city will grow sustainably—socially, ecologically, culturally, and economically. So far, the Umeå region ranks high in social progress based on indicators of basic human needs and foundations of well-being [28]. In addition, Umeå is considered a relatively egalitarian municipality since it has no neighbourhoods defined as "socially vulnerable." The Swedish police authority has identified 61 socially vulnerable neighbourhoods in Sweden, characterised by high crime, insecurity, and social exclusion, and none of these neighbourhoods are located in northern Sweden [29]. As Umeå Municipality hosts one of the largest universities in Sweden, its population is relatively young (average age of 38) and highly educated. The municipality encompasses rural and urban areas with different characteristics, and about 16% of the population lives in rural areas. Urban neighbourhoods typically contain mixed settlements with rental and tenant-owned apartments and detached houses, while rural neighbourhoods are mainly villages typically containing villas and farms.

2.2. Study Design

This study was conducted within a broader research project about the role of social capital in the design of health-promotive and socially sustainable neighbourhoods [30]. We designed this study as an ecological study across neighbourhoods in Umeå Municipality. We defined neighbourhoods as the residential environments where people interact daily by sharing a common service area and local identity. This definition implies a specific geographical area with a locally used name and geographical borders. The residential subdivision follows officially recognised neighbourhoods, as defined by the municipality and people in general based on local knowledge (i.e., by defined neighbourhood and village names). We identified the geographical borders of each neighbourhood by using postcode areas and municipal area maps. Since postcode sectors are small administrative units, we merged several geographically close postcode sectors to fit the geographical borders of the larger neighbourhood areas. In total, 46 geographic neighbourhood areas were identified, of which 26 were urban districts, and 20 were rural villages.

2.3. Data Sources

2.3.1. Social Capital Surveys in 2006 and 2020

We generated information on neighbourhood social capital from two social capital surveys conducted in 2006 (n = 5768 individuals) and 2020 (n = 5881) in Umeå Municipality. In 2006, Statistics Sweden randomly selected individuals who lived in the 46 neighbourhoods. In 2020, the number of individuals randomly selected and invited to the survey in each neighbourhood was proportional to the size of the population in each of the neighbourhoods. The survey was sent to 10,000 individuals in 2006 and 16,000 individuals in 2020 and yielded a response rate of 58% and 37%, respectively. We utilised the same neighbourhood divisions in both surveys. In 2020, we added new postcode areas to the existing neighbourhoods and removed expired postcodes from respective neighbourhoods. As we had information on postcodes for each survey respondent, we could aggregate the survey responses for each neighbourhood to obtain the neighbourhood-specific social capital level.

2.3.2. Outcome Variable: Neighbourhood-Level Social Capital

Both surveys utilised the same seven questions presented below to measure social capital, which facilitated the comparison of the levels of social capital across neighbourhoods over time. We recorded the individual responses to these questions so that low values signified low and high values signified high on each neighbourhood social capital indicator. We replaced no opinion values with the mean value of the respective variables.

1. "Is it common in this neighbourhood that neighbours talk to each other?" (yes, very common; yes, rather common; no, rather uncommon; no, very uncommon; no opinion.)
2. "In my neighbourhood, people are ready to help each other." (About enough; too much; too little; no opinion.)
3. "In my neighbourhood, one is expected to be involved in issues that concern this place." (About enough; too much; too little; no opinion.)
4. "In my neighbourhood, people care for each other." (About enough; too much; too little; no opinion.)
5. "Did you vote in the last (2006 and 2018) election?" (Yes; no.)
6. "During the last 12 months, have you participated in any social events?" (Yes; no.)
7. "Do you feel that you can trust people in general?" (Yes; no.)

We conducted a principal component analysis to reduce the dimension of these seven correlated questions into a smaller number of uncorrelated components [31,32]. We retained the first two components, which had an Eigenvalue larger than one. Both components accounted for 48% of the total observed variance in the data. We used the cut-off of 0.3 or greater for factor loading to consider an item relevant to each component. The first four items, which reflected place-related collective social capital, had high loading in the first component. In comparison, the remaining three items loaded higher in the second component. We generated composite scores on a continuous scale for each of the components. To align with the focus of this study on neighbourhood social capital, we used only the first component for subsequent analysis since that component reflects the local environment. The individual-level composite scores were then aggregated to the neighbourhood level. The average composite scores were used as the proxy of the neighbourhood social capital for each of the 46 neighbourhoods. Neighbourhoods with high composite scores represent neighbourhoods with high social capital and vice versa. We constructed the neighbourhood social capital scores separately for urban and rural neighbourhoods in 2006 and 2020 since previous research indicates that social capital might operate differently in urban versus rural areas [23]. Finally, we ranked the neighbourhoods in each urban and rural area. We divided neighbourhoods into five groups in urban and rural areas, from very low to very high social capital levels.

Previous research [10,33] has indicated that neighbourhood social capital is a stable construct that does not change dramatically over time. Based on this assumption, we assigned the same scores for neighbourhood social capital over 2006–2013 (based on the 2006 survey results) and 2014–2017 (based on the 2020 survey results).

2.3.3. Independent Variables

Statistics Sweden constructed aggregated neighbourhood-level socioeconomic and demographic characteristics, available annually during 2006–2017. These variables were extracted from Statistics Sweden's longitudinal database, LISA, which comprises detailed data on, e.g., health and unemployment insurance at the individual level. Data for all persons aged 16 or older who are registered in the population in Sweden are available and updated yearly in the LISA database [34].

For each neighbourhood, we obtained information about the proportion of households in the neighbourhood with:

(i).   at least one adult (aged 18+) with a higher level of education (i.e., at least three years of post-secondary education);
(ii).   at least one adult (aged 18+) foreign-born member (i.e., born outside Sweden);
(iii).   receipt of cash welfare benefits during the last year (i.e., economic support from the social services, indicating economic strain);
(iv).   single parents;
(v).   at least one child under 12 years old;
(vi).   at least one adult (aged 18+) receiving unemployment benefits during the last year (i.e., indicating periods of unemployment);
(vii).   at least one family member being an older pensioner (i.e., being at least 60 years and receiving a pension); as well as
(viii).the mean disposable household income.

2.4. Statistical Analysis

We conducted the descriptive analysis for sociodemographic and socioeconomic characteristics in urban and rural areas for 2006 and 2020. We presented the level of social capital for each neighbourhood based on the 2006 and 2020 surveys in a descriptive table. We also visualised neighbourhood social capital in 2006 and 2020 and its changes in a choropleth graph created in ArcGIS. To estimate the association between the sets of neighbourhood's sociodemographic and socioeconomic characteristics and the neighbourhood-level of social capital using the panel data during 2006–2017, we built a random-effect ordinal logistic regression model with time (12 years during 2006–2017) at level 1 and with the 46 neighbourhoods at level 2. The regression model included the interaction terms between sociodemographic and socioeconomic characteristics and the residential area (urban/rural). We estimated the probability of belonging to different neighbourhood social capital levels based on different sociodemographic and socioeconomic characteristics. All analyses were done using Stata 16.0.

3. Results

Table 1 summarizes the sociodemographic, socioeconomic, and social capital characteristics in urban and rural neighbourhoods in Umeå Municipality in 2006 and 2017. Over 11 years, the proportion of households in urban and rural areas with at least one adult with a high level of education, one foreign-born member, older pensioners, and the mean disposable income increased. There were no marked changes in the proportion of households with children under 12 years old. In contrast, the proportion of households that received cash welfare benefits or unemployment benefits, and those with single parents, decreased during the same period. On average, the proportion of individuals in urban and rural neighbourhoods who reported that neighbours talk to each other, help each other, are involved in issues concerning the place, and care for each other changed only marginally during the same period. The proportion of individuals in urban and rural areas responding that neighbours care for and help each other was higher in 2017 compared to 2006, indicating an overall increase in some of the social capital indicators in Umeå Municipality over time.

Table 1. Socioeconomic/demographic and social capital characteristics of urban and rural neighbourhoods in Umeå Municipality in 2006 and 2017.

Characteristics	Urban Areas		Rural Areas	
	2006	2017	2006	2017
	Mean (95% CI)	Mean (95% CI)	Mean (95% CI)	Mean (95% CI)
Socioeconomic and demographic characteristics				
% households with at least one adult with higher education	26.9 (23.3–30.5)	32.1 (28.7–35.4)	24.1 (21.2–27.1)	31.5 (27.9–35.1)
% households with at least one foreign-born adult member	11.7 (9.6–13.8)	15.7 (12.7–18.7)	6.7 (5.7–7.8)	8.6 (7.3–9.9)
% households with cash welfare benefits	4.7 (3.4–6.0)	3.0 (2.2–3.8)	1.9 (1.2–2.5)	1.0 (0.4–1.5)
% households with a single parent	6.8 (5.5–8.2)	6.2 (5.2–7.2)	6.8 (6.2–7.4)	6.6 (5.9–7.2)
% households with children under 12 years old	13.9 (10.4–17.4)	14.3 (11.0–17.5)	23.5 (21.3–25.6)	23.9 (21.5–26.4)
% households with unemployment benefits	11.7 (11.1–12.3)	3.4 (3.0–3.7)	10.5 (9.7–11.3)	3.6 (3.2–4.0)
% households with older pensioners	15.9 (12.7–19.2)	18.9 (15.7–22.1)	23.3 (21.1–25.6)	25.8 (23.5–28.2)
Mean disposable income (in thousand SEK)	1703 (1577–1829)	2474 (2292–2656)	1972 (1770–2175)	2821 (2724–2917)
Social capital characteristics *				
% who reported that neighbours talk to each other	74.9 (67.0–82.7)	74.4 (66.9–81.9)	97.3 (95.8–98.8)	96.7 (95.4–98.0)
% reported that people ready to help each other	83.2 (80.1–86.3)	85.0 (82.1–87.9)	92.5 (90.6–94.4)	94.0 (92.5–95.5)
% reported involvement in issues concerning place	83.3 (80.8–85.8)	82.0 (79.1–84.8)	85.8 (83.1–88.5)	88.2 (86.6–89.8)
% reported people care for each other	81.2 (77.4–85.1)	83.2 (79.5–86.9)	93.5 (91.9–95.1)	94.3 (92.8–95.9)

Note: * All the percentages are unweighted.

Figure 1 shows the spatial distribution of the five levels of neighbourhood social capital in urban and rural areas in Umeå Municipality in 2006 and 2017. Almost half of the neighbourhoods, and especially those in the urban area, were categorised into the same quintile in terms of their level of social capital at the two different points of observation (Appendix A). Only one neighbourhood in the urban area, Umedalen, moved by two quintiles, from being at the lowest quintile of (very low) social capital in 2006 to the third quintile (medium social capital) in 2017. Otherwise, the other neighbourhoods with changes in their levels of social capital either moved up or moved down by one quintile of the level of social capital. In contrast, in the rural area, some neighbourhoods moved up by two or even four quintiles (Botsmark and Sörmjöle moved from being in the lowest quintile with very low social capital in 2006 to the highest quintile, very high in 2017). Some neighbourhoods, such as Täfteå and Hissjö, had lower levels of social capital in 2017 (they moved from the fifth quintile, high, in 2006 to the 2nd quintile, low, in 2017).

In total, four of the urban neighbourhoods increased their levels of social capital, while five neighbourhoods decreased their social-capital levels. In the rural areas, five villages increased their social-capital levels, while ten villages decreased their social-capital levels (Appendix A).

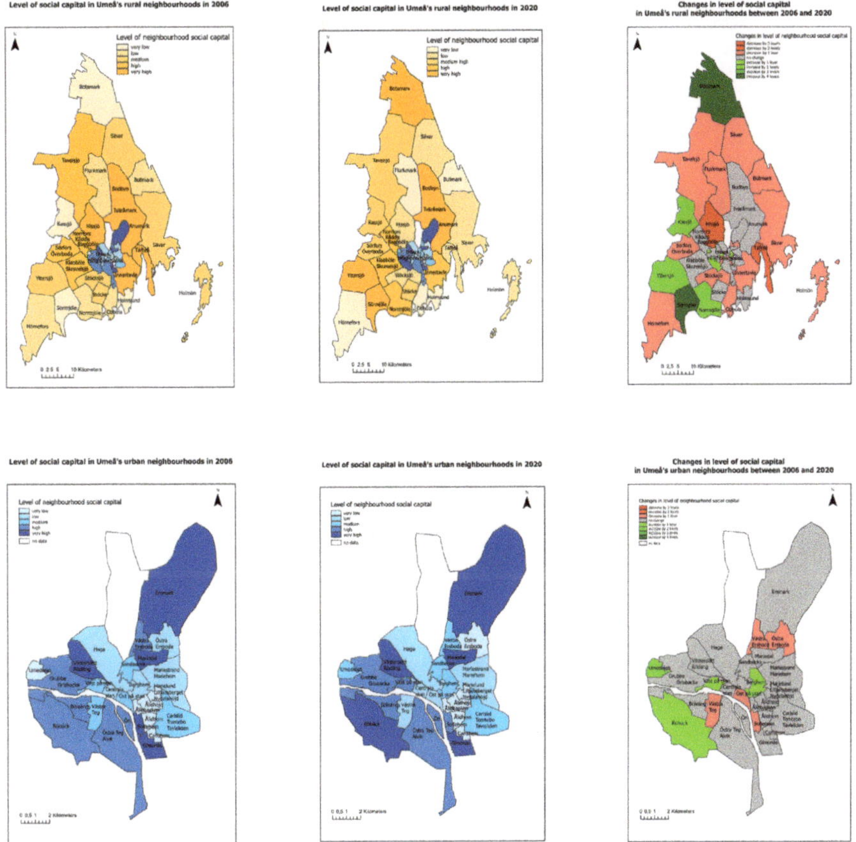

Figure 1. Levels of social capital across 46 neighbourhoods in 2006 and 2017 in Umeå Municipality, as well as changes in the levels of social capital.

Appendix B shows the five levels of social capital in urban and rural neighbourhoods in 2006 and 2017, with the corresponding distributions of responses to the four social capital indicators (neighbours talk, neighbours help, expected to be involved, and neighbours care). The darker gradient colours represent a higher proportion of responses to the questions. As expected, a higher proportion of people in rural neighbourhoods responded positively to these indicators. It is important to note that even if a neighbourhood moved up or down in the overall social capital level (which was derived as a composite estimate of all the indicators), average responses to a single indicator might show a reverse pattern. For example, Ö Ersboda changed from being categorised as a neighbourhood with low social capital in 2006 to very low social capital in 2017. Looking at the single indicator, however, the proportion of respondents who reported that neighbours talked to each other increased from 73% in 2006 to 82% in 2017.

Figure 2 presents the odds ratio of an urban or rural neighbourhood having a higher level of social capital based on their socioeconomic and sociodemographic characteristics. In the urban area, having a higher proportion of households with older pensioners (OR = 1.16, 95% CI = 1.01–1.34) or children under 12 years old (OR = 1.38, 95% CI = 1.11–1.70) were significantly associated with the probability of having a higher level of social capital. In contrast, the higher the proportion of households in a neighbourhood with at least one foreign-born member, the lower the probability of the neighbourhood being classified as having a higher level of social capital. A one percentage point increase in the proportion of households with at least one foreign-born member was associated with a 41% lower probability of the neighbourhood being classified as having a higher level of social capital. None of the sociodemographic and socioeconomic variables in the rural area was associated with the probability of a higher level of social capital, except for the proportion of single-parent households. A one percentage point increase in the proportion of single-parent households was associated with a 44% higher odds for the neighbourhood to be classified at a higher level of social capital (OR = 1.44, 95% CI = 1.04–1.98). As shown in Table 1, the proportion of households with older pensioners and at least one adult not born in Sweden increased by several percentage points in urban and rural areas between 2006 and 2017. In contrast, there were only marginal changes in the proportion of households with children under 12 and single-parent households.

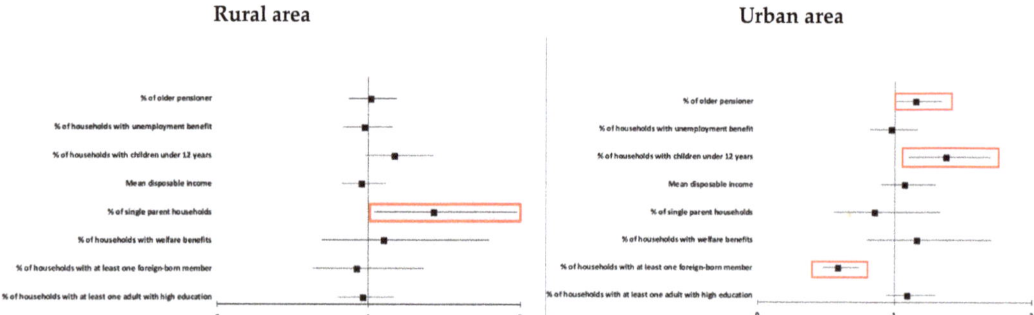

Figure 2. The odds ratio of a neighbourhood having a higher level of social capital based on their socioeconomic and sociodemographic characteristics in Umeå Municipality during 2006–2017. Note: The odds ratios for urban and rural areas were derived from the interaction terms between areas and each characteristic in a multivariable ordinal logistic regression, adjusted for all the other characteristics.

While Figure 2 indicates the odds ratio of a neighbourhood having a higher level of social capital, it does not provide details for each quintile, presented in Figure 3 and Appendix C. Figure 3 illustrates the probability of belonging to the fifth quintile of neighbourhood social capital (very high) based on different sociodemographic and socioeconomic characteristics. A more complete version of the graph with a probability of belonging to all the five different quintiles is presented in Appendix C. None of the results of the fifth quintile in rural areas were significant. In urban areas, the higher the proportion of households with at least one adult with a high level of education, receiving cash welfare benefits, with children under twelve, with older pensioners, or with higher household disposable income, the higher the probability that the neighbourhoods were classified in the highest quintile of social capital, i.e., very high social capital. Meanwhile, the higher the proportion of households with at least one foreign-born member or of single-parent households, the lower the probability of a neighbourhood being classified in the highest quintile of social capital.

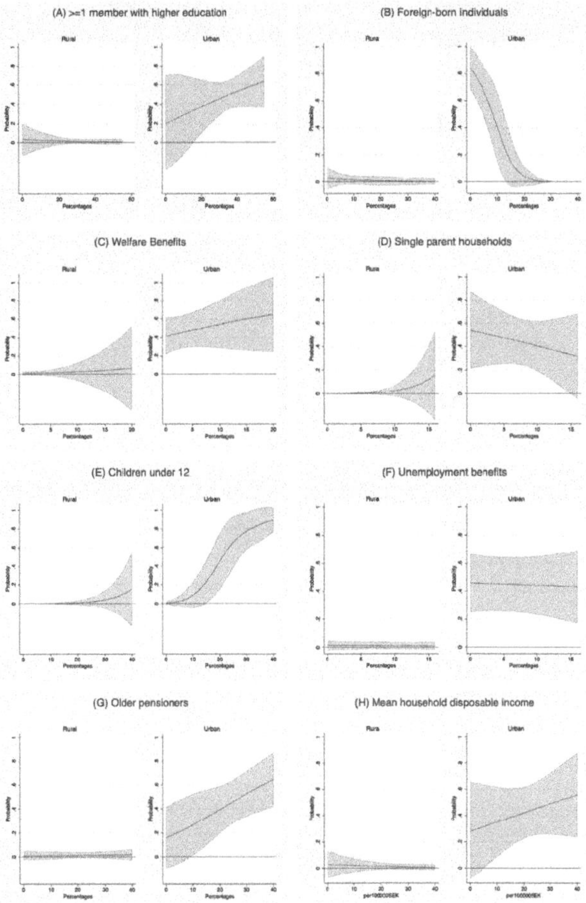

Figure 3. The probability of belonging to the fifth quintile of neighbourhood social capital—(y-axis) based on different sociodemographic and socioeconomic characteristics (x-axis). The characteristics include the proportion of households in the neighbourhood with: (**A**) at least one adult member with higher education, (**B**) at least one foreign-born adult member, (**C**) receipt of cash welfare benefits during the last year, (**D**) single parent, (**E**) at least one child under 12 years old, (**F**) at least one adult member receiving unemployment benefits during the last year, (**G**) at least one family member being an older pensioner, and (**H**) the mean disposable household income.

4. Discussion

This study examined changes in social capital in 46 urban and rural neighbourhoods in Umeå, a fast-growing municipality in northern Sweden, as a proxy for social sustainable development in the study setting. We analysed whether neighbourhood-level sociodemographic and socioeconomic changes were correlated with changes in neighbourhood social capital. To our knowledge, only a few studies have used repeated measures of social capital in local areas. Thus, our study contributes insights into how social capital changes over time in local areas and the factors influencing its development. This knowledge can guide local policies that aim to strengthen social capital to ensure sustainable social development. However, further research from various cultural contexts on how neighbourhood social capital and social sustainability change over time in local areas are needed, in order to examine if our results are valid in other settings.

4.1. Changes over Time in Social Capital and Socioeconomic and Sociodemographic Factors

Overall, our results show a positive development of social capital in Umeå Municipality during 2006–2017. On average, in both urban and rural areas, the proportion of people responding that neighbours care for and help each other was higher in the 2020 survey compared to the 2006 survey. The other social capital components changed only marginally over time. These findings mirror the results in the European Social Progress Index, showing that the region of northern Sweden (in which Umeå Municipality is located) had the highest score with regards to social development and quality of life, compared to other regions in Europe, in 2016 as well as in 2020 [35]. However, the EU Social Progress Index measures social progress at the overall regional level. At the same time, our results illustrate changes in social progress (social capital) over time at a lower hierarchical spatial level (see [36]). At the municipal level, this overall positive trend suggests that municipal strategies to ensure social sustainable growth have been relatively successful. Among other things, the municipality set up a commission for a socially sustainable Umeå, tasked with analysing differences in living conditions between groups and geographic areas and providing concrete measures for sustainable social development throughout the municipality [37]. On the other hand, the results also show how positive socioeconomic development accompanies the overall positive development of social capital. Over the study period, the proportion of households with at least one adult with higher education and mean disposable income increased across both urban and rural neighbourhoods. During the same period, the proportion of households receiving cash welfare benefits and unemployment benefits decreased. Our findings indicate an overall positive association between economic security and social capital at the municipal level, as suggested by others [10,38].

However, the overall positive trend in social capital over time in Umeå Municipality was not consistently observed in all neighbourhoods. Our results show that 15 out of the 46 neighbourhoods (five urban and ten rural neighbourhoods) had a negative development of social capital over the 14 years. These findings indicate that municipal strategies to strengthen social capital to ensure social sustainability should balance municipal and neighbourhood needs, which may vary. Thus, interventions might need to be designed differently in different neighbourhoods to achieve sustainable social development in the whole municipality.

Overall, the levels of social capital were higher in rural compared to urban neighbourhoods. The proportion of individuals reporting that neighbours talk, help each other, are involved in issues concerning their place, and care for each other was higher in rural than urban neighbourhoods. More than 95% of the participants in rural areas reported that neighbours talked to each other compared to around 75% in urban areas. Population density has been previously linked to lower probability of neighbours interacting with one another [39]. In smaller rural areas, it may be more natural for people to know each other and talk to each other (or to depend on each other for daily support), whereas denser urban settlements tend to produce fewer neighbourly interactions. Other studies have also indicated that social capital is generally higher in rural compared to urban areas [10,23,40], especially bonding social capital [41]. However, higher levels do not necessarily mean that people living there benefit from it [23].

Further, our results support previous research suggesting that social capital is a relatively stable characteristic in local areas that does not fluctuate too much over time, at least not in a relatively stable society [10,33,42]. Almost half of the 46 neighbourhoods retained their same social capital rank over 14 years, while 9 neighbourhoods increased in social capital rank, and 15 neighbourhoods decreased in social capital rank. Still, it is not unreasonable to believe that social capital fluctuates more in municipalities with rapid population growth, such as Umeå: however, this warrants further investigations.

We completed our follow-up survey in early spring 2020, just before the outbreak of the COVID-19 pandemic. To understand how the pandemic situation—with social restrictions such as staying at home and avoiding physical contact with other people—might have affected local social capital, we conducted a subsequent telephone survey (during

June–November 2020) among a sub-sample (i.e., non-responders to the 2020 survey) of those who participated in the baseline survey in 2006 [43]. Contradicting other studies that argued that the COVID-19 pandemic erodes social capital [44], our results showed that neighbourhood social capital increased during the pandemic, particularly in high-social-capital neighbourhoods. Our results are thus in line with previous research showing that neighbourhoods and societies with high levels of social capital tend to be more resilient, facilitate adaptation processes, and recover more easily (see, e.g., [45–47]). However, the spatial scale is important to consider. It is not unlikely that a societal crisis such as a pandemic could lead to a decrease in trust in authorities and politicians at the national level and at the same time increase trust, help, and support between neighbours in local areas, because people feel that one need to "stick together" to protect each other.

4.2. Factors Associated with Positive Changes in Neighbourhood Social Capital

Our main finding in this study was that higher levels of neighbourhood social capital are associated with specific sociodemographic factors, but these factors differ in urban and rural areas. In urban neighbourhoods, the probability of having higher social capital increased significantly with a higher proportion of households with older pensioners and households with children under 12 years in the neighbourhood. These results indicate that social capital is higher in areas where people spend their time since one could assume that families with children under 12, and retired people, are more bound to their living area and thus spend more time in their neighbourhoods compared to people of working age. It is reasonable to believe that people who spend time in their neighbourhood are also more likely to contribute to the neighbourhood's social climate. In line with this, a Dutch study about social capital, neighbourhood attachment, and participation concluded that older residents were more likely to participate in civic activities [48].

In rural neighbourhoods, the proportions of retired people and children under 12 were not associated with higher levels of social capital. Instead, a higher proportion of single-parent households was associated with higher levels of neighbourhood social capital in the rural areas, which is harder to interpret. This contradicts the conventional view that single parents tend to be socially isolated from the mainstream of society. However, Sweden has the highest proportion of single-parent households within the EU, with 34% of all households with children [49]. Stavrova and Fetchenhauer [50] compared well-being among single and partnered parents in 43 European countries. They found that, in individualist countries in which single-parent families were a socially acceptable practice (e.g., the Scandinavian countries), single parents did not report a lower level of well-being than partnered parents. Thus, it might be that the high proportion of single-parent households in Sweden, in combination with the overall organisation of the welfare state with, e.g., proportional costs for childcare, makes single parents less excluded from social life. Further, the proportion of households with children under 12 was considerably higher in rural areas (around 24%) compared to urban areas (around 14%). Therefore, social interaction in rural areas could potentially be more centred around family relations, with less room for engagement in the neighbourhood/village as a whole. Thus, single-parent households in rural areas might need to reach out to the broader village to mobilise help and support. These actions, in turn, might create a social environment that positively influences social capital in the whole village. A study from Austria [51] on urban–rural differences in social capital found that people in rural areas reported more family contacts, while people in urban areas reported more contact with friends. In sum, our study shows that social capital might operate differently between different spatial areas in the same municipality. Thus, interventions aiming to strengthen social capital to ensure social sustainability in local areas need to carefully consider the specific local context in order to adjust and plan the intervention in line with the specific needs and conditions of the intervention setting.

Further, our findings demonstrate that higher levels of neighbourhood social capital are mainly associated with sociodemographic rather than socioeconomic factors. None of the socioeconomic neighbourhood factors, i.e., mean of household income, reception of welfare or unemployment benefits, or education level, were significantly associated with having higher levels of neighbourhood social capital. It might be a characteristic of the Swedish welfare state that the overall high level of social security translates to the absence of a linkage between socioeconomic disadvantage and social exclusion, unlike less-generous welfare states such as the USA or Japan.

However, when we estimated the subsequent probability of belonging to different quintiles of neighbourhood social capital, we observed some significant (and unexpected) results in some of the factors in the urban area. For example, the higher proportion of households receiving cash welfare benefits, the higher the probability of belonging to the fifth (very high) quintiles of neighbourhood social capital. Thus, our results do not fully support the idea that high levels of neighbourhood social capital require a socioeconomic prosperous environment, as suggested by others [21]. Even neighbourhoods with a high proportion of households receiving cash welfare benefits (indicating a very low income level) had very high social capital. In our study setting, the sociodemographic composition of people in a neighbourhood, rather than their socioeconomic position, influenced neighbourhood interactions, help, and support (i.e., social capital).

4.3. Factors Associated with Negative Changes in Neighbourhood Social Capital

In rural neighbourhoods/villages, none of the sociodemographic and socioeconomic variables were negatively associated with levels of social capital. On the contrary, in urban neighbourhoods, an increase in the proportion of households with foreign-born adult members was significantly associated with a decrease in the likelihood of having higher social capital, even after controlling for all other sociodemographic and socioeconomic variables. This finding is in line with other (contested) studies that have found that ethnic diversity obstructs social capital development [20,21,52]. One proposed explanation is that people tend to have lower trust in people dissimilar to themselves in terms of income, religion, or ethnicity [53], thus eroding social capital. Consequently, neighbourhood interactions could be harder to develop when people differ regarding ethnicity since "birds of a feather flock together" [54]. Putnam's [52] contested study, based on data from 41 different communities in the US, found that people in ethnically diverse communities tended to withdraw from both bonding (with similar people) and bridging social networks, thus eroding social capital in general. Putnam's claim, known as the "constrict claim," was later tested on the country level in Europe. Gesthuizen, van der Meer, and Scheepers [55] found no support for Putnam's hypothesis in European societies. Instead, the authors concluded that economic inequality and the national history of democracy in European societies were more critical for explaining cross-national differences in social capital in Europe.

Results about the negative influence of ethnic diversity on social capital and cohesion have been criticised for not considering how low socioeconomic status influences social interactions [56]. Another concern is that these results could be used to encourage homogeneity and anti-immigrant policies rather than policies to encourage "strength in diversity" [56]. These conclusions and solutions could have detrimental effects since there is strong evidence for how increased opportunities for interethnic contact facilitate interethnic interactions, which stimulate both out- and in-groups' trust and trust in neighbours [57].

Van de Meer and Tolsma reviewed 90 studies from different countries about the effect of ethnic diversity on social cohesion [57]. In line with our results, they found a consistent association between high ethnic heterogeneity and lower levels of within-neighbourhood social cohesion. In contrast, they did not find any support for a negative association between ethnic heterogeneity and inter-ethnic cohesion. They [57] discuss that this finding supports the notion that ethnic diversity increases opportunities for interethnic contacts, which further increases interethnic trust and thus social cohesion. Further, beyond the spatial level of neighbourhoods, e.g., on a country level, they did not find any consistent evidence that ethnic heterogeneity is negatively associated with social cohesion. The authors [57] discuss that this could potentially be explained by ethnic heterogeneity obstructing the sense of shared social norms in the neighbourhood, which creates uncertainty about how to socially interact with neighbours. However, this uncertainty does not spill over to decrease social cohesion in the overall society since people, in general, are able to distinguish between how they view their immediate environment and how they view the world as such. In addition, van de Meer and Tolsma [57] found that the negative association between ethnic heterogeneity and intra-neighbourhood social cohesion was particularly strong in studies from the US (the only country with some additional evidence for negative spill-over effects on general trust in society). Van de Meer and Tolsma [57] discuss that this could be understood in light of the relatively high levels of heterogeneity in the US, combined with the pronounced segregation of cities and the persistence of ethnic inequalities. Segregation and ethnic inequalities obstruct interethnic contact opportunities and are more likely to occur in a context where multicultural policies are lacking. Taken together, this indicates that segregation, rather than ethnic diversity per se, is the problem [58,59]. Hence, it might not be the percent immigrants that matters for social cohesion; rather, it is the segregation of immigrants.

These results and arguments are important to consider in light of our current results. Sweden has traditionally had a generous immigration policy. However, this changed dramatically in 2015 (during the refugee crisis when Sweden received more than 160,000 asylum seekers), with the launching of a new, far more restrictive immigration policy (first launched as a temporary law but made permanent in 2021). These changes towards very restrictive migration legislation were accompanied by changes in public opinions about immigrants, mirrored by the fact that the ultra-nationalistic and anti-immigrant political party Sweden Democrats became the third-largest party in the parliament elections in 2018. Further, segregation and social inequality have increased significantly in Sweden during the last few decades [60] and are now viewed as huge societal challenges. This indicates that the patterns previously observed in the US with segregation, and ethnic inequalities, could now be evolving in the Swedish context. Thus, we agree with van der Meer and Tolsma [57] that policymakers need to consider the combination of heterogeneity, segregation, and inequality to understand and target the potential adverse effect of ethnic diversity for social capital and sustainable social development. Rather than concluding that homogeneity is "good" for social capital and sustainable social development, policies should focus on actions that can increase opportunities for interethnic contacts and, at the same time, fight inequality and segregation.

Further research is needed on how to stimulate opportunities for inter-ethnic interactions in local communities in various cultural settings. In addition, there is a need for more studies on how to simultaneous strengthen social capital and ensure social sustainable development at different spatial levels, such as overall municipal/city and neighbourhood levels. Strengthening within-neighbourhood ties and cohesion (i.e., bonding social capital) might promote a social sustainable neighbourhood but at the same time lead to polarizations and tensions at the municipal/city level, thus eroding social sustainability at a higher spatial level. Qualitative studies exploring how different social groups experience social capital, social inclusion, and social sustainability in various living environments (e.g., urban versus rural neighbourhoods) are also needed, as well as further investigations on the role of social capital and social sustainability during societal crises such as a pandemic.

4.4. Strengths and Limitations

Using a similar tool and analytical approach in the social capital survey in 2006 and 2020 allowed us to have comparable data in comparing the changes in social capital levels across neighbourhoods in Umeå Municipality. As we only had two data points of measurement for social capital, we had to assume that the levels of neighbourhood social capital did not change swiftly in between the surveys. We believe this assumption is valid, considering the stable nature of social capital in a relatively stable community in northern Sweden. The lack of access to neighbourhood registry sociodemographic and socioeconomic data for the years 2017–2020 limited the analyses to the period time of 2006–2017.

5. Conclusions—Implications for Local Policies Aiming to Strengthen Social Capital for Socially Sustainable Development

Sociodemographic, rather than socioeconomic, factors were associated with levels of social capital at the neighbourhood level. Local policies aiming to strengthen social capital for sustainable social development should therefore strive for neighbourhoods with mixed sociodemographic groups, i.e., designing neighbourhoods that are attractive for families with children, pensioners, and single-parent households. Ensuring that people (want to) spend time in their neighbourhoods is essential for social capital and sustainable social development. This could, e.g., imply planning for meeting places that attract different groups of people, such as libraries, safe and enjoyable playgrounds for children, youth centres for adolescents, and attractive recreation areas for older people. Local access to shops, cafés, and restaurants could also increase social interactions and people's interests in spending time in their neighbourhoods.

Neighbourhood social capital operates and is associated with different factors in urban and rural areas. Thus, policies to strengthen social capital as a means for ensuring sustainable social development need to consider the specific local area and adjust interventions to the local needs. Hence, it is important to consider both the municipal and neighbourhood conditions to ensure sustainable social development in the whole municipality. Overall positive development in the municipality might not benefit all neighbourhoods and vice versa. This requires careful mapping of local needs before implementing any intervention. In some neighbourhoods, interventions to increase neighbour interactions and social cohesion might be needed (i.e., strengthening bonding social capital), to ensure social sustainability. In other neighbourhoods, it might instead be important to increase a sense of inclusion in the municipality/city as a whole to ensure social sustainable development. This could be done by, e.g., increasing involvement and representativeness by the neighbourhood in municipal processes (i.e., building bridging and linking social capital).

Policies to strengthen social capital for sustainable social development should include interventions that increase possibilities for inter-ethnic social contacts, since opportunities for interethnic interactions increase interethnic relations, which in turn increases trust and social cohesion. Supporting opportunities for inter-ethnic interactions requires conscious municipal actions on how to fight segregation and discrimination and creating inclusive and ethnically mixed school-settings, meeting places, and leisure activities. This could, e.g., imply the strategic location of attractive leisure activities and schools to stimulate the flow of people between neighbourhoods, which could then increase opportunities for inter-ethnic contacts.

Author Contributions: Conceptualization, M.E., A.S., L.Z., I.K. and N.N.; methodology, M.E., A.S., L.Z., I.K. and N.N.; software, A.S. and N.N.; validation, M.E., A.S., L.Z., I.K. and N.N.; formal analysis, A.S., M.E., L.Z. and N.N.; investigation, M.E. (PI), A.S. (CO-PI), and N.N. (CO-PI); data curation, M.E., A.S. and N.N.; writing—original draft preparation, M.E. and N.N.; writing—review and editing, M.E., A.S., L.Z., N.N. and I.K.; funding acquisition, M.E., N.N. and A.S. All authors have approved the submitted version of the manuscript and agree to be personally accountable for their own contributions.

Funding: This project was funded by Formas, the Swedish Research Council for Sustainable Development (Dnr. 2018-00262 and Dnr. 2018-00076).

Institutional Review Board Statement: The study was conducted in accordance with the Declaration of Helsinki, and the studies involving human participants were reviewed and approved by Swedish Ethics Review Authority (Dnr: 2019-04395, 2019-10-28; Dnr: 2020-00160, 2020-02-18, 2020-06-08; Dnr 2020-02757). The participants provided their informed consent to participate in this study.

Informed Consent Statement: Informed consent was obtained from all participants involved in the study.

Data Availability Statement: The data presented in this study are available on request from the corresponding author. The data are not publicly available due to restrictions in the ethical approval for this project. Questions related to the data in this study should be directed to the corresponding author.

Acknowledgments: The authors would like to acknowledge Cecilia Falgas-Ravly, for developing the maps of neighbourhood social capital in ArcGIS.

Appendix A

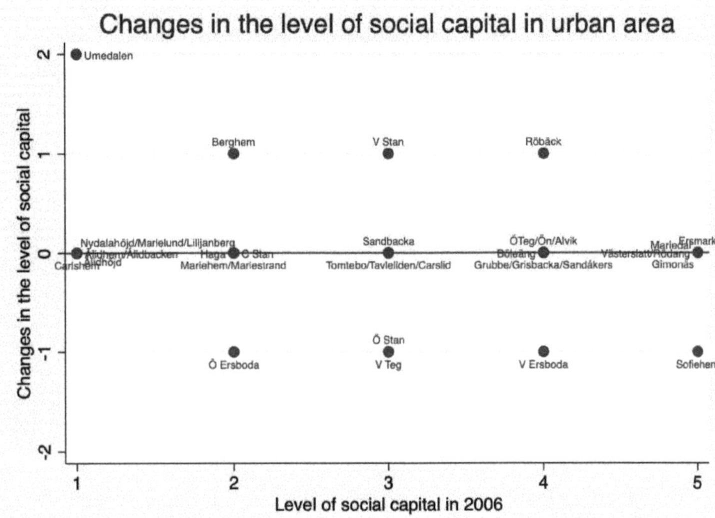

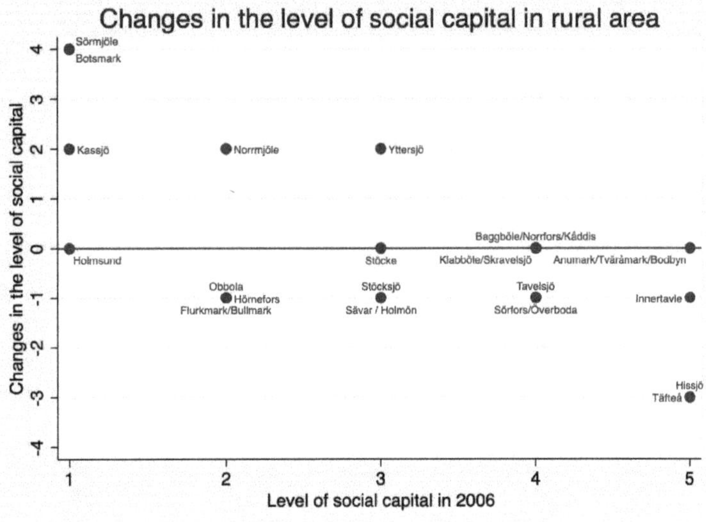

Figure A1. Changes in neighbourhood-level social capital in 26 urban and 20 rural neighbourhoods in Umeå Municipality between 2006 and 2017.

Appendix B

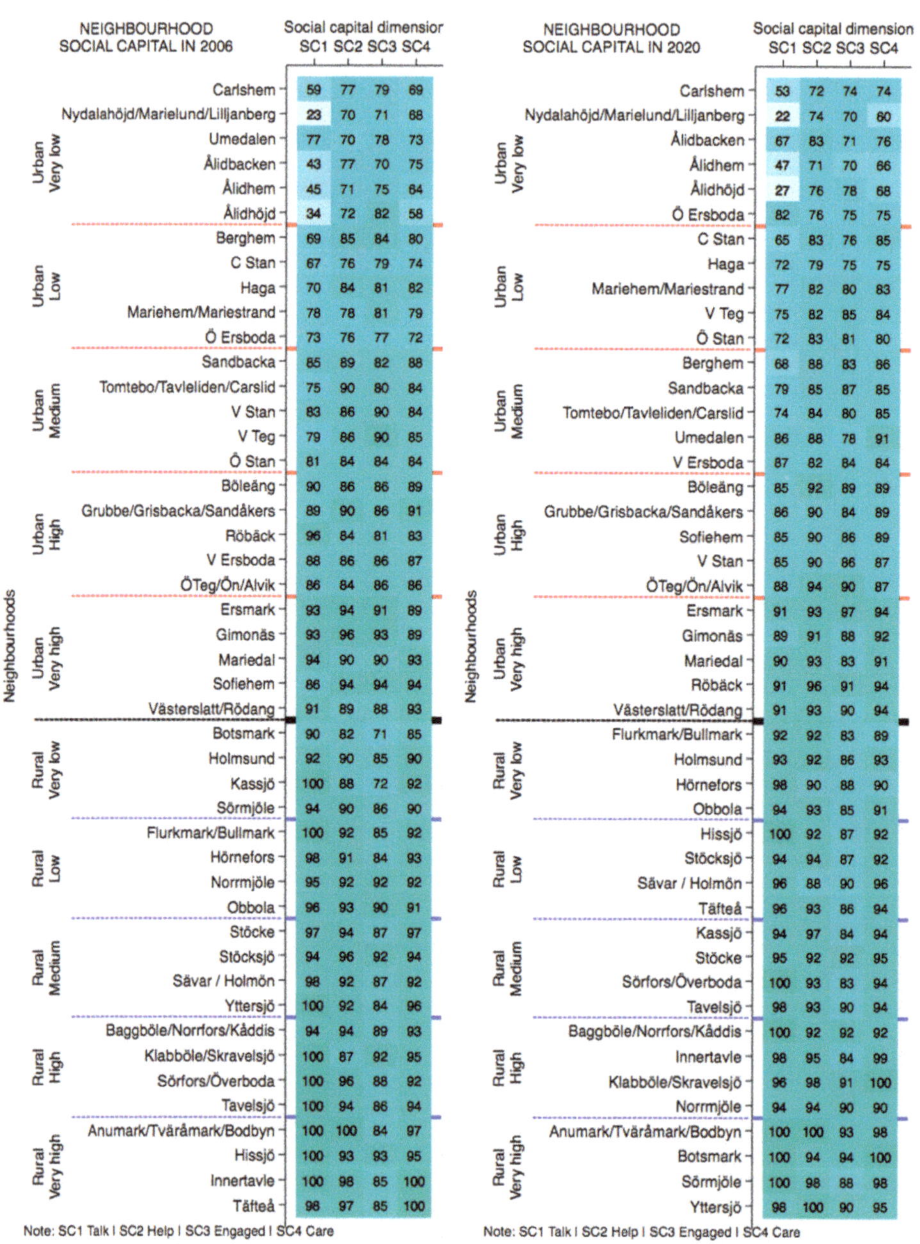

Figure A2. Social capital indicators across different social capital quintiles in urban and rural areas in 2006 and 2017.

Appendix C

Table A1. Probability of belonging to different quintiles of neighbourhood social capital (y-axis) based on different sociodemographic and socioeconomic characteristics (x-axis).

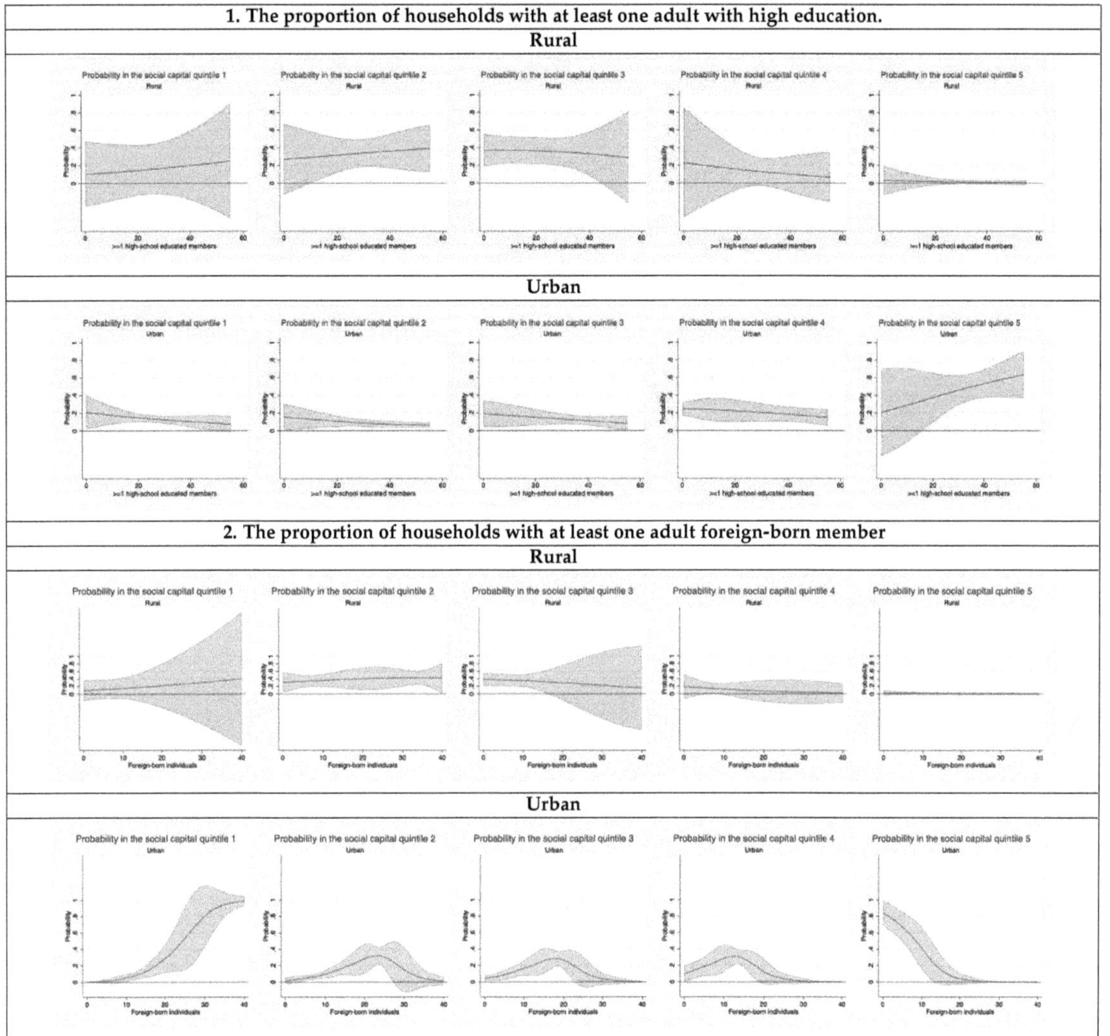

Table A1. *Cont.*

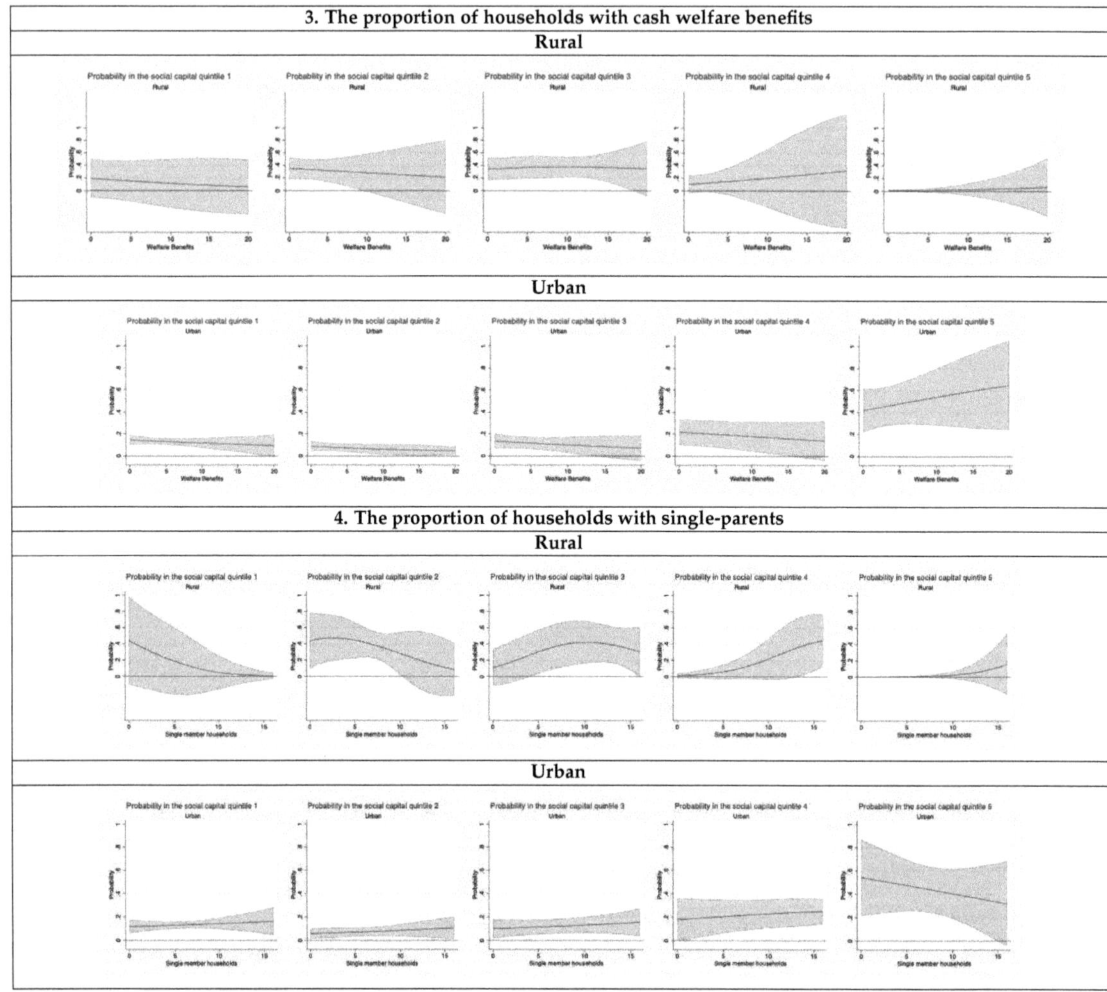

3. The proportion of households with cash welfare benefits

Rural

Urban

4. The proportion of households with single-parents

Rural

Urban

Table A1. *Cont.*

5. The proportion of households with children under 12 years old
Rural

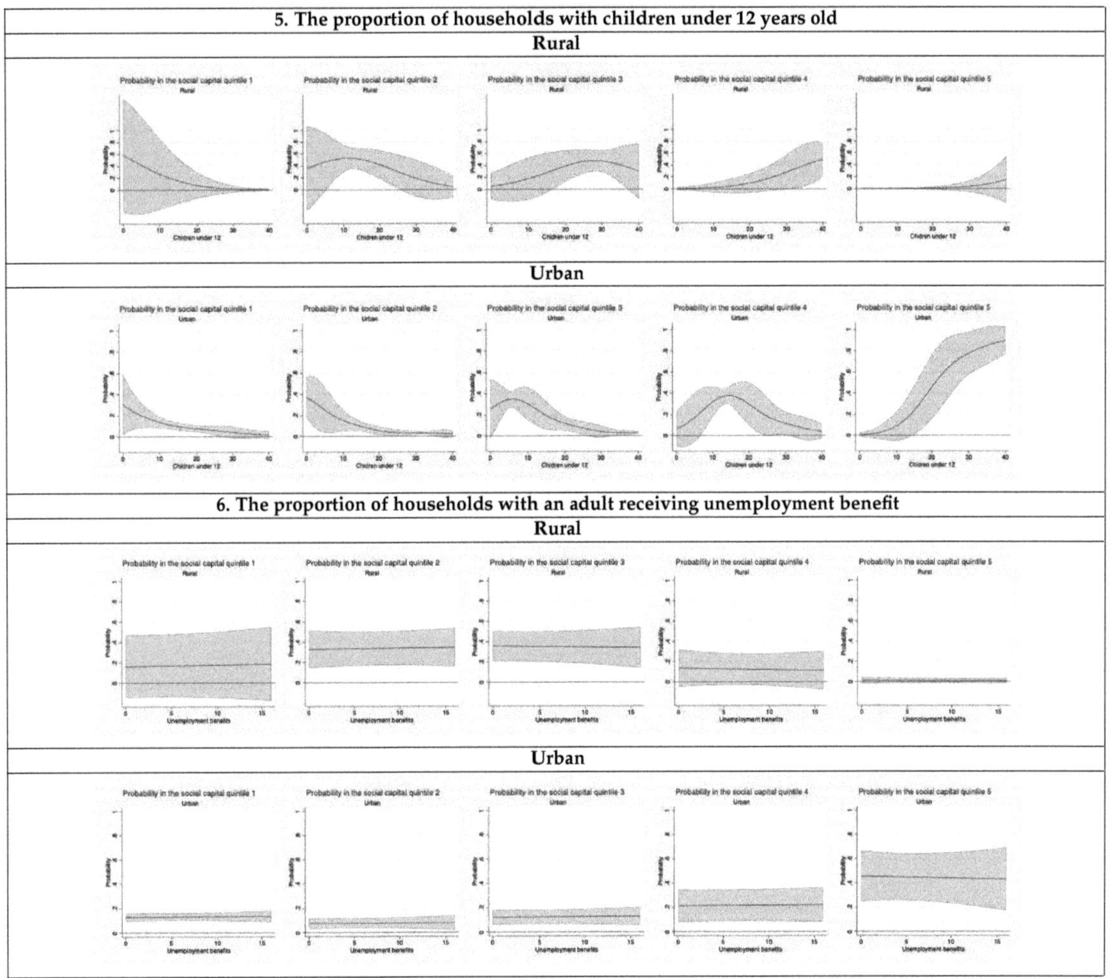

Urban

6. The proportion of households with an adult receiving unemployment benefit
Rural
Urban

Table A1. *Cont.*

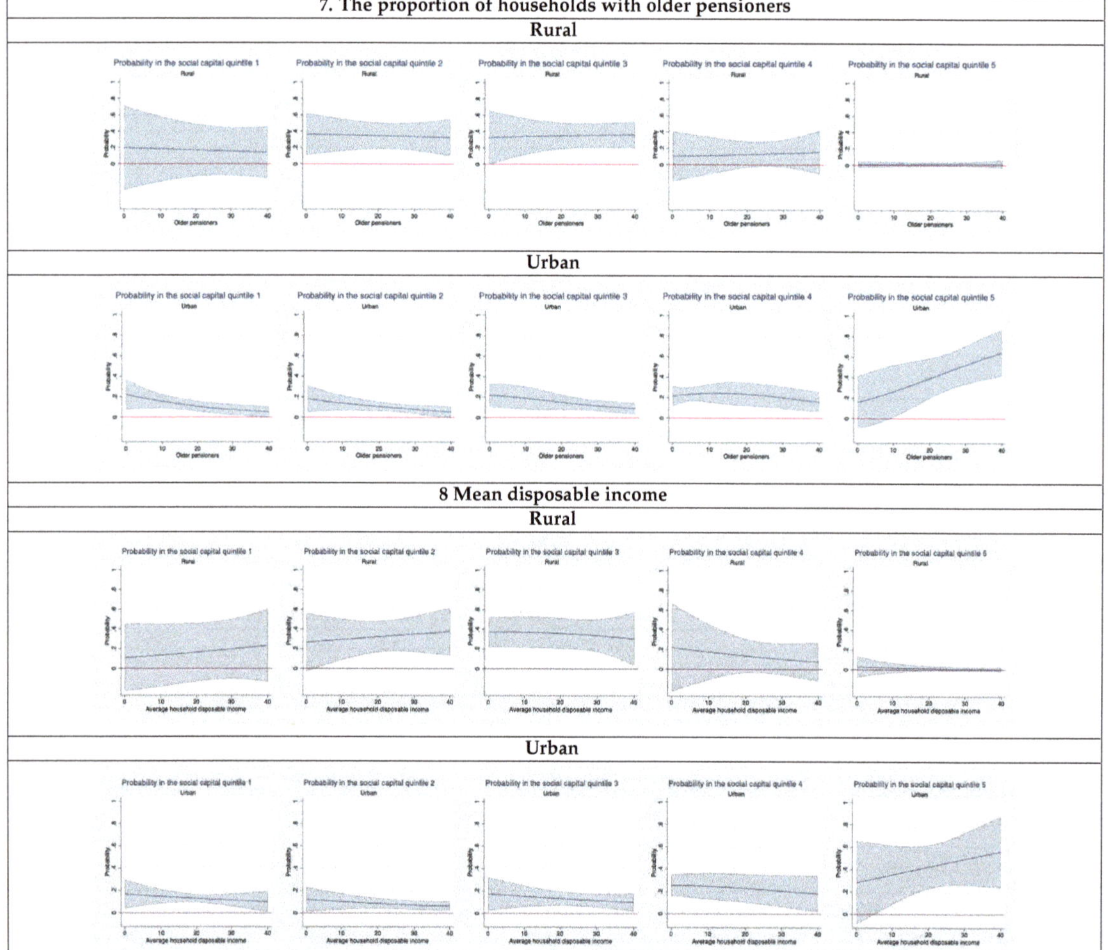

References

1. Shirazi, M.R.; Keivani, R. The triad of social sustainability: Defining and measuring social sustainability of urban neighbourhoods. *Urban Res. Pract.* **2019**, *12*, 448–471. [CrossRef]
2. Boström, M. A missing pillar? Challenges in theorizing and practicing social sustainability: Introduction to the special issue. *Sustain. Sci. Pract. Policy* **2012**, *8*, 3–14. [CrossRef]
3. Dempsey, N.; Bramley, G.; Power, S.; Brown, C. The social dimension of sustainable development: Defining urban social sustainability. *Sustain. Dev.* **2011**, *19*, 289–300. [CrossRef]
4. Schuller, T.; Baron, S.; Field, J. Social capital: A review and critique. In *Social Capital: Critical Perspectives*; Oxford University Press: Oxford, UK, 2000.
5. Woodcraft, S. Social Sustainability and New Communities: Moving from Concept to Practice in the UK. *Procedia-Soc. Behav. Sci.* **2012**, *68*, 29–42. [CrossRef]
6. Yoo, C.; Lee, S. Neighborhood Built Environments Affecting Social Capital and Social Sustainability in Seoul, Korea. *Sustainability* **2016**, *8*, 1346. [CrossRef]
7. Weingaertner, C.; Moberg, Å. Exploring Social Sustainability: Learning from Perspectives on Urban Development and Companies and Products. *Sustain. Dev.* **2014**, *22*, 122–133. [CrossRef]
8. Moore, S.; Kawachi, I. Twenty years of social capital and health research: A glossary. *J. Epidemiol. Community Health* **2017**, *71*, 513–517. [CrossRef]

9.   Putnam, R.D. *Making Democracy Work: Civic Traditions in Modern Italy*; Princeton University Press: Princeton, NJ, USA, 1993.
10. Putnam, R.D. *Bowling Alone: The Collapse and Revival of American Community*; Simon & Schuster: New York, NY, USA, 2000.
11. Portes, A. Social Capital: Its Origins and Applications in Modern Sociology. *Annu. Rev. Sociol.* **1998**, *24*, 1–24. [CrossRef]
12. Hooghe, M. Introduction: Generating social capital. In *Generating Social Capital: Civil Society and Institutions in Comparative Perspective*; Hooghe, M., Stolle, D., Eds.; Palgrave Macmillan: New York, NY, USA, 2003; p. 118.
13. Fukuyama, F. Social Capital and Development: The Coming Agenda. *SAIS Rev.* **2002**, *22*, 23–37. [CrossRef]
14. Semenza, J.C.; Krishnasamy, P.V. Design of a health-promoting neighborhood intervention. *Health Promot. Pract.* **2007**, *8*, 243–256. [CrossRef]
15. Semenza, J.C.; March, T.L.; Bontempo, B.D. Community-initiated urban development: An ecological intervention. *J. Urban Health* **2007**, *84*, 8–20. [CrossRef]
16. Farquhar, S.A.; Michael, Y.L.; Wiggins, N. Building on leadership and social capital to create change in 2 urban communities. *Am. J. Public Health* **2005**, *95*, 596–601. [CrossRef]
17. Michael, Y.L.; Farquhar, S.A.; Wiggins, N.; Green, M.K. Findings from a community-based participatory prevention research intervention designed to increase social capital in Latino and African American communities. *J. Immigr. Minor. Health* **2008**, *10*, 281–289. [CrossRef]
18. Ferlander, S.; Timms, D. Social Capital and Community Building through the Internet: A Swedish Case Study in a Disadvantaged Suburban Area. *Sociol. Res. Online* **2007**, *12*, 1–17. [CrossRef]
19. Vårheim, A. Gracious space: Library programming strategies towards immigrants as tools in the creation of social capital. *Libr. Inf. Sci. Res.* **2011**, *33*, 12–18. [CrossRef]
20. Costa, D.L.; Kahn, M.E. Civic Engagement and Community Heterogeneity: An Economist's Perspective. *Perspect. Politics* **2003**, *1*, 103–111. [CrossRef]
21. Hutchinson, R.N.; Putt, M.A.; Dean, L.T.; Long, J.A.; Montagnet, C.A.; Armstrong, K. Neighborhood racial composition, social capital and black all-cause mortality in Philadelphia. *Soc. Sci. Med.* **2009**, *68*, 1859–1865. [CrossRef]
22. Harpham, T.; Grant, E.; Thomas, E. Measuring social capital within health surveys: Key issues. *Health Policy Plan* **2002**, *17*, 106–111. [CrossRef] [PubMed]
23. Mohnen, S.M.; Groenewegen, P.P.; Völker, B.; Flap, H. Neighborhood social capital and individual health. *Soc. Sci. Med.* **2011**, *72*, 660–667. [CrossRef] [PubMed]
24. Poortinga, W. Social relations or social capital? Individual and community health effects of bonding social capital. *Soc. Sci. Med.* **2006**, *63*, 255–270. [CrossRef] [PubMed]
25. Snelgrove, J.W.; Pikhart, H.; Stafford, M. A multilevel analysis of social capital and self-rated health: Evidence from the British Household Panel Survey. *Soc. Sci. Med.* **2009**, *68*, 1993–2001. [CrossRef] [PubMed]
26. Eriksson, M.; Ng, N.; Weinehall, L.; Emmelin, M. The importance of gender and conceptualization for understanding the association between collective social capital and health: A multilevel analysis from northern Sweden. *Soc. Sci. Med.* **2011**, *73*, 264–273. [CrossRef]
27. Municipality, U. Demografen [The Demographer]. Available online: https://demografi.umea.se/ (accessed on 22 November 2021).
28. European Social Progress Index. Available online: http://ec.europa.eu/regional_policy/en/information/maps/social_progress (accessed on 22 November 2021).
29. Swedish Police. *Utsatta Områden. Social Ordning, Kriminell Struktur Och Utmaningar för Polisen. [Vulnerable Areas. Social Order, Criminal Structure and Challenges for the Police]*; Swedish Police: Stockholm, Sweden, 2017.
30. Santosa, A.; Ng, N.; Zetterberg, L.; Eriksson, M. Study Protocol: Social Capital as a Resource for the Planning and Design of Socially Sustainable and Health Promoting Neighborhoods- A Mixed Method Study. *Front Public Health* **2020**, *8*, 581078. [CrossRef]
31. Fabrigar, L.R.; Wegener, D.T.; MacCallum, R.C.; Strahan, E.J. Evaluating the use of exploratory factor analysis in psychological research. *Psychol. Methods* **1999**, *4*, 272–299. [CrossRef]
32. Vyas, S.; Kumaranayake, L. Constructing socio-economic status indices: How to use principal components analysis. *Health Policy Plan* **2006**, *21*, 459–468. [CrossRef] [PubMed]
33. Woolcock, M.J.V.; Narayan, D. Social capital: Implications for development theory, research, and policy. *World Bank Res. Obs.* **2000**, *15*, 225–249. [CrossRef]
34. SCB. Longitudinal Integrated Database for Health Insurance and Labour Market Studies (LISA). Available online: https://www.scb.se/en/services/ordering-data-and-statistics/ordering-microdata/vilka-mikrodata-finns/longitudinella-register/longitudinal-integrated-database-for-health-insurance-and-labour-market-studies-lisa/ (accessed on 15 November 2021).
35. European Commission. European Social Progress Index. Available online: https://ec.europa.eu/regional_policy/en/information/maps/social_progress2020/ (accessed on 22 November 2021).
36. Westlund, H.; Rutten, R.; Boekema, F. Social capital, distance, values and levels of space. *J. Econ. Behav. Organ.* **2010**, *18*, 965–970.
37. Municipality, U. Umeå Kommuns Bostadsförsörjningsprogram 2017–2024 [Umeå Municipality Housing Supply Programme 2017–2024]. Available online: https://www.umea.se/download/18.2bd9ced91726ea4d7b4484/1592481137809/Bostadsf%C3%B6rs%C3%B6rjningsprogram%202017-2024.pdf (accessed on 22 November 2021).
38. Whiteley, P.F. Economic Growth and Social Capital. *Political Stud.* **2000**, *48*, 443–466. [CrossRef]

39. Raman, S. Designing a Liveable Compact City: Physical Forms of City and Social Life in Urban Neighbourhoods. *Built Environ.* **2010**, *36*, 63–80. [CrossRef]
40. Ziersch, A.M.; Baum, F.; Darmawan, I.G.; Kavanagh, A.M.; Bentley, R.J. Social capital and health in rural and urban communities in South Australia. *Aust. N Z J. Public Health* **2009**, *33*, 7–16. [CrossRef] [PubMed]
41. Sørensen, J.F.L. Rural–Urban Differences in Bonding and Bridging Social Capital. *Reg. Stud.* **2016**, *50*, 391–410. [CrossRef]
42. Fukuyama, F. Social capital, civil society and development. *Third World Q.* **2001**, *22*, 7–20. [CrossRef]
43. Zetterberg, L.; Santosa, A.; Ng, N.; Karlsson, M.; Eriksson, M. Impact of COVID-19 on Neighborhood Social Support and Social Interactions in UmeÅ Municipality, Sweden. *Front. Sustain. Cities* **2021**, *3*, 685737. [CrossRef]
44. Pitas, N.; Ehmer, C. Social Capital in the Response to COVID-19. *Am. J. Health Promot.* **2020**, *34*, 942–944. [CrossRef] [PubMed]
45. Aldrich, D.; Meyer, M. Social Capital and Community Resilience. *Am. Behav. Sci.* **2015**, *59*, 254–269. [CrossRef]
46. Helliwell, J.F.; Huang, H.; Wang, S. Social Capital and Well-Being in Times of Crisis. *J. Happiness Stud.* **2014**, *15*, 145–162. [CrossRef]
47. Nakagawa, Y.; Shaw, R. Social Capital: A Missing Link to Disaster Recovery. *Int. J. Mass Emerg. Disasters* **2004**, *22*, 5–34.
48. Dekker, K. Social Capital, Neighbourhood Attachment and Participation in Distressed Urban Areas. A Case Study in The Hague and Utrecht, the Netherlands. *Hous. Stud.* **2007**, *22*, 355–379. [CrossRef]
49. Eurostat. How Many Single-Parent Households Are there in the EU? Available online: https://ec.europa.eu/eurostat/web/products-eurostat-news/-/edn-20210601-2 (accessed on 22 November 2021).
50. Stavrova, O.; Fetchenhauer, D. Single Parents, Unhappy Parents? Parenthood, Partnership, and the Cultural Normative Context. *J. Cross-Cult. Psychol.* **2014**, *46*, 134–149. [CrossRef]
51. Glatz, C.; Bodi-Fernandez, O. Individual social capital and subjective well-being in urban- and rural Austrian areas. *Osterr. Z. Für Soziologie* **2020**, *45*, 139–163. [CrossRef]
52. Putnam, R.D. E Pluribus Unum: Diversity and Community in the Twenty-first Century The 2006 Johan Skytte Prize Lecture. *Scand. Political Stud.* **2007**, *30*, 137–174. [CrossRef]
53. Vermeulen, F.; Tillie, J.; van de Walle, R. Different Effects of Ethnic Diversity on Social Capital: Density of Foundations and Leisure Associations in Amsterdam Neighbourhoods. *Urban Stud.* **2011**, *49*, 337–352. [CrossRef] [PubMed]
54. Rostila, M. Birds of a feather flock together–and fall ill? Migrant homophily and health in Sweden. *Sociol. Health Illn.* **2010**, *32*, 382–399. [CrossRef] [PubMed]
55. Gesthuizen, M.; Van der Meer, T.; Scheepers, P. Ethnic diversity and social capital in Europe: Tests of Putnam's thesis in European countries. *Scand. Political Stud.* **2009**, *32*, 121–142. [CrossRef]
56. Letki, N. Does Diversity Erode Social Cohesion? Social Capital and Race in British Neighbourhoods. *Political Stud.* **2008**, *56*, 99–126. [CrossRef]
57. Van der Meer, T.; Tolsma, J. Ethnic Diversity and Its Effects on Social Cohesion. *Annu. Rev. Sociol.* **2014**, *40*, 459–478. [CrossRef]
58. Laurence, J. Wider-community Segregation and the Effect of Neighbourhood Ethnic Diversity on Social Capital: An Investigation into Intra-Neighbourhood Trust in Great Britain and London. *Sociology* **2017**, *51*, 1011–1033. [CrossRef]
59. Uslaner, E. Trust, Diversity, and Segregation in the United States and the United Kingdom1. *Comp. Sociol.* **2011**, *10*, 221–247. [CrossRef]
60. Delegationen mot Segregation. *Segregation i Sverige–Årsrapport 2021 Om Den Socioekonomiska Boendesegregationens Utveckling [Segregation in Sweden–Annual Report about Development in Socioeconomic Housing Segregation]*; Delegationen mot Segregation [The Delegation against Segregation]: Stockholm, Sweden, 2021.

 *sustainability*

MDPI

Article

The Lived Experience of Residents in an Emerging Master-Planned Community

Laurie Buys [1,*], Cameron Newton [2] and Nicole Walker [1]

[1] Health and Behavioural Sciences Faculty, University of Queensland, St Lucia, QLD 4072, Australia; n.walker4@uq.edu.au

[2] Faculty of Business and Law, Queensland University of Technology, Brisbane, QLD 4000, Australia; cj.newton@qut.edu.au

* Correspondence: l.buys@uq.edu.au

Abstract: Master-planned communities around the world are developed and purposefully planned to address housing sustainability and community connectivity; they often have a distinctive look, and appeal to a particular customer base desiring a strong, utopian-esque community. However, the lived experience of new residents joining master-planned communities has not been explored. This paper examines the lived experience of new residents within an emerging Australian master-planned estate, and reports on the first two stages of a longitudinal study focusing on the results of an online forum. This unique study presents real-life findings on a culturally diverse community. The findings reveal how the purposeful development of community identity in the early stages of the MPCommunity has not led to satisfactory levels of social infrastructure or social connectedness for the pioneering residents. The physical and social environment, as interpreted by residents against the developers' imagined vision and marketing testimonies, has not been entirely satisfactory. Infrastructure issues—such as transport, and access to daily activities such as shopping, work, and school—were points of frustration and dissatisfaction. The findings provide insight into the challenges and opportunities for residents in a developing MPC, and further our understanding of the specific factors that inform us as to how social infrastructure can best encourage and support connection within existing and future MPC developments.

Keywords: master-planned estate; community; community identity; social connectedness; social infrastructure: physical infrastructure; housing developments; longitudinal study

Citation: Buys, L.; Newton, C.; Walker, N. The Lived Experience of Residents in an Emerging Master-Planned Community. *Sustainability* **2021**, *13*, 12158. https://doi.org/10.3390/su132112158

Academic Editor: Vida Maliene

Received: 30 September 2021
Accepted: 2 November 2021
Published: 4 November 2021

Publisher's Note: MDPI stays neutral with regard to jurisdictional claims in published maps and institutional affiliations.

1. Introduction

Creating residential communities that provide housing and accommodation along with an engaging environment that facilitates community connections has been a priority for many cities and communities. There have been a variety of approaches to addressing this challenge, with recent development focusing on master-planned communities (MPCs) [1]. MPCs are capital-intensive forms of master-planned estates (MPEs), purposefully designed to encourage the development of community identity and connectivity [2]. Given recent global events, including COVID-19, providing insight into how MPCs are experienced in the initial stages of their development, by people from diverse backgrounds, is important to delivering sustainable communities [3]. This study explores the lived experiences—specifically social connectedness—in an attempt to gain a better understanding of the issues faced by a group of pioneer residents in relation to their everyday community lives together in the early stages of MPC development; it does this by first examining the notion of MPCs, and the characteristics that are designed to influence the development of community within them, and then investigates an incipient MPCommunity (the label for the case study MPC) in Australia.

Sustainability **2021**, *13*, 12158. https://doi.org/10.3390/su132112158

https://www.mdpi.com/journal/sustainability

1.1. Community

The term "community" is part of common parlance, and is entrenched in urban planning and design, yet it lacks a fixed meaning, either stated or contested. It must be acknowledged that communities are not static entities, but are dynamic, and self-organise over time. Structures of interaction between residents change both as an outcome of changes in local interactions, and/or in response to external social challenges, such as COVID-19 [4].

Community has positive connotations as an adjective elevating nouns such as "development", "policing", and "garden" to a level of ethical superiority that does not exist without its use as a modifier [5]. While the word "community" is often considered to be variable and ambiguous, there is agreement among researchers that any definition should include three basic elements [6]: (1) shared territory, (2) meaningful social interaction, and (3) significant social ties [6]. While communities as networks exist beyond localised scales, the primary concern among academics and the urban planning and design professions is the importance of place to notions of community [3]. This paper looks at communities at the localised neighbourhood scale within an MPC.

1.2. Social Connectedness

The interaction between individuals is multifaceted, at the individual, community, and societal levels. In the case of MPCs, the developer consciously constructs an appealing idyllic lifestyle vision encompassing consumer preferences, community ethos, and lifestyle needs, with marketing strategies therefore attracting a specific consumer group. However, this may only account for one layer of what defines a community/neighbourhood. This paper examines the lived experiences and the role in the development of social infrastructure in socially connected communities.

1.3. Master Planned Community

MPCs in Australia are usually large, bounded, greenfield developments on city fringes, with significant planning challenges. These challenges include assimilating the development into existing urban landscapes, rolling out suitable support services, and infrastructure (e.g., telecommunications, schools, shops, transport, etc.) [7]. To address these planning issues, MPCs incorporate complementary mixed-use development as part of their multiphase master-planned outcomes [7]. In the Australian context, MPCs are the most capital-intensive type of modern planned private housing estate, integrating different levels of physical and social infrastructure [8]; they are designed to have an individual look, differentiating the MPC as a unique community distinctive from neighbouring residential areas, and thereby creating appeal to a market segment seeking such a community [9]. It has been suggested that MPCs seem to meet a real need in the market for community [3].

The strategic intention of an MPC is strengthened through place-making approaches and a "community compact" [2]. These are designed to facilitate social interaction, foster a collective identity/sentiment, and encourage the development of relational community, thereby building links/bonds, trust, and shared values between residents [7]. The community compact seeks to create immediate familiarity in order to develop "rapid" trust between residents, and between the developer and residents, through establishing homogeneity of values, aspirations, and lifestyles [2]. This immediate familiarity is in contrast to the usual habitual familiarity established over time within neighbourhoods or estates. The MPC community compact and place-making approaches are often complemented with community development programs designed to tie the developer and residents to the MPC vision and establish a localised practice of "community" with a social attachment to place [8].

Planners/developers' community development programs often initiate community sanctioning activities such as community events, newsletters, and a designated website to establish and kindle the community ethos [10]. MPC residents also initiate activities that serve to legitimise the community ethos, the common social code, and estate conventions within an amiable setting [2]. It is through the community compact and community activities that residents' expectations are set for behaviour that aligns with belonging to a great community [10].

While MPCs are usually defined as geographically bounded sites [3], they are designed to be socially and community bounded spaces, with developing a strong, place-based sense of community as the primary goal. The effectiveness of the social space is critical to developing community in the MPC [2,11]. Social space provides residents with the opportunity for regular and spontaneous face-to-face interactions in their everyday living, thereby enabling communal/neighbourhood connection and community development [12]. This is achieved through a pedestrian-friendly environment of wide footpaths with walkable destinations from mixed-use development (e.g., shops, bars, eating places), pocket parks, and sporting facilities freely and easily available to the community [13]. An MPC with such an environment establishes a distinct social identity and provides casual meeting places that are prerequisites to social connectivity and community development [13].

1.4. Context

Australia, like many developed nations, has begun to embrace MPC development. MPCs are seen as a viable strategy to address housing shortages in cities and communities across Australia [14]. MPCs have place-making and community-making approaches purposefully designed to target a particular market segment, which promote the concept of community wellbeing and social connectivity.

Research of the lived and dynamic nature of community life in Australian MPCs is underdeveloped [8], and it is not clear how MPCs are experienced by new residents in the early stages of their development. The present study explores the MPC phenomenon through the lived experience of a diverse group of pioneer residents from a growing MPC in Australia. This study seeks to understand the lived experience of moving to a new MPCommunity.

2. Materials and Methods

One of the biggest challenges facing communities is attempting to measure outcomes that are changeable and inconsistent. Applying one measure fails to capture the dynamic needs and preferences of the individuals and the community. Therefore, in this research project, both quantitative and qualitative techniques were employed to uncover deeper insights from the residents' lived experiences.

Qualitative research reveals subjective perceptions and meanings of the social reality; however, it does not identify trends or suggest any generalisation of results across populations. This research aimed to unearth phenomenology to provide a deeper understanding by highlighting meaningful responses from residents [15].

Qualitative research is a recognised and rigorous social research method that is used to gain in-depth knowledge and understanding of phenomena, issues, or questions [16]. This method is often used to explore a topic for the first time, and is quite often used in building and environmental research [17].

To explore the lived experiences of residents over time, a larger longitudinal study was designed, employing mixed methods, and collecting quantitative and qualitative data at 9–10-month intervals over a 5-year period commencing in July 2018. The present study focuses on data collected from the first two stages (July 2018 and May 2019), thereby capturing the lived experience of residents in the early stages of community development within the MPCommunity. A single case study was utilised in order to generate an in-depth, multifaceted understanding of this complex real-life context [18]. This study received ethical approval from the Queensland University of Technology's Human Research Ethics Committee (approval number 1600001085).

2.1. The Case Study

The case study site is a developing MPC located approximately 29 km from an Australian capital city. The MPCommunity was developed by two major Australian residential estate developers. The estate has access to a major highway and a train into the capital city. The developers' aim was to design a community for easy living, with access to local shops and services, green pedestrian connections, wide open spaces, and public transport. The developers' marketing campaign described living at the MPCommunity as being highly desirable in terms of amenities, access to the capital city, and available green space. It is anticipated that the development and delivery of major infrastructure will take around 12–14 years. It is proposed that the MPCommunity will include approximately 7000 homes of varied housing types upon its completion.

The MPCommunity is fast-selling. When the first package was marketed, a countdown timer for registration was initiated on the MPCommunity's website; there were ~1000 interested buyers waiting to put their names down. The capacity of 250 buyer registrations was saturated within 45 s. Buyers were invited to visit the site, and the 75 available lots were sold the following weekend.

The real-estate market in the capital city is tight, with buyers often having to wait years to secure land. The MPCommunity's developers tried to make the process of buying fair for everyone. Interested buyers participated in a lotto-style ballot, registered with AUD 1000, for the right to be part of the event.

2.2. Procedure

Over two time periods—a week in each of July 2018 (Time 1) and May 2019 (Time 2)—all residents and buyers of the MPCommunity were invited by the researchers to engage in an online forum. Initially, participants created a unique account that enabled them to enter the online site during the "open days".

2.3. Online Forum

This study used quantitative and qualitative measures to deliver an engagement approach that taps into the pulse of community feelings and experiences. All residents and buyers of the MPCommunity were invited to participate in both the online survey and the online forum.

Initially, demographic data were collected from participants, and they were invited to complete the Australian Unity Personal Wellbeing Index and Regional Community Wellbeing Survey. These instruments were used as a framework to identify self-perceived wellbeing in Australian adults in domains such as material wellbeing and personal and community relationship wellbeing [19], examining age, income, relationship status, and other factors that affect personal wellbeing. Since 2001, 37 nationally representative surveys have been conducted using these tools, with more than 65,000 Australian adults participating. The Australian Unity Wellbeing Index is a robust and established instrument that gives a "more comprehensive picture of wellbeing for the individual, the community and society" [19] (p.16).

Qualitative data were gathered over a week in each of July 2018 and May 2019 by using an online forum that was moderated by an experienced online forum facilitator. Online forum conversations were gathered, centred on residents' experiences and perceptions of the topics of place, people, and pursuits within the new MPCommunity. Three questions were asked during the forum: (1) Who are you connecting with in your village? (people); (2) Where are you going to in the village and your local area? (place); and (3) what are you doing in your village and surrounding areas? (pursuits). Participants used words to describe their experiences, and could upload photos or videos. Once the participants had given their responses, they joined the group for an online discussion (if they desired). To facilitate participant engagement, the format of the online discussion was similar to popular social media platforms.

2.4. Analysis

Transcripts of online conversations were analysed using a data-driven approach, where data were examined and organised, manually interpreting and identifying emerging themes. Nvivo12 software was used to support our analyses. Strategic decisions, based on the data analysis and interpretations, determined patterns or themes pertaining to commonalities and the range of variation in participants' experiences and perceptions [20]. This process included iterative data reviews and inter-rater agreement (reliability), establishing a nuanced understanding of the data in terms of emerging themes and meaningful categories [20]. The data were manually coded with the emerging key themes of hope and expectations, sense of community, feelings of exclusion, physical and social infrastructure, and developer-initiated community activities. This process captured conceptual categories that are empirically grounded, revealing the lived experiences of the participants [21].

3. Results

3.1. Demographic Characteristics of Participants

The demographic profiles of the participants are described in Table 1. Of the 154 participants, there were more males than females. Most participants were younger than 45, married/partnered, not Australian-born, educated to a graduate/postgraduate level, had an income of less than AUD 120,000, and reported their cultural origins as Australian, Indian, or "other".

Table 1. Demographic information.

Topic/Area	Category	Participants Time 1 ***	Participants Time 2 ***
Gender	Male	45	42
	Female	35	32
Age	Younger than 35	39	33
	Between 35 and 45	29	33
	Older than 45	8	5
	Omitted response	4	3
Relationship status	Married/Partnered	71	66
	Single	6	4
	Divorced	2	2
	Omitted response	1	1
Education	High school	7	10
	Postgraduate	21	24
	Trade/post-secondary	19	14
	University	31	24
	Omitted response	2	2
Income	More than AUD 140,000	13	18
	AUD 120,000–140,000	9	20
	Less than AUD 120,000	44	23
	Omitted response	14	13
Born in Australia	Yes	27	26
	No	53	47
	Omitted response		1

Table 1. *Cont.*

Topic/Area	Category	Participants Time 1 ***	Participants Time 2 ***
	Australian	27	21
	Indian	^	20
Culture	Mixed culture *	^	7
	Other **	^	24
	Omitted response		2
	Total participants	80	74

NB: ^ Time 1 did not capture any other culture apart from Australian. * Participants identified with more than one culture, i.e., Australian and Indian. ** "Other" includes Asian, Sri Lankan, Maltese, Italian, Filipino, Vietnamese, Egyptian, African, etc. *** Times 1 and 2 were 10 months apart.

3.2. Quantative—Survey Findings

Analysis revealed differences between Time 1 and Time 2. In particular, participants reported increased satisfaction with safety and security at Time 2 compared to Time 1; in contrast, participants reported reduced satisfaction with the sense of being a part of a community (found at Time 2 compared to Time 1)—that is, participants grew more confident over time about safety and security at the MPCommunity, but at the same time felt less connected to the community. In other words, participants reported high internal life satisfaction, which contrasts with possible external social connectedness. At Time 1, residents reported high overall personal satisfaction across most domains. Of note, at Time 1, scores in the domains of personal relationships and overall life satisfaction were very high (88.2 and 82.5, respectively) in comparison to the national average (see Table 2).

Table 2. Survey questions and results at Time 1 and Time 2.

	Question	Average (Mean) Score MPCommunity Time 1	Average (Mean) Score MPCommunity Time 2	Average (Mean) Scores 2018 *
1	How satisfied are you with your life as a whole?	82.5	79.6	75.9
2	How satisfied are you with your standard of living?	83.1	81.6	79.4
3	How satisfied are you with your health?	78.6	76.3	73.2
4	How satisfied are you with what you are achieving in life?	81.4	79.5	72.6
5	How satisfied are you with your personal relationships?	88.2	84.5	78.6
6	How satisfied are you with how safe you feel?	67.8	75.9	80.3
7	How satisfied are you with feeling part of your community?	78.2	70.0	72.4
8	How satisfied are you with your future security?	70.0	70.3	71.02
9	Personal Wellbeing Index (PWI)	78.87	77.2	75.5

* Average (mean) level of wellbeing in 2017 [22].

Furthermore, results revealed decreases in satisfaction in the domains of standard of living, health, life achievements, and general personal wellbeing (PWI) from Time 1 to Time 2 (81.6, 76.3, 79.5, and 77.2, respectively).

Interestingly, when examining participant responses to question 6 regarding safety, at Time 1, the median score was 7; however, over half of the participants scored below 7 (see Figure 1). Therefore, it is plausible to infer that many residents felt dissatisfied with aspects of safety within the MPCommunity.

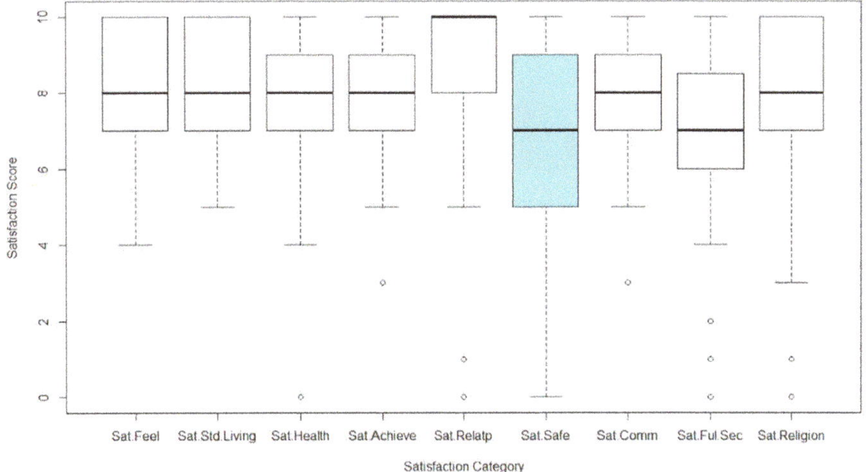

Figure 1. Time 1 (Round 1) satisfaction score distribution.

Further when comparing the same question (Sat.Safe) at Time 2, it was evident that the score range had shrunk, spanning from 4 to 10, which may indicate that those scoring below 7 at Time 1 were more assured over time (see Figure 2). However, as stated above, the scores of this domain remained lower than the national average (see Table 2), which possibly indicates that participants were not fully satisfied with their safety at the MPCommunity. Indeed, there was a great deal of discussion in the qualitative data regarding safety and security.

The distribution analysis for Time 2 evidenced a drastic change in the respondents' sentiments with regard to the level of satisfaction with being part of a community (Sat.Comm). As shown below, the score distribution ranged from 1 to 10 (see Figure 2). When compared with the scores for Time 1, 'Sat.Comm' showed negative growth; therefore, it is possible to argue that MPCommunity respondents were less satisfied with their community than in the previous year.

Analyses of the data using *t*-tests confirmed the sentiments of MPCommunity participants, revealing that the changes across seven of the nine domains from Time 1 to Time 2 did not reach significance, therefore indicating little change in the participants' emotional experiences across time in those domains (see Table 3). However, further substantiating the considerable mean changes reported above for satisfaction with safety (Sat. Safe) and satisfaction with being part of a community (Sat.Comm), the *t*-tests affirmed the changes, with both reaching significance ($p = 0.017$ and $p = 0.010$, respectively) from Time 1 to Time 2. Specifically, satisfaction with safety increased over time. In contrast, satisfaction with being a part of the community declined over time. The qualitative analysis further examines the intersection between safety and the sense of belonging to the community.

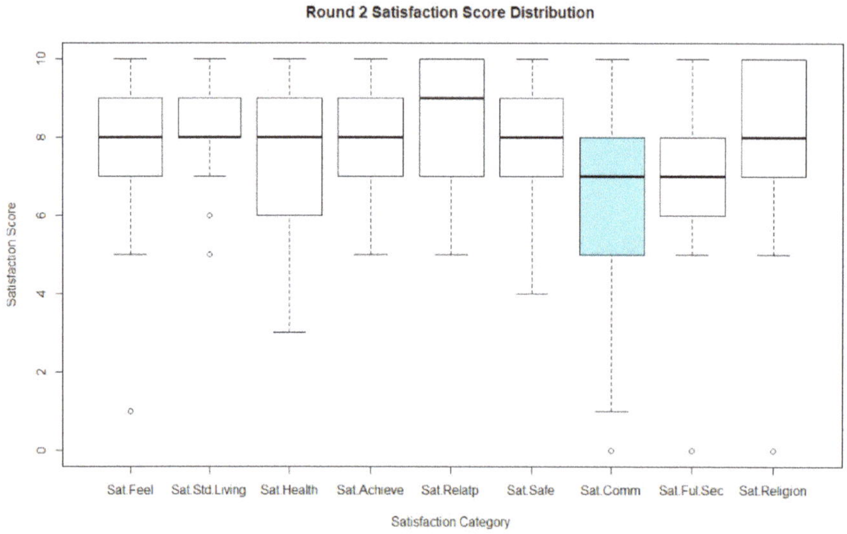

Round 2 Satisfaction Score Distribution

Figure 2. Time 2 satisfaction score distribution.

Table 3. *t*-Test satisfaction results.

Satisfaction Domains	Time 1 Mean Scores	Time 2 Mean Scores	*t*-Test Results *p* Values
Sat.Feel	8.25	7.96	0.245
Sat.Std.Living	8.31	8.16	0.513
Sat.Health	7.86	7.63	0.427
Sat.Achieve	8.13	7.95	0.448
Sat.Relatp	8.82	8.45	0.187
Sat.Safe	6.78	7.59	0.017 ***
Sat.Comm	7.83	7.00	0.010 ***
Sat.Fut.Sec	7.00	7.03	0.933
Sat.Religion	7.87	8.32	0.198

NB: *** $p \leq 0.05$ indicates statistical differences from Time 1 to Time 2; $p \geq 0.05$ indicates no statistical differences from Time 1 to Time 2.

3.3. Qualitative—Forum Findings

The forum extended the understanding of the data, and provided further clarity to distinguish differences between Time 1 and Time 2. The robust discussions illustrated the magnitude and earnestness of the connections participants had with the MPCommunity. Interestingly, many participants expressed concern about the consolidation of community within the MPCommunity. However, this could be the logical and even expected progression of community development in the MPCommunity. Importantly, it could be argued that, at Time 1, results revealed the anticipation and excitement—a type of "honeymoon" period—reflecting residents' hopes and expectations, rather than any real sense or experience of community participation. Meanwhile, Time 2, after a period of 10 months, revealed the actual incipient community experiences of residents after the "honeymoon" period of the development, inclusive of the length of time it takes for promised planned infrastructure to materialise. The qualitative results describe this transition under the theme headings discussed below.

3.3.1. Theme 1: Hopes and Expectations

Participants overwhelmingly reported loving living at the MPCommunity and feeling a personal sense of renewal when first becoming residents. Moving into the MPCommunity provided residents with the chance for a new start. Most of the participants felt lucky to have secured a piece of the Australian dream of home ownership, often for the first time. New residents resembled pioneers or honeymooners, with high hopes for their future lives within the MPCommunity. Participants indicated that they expected a high standard of living at the MPCommunity. The findings suggest that participants had bought into the marketing hype around the development, and that their expectations were piqued as a result.

> *"True, peaceful and quite (sic) life which me and my family looking after . . . and I think we can get in the MPCommunity."* (R1, Male, 32)

> *"Moved to* [MPCommunity] *from Sydney with so much expectations. We are newbies in the community and would love being active in this growing community as I know* [MPCommunity] *people would listen and would be caring for all our thoughts and views."* (R2, Female, 35)

The discussion had a nostalgic feel of desire for a utopian, more traditional neighbourhood, community, and set of values.

> *"I want to know who my neighbours are so we can look out for each other and create friendships like they do in the old days.".* (R2, Female, 30)

> *"I want neighbours to know one another, to be eating in one another's homes, to see kids playing with each other outside.".* (R2, Male, 32)

3.3.2. Theme 2: Sense of Community

Despite the decline in terms of satisfaction with being part of a community, participants talked about *"loving living at* [MPCommunity]*"* (R2, Female, 59). There was great optimism discussed about the case study community, discussed at Time 1 and Time 2.

> *"*[MPCommunity] *has the vibe of a country town with its sense of community while conveniently being close to the CBD. Strangers welcome each other and talk which is really nice and different to where I am moving from."* (R1, Male, 28)

> *"I love the growing community feel (even if this means we still have a long way for dirt flying around, construction on going, constant flyers for blinds and pavers filling up the mailbox)."* (R2, Female, 35)

At Time 2, participants identified a number of issues affecting positive community development. They acknowledged that they wanted more frequent and deeper interactions with their neighbours—to create a sense of community/togetherness—*"to feel connected and inclusive"* (R2, Female, 35).

While they wanted more meaningful and frequent interactions, not all contact was well received.

> *"*[I] *look them in the eye and say 'hello, how are you?' They turn their back on you and walk away. I guess some people just don't like to be polite and courteous".* (R2, Male, 35)

However, at Time 2, participants recognised the importance of individual responsibility in establishing a sense of community, and working with the developer in making this happen.

> *"Given that* [the developer] *is trying so hard to build a community, I'd like to also see people responding to that positively."* (R2, Male, 32)

Participants acknowledged the "need to take initiative at least rather than waiting for [an]other to start first" [to] "invite your neighbour to your house" (R2, Male, 45).

3.3.3. Theme 3: Feelings of Exclusion

The forum revealed feelings of exclusion within the general community. These included couples without children, who reported that they felt outside the group norm for most organised community activities.

> *"We have found that a lot of the activities are aimed at families and children which is great for the kids (unfortunately we have been unable to have kids of our own) . . . {and} . . . feel a little left out of the community feel.".* (R2, Female, 35)

Participants without children often had pets, and they indicated that there were a lot of developer-planned activities within the MPCommunity that excluded four-legged family members. Those participants felt that centrally organised community activities that included their pets would be a useful way for them to connect and keep connected to the MPCommunity.

> *" . . . we have tried to get involved with some other activities however we were turned away as we had our little dog with us, it would be great for* [the developer] *to run activities that are inclusive for all as we would love to feel connected to our community".* (R2, Female, 35)

The perceived cultural differences within the MPCommunity population also appeared to trigger feelings of exclusion. These cultural differences appeared to discourage social interaction, and resulted in cultural and ethnic-based tensions.

> *"I also think there is a small amount of racism particularly toward our Indian population. People have made comments that they don't want to attend particular community events or join particular sporting clubs because it's only Indians that attend these events".* (R2, Female, 32)

Some participants indicated that there were a number of resident-planned neighbourhood activities in the MPCommunity areas that were culturally based and by invitation only, which resulted in some of the non-invitees feeling excluded.

> *"Less division from the Indian community in the estate. Shown through their many Indian events over the last couple months . . . it feels very segregated".* (R2, Female, 26)

3.3.4. Theme 4: Physical and Social Infrastructure

After the honeymoon period comes the reality of the day-to-day living experience in an incipient MPC. The findings suggest that participants became impatient with the time it was taking to deliver the promised, master-planned infrastructure. MPCommunity residents often have no choice but to travel long distances for everyday activities such as shopping, work, school, etc. This results in residents spending a lot of time commuting outside of the MPCommunity, which could be spent within the MPCommunity if that physical and social infrastructure was available to them locally.

> *"grocery is the biggest frustration, I spent around 3 to 4 h traveling to other suburb weekly to get the basic groceries".* (R1, Male, 32)

Residents purchased property in the estate because of the promised and planned infrastructure; however, resident patience started to wane with the extent of commuting required in their daily activity. Most residents use their private cars, as public transportation was found to be non-existent or inconvenient due to frequency or accessibility to transport hubs.

> *"I'd also like to see the public transport like a bus service to the station and nearby shops as well as an easier pathway to get to the station."* (R2, Male, 32)

> *"Bus service* [hasn't] *started yet, I know shuttle is in service but it has very limited times and at weekends nothing at all."* (R2, Female, 35)

However, while participants voiced their frustration, they remained hopeful that the services and infrastructure would eventually be available.

"Of course things will only get better." (R2, Female, 40)

"I think It will be much easier getting around once more infrastructure is in place." (R1, Female, 36)

There was considerable discussion around the desirability/need for additional infrastructure, facilities, and services, including:

- A shopping/town centre that includes at least one of the major grocery chains, *"good eat out places eg pizza shop hungry jacks and local market"* (R2, Male, 36), a post office, library, and a medical centre that includes dental and allied health services.

  "It would be much better if we had a town centre asap". (R2, Female, 40)

  "I want the Town Centre to be busy, with people walking or riding to the shops and stopping to stay a while and chat to friends and eat in the cafes". (R2, Male, 32)

- Activity rooms *"for people to rent out to schedule yoga class, zumba class, etc."* (R2, Female, 35)
- Facilities for all residents, especially children and teenagers, including public pool, dance studio, martial arts, gym, etc. Several participants identified the need *"to bring teenagers together in the community"* (R2, Female, 40).
- An off-leash dog park, because currently the MPCommunity was not seen as a dog-friendly space. Residents with dogs would really appreciate the addition of an off-leash dog park within the MPC.

  "it would be great to see a park with a dedicated off leash dog zone." (R2, Female, 30)

3.3.5. Theme 5: Developer-Initiated Community Activities

Participants recognised the important role the developer plays in getting people to meet one another. The importance of developer-initiated community activities was clearly identified by participants across the two data collections. Such estate-wide community-unifying "get-togethers" organised by the developer at the MPCommunity included a food-truck festival with children's rides, characters, and a DJ; moonlight cinema; tree-planting days followed by sharing pizza together; night markets; a residents' Christmas party; and other resident-only events.

The importance of such events in unifying the diverse residents was highlighted by one participant:

"the turning point was when the movie night happened last summer. It was the mix of cultures and people so warm for each other. We just want our kids to grow up to tolerate and respect each other no matter where their roots are. We could easily picture our family fit in". (R2, Female, 36)

These organised activities also appeared important in "bedding in" the residents–developer relationship. Residents appreciated the developer-initiated community activities and events, and described the relationship between the community and the developer as good.

"Yes. It definitely has that tight knit community feel about it. I think many factors play into this. The developers putting time and effort in creating that type of community through organised events so neighbours can meet and mingle such as 'stage bbq meet ups', just this weekend there was a clean-up Australia day event and pizza night etc. It brings the community together and in turn forms a bond between neighbours. The resident [Facebook page] also fosters community togetherness." (R1, Female, 36)

However, it was also noted that the events:

" ... have been good but I find they are 'one off' things and it can be difficult to keep connected." (R1, Female, 36)

3.3.6. Theme 6: Safety and Security

While safety and security scores were less significant at Time 2 than at Time 1, it was a very intense topic of discussion. The desire for a "safe and secure" community was an integral part of participants' utopian reality. Perceived threats to utopian living were factors that threatened the residents' security. The data show that residents are seeking to support their physical and economic security.

3.4. Physical Security

Participants were concerned that law and order issues experienced in areas immediately surrounding the MPCommunity could extend into their MPC.

> *"what is important to me is community safety . . . this is important to me as you should feel safe in your home at all times and with the recent events of home invasions and gang meet ups in surrounding suburbs, it won't be long until* [MPCommunity] *will become a "hangout" suburb"* (DP)

Residents said that they were mostly happy with their home environment, but that they did not feel safe in the neighbouring suburbs—posing an immediate challenge, as the promised physical and social infrastructure has not yet been realised, and residents need to go into these surrounding areas to undertake everyday activities such as shopping, entertainment, schools, etc.

> *"[I] actually feel more unsafe outside of* [MPCommunity] *. . . I'd rather starve than shop at* [a particular suburb] *these days; it feels SO unsafe to me"*. (R1, Female, 36)

Many participants want security infrastructure incorporated within the MPCommunity, with some suggesting that *"CCTV all around MPCommunity would be nice"* (EL), as well as fencing around the estate and secured gates for the entrance. Some participants suggested something similar to *"how* [another MPC] *is designed"* (SS).

Participants also discussed the possibility or need for *"visible security guards and police patrols"* (DCa).

> *"I would suggest roaming security around the suburb"*. (DP)

> *"visibility of police car driving through the estate at random times would make residence feel safe"*. (M)

However, participants also recognised the importance of connecting with one another to feel safe and secure, as well as establishing a neighbourhood watch group within the estate as a strong safety and security measure.

> *"Know who your neighbours are. Ensure unsafe activities are reported. Appropriate action from law enforcement agencies. Neighbourhood watch"*. (PW)

> *"What is important as a community is to keep in the loop with everyone like a forum sort of thing; the old NABO app for like community events and reporting on security issues. Just things to keep everyone safe in the community"*. (C)

3.5. Economic Security

Of direct consequence to economic security is the lack of promised physical and social infrastructure, including places of employment, shopping, schools, leisure, etc. Participants indicated their disquiet with the time it was taking to deliver on the promised master-planned infrastructure and services.

Participants were frustrated with having to leave the MPCommunity for nearly every activity outside of their home. Residents of the MPCommunity have no choice in terms of shopping but to travel elsewhere. This puts a lot of time and pressure on residents, because shopping is one of the main chores in daily life. The amount of time spent outside of the MPCommunity is time not spent in activities with family and community within the MPCommunity, reducing the chance to build social networks closer to home.

"A lot more driving now, an extra 20 min each way to work for me. Further to drive for shops. 13–14 h per week minimum driving. Hopefully that will be less when shops open in [MPCommunity]*."* (R1, Female, 26)

"If I don't have my car . . . I think I can't do anything[sad emoji] *stuck at home."* (R1, Male, 32)

"Access to good health services and shops are also important to me". (San)

Participants discussed the lack of employment opportunities in the MPCommunity or the local area, resulting in long commutes and less time in the local community.

"I actually resigned from my job . . . Moving to [the MPCommunity] *definitely had an impact on that decision. I was spending over 1 hr to get to work as I had 2 drop offs in the morning—dropping one child to childcare and the other to school. I am looking for something closer to home now so I have a better work/life balance."* (R1, Female, 36)

4. Discussion

This research employed a mixed-methods approach to provide the contextual framing necessary to understand the lived experiences of pioneer residents moving into a new MPCommunity. The results demonstrate how the purposeful development of community identity in the early stages of the MPCommunity has not led to satisfactory levels of social infrastructure or social connectedness for the pioneering residents—that is, the physical and social environment, as interpreted by residents against the developers' imagined vision and marketing testimonies, was not entirely satisfactory. Specifically, master-planned infrastructure issues—such as transport, and access to daily activities such as shopping, work, and school—were common and significant points of frustration and dissatisfaction.

The central lifestyle marketing themes for this MPCommunity—such as safety, security, and community infrastructure and connectedness—were captured at Times 1 and 2. In terms of changes from Time 1 to Time 2, Time 1 could be characterised as the "honeymoon" period, when pioneering residents were full of anticipation and expectation for the idyllic lifestyle marketed by the developer. Meanwhile, Time 2 provides an insight into the realities of the day-to-day lived experience, capturing the teething problems as residents became discontented with delays in the rollout of the planned infrastructure. Results revealed a decline in satisfaction over time in terms of the residents' sense of belonging to a community. Nevertheless, most of the participants maintained hope of finding a place where they felt they belonged, reminiscent of their childhood, or an imagined, more secure time representing a more traditional set of family and community values.

While the MPCommunity, as a new development, promises a high standard of housing—which is important for quality of life [22]—the results show that the sense of social connectedness in this new estate is not yet developed. The findings indicate that participants are becoming increasingly impatient with the time that it is taking to deliver promised, master-planned infrastructure, and this appears to have fuelled some of their disquiet in terms of satisfaction with the community. The development of community needs engagement, which requires opportunities and time spent within the community [3]. Currently, in this MPCommunity, this is not possible, as residents need to spend large amounts of time outside the MPCommunity for everyday activities such as shopping and work.

The findings revealed another reason for the decline in satisfaction with being part of the community, which appeared to be related to the cultural heterogeneity of the MP-Community population. Greater degrees of homogeneity and compatibility have been shown to increase the probability of more intensive relationships that extend beyond a simple exchange of greetings [23]. Similarly, it has been found that ties between non-similar individuals disband more quickly, thus triggering the creation of enclaves within the social environment [23,24]. In this instance, the heterogeneity of this MPCommunity appeared to discourage or deter interactions.

Previous research suggests that there are parts of Australia that experience "multicultural anxiety", founded in safety and security concerns. Many developers strategically position their marketing to promote safe and secure developments, thereby protecting against this anxiety [25]. This MPCommunity has a culturally heterogeneous resident population, as seen in Table 1, who clearly were not looking to escape "multicultural anxiety"—and if they were, their attempts were completely misguided and unsuccessful. However, while not motivated by "multicultural anxiety" this particular resident population was very concerned about safety and security.

4.1. Sense of Community

MPC developers know their markets well, and promote more traditional concepts of community [24]. The participants in this study were full of hope and expectation for a "close-knit community"—a community utopia. Most of the participants described being interested in finding a place where they felt they belonged, reminiscent of their childhood, or an imagined, friendlier time representing a more utopian idea of community and community values. Participants reported feeling good about their MPCommunity, but wanted to feel more like a part of the resident community—what has previously been identified as desiring localised neighbourhood [3].

As evidenced in the findings, the community utopia sought by the participants was challenged by their lived experience in the MPCommunity after the "honeymoon" period. Importantly, the participants placed a great deal of value on "friendliness" and "community ethos", which they thought was absent from their MPCommunity, but continued to set as an expectation and desire for its future. Conversely, participants reported they felt lucky to be living in the MPCommunity, and said that they saw the potential for a community utopia, and recognised their own responsibility in making this a reality. Some suggested taking the initiative to be inclusive and engage in more meaningful ways with neighbours, as well as establishing community norms of cooperation, consideration, and kindness to one another; meanwhile, others reminisced about their enjoyment and importance of centrally organised community activities in establishing and strengthening a sense of community. These findings further highlighted the fact that the MPCommunity residents were at different life stages and, therefore, had different needs. Previous research has noted the importance of understanding the different life stages that may determine specific individual demands [3]. Beyond the sense of community connectedness and belonging, participants identified a number of challenges that hindered community utopia—for example, the lack of the promised physical and social infrastructure, which directly impacted organised community activities.

4.2. Physical and Social Infrastructure

This MPCommunity was predominantly in the early stages of development—specifically, master-planned physical and social infrastructure (e.g., adequate public transport; pedestrian and bike access to the local train station; shopping/town centre; facilities specifically designed for children, teenagers, pets, etc.) had not yet been realised. Principally, design aspects such as shopping centres, public transport, walkways, bike paths, and parks encourage people out of their homes and into public spaces, where they can meet others engaged in similar activities [12]. Participants with pets and no children reported a sense of exclusion from the community, with the absence of a dog park or organised pet-friendly community events. Many participants saw this lack of infrastructure as inconvenient, and as an impediment to incidental and spontaneous social gatherings or everyday activities. This finding is consistent with previous research that recognises the importance of well-designed physical and social infrastructure in supporting unprompted, natural social connection, facilitating a sense of belonging and trust between residents [3]. Notably, where infrastructure is inadequate, residents are more likely to be lonely, poorly supported, and have limited social agency for change of their circumstances, which are felt by the wider community [3].

Moreover, participants with children reported concerns about a lack of child-related areas for activity and play, despite the availability of parks. Parents wanted specific spaces for organised as well as free activities for children, including spaces/rooms for dance, martial arts, a swimming pool, etc. Furthermore, this specific cohort identified the lack of public transport as something that placed significant strain on families, specifically impacting the daily transport needs of the children within the MPCommunity. Having local access to schools through safe walking and bike paths between homes and schools provides children and parents with the opportunity to mix with other local children and adults undertaking the same activity. The critical nature of co-location of schools and homes has been noted in a previous Australian study [3].

The issues with children extended to adolescents. Adolescents are similar to other residents in terms of access to public transport for recreational and entertainment purposes [3], and adolescents can become more isolated from surrounding areas when there are public transport inadequacies. Targeted physical and social infrastructure to link adolescents with activities of interest within MPCs, and provide public transport access to external activities, is considered essential [3]. The participants in this study had strong desires for adolescents to have shared spaces and spend time within the MPCommunity.

A recurring theme and frustration centred on this lack of physical and social infrastructure for everyday activities within the MPCommunity, which consequently required residents to go to neighbouring suburbs for shopping and other services. Participants reported a discomfort within the neighbouring suburbs, leading to distress and disquiet in terms of personal safety and security. Encouragingly, participants expected that this problem will be overcome once the promised infrastructure is fully realised.

It should be noted that participants tabled several solutions to safety and security concerns. These solutions consisted of installing CCTV, security fencing, roving security, or police patrols. However, participants identified the importance of connecting and cooperating with one another in terms of neighbourhood safety and security concerns, including involvement in activities/committees such as neighbourhood watch. Therefore, participants were not simply looking for a "fortress solution"; they recognised the importance of community development, involvement, and engagement in bolstering residents' sense of safety and security.

4.3. Community Activities

Beyond the bounds of physical and social infrastructure, participants acknowledged the importance of community groups and events within the MPCommunity. The developer initiated a number of community activities, which were very well received by residents. Participants overwhelmingly spoke positively about developer-organised community events. Participants enjoyed these events, felt that they contributed to the sense of community, and had the desire for more formal activities organised by the developer. In particular, one participant reported that one movie night was a turning point, with the mix of cultures coming together in a warm and welcoming atmosphere. Community activities, facilitated by the developer, have been shown to be an integral component of respondents' symbolic conception of the community in an MPC [2]. Organised community activities allow residents to intermingle, which enables social connectivity to develop. Such events also provide residents with the opportunity to build a new identity as part of the MPC and the developer's ideology [2]. Furthermore, these activities provide the opportunity to monitor interactions with other residents in order to determine how well they "fit" with the wider community [2]. While it seems difficult to know how these singular events affect community development—especially when residents come together in an amplified anticipation of celebration [25]—participants reported remembering the events fondly, feeling real connection between residents at the events, and having a real appetite for more such events. The results show that these one-off events further support the development of a community ethos and the growth of community connectedness.

4.4. Safety and Security

There was an improvement in participants' reported levels of satisfaction in terms of the safety and security scores between Time 1 and Time 2. This may be a reasonably expected progression in the development phases of MPCs—as more homes are built, more people take up residency, thereby creating the feeling of increased safety in numbers. The improvement in satisfaction with safety and security scores at Time 2 was not obvious from the conversations and the survey question results (see Table 2). Furthermore, the level of satisfaction with safety and security remained lower than the national average. However, despite the score improvement, participants were notably vocal in the forum about safety and security issues in the MPCommunity. It could be argued that this reflects unmet expectations derived from marketing promises, as well as a timing issue, evolving from unrealised infrastructure commitments by the developer [26].

The unrealised status of promised physical and social infrastructure also affects economic security, and is an obstruction to authentic feelings of community utopia [27]. Occupational opportunities within the MPCommunity were non-existent. Lack of work within or close to residential communities reduces the opportunity to build social networks closer to home [3]. This structural issue puts community design and work in relation to one another, as the physical and temporal separation of community and care activities from work activities is particularly problematic, and its impact is felt across life-stage groups [3,28].

The most common tool utilised by developers for economic security is "community compact" [13]. Community compact establishes and enforces restrictive covenants (formal) and social norms (informal) designed to establish common social goals of a pecuniary code of beauty for the MPC and the community ethos [7,13]. The MPCommunity has a design review panel (DRP) that approves all components of the house design and construction, landscaping, and building siting. In the MPCommunity, the DRP attracted criticism regarding the consistency of its decisions and the length of time taken to reach a decision. Participants reported that these problems resulted in cost overruns and a lack of confidence and trust in the consistency of the decision-making process. There was also recognised disquiet amongst participants with regard to the state of the public spaces within the MPC. Like other MPCs, the common social norms include highly maintained gardens, cars parked in garages rather than on the road or footpath, and a positive, friendly demeanour to other residents of the MPC [2,8]. Participants complained about the state of public spaces and the disruptive effect of residents and visitors parking on roads and footpaths, sometimes blocking thoroughfares.

Participants were critical of the physical state of the community facilities in the MPCommunity, and stressed the importance of these formal and informal covenants. This type of maintenance affects property values, and goes to the heart of economic security [2].

Identity and belonging are strongly linked to home and neighbourhood, with a "sense of home" contributing to security and a sense of order, meaning, continuity, and agency [29,30]. As evidenced in the findings, the community utopia sought by the participants was not delivered by their lived experience in the MPC. In general, participants placed a great deal of value on "friendliness" and "community ethos", which was non-existent. Nevertheless, participants were keen to establish a strong, friendly, cooperative community ethos that was nurtured within the MPCommunity.

The formal and informal covenants within the MPCommunity, while not always adhered to, appeared to offer residents a sense of coherence, community ethos, and social order, as well as a degree of control over the physical and social environment. An important part of this was the community events organised and promoted by the developer, which appeared to go some way toward bringing this community together. These unifying events strengthened residents' sense of security by facilitating social connections between MPCommunity neighbours, creating a strong sense of neighbourhood identity and belonging. Residents were keen to have more of these types of events. Moreover, through these events, as well as the formal and informal covenants and norms, the MPCommunity appeared to offer residents a sense of active engagement, rules of belonging, and a code of behaviour, providing an anchor for communal living. Consequently, individual identity and a feeling of predictability were derived, allowing residents to confidently make life plans that strongly contributed to their security.

5. Conclusions

Master-planned community estates are promoted, sold, and purchased on the basis of powerful symbols of identity, community, safety, and security [1–3], and it is unclear how this manifests in the lived experience of pioneer residents moving to a new MPC [2,4]. This study addressed this gap and explored—through the lived experiences of residents—the link between how residents felt about the MPCommunity, their wellbeing, and their daily lives. This case study explores an MPC resident population and the evolving community in its early stages of development.

The MPCommunity promises a high standard of housing and infrastructure, which is important for daily life and wellbeing. The social connectedness in this incipient MPC is not yet fully realised in terms of infrastructure and social connections between residents. This case study offers valuable insight into concerns voiced by residents with different perspectives and challenges, living within a new and developing local Australian community. The desire for safety and security are prominent concerns of housing consumers, and enhanced community safety and security increase both the perceived and real value of the community [31]. This is not lost on MPC developers, with safety and security being central to the formation of MPCs, and marketing campaigns playing on prospective residents' security fears [27,32]. MPC developers aim to deliver on identity, community, and the inter-related factors of physical and economic security [2,33]. Through capturing participants' lived experiences, it was clear that while the MPCommunity residents have positive feelings toward the estate, there were issues of concern in terms of security. Participants acknowledged that in order to support and strengthen these areas of security, the developer and residents would need to simultaneously work on strategies that addressed the "sense of security" and the "sense of community".

This research addresses significant gaps in previous studies, and contributes to a growing body of knowledge around MPCs and community wellness. One limitation of this research is that results from case studies cannot be generalised to the broader population. However, this study provides "real-life data" of the lived experience of an incipient MPC community. While not definitive, this study offers a level of insight into the lived experiences of community wellness, safety, and security of an emerging heterogeneous MPC. This case study offers valuable insight into how social anxieties associated with a cultural mix of people with different languages and values can overcome the idea of threats to more established ways of life within a local community. By highlighting issues that impact on MPC wellness, safety, and security—especially for heterogeneous resident populations—this research furthers our understanding of the specific factors that make MPCs more desirable environments for existing and future MPC developments.

5.1. Limitations

A limitation of this study is that the results of the case study cannot be generalised to a wider population. This case study provides valuable insights into how the cultural combination of people with different languages and values can overcome threats to the more mature lifestyles of local communities. However, for the protection of community health, safety, and life experience, the results of this research cannot be confirmed at present.

5.2. Future Research

Based on this exploratory research, future research could develop in a range of theoretical and practical areas. Future theoretical research could focus on social networks in MPCs to explore and test the range of characteristics that explain patterns of inclusion/exclusion and possible structural barriers to communication and cooperation. In addition, future research to extend the understanding of community impacts could include social network analysis in order to map changes in the structure of interaction over time, barriers to communication/cooperation, and other relevant topics as they arise over time,

To assist in the future development of sustainable MPCs, additional research is required in order to develop a specific tool or instrument to use with residents, so as to identify issues arising as residents populate new communities.

Author Contributions: Conceptualization, L.B.; methodology, L.B.; formal analysis, L.B. and N.W.; investigation, L.B.; writing—original draft preparation, review, and editing, L.B., C.N. and N.W.; project administration, L.B. All authors have written and reviewed the paper. All authors have read and agreed to the published version of the manuscript.

Funding: Funding for the study was provided by the Developer of the MPCommunity. For confidentiality reasons, the Developers's name has not been included as it would identify the MPC location and potentially the participants.

Institutional Review Board Statement: The study was conducted accordance with the Australian National Health and Medical Research Council guidelines and approved by the Human Ethics Committee at Queensland University of technology (approval code 1600001085, June 2017)

Informed Consent Statement: Informed consent was obtained from all participants involved in the study.

Data Availability Statement: Data are not publicly available due to the potential of identifying individual participants.

Conflicts of Interest: L.B. is an editor for *Sustainability*. The funders had no role in the design of the study, in the collection, analyses, or interpretation of data, in the writing of the manuscript, or in the decision to publish the results.

References

1. Terwilliger Center for Housing, Urban Land Institute. *Residential Futures: Thought-Provoking Ideas on What's Next for Masterplanned Communities*; Urban Land Institute: Washington, DC, USA, 2012.
2. Gwyther, G. Paradise Planned: Community Formation and the Master Planned Estate. *Urban Policy Res.* **2005**, *23*, 57–72. [CrossRef]
3. Williams, P.; Pocock, B. Building 'community' for different stages of life: Physical and social infrastructure in master planned communities. *Community Work Fam* **2010**, *13*, 71–87. [CrossRef]
4. Bento, F.; Couto, K.C. A behavioral perspective on community resilience during the COVID-19 pandemic: The case of Paraisópolis in São Paulo, Brazil. *Sustainability* **2021**, *13*, 1447. [CrossRef]
5. Thompson, J. Beyond the Neighborhood: Community at the Urban Scale. *Int. J. Constr. Environ.* **2013**, *3*, 167–179. [CrossRef]
6. Vine, D.; Buys, L.; Aird, R. Conceptions of 'community' among older adults living in high-density urban areas: An Australian case study. *Australas. J. Ageing* **2013**, *33*, E1–E6. [CrossRef]
7. Cheshire, L.; Wickes, R.; White, G. New Suburbs in the Making? Locating Master Planned Estates in a Comparative Analysis of Suburbs in South-East Queensland. *Urban Policy Res.* **2013**, *31*, 281–299. [CrossRef]
8. McGuirk, P.M.; Dowling, R. Understanding Master-Planned Estates in Australian Cities: A Framework for Research. *Urban Policy Res.* **2007**, *25*, 21–38. [CrossRef]
9. Thompson, C. Master-Planned Estates: Privatization, Socio-Spatial Polarization and Community. *Geogr. Compass* **2013**, *7*, 85–93. [CrossRef]
10. Walters, P.; Rosenblatt, T. Co-operation or Co-presence? The Comforting Ideal of Community in a Master Planned Estate. *Urban Policy Res.* **2008**, *26*, 397–413. [CrossRef]
11. Rosenblatt, T.; Cheshire, L.; Lawrence, G. Social interaction and sense of community in a master planned community. *Hous. Theory Soc.* **2009**, *26*, 122–142. [CrossRef]
12. Jacobs, J. *The Death and Life of Great American Cities*; Modern Library Edition, Random House: New York, NY, USA, 1993.
13. Dowling, R.; Atkinson, R.; McGuirk, P. Privatism, Privatisation and Social Distinction in Master-Planned Residential Estates. *Urban Policy Res.* **2010**, *28*, 391–410. [CrossRef]
14. Bajracharya, B.; Donohue, P.; Baker, D. Master Planned Communities and Governance. Available online: http://soac.fbe.unsw.edu.au/2007/SOAC/masterplannedcommunitiesandgovernance.pdf (accessed on 30 September 2021).
15. Lincoln, Y.S.; Guba, E.G. *Naturalistic Inquiry*; SAGE Publications Inc.: Beverly Hills, CA, USA, 1985.
16. Curry, L.A.; Nembhard, I.N.; Bradley, E.H. Qualitative and mixed methods provide unique contributions to outcomes research. *J. Am. Heart Assoc.* **2009**, *119*, 1442–1452. [CrossRef]

17. Hamza, N.; Greenwood, D. Energy conservation regulations: Impacts on design and procurement of low energy buildings. *Build Environ.* **2009**, *44*, 929–936. [CrossRef]
18. Yin, R.K. *Applications of Case Study Research*; SAGE: Thousand Oaks, CA, USA, 2012.
19. Australian Unity. Evaluating 20 Years of the Australian Unity Wellbeing Index. Available online: https://www.australianunity.com.au/wellbeing/what-is-real-wellbeing/20-years-of-wellbeing-research (accessed on 30 September 2021).
20. Flick, U. *An Introduction to Qualitative Research*; SAGE Publications: London, UK, 2014.
21. Corbin, J.M.; Strauss, A.L. *Basics of Qualitative Research*; SAGE Publications: London, UK, 2015.
22. Nelson, G.; Sylvestre, J.; Aubry, T.; George, L.; Trainor, J. Housing choice and control, housing quality, and control over professional support as contributors to the subjective quality of life and community adaptation of people with severe mental illness. *Adm. Policy Ment. Health Ment. Health Serv. Res.* **2006**, *34*, 89–100. [CrossRef] [PubMed]
23. Gans, H.J. The balanced community: Homogeneity or heterogeneity in residential areas? *J. Am. Inst. Plan.* **1961**, *27*, 176–184. [CrossRef]
24. McPherson, M.; Smith-Lovin, L.; Cook, J.M. Birds of a feather: Homophily in social networks. *Ann. Rev. Sociol.* **2001**, *2001 27*, 415–444. [CrossRef]
25. Kenna, T.E. Consciously Constructing Exclusivity in the Suburbs? Unpacking a Master Planned Estate Development in Western Sydney. *Geogr. Res.* **2007**, *45*, 300–313. [CrossRef]
26. Bauman, Z. *Community: Seeking Safety in an Insecure World*; John Wiley & Sons: Hoboken, NJ, USA, 2013.
27. Cheshire, L.; Walters, P.; Wickes, R. Privatisation, Security and Community: How Master Planned Estates are Changing Suburban Australia. *Urban Policy Res.* **2010**, *28*, 359–373. [CrossRef]
28. Heid, J.; Urban Land Institute. Greenfield Development without Sprawl: The Role of Planned Communities. Available online: http://europe.uli.org/wp-content/uploads/ULI-Documents/GreenfieldDev.ashx_.pdf (accessed on 30 September 2021).
29. Aidala, A.A.; Sumartojo, E. Why housing? *AIDS Behav.* **2007**, *11*, 1–6. [CrossRef] [PubMed]
30. Padgett, D.K. There's no place like (a) home: Ontological security among persons with serious mental illness in the United States. *Soc. Sci. Med.* **2007**, *64*, 1925–1936. [CrossRef] [PubMed]
31. Halter, V.R. Rethinking Master-Planned Communities. In *Trends and Innovations in Master-Planned Communities*; Schmitz, A., Bookout, L.W., Eds.; Urban Land Institute: Washington, DC, USA, 1998; pp. 1–9.
32. Johnson, L.C. Master Planned Estates: Pariah or Panacea? *Urban Policy Res.* **2010**, *28*, 375–390. [CrossRef]
33. Costley, D. Master planned communities: Do they offer a solution to urban sprawl or a vehicle for seclusion of the more affluent consumers in Australia? *Hous. Theory Soc.* **2006**, *23*, 157–175. [CrossRef]

Making the Gold Coast a Smart City—An Analysis

Isara Khanjanasthiti *, Kayalvizhi Sundarraj Chandrasekar and Bhishna Bajracharya

Faculty of Society & Design, Bond University, Gold Coast 4226, Australia; kayal@bond.edu.au (K.S.C.);
bbajrach@bond.edu.au (B.B.)
* Correspondence: ikhanjan@bond.edu.au

Abstract: In recent times, there has been a worldwide trend towards creating smart cities with a focus on the knowledge economy and on information and communication technologies. These technologies have potential applications in managing the built and natural environments more efficiently, promoting economic development, and actively engaging the public, thus helping build more sustainable cities. Whilst the interest in smart cities has been widespread predominantly amongst metropolitan cities, several regional cities such as the Gold Coast in Australia have also recently endeavoured to become smart cities. In response to this emerging trend, this study aimed to investigate key opportunities and challenges associated with developing regional cities into smart cities using the Gold Coast as a case study. It identified key factors critical to the planning and development of smart cities. These factors fall under five broad themes: cultural and natural amenities, technology, knowledge and innovation precincts, people and skills, and governance. The factors were applied to the Gold Coast to analyse the key opportunities and challenges for its development into a smart city. Finally, key lessons, which are potentially applicable to other regional cities seeking to develop into smart cities, are drawn from the case study.

Keywords: smart cities; regional cities; technology; governance; knowledge workers; knowledge precincts; open data; Gold Coast

Citation: Khanjanasthiti, I.;
Chandrasekar, K.S.; Bajracharya, B.
Making the Gold Coast a Smart
City—An Analysis. *Sustainability*
2021, *13*, 10624. https://doi.org/
10.3390/su131910624

Academic Editor: Tan Yigitcanlar

Received: 18 August 2021
Accepted: 18 September 2021
Published: 24 September 2021

Publisher's Note: MDPI stays neutral
with regard to jurisdictional claims in
published maps and institutional affiliations.

1. Introduction

Many cities worldwide are now facing several ongoing challenges, such as increasing traffic volume and congestion, diminishing quality of life, urban sprawl, and a degrading natural environment. To address these challenges, several cities are now seeking to establish themselves as smart cities. Several metropolitan cities such as London, Barcelona [1], Seoul, Hong Kong, Tokyo, and Singapore [2] have been at the forefront of these trends [3]. Recently, however, a growing number of regional cities in Australia, such as the Gold Coast, the Sunshine Coast, Launceston, and Gawler, have also shown interest in establishing themselves as smart cities. This trend became highly evident following the $50 million Smart Cities and Suburbs program by the Australian Federal Government launched in 2016 [4], which is delivered through City Deals, another Federal initiative. Throughout the two application rounds to date, several local government agencies of non-metropolitan cities have successfully secured funding support for their smart city projects, thus illustrating the growing level of interest in smart cities amongst regional cities in Australia.

The Gold Coast is Australia's sixth most populated and largest non-capital regional city and yet is often regarded as an "overgrown resort town" [5]. The city has traditionally been "Australia's playground" [6] with its beautiful surf beaches, rainforests, and theme parks that attract domestic travellers and, lately, an increasing number of international tourists [7]. Contrary to its early days, the cultural and creative identities of the Gold Coast have continued to evolve, especially since hosting the 2018 Commonwealth Games [8]. The 2016–2017 budget plan of the City of Gold Coast (CoGC), the local government agency responsible for managing the Gold Coast, has specifically allocated $3.6 million to Digital City, a smart city project. The project seeks to identify, investigate, and provide emerging opportunities, including smart lighting, smart travel, and smart health initiatives [9]. Since then, a few projects have been underway or in the pipeline under the Digital City platform.

Given this context, this study aimed to explore major opportunities and challenges that regional cities encounter in their endeavour to become smart cities. It identified key factors for the successful development of smart cities and analysed a case study of the Gold Coast using these factors. This article comprises two sections. The first section is a literature review of smart city concepts, which provides a basis for an analytical framework of smart cities. The second section applies the framework to the Gold Coast to identify key opportunities and challenges the city is facing in its development into a smart city. The case study analysis is based on a review of relevant policy documents and web-based resources related to the Gold Coast. Lastly, the article concludes with key lessons from the case study, which are potentially applicable to other regional cities seeking to establish themselves as smart cities.

2. Literature Review

The literature on smart cities has grown exponentially in the last decade, e.g., [10–13]. However, despite the growing body of knowledge, no universally accepted definition currently exists for smart cities [14–18]. Several terms such as "smart," "smarter," "digital," "intelligent," "knowledge-based," and "ubiquitous" are used interchangeably in the smart city literature [10,19].

2.1. What Are Smart Cities?

According to Townsend [13] (p. 15), smart cities are "places where information technology is combined with infrastructure, architecture, everyday objects, and even our bodies to address social, economic, and environmental problems." He also proposes that these cities have the prospect of being efficient, transparent, resilient, secure, and sociable. Smart cities rely upon an appropriate mix of human capital (e.g., skilled labour), infrastructure capital (e.g., telecommunication infrastructure), social capital [20], and entrepreneurial capital [21]. Recently, Alizadeh [22] (p. 71) has defined smart cities as "urban settlements that capitalise on telecommunication technologies to enhance liveability, workability and sustainability." This definition highlights the importance of telecommunication infrastructure in creating smart cities. Several information and communication technologies (ICTs) (e.g., smartphones, radio frequency identification, sensors, and drones) have enabled crowdsourcing of data from the public to be conducted efficiently without compromising privacy and security [12,22]. There is an opportunity for open data to be harnessed for addressing the long-term challenges that cities are facing [23]. Such data can be used by different stakeholders to provide a variety of useful information products and services for local communities.

Although smart cities are structurally "wired" and supported by an integrated system of ICTs, the use of technologies alone does not automatically make cities "smart." Hollands [14] (p. 315) stated that "smart cities must seriously start with people and the human capital." Therefore, ensuring "the balance between using advanced technology and still maintaining the humanness aspect" is an essential consideration for smart cities [24] (p. 30). Traditionally, smart cities primarily focused on corporate marketing needs on an incremental and piecemeal basis [25,26]. However, contemporary smart cities should invest in intellectual capital by accumulating a funding policy strategy that promotes social, cultural, and environmental development [3,27], which is often disregarded in smart city frameworks [28]. This process can also help cities achieve socio-economic equality, a critical element of smart cities, by engaging all citizens in a problem-solving process to build a socially inclusive society [29–31].

In addition to technology and people, institutions form another key component of smart cities, and there is a need for governance to encourage public participation and build institutional capacity [32,33]. Paskaleva [34] highlighted the importance of using knowledge networks and e-governance to boost a city's competitiveness. Strategic plans to promote e-governance and integration of ICTs are also vital for developing smart cities [35].

2.2. Critique of Smart Cities

According to Yigitcanlar [36], although the smart city movement has a prospect, the topic is still under exploration and, therefore, not well-developed. Thus, the smart city concept is not an effective urban development and management model in its current stage. Moreover, "local characteristics, priorities and the needs" of the city often influence the smart city landscape [36] (p. 32). Several cities around the world have implemented strategies to become smarter over the last decade. Nevertheless, in doing so, the simplistic use of such terms as "smart" and "intelligent" has emerged primarily for marketing purposes, resulting in the failure to specify which aspects of their intelligence are being enhanced and how they intend to achieve a high level of intelligence [37–39]. Cities seeking to establish themselves as smart cities can also appear elitist [18,40]. Many cities, through their desire to join the "smart city elite," are adding several embellishments to the "smart" definition of smart cities beyond the technology dimension, thus making the smart city concept more convoluted [41] (p. 2).

According to Saunders and Baeck [42], smart city ideas have often been criticised in three ways. Firstly, they have been overly focused on hardware and hard infrastructure rather than people, even though cities should be for people [43]. Secondly, given the rapid advancement in technologies in recent years, there has been an overemphasis on finding uses for new technologies rather than finding appropriate technologies to solve major urban problems. Thirdly, similarly to Komninos' [39] argument above, several smart cities have concentrated on marketing and promoting themselves at the expense of testing different solutions in the real world.

In his review of current smart city practices, Yigitcanlar [29] discovered that cities worldwide are trialling different technologies to improve their operation. Cities are employing technologies to achieve smart city objectives in vastly different ways, even in the same country. This trend illustrates that a one-size-fits-all approach is not appropriate for implementing smart city initiatives due to each city's unique characteristics, context, and issues [24]. A one-size-fits-all narrative, which does not adequately consider socio-economic, spatial, and political variables in the local context, has been identified as one of the significant shortcomings in the current smart city debates [44–46]. As regional cities may not possess the sufficient financial resources required for smart city initiatives, there is a need to better align smart city strategies with existing government policy and priorities to improve their funding capabilities [47,48]. By carefully evaluating local context before implementing smart city ideas, critical issues can be prioritised and addressed accordingly in a more cost- and time-efficient manner.

This review of the literature is fundamental to understanding the key themes and debates relating to smart cities. The literature review findings will be used as a basis for analysing a case study of the Gold Coast in terms of its key challenges and opportunities for developing into a smart city.

3. Case Study of the Gold Coast

As a coastal city with approximately half a million population, the Gold Coast is located on the south-eastern corner of the South East Queensland (SEQ) region, which also comprises Brisbane, Logan, and the Sunshine Coast, all of which have recently demonstrated an intent to develop into smart cities. Figure 1 below displays the location of the Gold Coast in relation to the nearby cities and regional areas from both the Queensland and New South Wales states.

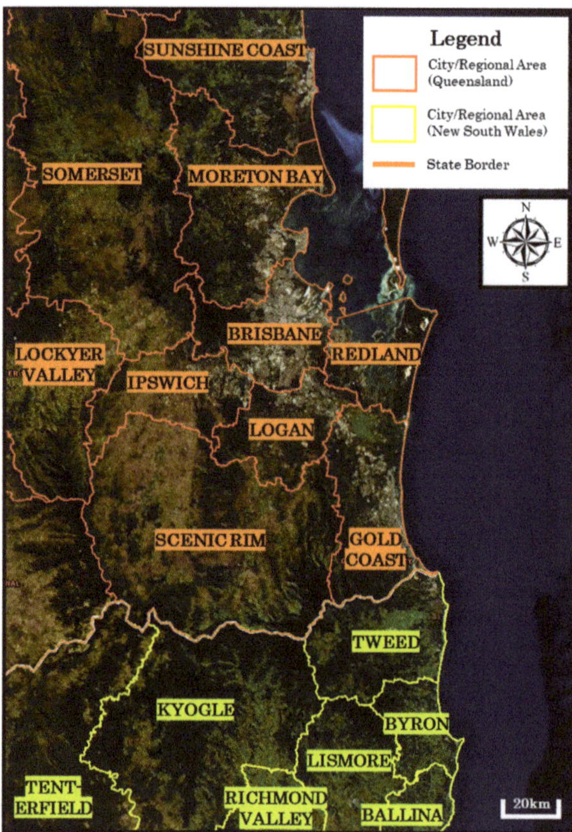

Figure 1. Map of the Gold Coast and its surrounding cities and regional areas, created by the authors using satellite imagery from Queensland Globe [49] under the CC BY 3.0 AU license [50].

The Gold Coast is the sixth-largest city in Australia, with an expected population figure of about three-quarters of a million by 2026 [51]. The city has a subtropical climate with more than 80 kilometres of beaches on the eastern end and hinterland on the western side. Characterised by a cluster of high-rise residential towers along the beaches, most of the Gold Coast's urban development and population are concentrated close to the coastline. Given the increasingly limited availability of land in coastal suburbs, the Gold Coast is currently experiencing suburban growth in inland outskirts, which is progressively constrained by the hinterland. With an extensive network of waterways acting as barriers to walking, cycling, and public transport, the Gold Coast is a car-dependent city. In recent years, there has been a constant need for expanding the Gold Coast's road infrastructure to accommodate its rapidly growing volume of vehicular traffic [52]. As such, the Gold Coast is facing a significant challenge for accommodating future growth in population, traffic congestion [53], and housing demand [54] in a sustainable manner. In response to this long-term challenge, the CoGC has specified an objective to transform the Gold Coast into an "active digital city" where "data and real-time information [are used] to shape [the] city" [55] (p. 14). As part of this objective, a sensor network will be installed to monitor and manage the local transport infrastructure. Additionally, real-time data will be increasingly used to support disaster responses, public safety, and other initiatives in the future.

The entire Gold Coast region is governed by the CoGC, making the council the second-largest local government agency in Australia behind Brisbane regarding jurisdictional population size. Whilst tourism has historically been an important economic driver of the Gold Coast, it also has a foothold in other industries, including education, sports, film, marine, and ICTs. Accordingly, the council has recently displayed an intent to develop these industries further to improve the Gold Coast's profile as a major event destination [56]. Driven by the prospect of hosting the Commonwealth Games in 2018, the Gold Coast became the first city in Queensland to incorporate a light rail system into its public transport services. Partly funded by both the Federal and state governments, the project is identified as a successful example of the "Smart Infrastructure" pillar within the Smart Cities Plan launched in 2016 by the Federal Government, with the other two pillars underpinning the policy being "Smart Policy" and "Smart Technology" [57]. The Smart Cities Plan considers delivering infrastructure to not just capital cities but also for regional cities to support long-term growth. With the 2032 Olympic and Paralympic games confirmed to be hosted in SEQ and seven venues planned throughout the Gold Coast for the events [58], the city needs to utilise this major opportunity to address the planning challenges it is currently facing. Especially given that the SEQ City Deal is still under negotiation with the Federal Government after it was signed in March 2019 [59], the strategic priorities need to be carefully thought out for future-proofing the city in this context as interconnected smart cities at the regional level can be beneficial for stimulating the development of innovative and inclusive urban infrastructure networks [60]. City Deals involve a partnership between all government and community levels in driving long-term investment and economic growth for eight Australian cities/regions, one of which is SEQ.

Smart city initiatives by major Australian metropolitan cities, such as Sydney, Melbourne, and Brisbane, garner greater attention and benefit from legacy planning and federal funding [61]. Brisbane was recognised as one of the earliest proponents of smart city initiatives in the city-building process. Recently, Brisbane Vision 2031, the planning scheme for Brisbane, identifies "smart, prosperous city" as one of its eight themes [62]. In addition, the Smart, Connected Brisbane Framework focuses on Brisbane City Council's dedicated approach to enabling (a) user-centred design with clear goals and aspirations, (b) pathways to collaboration, (c) fit-for-purpose and future-proofed infrastructure, and (d) data usage for informed decision making [63]. Initiatives implemented under the framework include projects on smart poles and road safety, and innovative technology exploration in partnership with businesses, community, and universities.

On the other hand, the Gold Coast's smart city journey commenced with its participation in IBM's Smarter Cities Challenge as a grant recipient in 2013, but, beyond this milestone, the various smart city initiatives implemented for the city since then have been ad-hoc compared to other cities [56]. Similar regional cities in Australia, such as the Sunshine Coast and Newcastle, are early adopters of smart city strategies and have charted a more mature and unique approach. In particular, Newcastle has emerged with a strategic vision that does not focus on corporate images of smart cities but prioritises solutions for the local community's needs [64]. In comparison, the Gold Coast is still in a state of infancy in its transition to a smart city. Though, in recent times, the CoGC has been investing in several smart city initiatives under the "Our sustainable city" program [65]. However, there is still scope for a more co-ordinated approach. The SEQ City Deal proposition document places greater emphasis on Brisbane, while the Gold Coast light rail is the only project that the document highlights [66]. Additionally, in the case of Brisbane, advocacy groups such as the Committee for Brisbane, which includes representatives from businesses, industry bodies, and universities, publishes policy papers to position Brisbane as a highly liveable city [67]. A similar stakeholder arrangement is lacking for the Gold Coast. Therefore, more research and planning are needed to understand the unique challenges and opportunities for the Gold Coast's smart city transformation, thus motivating this study.

As previously shown in the literature review, the current concepts of smart cities are transitioning away from just a technology-centred approach towards a more holistic approach. Despite the lack of a universal definition, several researchers have proposed smart city frameworks to explain the emerging city model, e.g., [15,68–70]. Six key dimensions, including governance, economy, mobility, environment, people, and living, are recurring themes in some frameworks. For this article, we identified the following five thematic areas, developed in a previous study on the Gold Coast [71] and further expanded in this research, as essential considerations for smart cities:

1. Cultural and natural amenities—the various quality of life factors, such as public spaces, natural environment, events, and cultural activities and facilities.
2. Technology—implementation of ICTs and advanced telecommunication infrastructure for the improvement of city systems and services.
3. Knowledge and innovation precincts—facilities for attracting, creating, and retaining a knowledge workforce.
4. People and skills—attraction, creation, and retention of knowledge/creative workers and businesses.
5. Governance—e-governance, strategic plans, stakeholder collaboration, and open data, which are accessible to the public in real-time.

The themes above collectively form a conceptual framework of smart cities, which is visually displayed in Figure 2 below. The figure also illustrates some of the key factors underpinning each theme.

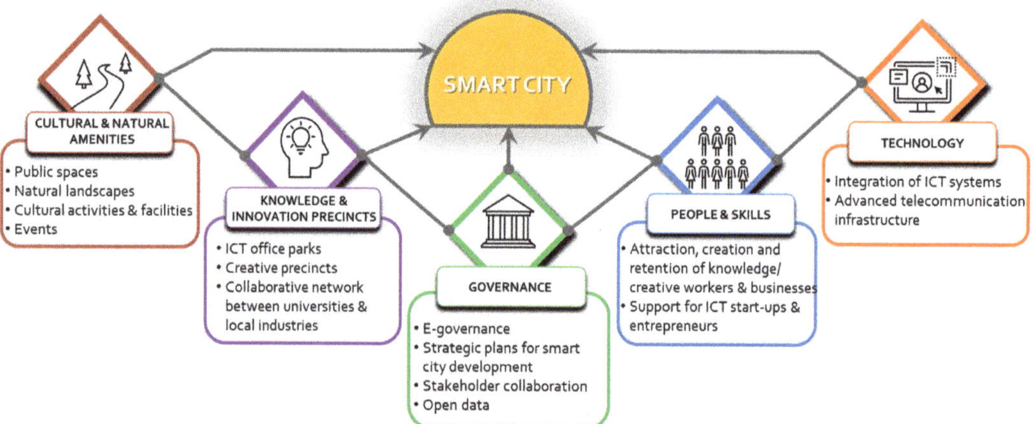

Figure 2. Key themes and factors for analysing the Gold Coast case study, adapted from [71].

The study applied the framework to the Gold Coast. Based on the five key themes articulated in the framework, opportunities and challenges for the Gold Coast to develop into a smart city were identified. The discussion below is primarily based on a critical analysis of the city's existing context and relevant policies and initiatives across both the public and private sectors.

3.1. Cultural and Natural Amenities

The first theme is related to the quality of life, which can be enhanced by factors such as natural environment, public spaces, events, and cultural activities and facilities. Importantly, a high quality of life can act as a key catalyst for attracting knowledge/creative workers, who are critical to creating a smart city.

The Gold Coast is promoted as a unique location with natural beauty and as an ideal place to live, work, and play. As one of Australia's most biologically diverse cities, the city is home to several World-Heritage-listed rainforests, coastal ecosystems, and world-renowned beaches, which are frequented by both locals and visitors alike. The Gold Coast also comprises a range of entertainment facilities, such as theme parks and wildlife sanctuaries, sporting and exhibition facilities, recreational facilities, and outdoor activities. Nevertheless, the unique ecosystems within coastal cities like the Gold Coast are prone to natural hazard events. The CoGC has implemented some initiatives to mitigate these effects, which are further discussed in Section 3.2. However, there is still scope for introducing some of the emerging smart technologies and practices for coastal disaster management. Some of the prominent uses of technology for disaster management include ICT-based global and regional disaster alert systems, water-level monitoring through the use of sensors, early flood detection and processing closed-circuit television (CCTV) imagery for machine learning, foldable flood barriers, and transmission of disaster information to smartphones [72].

As part of the city's Culture Strategy 2023 vision, the first stage of Home of the Arts (HOTA), a new Cultural Precinct seeking to showcase the Gold Coast community's culture, arts, and creativity to the world, has been recently completed [73]. In addition to HOTA, several art galleries and museums are located throughout the city and have served as the city's popular attractions. Along with smaller events that are either free or low-cost, the city also regularly hosts several major events. The rich diversity and availability of natural environment, events, and entertainment and cultural facilities, which have all contributed to making the city one of Australia's most popular destinations, can play a major role in attracting knowledge/creative workers.

However, the Gold Coast's image and branding are often perceived poorly by the rest of the country due to issues with safety, drugs, bikers, and violence, which are also frequently overplayed by the media [74]. Meanwhile, swift population growth and the increasingly limited availability of greenfield land supply on the Gold Coast have led to high demand for new residential developments and rapidly rising housing prices [75]. Likewise, a recently conducted study on the supply of affordable rental housing revealed that the Gold Coast had the greatest shortage of affordable housing amongst large regional satellite cities in Australia [54]. Throughout the COVID-19 pandemic, this situation has further worsened with a large influx of interstate migration to the Gold Coast, adding to the strain on the supply of affordable rental housing [76]. The Gold Coast's negative image and lack of affordable housing could be a significant barrier to attracting knowledge/creative workers.

3.2. Technology

The technology theme is related to the use of ICTs to improve how the city operates and the availability of advanced telecommunication infrastructure. In addition to improving the efficiency of several city services, these technologies can play a vital role in attracting knowledge workers to a city. They can also support land use intensification policies to address land supply shortages and their associated issues such as traffic congestion and lack of housing affordability.

The CoGC, through its Local Government Infrastructure Plan and Economic Development Strategy 2013–2023, promotes smart infrastructure development in three key areas, namely, climate change response infrastructure, transport infrastructure, and ICT transformation [77]. As a city prone to a rising sea level, coastal erosion, and other climate change issues, the CoGC utilised the IBM Smarter Cities Challenge grant as a measure to improve its climate change resilience and response capabilities. The establishment of the Transport Coordination Centre during the 2018 Commonwealth Games [78] has improved the efficiency and attractiveness of the local public transport infrastructure by providing real-time information. Likewise, the recently implemented ICT Transformation Program has enhanced the council's digital customer service capabilities [77].

The CoGC has recently implemented a range of ICTs to improve the operational efficiency of the city's infrastructure. Wireless sensors have been installed at Springbrook National Park, a 6725-hectare World-Heritage-listed rainforest, to monitor environmental variables and track biodiversity restoration [79]. The Gold Coast Waterways Authority manages and protects the 260-kilometre stretch of navigable inland waterways throughout the city. In early 2020, the agency collaborated with the Queensland University of Technology and invested in a smart camera trial to study the waterways. The trial will use machine learning and statistical analysis to investigate the type of vessels using the waterways and to document marine incidents and weather conditions [80]. The CoGC has invested in several water infrastructure solutions to improve the city's water resilience. Drinking water quality and safety are ensured through regular testing and sampling, the results of which can be accessed by the public using the online interactive mapping tool [81]. Since June 2020, the council has been upgrading existing water meters with smart metering devices that can collect, transmit and analyse water-usage data. The CoCG envisages "access to near real-time water-use data" for non-residential users in the first stages of the smart water meter project [82].

The mid-life review of Gold Coast Transport Strategy 2031 identified traffic congestion management as one of the key priorities. In addition to the major road and intersection upgrades proposed, the review recognised opportunities for employing ICTs to gather and share real-time traffic information, monitor incidents, and coordinate traffic control [52]. As part of the strategy, the CoGC has implemented two fully automated parking-enforcement units. The vehicle-mounted units employ cloud-based, preloaded parking rules and include automatic license-plate-recognition cameras.

In December 2020, the Gold Coast became the first city in Queensland to incorporate smart ticketing into its public transport system. The project has enabled payments for trips to be made using smartphones, smartwatches, or debit/credit cards, effectively replacing existing card- and ticket-based systems [83]. By increasing the convenience of using public transport, the smart ticketing system will expectedly drive the city's public transport patronage, which has substantially expanded following the recent introduction of the light rail [84]. Stage 2 of the light rail was completed in December 2017 ahead of the 2018 Commonwealth Games, and two additional stages are planned for further expansion of the infrastructure. Enhancing the use of public transport will aid in reducing the amount of land and investment required for accommodating vehicular traffic in the future, thus assisting in land-use intensification for the city. In addition, the CoGC has invested in renewable energy technology for sustainable transport options. The installation of ten electrical charging stations in 2021 [85], an example of which is shown in Figure 3 below, demonstrates the council's intent to promote the use of alternative fuel vehicles on the Gold Coast.

Figure 3. An electric vehicle charging station at HOTA (source: authors).

A major telecommunication infrastructure in Australia, which has received significant political, public, and media attention over the past decade, is the National Broadband

Network (NBN). As at September 2021, most of the urban areas of the Gold Coast are serviced by fixed-line NBN infrastructure [86]. The NBN was expected to provide high-speed broadband access to all Australian households [87], yet its average internet speed is one of the lowest amongst OECD countries [88]. Most Australian cities, including the Gold Coast, have experienced a high level of internet congestion and the resulting lower internet speed throughout the COVID-19 pandemic [89] due to the increased adaptation of remote working from home induced by government-imposed lockdowns [90]. Cities like Adelaide have invested in enterprise-grade fibreoptic networks to establish the "Ten Gigabit Adelaide" network for seamless and high-speed internet connection [91]. Likewise, several other regional cities in Australia have made similar investments to attract businesses (e.g., Bendigo, with its 100-gigabit network [92] and the Sunshine Coast, through the international submarine cable connected to its "new Gigabit Maroochydore City Centre" project [93]). As part of its objective to transform the Gold Coast into an active digital city, the CoGC has recently invested in a range of infrastructure to facilitate digital connectivity for the city. Table 1 below outlines the range of existing and developing digital infrastructure on the Gold Coast.

Table 1. Summary of digital infrastructure on the Gold Coast, adapted from [94].

Infrastructure Type	Details
Internet of Things	• Installation of a commercial-grade low-power wide-area network (LPWAN), which enables large-scale Internet-of-Things (IoT) applications and encompasses more than 1300 km^2.
Telecommunications	• Council-invested telecommunications network, which is expected to enhance local internet speeds and enable the council to keep rates low.
Fibreoptic network	• The CoGC is the first council in Australia to invest in an 864-core carrier-grade fibreoptic network, covering more than 95 km; • Provision of 1 gigabit-per-second internet speed in the Gold Coast Health and Knowledge Precinct, further discussed below, and additional locations in the future.
Free Wi-Fi	• Provision of free Wi-Fi for residents, visitors, and local businesses in the major urban centres of the city.
Data intelligence	• Digital City Insights (DCI)—a council initiative utilising data collection, research, and analysis to develop new insights for guiding the council and local businesses [95].
Open data portal	• Availability of Geographic Information System (GIS) data on the council's Open Data Portal for public use; • Published datasets include information on council administration, local environment, parks and recreation, planning and building, stormwater, transport, and water and sewerage services.
Mobile app	• City App—the CoGC's mobile application for access to online council services.

The CoGC's investment in telecommunications and fibreoptic networks will crucially support working-from-home arrangements both throughout and post the COVID-19 pandemic. As such, the infrastructure can potentially play an important role in land-use intensification in the future, thus addressing the city's land use supply issue previously discussed.

The CoGC's investment in a range of relevant technologies to create digital connectivity for the Gold Coast illustrates the council's proactive approach to ensuring adequate telecommunication infrastructure in line with its corporate plan. However, with increasing competitiveness amongst other regional cities for digital connectivity dominance, it is relevant that the Gold Coast continues to upgrade its digital infrastructure to stay ahead. This can be challenging, however, as the council has been reducing its yearly rate increases since 2012. The city's rates are the second-largest source of income for the council, contributing to approximately one-third of the budget [96]. Additionally, limited funding could be available for future smart city initiatives, especially in the post-pandemic period.

3.3. Knowledge and Innovation Precincts

The third theme applied to the case study is associated with knowledge and innovation precincts, which play a critical role in attracting, nurturing, and retaining knowledge/creative workers.

Several ICT office parks are located throughout the Gold Coast and specifically cater to the needs of ICT businesses. As part of the CoGC's Pacific Innovation Corridor strategy, Varsity Lakes, a master-planned community adjoining Bond University, has been designated as a specialised ICT hub with offices for ICT businesses. In addition to Varsity Lakes, Southport, the city's central business district, has been designated as another ICT hub.

The Gold Coast is also home to Village Roadshow Studios, a creative precinct providing world-class film production facilities where several film companies are jointly located. With various film equipment and facilities available on-site, the precinct's primary competitive advantage is its relatively higher cost-effectiveness for film production than other major filming locations such as Los Angeles. The CoGC [97] (p. 11) describes the Gold Coast as "an enviable [filming] location for any production company" due to Village Roadshow Studios, easily accessible and diverse locations, and readily available production crews who are experienced in both film and television. The CoGC is also the only local government agency in Australia to offer financial and non-financial incentives to attract film and television productions to the city through its Screen Attraction Program. To date, several high-profile films and television series have been produced on the Gold Coast (e.g., Aquaman, Thor: Ragnarok, San Andreas, Peter Pan, and Terra Nova).

The Queensland Government has recently established the 200-hectare Gold Coast Health & Knowledge Precinct (GCHKP) in Southport. The mixed-use precinct comprises Gold Coast University Hospital, Gold Coast Private Hospital, a co-working and innovation hub, a residential precinct, and a Griffith University campus. In addition to the GCHKP, the other two universities on the Gold Coast are also co-located with hospitals. Bond University is situated in proximity to Robina Hospital. On the other hand, two major hospitals, namely, John Flynn Private Hospital and Tweed Hospital, adjoin Southern Cross University's Gold Coast campus. The three universities have been collaborating with their nearby hospitals to provide practical experience for their medical students. The co-location of the universities and the hospitals throughout the Gold Coast not only promotes the city as a desirable location for medical studies but also locally generates knowledge workers in the healthcare industry.

Another major knowledge precinct, which plays an important role in attracting and generating knowledge workers for the city, is centred around the Southern Cross University campus. The campus is located inside Gold Coast Airport, one of the fastest-growing airports in Australia. This co-location of an airport and a university campus is the first of its kind in Australia as of 2018 [98]. The knowledge precinct has significant knowledge creation implications for the Gold Coast as the airport attracts fly-in/fly-out students both domestically and internationally [99].

However, due to the historically car-oriented design of the Gold Coast, there is currently a lack of direct connectivity, particularly by walking, cycling, and public transport, between its knowledge precincts [100]. Enhancing the connectivity between the precincts could facilitate further training for students and additional collaborative initiatives between the universities and the hospitals. The recently implemented smart ticketing system for the local public transport and the light rail, both previously discussed, will play a critical role in promoting public transport connectivity between these knowledge precincts. The planned installation of a sensor network across the city's transport network will aid in alleviating traffic congestion and reduce travelling time, thus assisting in strengthening the transport connectivity between the knowledge precincts.

3.4. People and Skills

The article now examines the theme of people and skills associated with attracting, creating, and retaining knowledge/creative workers and businesses in a city.

Gold Coast TechSpace, a community innovation hub, regularly delivers workshops for local communities to learn about different technologies, such as robotics, green technologies, and a range of hardware and software. It also provides different memberships to cater to a range of interests and needs amongst community members. The organisation plays a significant role in educating local ICT workers for the city. The Gold Coast Innovation Hub, a not-for-profit organisation established in 2017 through a partnership of several stakeholders, provides physical and virtual co-working spaces to connect and support start-ups and foster the "innovation ecosystem" of the Gold Coast [101]. Likewise, in partnership with Telstra and other ICT companies, the CoGC launched the Mayor's Telstra Technology Award in 2014. The event, which has been running on an annual basis, invites high-school students to submit innovative ideas for new technology products. During the competition, students are mentored by representatives from Bond University, Telstra, and Startup Apprentice to transform their initial ideas into a concrete plan [102]. Startup Apprentice, a local organisation recently established in response to the high youth unemployment issue in the city, has been providing after-school entrepreneurship education to local students. It aims to equip students with entrepreneurship. The Mayor's Telstra Technology Award and Startup Apprentice play an important role in nurturing young, local entrepreneurs for the city.

In recent years, the CoGC has implemented several initiatives to attract international students to the Gold Coast. The council currently has active Sister City Agreements for student exchange programs with cities in China, the United Arab Emirates, the United States, Japan, Taiwan, and New Caledonia [103]. However, in contrast to metropolitan cities in Australia, the Gold Coast has a relatively limited range of employment options due to its dependence on tourism and associated industries. The relative scarcity of the Gold Coast's employment opportunities may constrain the city's ability to attract and retain tertiary students and knowledge/creative workers.

3.5. Governance

The article now examines the theme of governance, which relates to e-governance and strategic plans, stakeholder collaboration, and the availability of real-time open data.

Smart city policy at the local government level has been gaining prominence in Australia, with several cities implementing smart city strategies in addition to state-level policies [61]. For instance, Brisbane City Council launched its "Smart, Connected Brisbane Framework" [63], outlining its vision for achieving goals for Brisbane using innovation, technology, and data. The Sunshine Coast Regional Council [104] adopted the Sunshine Coast Smart city framework in 2016 and set out the Smart City Implementation plan 2016–2019. Likewise, in 2018, Logan City Council released its City Futures Strategy, a smart city plan which is aligned with the Smart Cities Plan and the council's other strategic plans [105]. In contrast, the CoGC has not yet implemented a specific smart city strategy or framework. The Corporate Plan—Gold Coast 2022, which highlights the council's strategic directions for the Gold Coast, identifies a few key plans and programs to realise the outlined objective of becoming an active digital city [55], without underlining specific strategies or implementation plans. This objective centrally focuses on the use of technology to shape and manage the city. The other themes critical to the creation of a smart city—namely, cultural and natural amenities, knowledge and innovation precincts, and people and skills—have not been addressed in the corporate plan for the purpose of transforming the Gold Coast into a smart city. As such, the objective does not sufficiently integrate the different elements of smart cities in its current state.

The smart city objective in the CoGC's corporate plan has not been adequately integrated into the strategic framework of the City Plan, the council's planning scheme regulating land uses throughout the Gold Coast. Without sufficient incorporation of smart city directions into the City Plan, future development driven by the private and community sectors may inhibit the Gold Coast's evolution into a smart city. In contrast, Brisbane's planning scheme articulates the establishment of the city as a "smart, prosperous city" as a central theme. The document specifies that "Brisbane's highly-skilled workers will be a major competitive advantage for local businesses and attract new businesses to the city" and that, by 2031, Brisbane will be "a major Australian study destination for international students" [62].

The Gold Coast is one of the first three Australian cities to participate in the Open & Agile Smart Cities (OASC) initiative, along with Brisbane and Springfield [106]. OASC is an international collaborative program comprising more than 150 cities from around the globe as of December 2020, according to the OASC's website. The initiative seeks to promote the development of smart cities using standard Application Programming Interfaces (APIs) and datasets amongst member cities. Through its Open Data project, the CoGC provides access to a broad range of GIS data through its online Open Data Portal. However, in an Open Data Forum event recently hosted by the CoGC, data ownership has been highlighted by local ICT businesses as a major barrier to creating innovative information products for the public. Whilst some datasets are under the council's ownership, others are exclusively owned by other government agencies, thus preventing adequate access. The CoGC has also been actively collaborating with several businesses to spearhead the development of the city's knowledge industry, infrastructure, and workers. For instance, the council has been working with CoastalCOMS, a local ICT firm, in monitoring the city's beach conditions and is a Founding Partner for TechSpace. The Open Data project, which can encourage innovative information products and services to be created by individuals and businesses, also heavily involves the private sector to ensure a comprehensive data range is available to the public.

The CoGC has implemented some e-governance initiatives for more efficient city management and to improve the accessibility of its services to the public. The council released a smartphone app named "City App" in 2019, which functions as a central point of communication for local citizens and visitors to report non-urgent issues throughout the city, such as graffiti, illegal dumping of rubbish, water leaks, and abandoned vehicles. It also allows photos taken with smartphones to be attached to a report. However, as of 10 September 2021, the app has attracted poor ratings on both Apple's App Store (2.1 out of the possible 5 stars) and Google's Play Store (1.9 out of 5), with reviews highlighting issues associated with the app's layout, navigation, and functions. Another e-governance initiative of the CoGC is related to its rates and water bills, which were previously only mailed to residents in a paper format. The council's website now enables residents to instantly update their contact details and view and pay their rates and water bills.

The CoGC has a community engagement program to facilitate and encourage public participation in generating ideas for shaping the city's future. The council also has an online platform titled "GC have your say," where residents can share their opinions on its draft strategies and policies. In addition, the platform allows residents to provide feedback on development applications requiring public notification before the council's decision is made. In recent years, local Gold Coast community members have been involved in providing feedback on several plans and policies, such as the Gold Coast City Transport Strategy 2031, the Culture Strategy 2023, and amendments to the City Plan. However, the majority of community engagement undertaken by the council has historically been initiated at the stage when a draft plan has been finalised. This process inherently limits the ability of local community to shape the council's plans and policies. Alternatively, if community members were engaged at an earlier stage of the plan-making process, their ideas would influence these documents and the future of their city to a greater extent [107].

Figure 4 below summarises the opportunities and challenges for developing the Gold Coast into a smart city.

The dotted lines between the five themes in Figure 4 above reflect the ad hoc nature of the Gold Coast's smart city development to date. The CoGC has been proactive in developing the city into a smart city, evident from its implementation of several initiatives that complement the different smart city themes in Figure 4. However, the existing strategic objective for the city to become a digital city primarily considers the technology dimension of smart cities. As such, there is an opportunity to holistically integrate all five smart city dimensions outlined in Figure 4 across the relevant strategic and statutory planning frameworks for the Gold Coast. There is also an opportunity to systematically assess the potential of smart city ideas and ICTs to contribute to each of the five themes and to the suggested integration of the themes. Capitalising on these opportunities will enable future initiatives, investment, and development on the Gold Coast to support, rather than inhibit, the city in its transformation into a smart city.

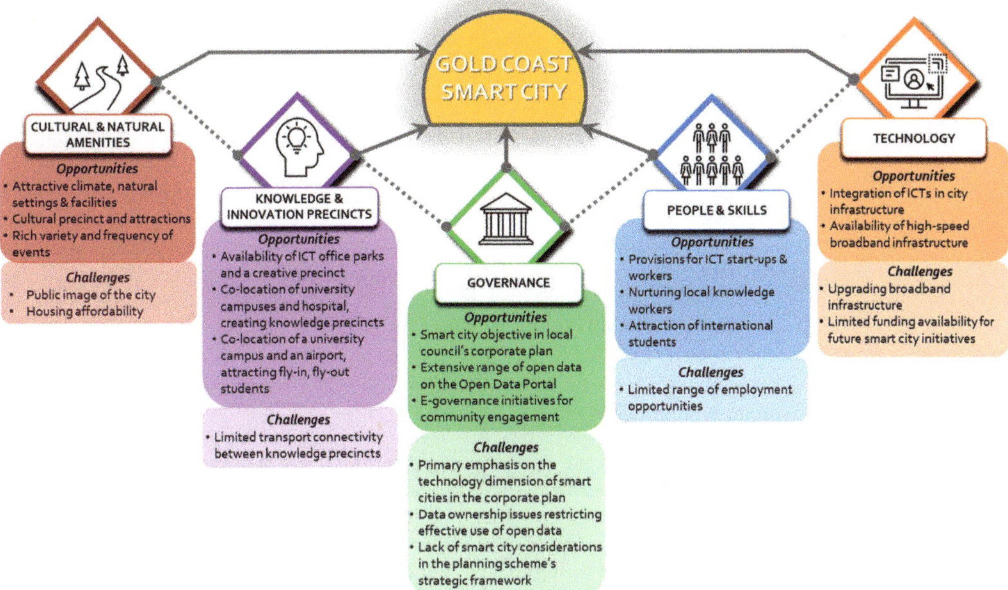

Figure 4. Opportunities and challenges for developing the Gold Coast into a smart city.

4. Conclusions

As cities continue to become increasingly globalised and competitive and move towards the knowledge and information economy, the concept of smart cities is gaining worldwide traction. This study examined the nature of smart cities through a literature review, which revealed five important dimensions of smart cities, including: cultural and natural amenities, technology knowledge and innovation precincts, people and skills, and governance. These factors were organised into themes, which collectively form a conceptual framework of smart cities. The framework was then applied to the case study of Australia's Gold Coast, a regional city still in its infancy in its smart city journey, to identify key challenges and opportunities for its development into a smart city. The following seven key lessons for developing regional cities into smart cities can be drawn from the case study:

(1) The Gold Coast's successful bid for hosting the 2018 Commonwealth Games helped the city to attract external investments in smart infrastructure such as IBM's Smarter Cities Challenge grant and the city's light rail infrastructure, which received state and federal funding support. The 2032 Olympics and Paralympics provide another excellent opportunity to attract funding in smart infrastructure for both the Gold Coast and the whole SEQ region. The Gold Coast's success in securing funding by leveraging a major international event is an example that other regional cities can follow in attracting external investment for smart city initiatives.

(2) Cultural and natural amenities play an important role in attracting and retaining knowledge workers, particularly for regional cities, as they may lack the "pull factors" that metropolitan cities often do. The nature of these amenities can be tangible (e.g., entertainment and lifestyle amenities, natural landscapes, and cultural precincts) or intangible (e.g., city's image and branding and housing affordability).

(3) Ensuring widespread availability of high-speed internet infrastructure is essential for supporting remote-working arrangements, which have become critically important throughout the COVID-19 pandemic, and for improving global competitiveness of cities. Future-proofing digital infrastructure is especially relevant before a major event such as the Olympic Games 2032, which has long-term planning implications for the Gold Coast.

(4) The partnership between a university and state and local governments to establish the Gold Coast Health & Knowledge Precinct has strengthened the Gold Coast's health-care and education industries. Other regional cities can apply a similar partnership arrangement to diversify their economic bases and attract knowledge workers.

(5) Government-collected data related to the natural and built environments should be made available to the public in real-time. However, when publishing this data, issues associated with data ownership should be addressed to allow local stakeholders to efficiently access and utilise the data and to create useful information products and services for the public.

(6) Planners should ensure that smart city directions articulated in strategic plans are adequately reflected in the city's land-use regulations. By doing so, future development will become more aligned with smart city objectives, which could assist in expediting the overall process of smart city development. The SEQ City Deal also provides a significant opportunity for coherently strategising the region's future priorities, including smart city developments for the Gold Coast and other cities in the region.

(7) Planning frameworks should holistically integrate the different elements of smart cities. The current strategic objective for the Gold Coast to become a digital city places a strong emphasis on the technology dimension of smart cities. However, other dimensions highlighted in this study, including cultural and natural amenities, knowledge and innovation precincts, people and skills, and governance, are also important considerations for developing smart cities.

These lessons have potential applicability to not only the Gold Coast but also other emerging regional smart cities around the world.

Author Contributions: Conceptualisation, I.K. and B.B.; methodology, I.K. and B.B.; validation, B.B.; formal analysis, I.K.; investigation, I.K.; data curation, I.K. writing—original draft preparation, I.K.; writing—review and editing, I.K., K.S.C. and B.B.; visualisation, K.S.C.; supervision, B.B.; project administration, I.K. and B.B. All authors have read and agreed to the published version of the manuscript.

Funding: This research received no external funding.

Conflicts of Interest: The authors declare no conflict of interest.

References

1. Bibri, S.E.; Krogstie, J. The Emerging Data–Driven Smart City and Its Innovative Applied Solutions for Sustainability: The Cases of London and Barcelona. *Energy Inform.* **2020**, *3*, 1–42. [CrossRef]

2. Joo, Y.M.; Tan, T.-B. *Smart Cities in Asia: Governing Development in the Era of Hyper-Connectivity*; Edward Elgar Publishing Limited: Cheltenham, UK, 2020; ISBN 978-1-78897-287-1.

3. Deakin, M. Smart Cities: The State-of-the-Art and Governance Challenge. *Theor. Chem. Acc.* **2014**, *1*, 1–16. [CrossRef]

4. Hu, R. Australia's National Urban Policy: The Smart Cties Agenda in Perspective. *Aust. J. Soc. Issues* **2020**, *55*, 201–217. [CrossRef]

5. Dedekorkut-Howes, A.; Bosman, C. The Gold Coast: Australia's Playground? *Cities* **2015**, *42*, 70–84. [CrossRef]

6. Dedekorkut-Howes, A.; Bosman, C.; Leach, A. Considering the Gold Coast. In *Off the Plan: The Urbanisation of the Gold Coast*; Dedekorkut-Howes, A., Bosman, C., Leach, A., Eds.; CSIRO Publishing: Clayton, Australia, 2016; pp. 1–16. ISBN 978-1-4863-0183-6.

7. Gold Coast Tourism Advisory Panel. *Gold Coast Global Tourism Hub: Community and Stakeholder Consultation*; Queensland Government, 2019. Available online: https://bit.ly/3Cuuv7Q (accessed on 20 July 2021).

8. Cunningham, S.; McCutcheon, M.; Hearn, G.; Ryan, M.D.; Collis, C. *Australian Cultural and Creative Activity: A Population and Hotspot Analysis: Gold Coast*; Digital Media Research Centre, 2019. Available online: https://bit.ly/3EJh31K (accessed on 20 July 2021).

9. CoGC. *Annual Plan 2016–17*; CoGC, 2016. Available online: https://bit.ly/3CB5mbA (accessed on 24 July 2021).

10. Allam, Z.; Newman, P. Redefining the Smart City: Culture, Metabolism and Governance. *Smart Cities* **2018**, *1*, 4–25. [CrossRef]

11. Angelidou, M. Smart City Policies: A Spatial Approach. *Cities* **2014**, *41*, S3–S11. [CrossRef]

12. Kitchin, R. The Real-Time City? Big Data and Smart Urbanism. *GeoJournal* **2014**, *79*, 1–14. [CrossRef]

13. Townsend, A.M. *Smart Cities: Big Data, Civic Hackers, and the Quest for a New Utopia*; W.W.Norton: New York, NY, USA, 2014; ISBN 978-0-3933-4978-8.

14. Hollands, R.G. Will the Real Smart City Please Stand Up?: Intelligent, Progressive or Entrepreneurial? *City* **2008**, *12*, 303–320. [CrossRef]

15. Albino, V.; Berardi, U.; Dangelico, R.M. Smart Cities: Definitions, Dimensions, Performance, and Initiatives. *J. Urban Technol.* **2015**, *22*, 3–21. [CrossRef]

16. Vanolo, A. Smartmentality: The Smart City as Disciplinary Strategy. *Urban Stud.* **2014**, *51*, 883–898. [CrossRef]

17. Barlow MMichael, A.; Lévy-Bencheton, C. *Smart Cities, Smart Future: Showcasing Tomorrow*; John Wiley & Sons, Incorporated: Newark, NJ, USA, 2018; ISBN 978-1-1195-1618-7.

18. Toli, A.M.; Murtagh, N. The Concept of Sustainability in Smart City Definitions. *Front. Built Environ.* **2020**, *6*, 1–10. [CrossRef]

19. Anthopoulos, L.G.; Vakali, A. Urban Planning and Smart Cities: Interrelations and Reciprocities. In *The Future Internet FIA 2012 Lecture Notes in Computer Science*; Álvarez, F., Cleary, F., Dara, P., Domingue, J., Galis, A., Garcia, A., Gavras, A., Karnourskos, S., Schaffers, S.K., Li, M.-S., et al., Eds.; Springer: Berlin/Heidelberg, Germany, 2012; p. 1. Available online: https://bit.ly/2UX8vmc (accessed on 28 August 2021).

20. Yigitcanlar, T.; Lee, S.H. Korean Ubiquitous-Eco-City: A Smart-Sustainable Urban Form or a Branding Hoax? *Technol. Forecast. Soc. Chang.* **2014**, *89*, 100–114. [CrossRef]

21. Ferraris, A.; Santoro, G.; Pellicelli, A.C. "Openness" of Public Governments in Smart Cities: Removing the Barriers for Innovation and Entrepreneurship. *Int. Entrep. Manag. J.* **2020**, *16*, 1259–1280. [CrossRef]

22. Alizadeh, T. An Investigation of IBM's Smarter Cities Challenge: What Do Participating Cities Want? *Cities* **2017**, *63*, 70–80. [CrossRef]

23. Bibri, S.E. On the Sustainability of Smart and Smarter Cities in the Era of Big Data: An Interdisciplinary and Transdisciplinary Literature Review. *J. Big Data* **2019**, *6*, 25. [CrossRef]

24. Chandrasekar, K.S.; Bajracharya, B.; O'Hare, D. A Comparative Analysis of Smart City Initiatives by China and India: Lessons for India. In Smart Cities for 21st Century Australia—How Urban Design Innovation Can Change our Cities. In Proceedings of the International Urban Design Conference, Canberra, Australia, 7–9 November 2016; pp. 339–357.

25. Shelton, T.; Zook, M.; Wiig, A. The 'Actually Existing Smart City'. *Camb. J. Reg. Econ. Soc.* **2015**, *8*, 13–25. [CrossRef]

26. Hollands, R.G. Critical Interventions into the Corporate Smart City. *Camb. J. Reg. Econ. Soc.* **2015**, *8*, 61–77. [CrossRef]

27. Masik, G.; Sagan, I.; Scott, J.W. Smart City Strategies and New Urban Development Policies in the Polish Context. *Cities* **2021**, *108*, 102970. [CrossRef]

28. Ahvenniemi, H.; Huovila, A.; Pinto-Seppä, I.; Airaksinen, M.; Pinto-Seppa, I.; Airaksinen, M. What Are the Differences between Sustainable and Smart Cities? *Cities* **2017**, *60*, 234–245. [CrossRef]

29. Yigitcanlar, T. *Technology and the City: Systems, Applications and Implications*; Taylor and Francis: Milton Park, UK, 2016; ISBN 978-1-1388-2670-0.

30. Thomas, V.; Wang, D.; Mullagh, L.; Dunn, N. Where's Wally? In Search of Citizen Perspectives on the Smart City. *Sustainability* **2016**, *8*, 207. [CrossRef]

31. Tan, S.Y.; Taeihagh, A. Smart City Governance in Developing Countries: A Systematic Literature Review. *Sustainability* **2020**, *12*, 899. [CrossRef]

32. Nam, T.; Pardo, T.A. Conceptualizing Smart City with Dimensions of Technology, People, and Institutions. In Proceedings of the 12th Annual International Conference on Digital Government Research, College Park, MD, USA, 12–15 June 2011; pp. 282–291.

33. Clement, D.J.; Crutzen, P.N. How Local Policy Priorities Set the Smart City Agenda. *Technol. Forecast. Soc. Chang.* **2021**, *171*, 1–11. [CrossRef]

34. Paskaleva, K. E-Governance as an Enabler of the Smart City. In *Smart Cities: Governing, Modelling and Analysing the Transition*; Deakin, M., Ed.; Routledge: London, UK, 2014; ISBN 978-0-4156-5819-5.

35. Angelidou, M. Strategic Planning for the Development of Smart Cities. Ph.D. Dissertation, Aristotle University of Thessaloniki, Thessaloniki, Greece, 2015. Available online: https://bit.ly/3j1Qw5P (accessed on 1 July 2021).

36. Yigitcanlar, T. Smart Cities: An Effective Urban Development and Management Model? *Aust. Plan.* **2015**, *52*, 27–34. [CrossRef]

37. Datta, A. New Urban Utopias of Postcolonial India: "Entrepreneurial Urbanization" in Dholera Smart City, Gujarat. *Dialogues Hum. Geogr.* **2015**, *5*, 3–22. [CrossRef]

38. Watson, V. The Allure of "Smart City" Rhetoric: India and Africa. *Dialogues Hum. Geogr.* **2015**, *5*, 36–39. [CrossRef]

39. Komninos, N. Intelligent Cities: Variable Geometries of Spatial Intelligence. *Intell. Build. Int.* **2011**, *3*, 172–188. [CrossRef]

40. Grossi, G.; Pianezzi, D. Smart Cities: Utopia or Neoliberal Ideology? *Cities* **2017**, *69*, 79–85. [CrossRef]

41. Glasmeier, A.K.; Nebiolo, M. Thinking about Smart Cities: The Travels of a Policy Idea That Promises a Great Deal, But So Far Has Delivered Modest Results. *Sustainability* **2016**, *8*, 1122. [CrossRef]

42. Saunders, T.; Baeck, P. *Rethinking Smart Cities from the Ground up*; Nesta, 2015. Available online: https://bit.ly/3lQz7yi (accessed on 17 June 2021).

43. Gehl, J. *Cities for People*; Island Press: Washington, DC, USA, 2010; ISBN 978-1-5972-6573-7.

44. Neirotti, P.; De Marco, A.; Cagliano, A.C.; Mangano, G.; Scorrano, F. Current Trends in Smart City Initiatives: Some Stylised Facts. *Cities* **2014**, *38*, 25–36. [CrossRef]

45. Kitchin, R. Making Sense of Smart Cities: Addressing Present Shortcomings. *Camb. J. Reg. Econ. Soc.* **2015**, *8*, 131–136. [CrossRef]

46. Esposito, G.; Clement, J.; Mora, L.; Crutzen, N. One Size Does Not Fit All: Framing Smart City Policy Narratives within Regional Socio-Economic Contexts in Brussels and Wallonia. *Cities* **2021**, *118*, 1–14. [CrossRef]

47. Alizadeh, T.; Sipe, N. Vancouver's Digital Strategy: Disruption, New Direction, or Business as Usual? *Int. J. E-Plan. Res.* **2016**, *5*, 1–15. [CrossRef]

48. Joss, S.; Sengers, F.; Schraven, D.; Caprotti, F.; Dayot, Y. The Smart City as Global Discourse: Storylines and Critical Junctures across 27 Cities. *J. Urban Technol.* **2019**, *26*, 3–34. [CrossRef]

49. Queensland Government. Queensland Globe. Available online: https://bit.ly/2KqWUTg (accessed on 7 September 2021).

50. Creative Commons. Attribution 3.0 Australia (CC BY 3.0 AU). Available online: https://bit.ly/2OmJIVj (accessed on 7 September 2021).

51. Howes, M.; Dedekorkut-Howes, A. The Adaptation Roller-Coaster: Planning for Climate Change on the Gold Coast, Queensland. In Proceedings of the State of Australian Cities Conference Adelaide: 8th State of Australian Cities National Conference, Adelaide, Australia, 28–30 November 2017; pp. 1–12. Available online: https://bit.ly/3C5rnzW (accessed on 17 June 2021).

52. CoGC. *Gold Coast Transport Strategy 2031—Mid-Life Review Summary*; CoGC, 2019. Available online: https://bit.ly/3kstlnl (accessed on 17 June 2021).

53. Farndon, D.; Burton, P. Avoiding the White Elephants: A New Approach to Infrastructure Planning at the 2018 Gold Coast Commonwealth Games? *Qld. Rev.* **2019**, *26*, 128–146. [CrossRef]

54. Hulse, K.; Reynolds, M.; Nygaard, C.; Parkinson, S.; Yates, J. *The Supply of Affordable Private Rental Housing in Australian Cities: Short-Term and Longer-Term Changes*; Australian Housing and Urban Research Institute, 2019. Available online: https://bit.ly/3o3LABs (accessed on 20 June 2021).

55. CoGC. *Corporate Plan—Gold Coast 2022*; CoGC, 2018. Available online: https://bit.ly/3fbHaDC (accessed on 17 June 2021).

56. Alizadeh, T.; Irajifar, L. Gold Coast Smart City Strategy: Informed by Local Planning Priorities and International Smart City Best Practices. *Int. J. Knowl.-Based Dev.* **2018**, *9*, 153–173. [CrossRef]

57. Australian Government. *Smart Cities Plan*; Australian Government, 2016. Available online: https://bit.ly/3nSiJ3b (accessed on 2 August 2021).

58. Nothling, L. Brisbane's 2032 Olympic Games Venues Will Be a Mix of New and Old. Here's How It Will Look. ABC News. Available online: https://ab.co/3lcqX4X (accessed on 24 July 2021).

59. Australian Government. *South East Queensland City Deal—Statement of Intent*; Queensland Government, 2019. Available online: https://bit.ly/3lP0g4J (accessed on 2 August 2021).

60. Ersoy, A. Smart Cities as a Mechanism towards a Broader Understanding of Infrastructure Interdependencies. *Reg. Stud. Reg. Sci.* **2017**, *4*, 26–31. [CrossRef]

61. Yigitcanlar, T.; Kankanamge, N.; Vella, K. How Are Smart City Concepts and Technologies Perceived and Utilized? A Systematic Geo-Twitter Analysis of Smart Cities in Australia. *J. Urban Technol.* **2020**, *28*, 135–154. [CrossRef]

62. Brisbane City Council. Our Smart, Prosperous City. Available online: https://bit.ly/2YI8jIT (accessed on 7 September 2021).

63. Brisbane City Council. *Smart, Connected Brisbane Framework*; Brisbane City Council, 2017. Available online: https://bit.ly/3E0BWFr (accessed on 7 September 2021).

64. Dowling, R.; McGuirk, P.; Sophia, M. Realising Smart Cities: Partnerships and Economic Development in the Emergenece and Practices of Smart in Newcastle, Australia. In *Inside Smart Cities: Place, Politics and Urban Innovation*; Karvonen, A., Cugurullo, F., Caprotti, F., Eds.; Routledge: London, UK, 2018; ISBN 978-1-3511-6620-1.

65. CoGC. Our Sustainable City. Available online: https://bit.ly/3E3weCN (accessed on 8 September 2021).

66. Coucil of Mayors. *Transforming SEQ: The SEQ City Deal Proposition*; Council of Mayors, 2019. Available online: https://bit.ly/38WUTL4 (accessed on 7 September 2021).

67. Committee for Brisbane. Advocacy. 2021. Available online: https://bit.ly/3zZUFi5 (accessed on 7 September 2021).

68. Giffinger, R.; Fertner, C.; Kramar, H.; Kalasek, R.; Pichler-Milanovic, N.; Meijers, E. *Smart Cities Ranking of European Medium-sized Cities*; University of Copenhagen, 2007. Available online: https://bit.ly/3lQ7422 (accessed on 20 July 2021).

69. Batty, M.; Axhausen, K.W.; Giannotti, F.; Pozdnoukhov, A.; Bazzani, A.; Wachowicz, M.; Ouzounis, G.; Portugali, Y. Smart Cities of the Future. *Eur. Phys. J. Spec. Top.* **2012**, *214*, 481–518. [CrossRef]

70. Cohen, B. Methodology for 2014 Smart Cities Benchmarking. Available online: https://bit.ly/3faZVa6 (accessed on 20 March 2021).

71. Bajracharya, B.; Cattell, D.; Khanjanasthiti, I. Challenges and Opportunities to Develop a Smart City: A Case Study of Gold Coast, Australia. In Proceedings of the REAL CORP 2014: Plan It Smart, Clever Solutions for Smart Cities, Vienna, Austria, 21–23 May 2014; pp. 119–129.

72. Tonmoy, F.N.; Hasan, S.; Tomlinson, R. Increasing Coastal Disaster Resilience Using Smart City Frameworks: Current State, Challenges, and Opportunities. *Front. Water* **2020**, 2, 1–13. [CrossRef]
73. CoGC. Get Set for Our Cultural Evolution. Available online: https://bit.ly/37bTSO6 (accessed on 7 March 2021).
74. Breen, S. The Gold Coast: The Sun-Drenched Sin City that Wants to Shine. Available online: https://bit.ly/3j7oq9c (accessed on 12 January 2021).
75. Armitage, L.; Khanjanasthiti, I.; Chand, S. Micro Houses: Trends and Implications on the Gold Coast. In Proceedings of the 4th Australian Regional Development Conference: Regional Development in a Changing World, Coffs Harbour, Australia, 11–12 September 2017.
76. Cansdale, D. Gold Coast Rental Market Described as "Hectic" as Vacancy Rates Drop. Available online: https://ab.co/3zPMvbx (accessed on 7 March 2021).
77. Alizadeh, T.; Irajifar, L. Towards Gold Coast Smart city: A Combination of Local Planning Priorities and International Best Practices. In Proceedings of the 8th State of Australian Cities National Conference, Adelaide, Australia, 28–30 November 2017. Available online: https://bit.ly/3BURB7q (accessed on 7 March 2021).
78. Bailey, M. Commonwealth Games Rapid Response Transport Centre Ready to Go. Available online: https://bit.ly/3txj6B8 (accessed on 10 September 2021).
79. Queensland Government. Springbrook Wireless Sensor Network—Information Sheet. Available online: https://bit.ly/3xeJzUe (accessed on 7 March 2021).
80. GCWA. Our Organisation. Available online: https://bit.ly/3xddyvL (accessed on 8 September 2021).
81. CoGC. Drinking Water Quality Results. Available online: https://bit.ly/2YJK4Ko (accessed on 8 September 2021).
82. CoGC. Smart Water Meter Solutions Project. Available online: https://bit.ly/3tugBiH (accessed on 9 September 2021).
83. Bailey, M. Smart Ticketing Touches on for the First Time on Gold Coast. Available online: https://bit.ly/3ldD81s (accessed on 22 January 2021).
84. Palaszczuk, A.; Trad, J. Stage Two of Gold Coast Light Rail in Track for Commonwealth Games. Available online: https://bit.ly/3ySW5cV (accessed on 30 March 2021).
85. CoGC. Our Energy Responsibilities. Available online: https://bit.ly/3ldzbIc (accessed on 10 September 2021).
86. NBN Co. NBN Rollout Map. Available online: https://bit.ly/2WBPHt3 (accessed on 23 July 2021).
87. Alizadeh, T. The Spatial Justice Implications of Telecommunication Infrastructure: The Socio-Economic Status of Early National Broadband Network Rollout in Australia. *Int. J. Crit. Infrastruct.* **2015**, *11*, 278–296. [CrossRef]
88. Acharya, M. Australia's Internet Speed Ranks below India, Kazakhstan or Latvia. Available online: https://bit.ly/3zPuzOb (accessed on 15 March 2021).
89. Hobday, L.; Sas, N. Coronavirus Affecting Internet Speeds, as COVID-19 Puts Pressure on the Network. Available online: https://ab.co/3ydjdDq (accessed on 12 January 2021).
90. Australian Institute of Family Studies. *Employment & Work—Family Balance in 2020*; Australian Institute of Family Studies, 2021. Available online: https://bit.ly/37GbJwY (accessed on 7 September 2021).
91. City of Adelaide. Ten Gigabit Adelaide. Available online: https://bit.ly/3gn900n (accessed on 13 August 2021).
92. Preiss, B. The Need for Speed: Bendigo Is out to Deliver the Fastest Internet in Australia. Available online: https://bit.ly/3g2hEB0 (accessed on 12 August 2021).
93. Sunshine Coast Regional Council. Digital Connectivity. Available online: https://bit.ly/3yOHW16 (accessed on 12 August 2021).
94. CoGC. Digital Connectivity. Available online: https://bit.ly/3BGmSei (accessed on 7 September 2021).
95. CoGC. Data Intelligence. Available online: https://bit.ly/3BHsBR6 (accessed on 7 September 2021).
96. CoGC. City Budget 20–21 Paves Roads to Recovery. Available online: https://bit.ly/3lQYsZj (accessed on 23 July 2021).
97. CoGC. *Film Gold Coast*; CoGC, 2019. Available online: https://bit.ly/3fcW8Jj (accessed on 13 August 2021).
98. Queensland Airports Limited. *Productivity Commission, Economic Regulation of Airports*; Productivity Commission, 2018. Available online: https://bit.ly/2LVRPTC (accessed on 7 September 2021).
99. Khanjanasthiti, I. Planning for Economic Development around a Second-Tier Airport: A Case Study of Gold Coast Airport. Ph.D. Dissertation, Bond University, Gold Coast, Australia, 2021. Available online: https://bit.ly/3BX3IAV (accessed on 7 September 2021).
100. O'Hare, D.; Bajracharya, B.; Khanjanasthiti, I. Transforming the Tourist City into a Knowledge and Healthy City: Reinventing Australia's Gold Coast. In Proceedings of the 7th International Forum on Knowledge Asset Dynamics, 5th Knowledge Cities World Summit: Knowledge, Innovation and Sustainability: Integrating Micro & Macro Perspectives (IFKAD-KCWS 2012), Matera, Italy, 13–15 June 2012; pp. 2488–2517.
101. The Gold Coast Innovation Hub. About Us. Available online: https://bit.ly/37bwZdC (accessed on 20 July 2021).
102. CoGC. The Mayor's Telstra Technology Award Finalists Announced. Available online: https://bit.ly/3xeeXCn (accessed on 15 July 2021).
103. CoGC. Sister City Relationships & International Partnerships. Available online: https://bit.ly/2XzvlRB (accessed on 11 March 2021).
104. Sunshine Coast Regional Council. Smart City Framework. Available online: https://bit.ly/3zzmNHL (accessed on 13 August 2021).

105. Logan City Council. *City Futures Strategy*; Logan City, 2018. Available online: https://bit.ly/3zUd4fx (accessed on 13 August 2021).
106. OASC Secretariat. Australian Cities Join Open & Agile Smart Cities. Open & Agile Smart Cities. Available online: https://bit.ly/3zSE7YU (accessed on 14 March 2021).
107. Leading Inside Out; Collective Impact Forum; FSG. *Community Engagement Toolkit*; Collective Impact Forum, 2017. Available online: https://bit.ly/3zPveix (accessed on 13 August 2021).

 *sustainability*

Article

Leveraging Smart and Sustainable Development via International Events: Insights from Bento Gonçalves Knowledge Cities World Summit

Larissa Diana Michelam [1,*], Tatiana Tucunduva Philippi Cortese [1,2], Tan Yigitcanlar [3], Ana Cristina Fachinelli [4], Leonardo Vils [1] and Wilson Levy [1]

[1] Graduate Program in Smart and Sustainable Cities, Nove de Julho University, Rua Vergueiro, 235/249, São Paulo 01525-000, Brazil; taticortese@gmail.com (T.T.P.C.); vilsleo@gmail.com (L.V.); wilsonlevy@gmail.com (W.L.)

[2] Institute of Advanced Studies, University of São Paulo, Rua Praça do Relógio, 109, Ground Floor, Cidade Universitária, São Paulo 05508-050, Brazil

[3] School of Architecture and Built Environment, Queensland University of Technology, 2 George Street, Brisbane, QLD 4000, Australia; tan.yigitcanlar@qut.edu.au

[4] Graduate Program in Administration, University of Caxias do Sul, Rua Francisco Getúlio Vargas, 1130, Caxias do Sul 95070-560, Brazil; acfachin@ucs.br

* Correspondence: larissamichelam@uni9.edu.br

Citation: Michelam, L.D.; Cortese, T.T.P.; Yigitcanlar, T.; Fachinelli, A.C.; Vils, L.; Levy, W. Leveraging Smart and Sustainable Development via International Events: Insights from Bento Gonçalves Knowledge Cities World Summit. *Sustainability* **2021**, *13*, 9937. https://doi.org/10.3390/su13179937

Academic Editor: Anna Visvizi

Received: 31 July 2021
Accepted: 31 August 2021
Published: 4 September 2021

Publisher's Note: MDPI stays neutral with regard to jurisdictional claims in published maps and institutional affiliations.

Abstract: During the last couple of decades, making cities smarter and more sustainable has become an important urban agenda. In this perspective, knowledge-based development is seen as a strategic approach for cities seeking to thrive through innovation and resilience. Accomplishing a knowledge-based development agenda is, however, challenging, and cities need support mechanisms to effectively develop and then incorporate such agendas into their decision-making processes. This study investigates the role of international events as one of these support mechanisms for the development and implementation of local knowledge-based development agendas. The study aims to address how international events contribute to the local knowledge-based development efforts. This study takes the Knowledge Cities World Summit (KCWS) series as the exemplar international event, and the Brazilian city of Bento Gonçalves as the case study city. The methodological approach of the study consists of semi-structured interview-based qualitative analysis and case study investigations. The findings of the study revealed the following: (a) international events can be fundamental drivers of local knowledge-based agendas; (b) these events contribute to host cities' development, especially at an institutional level, by generating outcomes such as engagement in cooperation networks and leveraging local actors' influence on the development process; and (c) KCWS was instrumental in placing the local university as a protagonist of the knowledge-based development movement of Bento Gonçalves. The study reported in this paper provides invaluable insights for cities seeking to use international knowledge-based development events for smart and sustainable city formation.

Keywords: knowledge-based development; knowledge-based urban development; smart and sustainable city; sustainable urban development; local development; urban development; knowledge cities world summit; international events; Bento Gonçalves; Brazil

1. Introduction

During the last decades, global economic, political, and environmental dynamics territorialized at the local level are changing the urban context and creating unprecedented challenges for cities [1,2]. Increased competition for investment capital, socioeconomic disparities, digital and knowledge divides, pandemics, escalating natural disasters, and other climate change effects are some of these global-scale issues currently facing cities [3–5].

Urbanization's extraordinary growth of the last decades, which will continue in the upcoming ones, especially in emerging economies [6,7], has turned global sustainability into an increasingly local agenda, and cities into an arena where the battle for sustainable development takes place [1,8–10]. The inclusion of a goal exclusively aimed at cities (SDG #11—Sustainable Cities and Communities) in the United Nations 2030 Agenda for Sustainable Development [11] is representative of how urban sustainability is now critical for the global development strategy [8].

For urban planners, policymakers, and city managers, urban development in this context involves challenges such as attracting and allocating resources for infrastructure implementation, expansion, and the management of services. Most importantly, it also involves the support of an institutional structure that converses with the different agents of the development process while ensuring adequate and effective governance of the city [7]. Moreover, the COVID-19 pandemic has highlighted the importance and urgency of evidence-based urban planning to provide efficient and effective responses to cities during crisis episodes such as this [12].

These are neither quick nor simple endeavors for any city, and they may be even more challenging for cities in developing nations, where most future urban growth is expected to occur [6]. For these cities, the challenges of fast-paced urbanization and sustainable development are even greater, as they are often faced with limited financial resources and small institutional capacity for urban planning and enforcement [7,13–16]. Furthermore, different studies have indicated that medium- and small-sized cities in developing countries may face even more challenges to achieve sustainable urban development [3,8,13,17].

Amidst these difficulties, new smart technologies, especially in the field of information and communication technology (ICT), have been regarded as the main instruments for solving complex urban problems, making the "smart city" emerge as the urban model to be achieved in the 21st century [2,18]. Initially, on a practical level, the smart city approach was strongly associated with applying data science and smart technology in the urban context [19]. Recently, this approach has been updated to the "smart and sustainable city" model, as it has become progressively clear that technology alone is not the solution to all cities' problems [2,20]. Instead, to be truly smart and sustainable, cities need a holistic approach that uses the opportunities provided by technology applications as a means of promoting all areas of urban development—economic, social, environmental, and institutional [2,8,20].

1.1. Knowledge-Based Development of Smart and Sustainable Cities

In the last two decades, knowledge-based development of cities or knowledge-based urban development (KBUD) has been increasingly applied as a strategic approach for the promotion of smart and sustainable cities [21–24]. Conceived in the 1990s [25], over the years, the KBUD framework has undergone significant updates [23], becoming consolidated as a prevalent policy for cities and regions seeking to thrive through the paths of innovation and sustainability.

KBUD has become increasingly popular as a planning and development approach for cities and regions interested in transforming knowledge resources into local smart and sustainable development [25–28]. By encouraging the attraction, development, and retention of intellectual and human capital, and by fostering innovation and knowledge dynamics, KBUD leverages urban transformation [21], advancing innovative capacities [29], diversifying the economic base [24], upgrading infrastructures [30], and improving quality of life [31]. Furthermore, KBUD can operate as a powerful, multidimensional, and integrated platform that facilitates the application of smart solutions at a practical level, without losing sight of all of the dimensions of sustainability [22].

In order to do that, KBUD's conceptual framework (Figure 1) draws upon a multidisciplinary and balanced perspective that considers urban development through four main elements, or policy domains, namely economic, sociocultural, spatial (or environmental), and institutional [23,28,29,32].

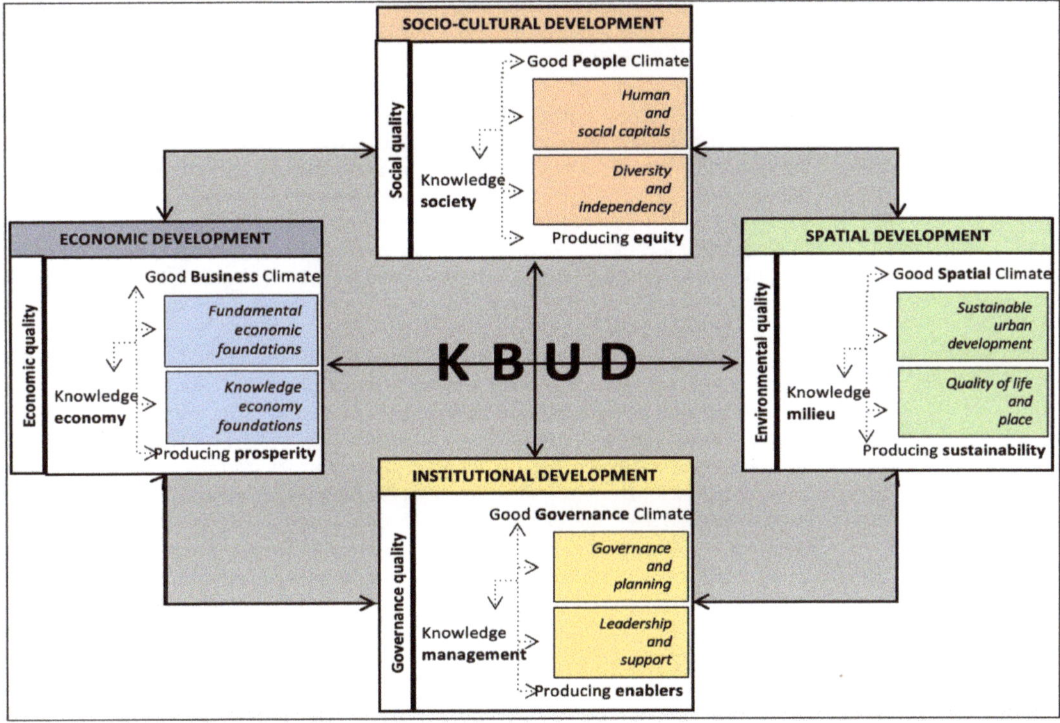

Figure 1. Conceptual framework of KBUD, derived from [29].

The economic development perspective places knowledge as a strategic resource and aims to achieve development by leveraging local endogenous knowledge assets and local research and development (R&D) and innovation processes [26,29].

The socio-cultural development perspective considers that citizens' skills and knowledge are critical for the community's development. This perspective advocates that socio-cultural development can only be achieved through social equity, diversity, independence, and strong human and social capitals [28,29].

The enviro-urban (spatial) development perspective values sustainable and environmentally sound alternatives for the spatial improvement of cities, emphasizing the importance of security and quality of life in urban development [23,33,34].

The institutional development perspective acts as an enabler of the former three through the application of strategic planning, institutional leadership, and partnership principles. In KBUD's framework, governance processes established through partnership and collaboration with all local development actors are elements that improve the city's management [26–29].

Integration and balance among the four perspectives are central to the success of KBUD initiatives, and cities must give equal attention to all policy domains if they want to achieve prosperous knowledge-based urban development [32].

Cities interested in planning and implementing KBUD approaches must start by forming a long-term KBUD strategy and adapting their planning mechanisms to it [27]. A comprehensive understanding of the unique characteristics; the identity differences; the diverse socioeconomic and socio-spatial forms; and, mainly, the valuable knowledge assets of the city must be at the heart of a KBUD process [28,31,35–37]. A central position is given to universities, seen as critical knowledge assets, deeply embedded in systems of innovation and knowledge training, generation, exchange, circulation, and commercialization [35,38].

Additionally, essential for the success of a KBUD process is the ability of city managers and policymakers to establish and cultivate collaboration through a partnership model, often a triple-helix model, in which the collaborative action of the university, government, and the private sector produces innovation and economic development [39,40]. The KBUD strategy raises universities and research institutes to a more prominent position, moving from a supporting to an entrepreneurial role [39,41]. The KBUD approach also recognizes the benefits of a quadruple-helix model, where the community joins forces with the university, government, and the business sector [35,41]. KBUD's success is heavily dependent on community support and support policies [26,40]. Research on this topic has indicated that local actors play a differential role in planning and implementing efforts, and, therefore, must participate in the orchestration of the local KBUD [7]. This requires awareness amongst all local development actors on the importance of supporting the KBUD frameworks [42].

Therefore, the effectiveness of KBUD depends on how the policy is formulated, implemented, and supported. However, long-term strategic vision and sound policies, plans, and actions, although fundamental aspects for a successful KBUD, are not a trivial arrangement for most cities and, as previously seen, can be even more challenging for cities in an emerging economy context [7]. Insofar KBUD is a robust strategy capable of leveraging urban development in a smart and sustainable way, it demands the use of different approaches and instruments for successful implementation [27]. In order to make KBUD their development strategy, cities need innovative solutions and affordable tactics and tools to successfully incorporate KBUD into their urban management processes and achieve effective, inclusive, and resilient local smart and sustainable development.

To date, several cities around the world have found a way to innovation and sustainability through KBUD, including, for example, Austin (USA), Barcelona (Spain), Delft (Netherlands), Helsinki (Finland), Melbourne (Australia), Montreal (Canada), and Stockholm (Sweden) [25,32,35]. A number of others have been engaged in planning and implementing KBUD strategies in the pursuit of sustainable economic growth and prosperity, e.g., Bento Gonçalves (Brazil), Brisbane (Australia), Dublin (Ireland), Florianopolis (Brazil), Istanbul (Turkey), Monterrey (Mexico), Shenzhen (China), Shanghai (China), and Tampere (Finland) [7,23,26,28,30,35,43].

A closer look at the development journey of these cities shows that each of them took a different pathway for KBUD implementation. Nevertheless, some similarities can be drawn. Some of these cities, for instance, shared the fact of having hosted, during their KBUD implementation process, a renowned international event specifically focused on the theme of knowledge-based urban development, namely, the Knowledge Cities World Summit (KCWS) series. This, and the fact that these host cities are succeeding in their KBUD endeavors, has raised interest in investigating the relationship between international events and knowledge-based development at a local level.

1.2. International Events and City Development

The connection between events and cities dates back to the beginning of urban history, as events sprung from the very fundamental human need for economic and social exchange [44]. In the last decades, cities and regions have been using events to generate a growing range of outcomes in different areas, including in the policy domain [45]. The theme of sustainability, for example, has gained much more visibility, as events such as the United Nations Climate Change Conferences (COP) and the UN-Habitat Conferences have gained importance and prominence worldwide [46–48].

Additionally, cities are progressively moving from a passive role, as merely a location or backdrop for events, towards a more proactive use, by drawing different policy agendas [49]. In the economic domain, events are commonly associated with the attraction of new investments; income generation; increases in retail activities, employment, and tax revenues; and opportunities to diversify the local economy [44,50–56]. The socio-cultural contributions assessed include knowledge exchange and the transmission of local cultural values and traditions, while impacts often comprise improvements in the standard of living and quality of life, strengthening of local or regional identity, and increase in community's self-esteem [55–57]. Events can play a significant role in the urban landscape as well, leading to urban revitalization, regeneration, and development [5,50,52,57–61]. At an institutional level, events can contribute directly to the learning and development of the cooperation capacity [61]. The theme of cooperation is a key notion in the literature on smart cities. Drawn from John Dewey's cooperative democracy, the concept is mobilized to deal with the governance challenges and democratic deficits of Brazilian cities [62]. In the policy field, researchers have identified that events may create shared understandings capable of motivating engagement in joint action [63], or act as catalysts of change in an institutional field [47]. For instance, events can bring together actors and partners that do not often interact, creating unusual discursive spaces that enable information flows and innovation [48]. In light of this, cities have progressively adopted a more integrated approach in which events can be part of broader policy frameworks, and policymakers can employ a range of different strategies to increase the benefits of events for different stakeholders [49]. The intensity and sustainability of these benefits are dependent on whether the event is the product of a dispersed and fragmented endeavor or part of a strategic development trajectory. To maximise the benefit from hosting events, cities need to use them to serve their long-term planning or development goals [5,50,60,61]. A clear strategy for the post-event period, designed and planned with the support and commitment of local and political stakeholders, is also critical to meeting broader urban objectives [57].

This paper is specifically interested in exploring the relationship between events and a KBUD strategy by analyzing how an international event on knowledge-based development can contribute to the local KBUD of the host city. In order to do this, the study places the Knowledge Cities World Summit (KCWS) under the microscope, an international annually-held event with the aim to shed light on the various dimensions of a city's KBUD [64].

Held for the first time in 2007 in Monterrey, Mexico, throughout its editions, 11 different cities around the world have hosted the summit, namely: Shenzhen (China, 2009), Melbourne (Australia, 2010), Bento Gonçalves (Brazil, 2011), Matera (Italy, 2012), Istanbul (Turkey, 2013), Tallinn (Estonia, 2014), Daegu (Korea, 2015), Vienna (Austria, 2016), Arequipa (Peru, 2017), Tenerife (Spain, 2018), and Florianopolis (Brazil, 2019). The 13th edition, in 2020, was meant to take place in the usual face-to-face mode in Tijuana (Mexico). However, due to COVID-19 pandemic restrictions, it was held in an online summit format, with an operational base in Monterrey (Mexico).

In this study, in order to gain an in-depth view and explore the topic in different dimensions, covering contextual settings, one of the host cities was selected as an object for a case study: the city of Bento Gonçalves, which hosted the fourth edition of the KCWS in 2011. Bento Gonçalves is a small city (with over 121,000 population) located in southern Brazil, yet it is recognized on the national scene as a relevant cultural, touristic, and economic hub. The city attracted international attention in 2019 after receiving the MAKCi Award (Most Admired Knowledge City Award) as an emerging knowledge city, which increased the interest in investigating how KCWS may have affected the city's KBUD trajectory. Furthermore, taking Bento Gonçalves as a case study meets the importance of developing more studies on KBUD in medium- and small-sized cities [65,66] in emerging countries [7].

The analysis of the collected data indicated that the KCWS's role in the KBUD of the host cities is that of an enabler, promoting the exchange of knowledge, increasing awareness, building networks, and highlighting development actors' relevance to the KBUD process. In Bento Gonçalves, the study identified several KBUD initiatives and achievements, some of them dating back to an older historical context. Others are related to events held in 2011, such as the city's MAKCi Award in 2019. Among the most relevant contributions of KCWS to Bento Gonçalves' KBUD are those associated with the university's role in local development. KCWS contributed to increasing the hosting university's influence over different regional development agents and placed it at the center of the local KBUD movement. From this, other effects and other KBUD initiatives were developed in the city, even in the region.

The remainder of the paper is structured as follows: In Section 2, we present the methodological procedures carried out in the development of the study. Section 3 reports the main research findings, which are then discussed and interpreted in Section 4. Finally, Section 5 summarizes the research content presented, highlighting lessons derived and future perspectives.

2. Materials and Methods

The methodological approach of this paper drew upon two main research strategies: a semi-structured interview-based qualitative analysis [67,68], which was adopted for empirical investigation of the annually held event, i.e., KCWS, and a case study strategy, which covered the contextual dynamics and particular settings on the specific case of the Brazilian city of Bento Gonçalves, which hosted the fourth edition of the KCWS in 2011.

2.1. Data Collection and Analysis

KCWS is promoted by the World Capital Institute (WCI), a non-profit professional association that operates as a network focused on professional community building and diffusion activities on knowledge-based development [69]. The research design included semi-structured interviews with key members of the WCI Executive Board to gain insight into KCWS as a whole and the role it plays on the KBUD of host cities. The selection of interviewees was done through purposive sampling and sought to include key people involved in the conception, development, and promotion of the KCWS since its outset. The purpose was to capture KCWS's creators and promoters' perspectives regarding the aims, achievements, and challenges of taking the event to the selected cities. This approach relied on interviewees' discursive accounts of their experiences and perceptions of the event throughout the years and how it has contributed to host cities' development.

Based on the research objectives, an interview script was developed and tested with people familiar with the research topic. After considering their feedback, the final interview guide consisted of the following five key questions: (a) KCWS's selection criteria and goals for host cities; (b) event's stakeholders and their aims for hosting KCWS; (c) achievements and impacts triggered as a result of KCWS; (d) tools, methods, and indicators to measure those impacts; and (e) challenges and opportunities regarding the summit series.

The interviews were conducted in May 2020 through Skype or Google Meet platforms, and each lasted between 45 and 100 minutes. They were digitally recorded and manually transcribed. During the interviews, some participants referred to documents and archival content to illustrate or amplify the comprehension of the topics discussed. This material was collected for analysis as a complementary source of evidence about KCWS.

Data analysis procedures were performed through the content analysis method [70,71]. A Computer-Assisted Qualitative Data Analysis Software (CAQDAS), namely NVivo, was used to support the coding process of the interviews' transcripts and later to facilitate iterations within the data coding and analysis. The process included two subsequent coding cycles [72]. For the first cycle, categories based on the key questions that guided the semi-structured interviews were defined and applied through the structural coding method [72]. Then, in the second coding cycle, the pattern coding method [72] was used to identify emergent topics, providing a second, and in some cases, even a third, level of codes (subcodes) derived directly from the interview's content. The data corpus was carefully read and analyzed multiple times in each cycle until a point of saturation [70] was reached, i.e., until no new additional topics could be identified.

2.2. The Case of Bento Gonçalves

In order to gain an in-depth view and explore the topic in different dimensions, covering contextual settings, we also applied a case study strategy for empirical inquiry on Bento Gonçalves. Single case design [73] was adopted, as it allowed for the observation of the unique characteristics of the case, and simultaneously provided a longitudinal understanding of the research topic. The selection of the case occurred at the beginning of the research, during the design phase. Among the 12 cities that have hosted a KCWS, Bento Gonçalves was purposively selected due to the city's particularly revelatory conditions regarding the research issue. With an estimated population of approximately 121,000 inhabitants, Bento Gonçalves is a good representation of a small town in an emerging country like Brazil that manages to stand out regionally in terms of development.

Formed by Italian immigrants in the 19th century, the city became an important regional industrial and touristic center in the Rio Grande do Sul (RS) state, known for its high-quality wine production and furniture industry, activities strongly associated with the city's Italian cultural identity. As we have mentioned before, the city hosted the KCWS in 2011. In 2019, eight years after holding the event, Bento Gonçalves received the Most Admired Knowledge City (MAKCi) Award. Promoted by the World Capital Institute (WCI), the award aims to identify and recognize communities worldwide engaged in formal and systematic KBUD processes [74]. In addition, illuminating studies in the knowledge-based development field were developed in the years after the event, focusing on Bento Gonçalves and the surrounding region as their object of study [75,76]. All of this was reason to believe that the activities and discussions developed during the event in 2011 hold a connection to Bento Gonçalves' KBUD trajectory.

Interviews were the primary data collection method for this case study. The goal was to capture key local actors' perceptions and narratives about the KCWS in 2011, and its connection to KBUD initiatives and the achievements of Bento Gonçalves. Again, we adopted a purposive sampling approach for the selection of interviewees. Informed by content collected on WCI board interviews, the selection covered key people directly involved with the organization of the fourth KCWS in 2011. In addition, considering literature references that emphasize the central role of the quadruple-helix in KBUD [23,41], the interviews included local representatives of the four sectors, i.e., university, civil society, the public sector, and the private sector.

Semi-structured interview scripts were developed and tested, following the case study protocol and research objectives. Preliminary findings from the WCI board interviews also contributed to improving the questionnaire. Although each interview had a different focus according to each interviewee's group, the main topics addressed in the interviews were as follows: (a) KCWS organizing process, (b) the event's stakeholders and their goals in organizing/sponsoring/supporting/participating in KCWS, (c) the event's contributions to the involved institutions and the city, (d) challenges and opportunities that could have been better explored, and (e) perceptions about what makes Bento Gonçalves a smart and sustainable city.

The interviewees signed an online informed consent form agreeing to participate in the research. In total, nine interviews were conducted, seven of which through the Skype or Google Meet platforms. At the request of the participants, the two other interviews were carried out via email, with a follow-up interview being subsequently conducted through the Google Meet platform to clarify and extend some topics. Each interview lasted from 40 to 120 minutes, and all of them were conducted, manually transcribed (when applicable), and analyzed in Portuguese. Content analysis was the approach to analyze case study's interviews. The same coding process used to analyze the WCI executive board interviews was applied in the Bento Gonçalves interviews. Again, two subsequent coding cycles, using the structural coding method and then the pattern coding method [72], were performed with NVivo software.

According to Krippendorff [71], every content analysis requires a context that gives meaning to the findings and serves as a conceptual justification for reasonable interpretations. In line with this, the case study's data collection included research and gathering of documents and indicators. Document analysis [77] comprised material collected from academic literature and grey literature, including technical reports, research reports not peer-reviewed, institutional websites, legislation, and policy reports. Likewise, official databases provided indicators and data sets for the indicator analysis. Indicators' selection was informed by the KBUD Assessment Model (KBUD/AM) [32] and included datasets and indexes related to the economic, socio-cultural, spatial, and institutional aspects of Bento Gonçalves' development. Indicator and document analyses served as a source of evidence about the facts, actions, and events regarding Bento Gonçalves' development, providing a profile of the city and the context within which we considered the interview analysis results. They were also instrumental in refining ideas, identifying conceptual boundaries, and corroborating the relevance of the categories derived from interview analysis. The documents and indicators collected also served to confront, corroborate, or augment the evidence from the interviews.

3. Results

3.1. Interviews with WCI Executives

This section presents the results of the interviews with members of the WCI executive board. Interviewees' selection considered their seniority and involvement in the KCWS conception, development, and promotion since its outset. Accordingly, five members of the WCI executive board were interviewed, namely:

- WCI President;
- WCI (Former) Executive Director of the Events Program;
- WCI Executive Director of the Awards Program;
- WCI International Advisory Board Member #1;
- WCI International Advisory Board Member #2.

Altogether, the interviews totaled about five recording hours. Each transcribed interview was carefully analyzed to identify the interviewees' perspectives about the goals, achievements, and challenges of taking KCWS to different cities. The interviews also provided an overview of KCWS's history and hosting process.

A total of 42 codes and subcodes, grouped in five categories ("goals", "stakeholders", "hosting", "contributions", and "challenges and opportunities"), were applied (Table 1).

Table 1. Distribution of coding references from the WCI executive board interviews.

Category	Code Level 1	Code Level 2	References
Goals (n = 83)	Host (n = 38)	To address institution-specific agenda	13
		To learn about KBUD	9
		To build profile	8
		To leverage local KBUD initiatives	5
		To create networks	3
	WCI (n = 34)	To help cities build or improve their local KBUD	11
		To further the study and application of KBUD	8
		To promote the socialization of the KBUD community	7
		To extend networks	6
		Not-for-profit activities	2
	Alignment (n = 11)	Mutual benefits	11
Stakeholders (n = 138)		University	35
		Government	34
		Private sector	25
		Multi-stakeholder partnership	15
		Experts and speakers	10
		Local community	7
		Civil society	6
		International audience	6
Hosting (n = 46)		Bidding motivation	19
		Selection criteria	16
		Hosting process	11

Table 1. *Cont.*

Category	Code Level 1	Code Level 2	References
Contributions (n = 47)		Network connections	13
		Enhancement of local initiatives	10
		Growth of KBUD awareness	7
		Profile building	7
		Development of academic agenda	5
		Knowledge exchange and skill training	5
Challenges and Opportunities (n = 58)	Opportunities (n = 31)	Technology and new online platforms	11
		Consolidated methods and tools	8
		Shifting of thematic focus	8
		Update of the conference format	4
	Challenges (n = 27)	Travelling and conferencing post COVID-19	7
		Continuity of initiatives	6
		Institutional memory	4
		Resource constraint	4
		Impact assessment	3
		Maintenance of the network	3

From WCI executives' perspective, the event's "goals" (83 references) are divided between "WCI's goals" (34 references) and "host's goals" (38 references). Among the former, the most cited one is "to help cities build or improve their local KBUD" (11 references), highlighting the event's commitment to promoting host cities' development. One of the interviewees explained that, sometimes, it is just a matter of bringing KBUD awareness to the city, "… because a lot of cities, they are aware that they have this history, they have these monuments, and they have heroes of the past, for instance, but they do not use their history to trigger some more movements of the present and future. (…) and the awareness is so that they use consciously, purposefully, their capital system for development". The other WCI's goals are of a more institutional nature, "to further the study and application of KBUD" (eight references), "to promote the socialization of KBUD community" (seven references), "to extend networks" (six references), and "not-for-profit activities" (two references).

On content coded as "host's goals", the most applied subcode is "to address institution-specific agenda" (13 references), as all interviewees pointed out that, in addition to the broad topics of KBUD and knowledge cities, particular focus is given to a theme significant to the host city's context in each summit. Therefore, according to the leading host partner of each event, the local host agenda may vary from city to city. Other host's goals, according to the interviewees, include "to learn about KBUD" (nine references), "to build profile" (eight references), "to leverage local KBUD initiatives" (five references), and "to create networks" (three references).

The overall perception is that there is an "alignment" (11 references) between the local host's objectives and those of WCI. This is corroborated by the coincidence of subcodes in each category, such as "extend networks" (WCI) and "create networks" (host), or "to help cities improve their KDB" (WCI) and "to leverage local KBUD initiatives" (host).

For this matter, another aspect of interest in the analysis was understanding who the event's primary stakeholders are. In the "stakeholders" category (138 references), the agents of the triple-helix, "university" (35 references), "government" (34 references), and "companies" (25 references) are the most cited, followed by "multi-stakeholder partnerships" (15 references). "Experts and speakers" (10 references) are also mentioned as they are responsible for delivering the event's value proposition. Finally, interviewees also cited "local community" (seven references), "civil society" (six references), and "international audience" (6 references) as relevant stakeholders of the event.

The "hosting" category (46 references) includes content about the circumstances, processes, and activities that enable the hosting of a KCWS by a specific city. Three key factors are coded on this category: "city's motivation" (19 references), WCI's "selection criteria" (16 references), and "hosting activities" (11 references). Like the goals, a city's motivation to host the event may vary according to the context and institution or group of institutions that lead the bid for hosting the summit. Interviewees cited, for example, a city's interest in showcasing its knowledge-based urban development on an international scale, as in Melbourne, Australia; or academic motivation, when universities are the main hosting partner, as in Arequipa, Peru; and sometimes it is the private sector that leads the efforts to bring the event to the city, as in Florianopolis, Brazil.

Another code in the "hosting" category is "selection criteria", which includes content about the factors considered by WCI to decide where to hold the next KCWS. Besides the practical and operational criteria, such as the organizing capacity and sufficient financial resources, interviewees indicated that a combination of stakeholders' engagement level, the current or desired KBUD level of the city, and the willingness and enthusiasm to host the event are the most relevant aspects to be considered. According to one of the interviewees, this combination is crucial for generating the proposed contributions to the host city.

The last code in the "hosting" category is "hosting activities", and this includes content about the production and execution of the event once the host city is defined. As the circumstances in which each event is held vary widely, different interviewees pointed out that there is not one concept for running a KCWS in one place. Each city's characteristics are significant, so the subjects and the sense of what is needed are very different in each place. A constant dialogue is what makes an intersection of interests possible. As highlighted by one of the interviewees, local concerns are essential, and dialogue makes it possible to format the event to fulfil each stakeholder's goals, including those of WCI.

Another codified category is "contributions" (47 references). At this point, it is necessary to clarify that during the interviews, the questions presented to the interviewees referred to the impacts of the event in the host cities. However, during the analysis procedure, we concluded that the answers were focused on the event's contributions, meaning the part played by KCWS in bringing about a result. Therefore, this content was coded as "contributions". This finding also influenced the data collection of the case study. In interviews with Bento Gonçalves representatives, the term used was "contributions".

As for the content coded in the "contributions" category, the most cited is "networking" (13 references). The interviewees pointed out that one of the most relevant event outcomes is that it triggers connections to formal and informal networks of leading global thinkers, experienced practitioners, and host city partners, enabling active engagement around KBUD. One interviewee highlighted that KCWS also brings local stakeholders together and enables them to see a way to work together in partnership and tackle their local issues around KBUD, which may produce several other positive results for the city.

The second most coded contribution is "enhancement of local initiatives" (10 references), as all interviewees indicated that the event helps improve the KUBD strategies or initiatives undertaken by the local stakeholders. Another contribution cited is "growth of KBUD awareness' (seven references), which is directly connected with the event's very reason, namely to promote KBUD and knowledge cities as a model of sustainable urban development. Interviewees also referred to the following contributions: "profile building" (seven references), as the event allows the host city to showcase itself to a highly qualified audience; "development of academic agenda" (five references), especially when the leading host institution is a university or research institute; and "knowledge exchange and skill training" (five references), as a result of the cutting-edge lectures and debates delivered by the experts and speakers brought by the event.

Finally, the last content category is "challenges and opportunities" (58 references). As a reflection of the recent COVID-19 pandemic, the most cited "challenge" (27 references) for the future is "travelling and conferencing after COVID-19" (seven references), followed by "continuity of initiatives" (six references), "institutional memory" (four references), "resource constraint" (four references), "impact assessment" (three references), and "maintenance of the network" (three references).

The continuity of initiatives is perceived as a challenge, mainly because it also depends on the host city's actors or institutions. In this sense, something that concerns some of the interviewees is how to continue contributing so that the ideas and initiatives developed during the event continue to be nurtured after the event. "Institutional memory" refers to the documentation of the events and activities carried out by WCI., and is perceived as a challenge due to the organizational nature of the institute, whose members are in different countries and dedicated to several other matters. As for the "resource constraint", the COVID-19 pandemic is expected to cause an economic crisis in the short term, causing cities to stop applying resources for events. "Impact assessment" is perceived as a challenge mainly because of the methodological aspects—for example, how to measure the intangible impacts of KCWS or how to keep track of it several years past the event. Considering that, as mentioned before, networking is one of the significant contributions KCWS brings to the city, some interviewees also mentioned that continually nurturing these networks is a challenge in the sense that it demands constant dedication.

On the other hand, the interviewees perceive different "opportunities" (31 references) for the future. They see the possibilities of "technology and new online platforms" (11 references) as a powerful instrument for responding to the challenges presented above. Virtual conferencing through online communication platforms is seen as an alternative to the difficulties brought about by the COVID-19 pandemic, and already contributed to the event's continuity in 2020. This is why the "update of the conference format" (four references) is also seen as an opportunity, preceded by "methods and tools" (eight references) and "shifting of thematic focus" (eight references). The interviewees agree that, over the years, the methodologies and frameworks used and disseminated in KCWS were well consolidated. Nevertheless, new technologies and the emergence of themes such as the Anthropocene, climate change, and the smart city phenomenon may have created some room for an update. The interviewees also see an opportunity to incorporate these themes into the smart and sustainable cities debate. Thus, through the perspective of the WCI's executive board members, it is possible to see that the event successfully involves the main KBUD agents, namely, the triple helix—university, government, and companies. However, there may be room for more civil society participation. Considering what the interviewees pointed out, the events' contributions resonate with the purposes of both the WCI and those of the host city. This can be verified by the coincidence of codes such as "to help cities build or improve their local KBUD" (goals/WCI's goals), "to leverage local KBUD initiatives" (goals/host's goals), and "enhancement of local initiatives (contributions); or "to extend networks" (goas/WCI's goals), "to create networks" (goals/host's goals), and "networking" (contributions); and also "to further the study and application of KBUD" (goals/WCI's goals), "to learn about KBUD" (goals/host's goals), and "growth of KBUD awareness" and "knowledge exchange and skill training" (contributions). This convergence seems to be connected to the event's sensitivity to the local context and host city interests, making it consistent with the local expectations.

3.2. Case Study Investigations and Interviews

3.2.1. Bento Gonçalves in a Nutshell

Bento Gonçalves is located in the Serra Gaúcha region, in the Rio Grande do Sul (RS) state, Brazil (Figure 2). The city has its origins as a colony settled to receive part of the Italian immigrants who arrived in the region at the end of the 19th century. This Italian ancestry would later become one of Bento Gonçalves' main knowledge assets, profoundly shaping its development processes [78]. Today, with an estimated population of 121,803 [79], the city is an important regional hub, compounding the Serra Gaúcha Metropolitan Region.

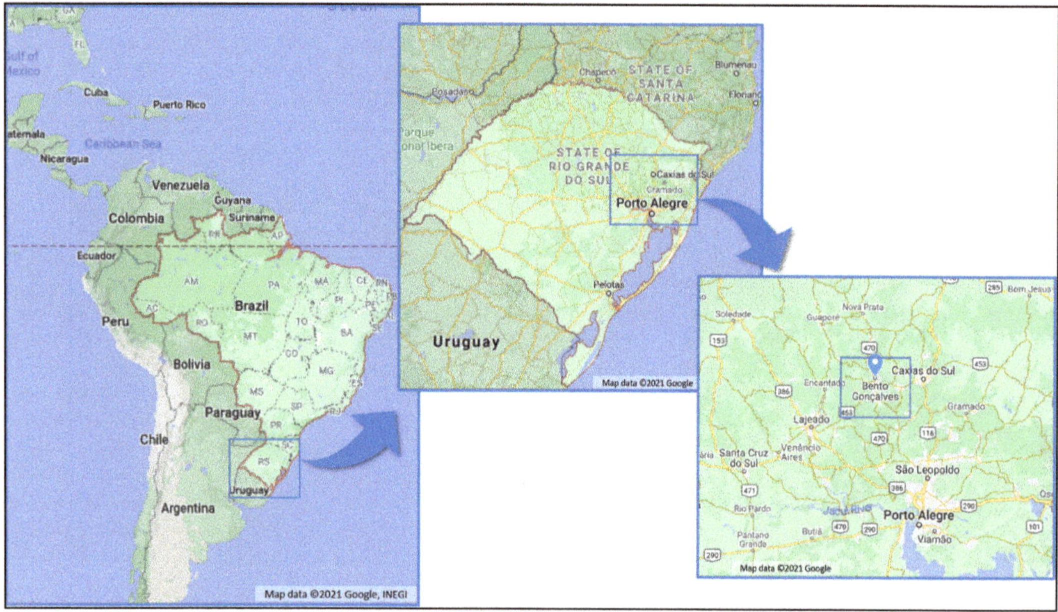

Figure 2. Location of Bento Gonçalves, drawn by the authors on a Google Map [80].

Regarding national averages, Bento Gonçalves sustains a good performance in terms of development, with a Human Development Index of 0.778, which is considered a high score and places the city in the 145th position among the 5565 Brazilian cities and 16th in the Rio Grande do Sul state [81]. Bento Gonçalves' advances in terms of development have been recognized even internationally, as in 2019, the city received the MAKCi award, taking its position among a select group of cities in the world that have been thriving under the KBUD flag.

In terms of economic development, Bento Gonçalves ranks as one of the largest economies in the Rio Grande do Sul state. In 2018 (most recent available data), the city's gross domestic product (GDP) was USD 1.54 billion, the 14th largest among the state's 497 cities [82]. Bento Gonçalves is listed in the national "Best Cities to Do Business" ranking, moving from the 84th position in 2014 to the 18th position in 2019 [83]. The industry sector represents the main economic activity, with a 59% share in the municipality's revenue, followed by the commercial (21.2%) and services (19.8%) sectors [84]. One interesting aspect of Bento Gonçalves' economic development is the presence of a strong cultural identity associated with the Italian pioneers and their entrepreneurial spirit [85]. The first companies were family-owned and started production to supply the local demand [76,86].

The timber and furniture industry is one of the most relevant segments in the Bento Gonçalves economy, accounting for the most jobs (13.4% of all formal jobs in 2018) and revenues (45% of total industrial sector revenues in 2018) for the city [84]. Bento Gonçalves also stands out on the national scene for its grape and wine production. The city is known as the "Brazilian Capital of Wine", and incorporates the largest and most important wine region in Brazil, i.e., the Serra Gaúcha region, which accounts for about 85% of the national wine production [87]. Bento Gonçalves' tourism sector has significantly benefited from the grape and wine industry's performance, which placed the region on the map of national and even international wine tourism [76].

Regarding sociocultural development, Bento Gonçalves stands out, together with other cities in the region, for its good human and social development levels. In the Rio Grande do Sul state, the government monitors the municipalities' societal development through IDESE, the Socioeconomic Development Index. Bento Gonçalves scored an IDESE of 0.834 in 2018, ranking as the first in the state among the municipalities with more than 100,000 inhabitants and 19th in the general ranking [88]. Notably, Bento Gonçalves has been investing in education as a development strategy. In Brazil, the Federal Constitution requires states and municipalities to invest at least 25% of their income into maintaining and developing education. In the case of Bento Gonçalves, spending on education has been exceeding the constitutional minimum in the last decade [89]. For instance, the city's investment ratio in 2018 (29.8%) was even higher than that of the state (26.7%) [89,90].

Bento Gonçalves also stands out as a regional hub of higher education. Some of the institutions located in the city are listed among the best in the country, such as the University of Caxias do Sul (UCS), 42nd in the national ranking, and the State University of Rio Grande do Sul (UERGS), ranked 163rd [91]. In particular, because of its community DNA and central role in the region's development, UCS has been a relevant institution in terms of Bento Gonçalves' KBUD. In addition to teaching and research, UCS promotes various initiatives to foster regional entrepreneurship and scientific and technological innovation. For instance, since 2015, the University has maintained a Science, Technology, and Innovation Park—the TecnoUCS. One of the park's most notable projects is UCSGRAPHENE, the first and largest industrial graphene production plant in Latin America installed by a university [92].

As for spatial development, Bento Gonçalves experienced a very intense increase in population in the last century (43,144.76% from 1876 to 2009) due to vegetative growth and both internal and external migratory attraction [93,94]. As in many other cities in Rio Grande do Sul and Brazil, Bento Gonçalves' urbanization process intensified since the 1950s due to expanding national industrialization programs and countryside mechanization [94]. Today, 47.7% of Bento Gonçalves' territory is of urban occupation [95], with the vast majority of the population being urban (92.3%) [96].

These fast urbanization processes, driven by rural exodus and migratory processes induced by economic development, brought some challenges to the city [93]. In Bento Gonçalves, whose geomorphology imposes limitations on urban growth, there are issues such as irregular occupations and settlements in risk areas [97,98]. On the other hand, an interesting aspect of the spatial organization of Bento Gonçalves concerns the relevance that the city gives to the protection of the cultural landscape and the preservation of the traditional rural zone, the locus of the wine and cultural tourism [99].

Finally, in terms of institutional development in Bento Gonçalves, it is possible to observe different participation spheres and groups of actors involved in urban development governance. The Regional Development Council (COREDE) is one of these spaces where strategic development plans are debated and voted, guiding the state budget application in the region. Bento Gonçalves is part of COREDE Serra, composed of 32 municipalities. COREDE Serra elaborated, in a participatory manner, through micro-regional assemblies, the Regional Development Strategic Plan 2015–2030 [100], which includes a portfolio of KBUD projects such as implementing technology parks by attracting national and foreign technology-intensive companies, promoting technology-intensive sectors, creating an innovation program, strengthening the TecnoUCS through the triple helix, identifying and developing the regional innovation ecosystem, creating local productive arrangements in technology-intensive segments, and strengthening the Serra Technological Modernization Pole. However, to date, only a few of the projects listed in the plan have made progress. Even so, the existence of the plan and the fact that it is being discussed by the community indicates a level of KBUD awareness, as well as the region's desire to move forward on this path.

Another notable local feature in terms of institutional arrangement is local public managers and government leaders' ability to congregate in regional organizations in order to strengthen their institutional capacity. Organizations such as AMESNE (Municipalities of Northeast Upper Hillside Association), composed of 36 municipalities represented by their respective mayors, and the Regional Parliament of Serra Gaúcha, with representatives of the City Councils of 24 cities in the region, are examples of this.

Bento Gonçalves' business community also stands out for its institutional leadership and participation in urban development. One of the most active local bodies is the Bento Gonçalves Centre for Industry, Commerce, and Services (CIC-BG), which acts directly in a series of planning and governance initiatives for sustainable development, such as Bento+20. In October 2020, Bento+20 delivered a masterplan [95] to the city, with the main purpose to make Bento Gonçalves a smart and sustainable city until 2040. Taking the United Nations Sustainable Development Goals and ISO 37120 and ISO 37122 standards as a guideline, the masterplan presents guidelines and a detailed action plan considering ten thematic areas (technology, innovation, and entrepreneurship; education; health; safety; industry, commerce, and services; tourism and culture; urbanism, urban mobility, and infrastructure; environment and sustainable development; citizenship; and rural development).

3.2.2. Case Study Interviews

This section reports the findings of the nine interviews carried out with Bento Gonçalves' representatives. The interviewees were selected considering their involvement with the fourth KCWS in 2011 and/or their local representativeness in one of the four quadruple-helix sectors, i.e., university, civil society, the public sector, and the private sector. Accordingly, the following representatives were interviewed (Table 2).

Table 2. Profiles of the Bento Gonçalves interviewees.

Relevance	Interviewee No	Position
Local organizing committee member	Interviewee #1	University of Caxias do Sul (UCS) professor and KCWS local chair in 2011
Local organizing committee member	Interviewee #2	UCS professor and member of the local organizing committee in 2011
Local organizing committee member	Interviewee #3	UCS professor, innovation centre director and member of the local organizing committee in 2011
Local organizing committee member	Interviewee #4	UCS professor and member of the local organizing committee in 2011
Local organizing committee member	Interviewee #5	UCS professor and member of the local organizing committee in 2011
Academic conference participant	Interviewee #6	UCS tele-diffusion director and KCWS attendee in 2011
Public sector conference participant	Interviewee #7	Former mayor of Bento Gonçalves
Civic society leader conference participant	Interviewee #8	President of the Rio Grande do Sul Regional Development Council
Private sector conference participant	Interviewee #9	Leading furniture company executive manager and KCWS attendee in 2011

Altogether, the interviews totaled over seven recording hours, plus 15 pages of written interviews. The complete content of the interviews, written or transcribed, was carefully analyzed to capture key local actors' perceptions and narratives about the KCWS in 2011, and its connection to KBUD initiatives and projects developed in Bento Gonçalves.

Bento Gonçalves Summit

Before entering the content analysis itself, a brief background about the fourth KCWS in Bento Gonçalves is presented below. This content is based primarily on the organizing committee interviewees' narratives, complemented by document analysis, and seeks to give insights about the particular settings and the context in which the KCWS was carried out.

Three main factors concurred to bring the event to Bento Gonçalves. The first of them concerns the historical role of the University of Caxias do Sul (UCS) in advancing the KBUD agenda in the region. According to the interviewees, since at least 2003, UCS had been cultivating institutional partnerships and connecting its professors and researchers with universities and institutes particularly dedicated to studying knowledge management and knowledge-based development disciplines, such as WCI and Ibero-American Community for Knowledge Systems (ICKS). These events are relevant because they connected local actors, especially UCS's researchers, to some of the most engaged and cutting-edge international communities in the KBUD discipline. By the time the KCWS was held, a UCS professor had integrated WCI's international advisory board, and there were two ICKS active cells in the region with regular meetings and debates joined by other professors, students, business people, and even community representatives. These initiatives were the first steps in raising local awareness about the potential of KBUD. Much of the local support that KCWS received in 2011 was due to these previous initiatives, which set the stage and contributed to the community's comprehension of the proposed debate.

The second factor concerns the involvement of these same UCS professors and researchers in local networks and organizations dedicated to regional development. These local networks played a fundamental role in making the event happen. For instance, one of the main host partners of the fourth KCWS was the Furniture Innovation Management Center (CGI), whose director, also a professor at UCS, was a member of the local organizing committee. CGI's participation was crucial in connecting the business sector, especially the timber and furniture segment, to the event. Representatives, entities, and companies of the industry not only attended the KCWS, but also sponsored the event. In addition, due to UCS professors' networks, the event had the support and sponsorship of government representatives, such as the Municipality of Bento Gonçalves and AMESNE (Municipalities of Northeast Upper Hillside Association), and the contributions of different local businesses and organizations.

Finally, the third factor concerns WCI's institutional connections themselves. In 2011, the fourth KCWS was a joint conference with the IX ICKS conference. Like WCI, ICKS also holds annual meetings to gather the knowledge systems community and discuss specific topics of the field. Bento Gonçalves was selected as the host city for the 2011 ICKS Conference the previous year, in 2010. Some WCI board members were simultaneously members of ICKS, and started to consider holding the KCWS and the ICKS conference together, as the potential for synergy between both was clear. When preparations for the ICKS event were already underway, the proposal was made to hold the two events together. Accordingly, the fourth KCWS and the ninth ICKS conference took place on 26 and 27 October 2011 in Bento Gonçalves. The general theme of the fourth KCWS was "Knowledge Cities for Future Generations", specifically addressing aspects such as cultural tradition, knowledge, and innovation in the community's future. Altogether, the event received about 200 participants—an expressive amount, according to the interviewees, considering the representativeness of those who attended.

Content Analysis

The findings of the content analysis on Bento Gonçalves' interviews are presented in this section. In total, five categories and 26 codes were applied (Table 3).

Table 3. Distribution of coding references from the Bento Gonçalves interviews.

Category	Code Level 1	Code Level 2	References
	Business sector (n = 5)	Learning about KBUD	5
Goals (n = 9)	Public managers (n = 2)	Learning about innovation and city cases	2
	University (n = 2)	Contribute to city development	2

Table 3. *Cont.*

Category	Code Level 1	Code Level 2	References
Stakeholders (n = 70)		University	13
		Local government	12
		Private sector	16
		Sectoral organisations and non-profit entities	11
		Experts, speakers	5
		Local community	5
		ICKS	8
Hosting (n = 14)		Local leaders	14
Contributions (n = 72)		Network connections	11
		KBUD awareness	16
		profile	5
		Academic agenda	12
		Knowledge exchange and skill training	13
		University's role	15
Challenges and opportunities (n = 14)	Opportunities (n = 8)	Local actors' commitment	5
		Broader audience	3
	Challenges (n = 6)	Balanced audience	2
		Funding	4

The "stakeholders" category (70 references) aggregates citations about the local agents who attended or were involved with the event in 2011. Similar to WCI members, Bento Gonçalves' interviewees indicated "private sector" (16 references), "university" (13 references), and "local government" (12 references) as the main stakeholders of the event. As the fourth KCWS was a joint event with the ICKS conference, "ICKS" (eight references) is also a frequently mentioned stakeholder. In Bento Gonçalves, the particularity is the involvement of different types of "sectoral organizations and non-profit entities" (11 references), which directly or indirectly contributed to the event, as mentioned above. The interviewees also pointed out "local community" (five references), e.g., local small merchants and suppliers, as playing an essential role in supporting the event, and the "experts and speakers" (five references) as one of the summit's main differentials.

The "goals" category (nine references) includes references to the multiple stakeholders' aims in participating in the event. Overall, learning from international experts was the main objective of those who attended the event. The "business sector" (five references) audience expressed an interest in learning about innovation, KBUD, sustainable development, and knowledge management. According to the private sector interviewee, many executives or managers of local companies were postgraduate students in a knowledge management program offered by UCS at the time. They became aware of KBUD through the course and therefore recognized the opportunity to learn and exchange experiences that the event would provide. The "municipalities representatives and public managers" (two references) were interested in learning from city cases and experiences worldwide about innovation and the future of cities. As for the university (two references), as highlighted by the academic interviewee, supporting the event was taken as an opportunity to contribute to the region's development and advance the commitment as a community university.

The "hosting" category (44 references) includes references about the context, circumstances, and activities related to the hosting of the KCWS. Most of its content corresponds to storytelling about how the event happened in Bento Gonçalves, what it took to organize the event in the city, how it was experienced, and so on, and has already been presented in the previous section. However, one topic in particular emerged from the speeches of different interviewees, namely, the crucial role of some "local leaders" (14 references) in making the event happen. In the case of the fourth KCWS, different interviewees coincided in pointing out the strong leadership of UCS professors as the critical factor in making the event feasible.

The "challenges and opportunities" category (14 references) includes content about the main difficulties observed by the interviewees regarding the event and the possibilities of improvement. "Funding" (four references) was emphasized by the organizing committee interviewees as the main challenge. As a small town, Bento Gonçalves and its companies and institutions have fewer resources available for sponsoring events and conferences. According to the interviewees, this difficulty was overcome with the involvement and sponsorship of several entities, local companies, UCS, and the Municipality of Bento Gonçalves, who collaborated to make the event happen. Some organizing committee interviewees also pointed out the challenge of achieving the right balance, in terms of "audience" (two references), between academics, government representatives, business leaders, and practitioners, so that the event could create a movement all across the city. According to the interviewees, this challenge was overcome thanks to the influence of those local leaders who activated their networks, ensuring the main local actors' participation.

As for the opportunities, the organizing committee and university's interviewees suggested that a greater effort could have been made to involve a "broader audience" (three references), such as universities and municipalities from the rest of the state or even the country. One of the interviewees stated that this would have allowed institutions directly involved with the event, such as UCS, to explore its strategic potential better. Meanwhile, different interviewees among those who do not belong to the organizing committee mentioned an untapped opportunity of using the event to induce local actors to "commit" (five references) themselves to local development projects. These interviewees coincided on the expectation that, after the event, some groups of actors or local leaders would have addressed concrete initiatives, directly assuming commitments, plans, or actions.

Regarding the content coded in "contributions" (72 references), the most significant number of citations are in "KBUD awareness" (16 references) and "university's role" (15 references). In all of the interviewees' groups, the perception is that one of the main contributions of the event in 2011 was to raise awareness on local actors about the themes of KBUD and knowledge cities. According to the civil society representative, "many regional and business leaders have been contaminated by the debate of the knowledge citie" s due to the event. Other interviewees also mentioned the effect on the business community, recalling that this awareness raised the local business sector's interest in monitoring the city's development progress through the KBUD methodology, which was later accomplished through a partnership between CIC-BG; the Bento Gonçalves Centre for Industry, Commerce, and Services; and UCS for the publication of the Bento Gonçalves' capital system [75,101].

The perception of how much this growth in KBUD awareness has unfolded into local development initiatives differs among respondents. Some interviewees indicated that it was not translated into concrete local initiatives after the event. Those who share this view, in general, referred to these initiatives as if they were the exclusive responsibility of the public authorities. On the other hand, other interviewees, including government and civil society representatives and others from the organizing committee, indicated that this growth in KBUD awareness paved the way for different local initiatives, especially those developed by or in partnership with local universities.

In line with this, recognition of the "university's role" (15 references) is the next codified contribution. Interviewees indicated that the KCWS in 2011 was fundamental in increasing UCS's influence over different development agents in the region and placing the university at the center of the local KBUD movement. One interviewee highlighted that due to the event, collaboration and data sharing processes could be established between the university and different institutions, including the municipal government. This, in turn, allowed UCS to advance and consolidate research on the topics of KBUD. The outcomes of these studies generated valuable insights that were shared with municipal management. According to the local government representative, the university acted as a translator, making the KBUD concepts more "palatable" so that the city administration could transform them into policies. The university representative pointed out that the KCWS consolidated and gave more visibility to this group of UCS researchers involved with the theme, who had become a KBUD reference even at a national level. According to one of the organizing committee interviewees, by bringing to the debate ideas such as the university's central role in a development process, the KCWS contributed to materialize UCS's protagonism in other spheres, expanding its influence on regional and even national levels.

Another mentioned contribution concerns "knowledge exchange and skill training" (13 references). Among the interviewees who attended the event, there is a consensus that one of the most positive aspects of the event was the high quality of the content transmitted, mainly due to the high level of international speakers, who brought research and experiences from different parts of the world. One of the organizing committee interviewees pointed out that KCWS is an accelerator capable of connecting attendees with the best in terms of knowledge, and "having accessed this cutting-edge knowledge created a movement in the community, in people who attended the event".

Contributions to the "academic agenda" (12 references) are also cited by interviewees linked to the university, whether from the organizing committee or not. Among the advances related to the fourth KCWS, the interviewees mentioned the following ones: the promotion of the postgraduate program in knowledge management, the consolidation of knowledge-based development as a line of research in the graduate program in administration, an increase in the number of master's and doctoral studies on the subject, the strengthening of partnerships with researchers from other countries, and publications in renowned journals of the field. After the fourth KCWS, further research on the capital system, as a value-based tool, began to be developed by researchers of the university (see [75,102–104]), culminating with the creation of the Brazilian Knowledge-Based Development Observatory in 2018. According to one of the organizing committee interviewees, this trajectory was inspired by the KCWS in 2011. Furthermore, the university representative highlighted that the event in 2011 expanded the KBUD debate within UCS itself to the point that it became a guideline for the university's current Institutional Internationalization Plan, which can also be considered a legacy of the event.

Among the organizing committee interviewees, another contribution cited with emphasis is creating or strengthening "networks" (11 references). The interviewees of this group mentioned that the dynamics of the event favored interaction, allowing for the connection between professionals, researchers, and speakers from all around the world. At an institutional level, the event provided the opportunity to strengthen the connections and develop joint projects between UCS and other institutions such as ICKS, WCI, the Municipal Government, and AMESNE. The creation or strengthening of these connections also expanded the possibilities for academic dialogue, allowing for collaborations that significantly accelerated research development and KBUD application in the region, according to an organizing committee interviewee.

The organizing committee interviewees also highlighted that the event gave "profile" (five references) to Bento Gonçalves before the international participants and speakers. In the analysis of some interviewees, having these experts visiting and getting to know some of the city's attributes boosted Bento Gonçalves reputation on KBUD and contributed to its nomination in the MAKCi Award in the following years.

4. Discussion

Looking at the results of the interviews with WCI executives and representatives of Bento Gonçalves, some insights into KCWS' contributions to the KBUD of host cities come to light. There is a significant overlap between the motivations to host the KCWS reported by Bento Gonçalves representatives and those pointed out by WCI executives. In general, regarding the host, the main goals correspond to learning and leveraging the city's initiatives, especially those of the hosting partner institutions. As for the event's main stakeholders, it is possible to see that the event successfully involves the main KBUD agents, namely, the triple helix (university, government, and private sector). Nevertheless, there may be room for more civil society participation, although references to "local community", "non-profit entities", and "civil society" also occurred in both cases. Regarding hosting requirements, it is clear that the type of motivation and commitment that WCI seeks to find in a host depends a lot on the leadership and involvement of local leaders in the event's organization, as pointed out by Bento Gonçalves' interviewees.

The category with the most matches is "contributions". There is a meaningful alignment between the contributions visualized by the event promoters and those perceived by the host city. On the tail of the event's goals, the interview analysis indicates that knowledge exchange is one of the main contributions of the event. The KCWS, as an international summit with the field's leading experts and researchers, acts as a facilitator for knowledge transfer on topics related to KBUD and knowledge cities. Consequently, KCWS also contributes to raising KBUD awareness among the local actors who attend the event. KBUD awareness, as defined by one of the interviewees, is this capability of understanding and employing local knowledge assets, and it is fundamental for the engagement of local actors in any KBUD initiative [23,42].

Events may act as institutional catalysts, facilitating the cooperation and collaboration among local actors [57,61]. In line with this, "networking" is the most cited contribution of the event among WCI interviewees, and is also highly referred by Bento Gonçalves representatives. In Bento Gonçalves, the KCWS strengthened pre-existing relationships and allowed for new connections between local actors and leading academics, professionals, and policymakers from all around the world in collaborative networks. In this sense, KCWS enhanced the local actors' ability to establish and nurture collaboration through partnerships, which, as we previously presented, is essential for the success of a KBUD process [40].

Another contribution pointed out by both groups of interviewees was the opportunity to gain an international profile as a knowledge city. In the case of an international conference such as KCWS, which draws together leading national and international specialists and practitioners in their fields, hosting the event conveys a message about the city's development intentions. In 2011, KCWS provided an opportunity for Bento Gonçalves to position itself as a city engaged in cutting-edge development strategies such as KBUD.

Both WCI's and Bento Gonçalves' interviewees highlight contributions and developments in the academic sphere generated by the event. This type of contribution is especially relevant for KBUD, which advocates that the knowledge produced by the universities is a substrate for the smart and sustainable development of cities. In the case of a community university like UCS, whose institutional identity is so deeply intertwined with the region, academic advances have the potential to project the development of the entire region.

Finally, one last contribution emerges from the interviews, however, it is pointed out only by Bento Gonçalves representatives, which is the elevation of the local university (UCS) to the role of a protagonist of local development. Since its foundation, UCS has been a university deeply involved with the region's development process. The fourth KCWS, however, contributed to increasing UCS's influence over different development agents in the region and placing the university at the center of the local KBUD movement. As seen in previous sections, having the university moving from a supporting role to a more entrepreneurial one is key in any KBUD strategy [39,41]. The KCWS not only conveyed this vision to the participants, but also presented UCS as an agent capable of leading this movement.

The fact that only Bento Gonçalves' interviewees mentioned this contribution suggests that its occurrence may be associated with a particular context. In fact, it is necessary to consider the context's effect on all of the results. The contributions are deeply connected to the event's central theme: KBUD and knowledge cities. Should the event be on a different topic, there might be different effects. The way these contributions developed in Bento Gonçalves is linked to the context in which the city is inserted. As presented before, UCS is a community university founded with the purpose of contributing to the development of the region. Years before the fourth KCWS, the university was already engaging in projects in the KBUD field. In fact, the event in 2011 was largely due to the institutional collaborative partnerships established years before between UCS, and other institutions dedicated to the subject. Furthermore, UCS professors and researchers were already actively engaged in projects on the topics of KBUD and knowledge cities by the time the event occurred. The outcomes could also be different if it were not a small city where the central university is already on a path to expand its relevance in the local development process.

The context is also relevant when considering how those contributions developed. Social and community identity can enhance the knowledge exchange process [105], and as a university with a community identity, UCS is highly distributed and embedded in the region. In this way, UCS's presence in different municipalities also makes it an integrating element that facilitates the exchange of knowledge between cities in the region. By connecting to an international event such as KCWS, UCS had the potential to take all the event's cutting-edge content to other development agents in the region, expanding the reach of these ideas.

UCS's role can also be perceived through the studies produced by this group of researchers directly involved with KCWS in 2011. Many of these studies were presented to the municipal government and supported the city's planning and policymaking. This is also a critical aspect for KBUD: providing city administrations, planners, and policymakers with data and knowledge for informed decisions and evidence-based policies [32,40]. However, data collection on Bento Gonçalves did not find evidence of a plan structured by the city with goals and measures to become a knowledge city, as occurred, for instance, in cities like Monterrey, Mexico, and Melbourne, Australia. Nevertheless, we identified that these studies developed by UCS's researchers supported, in the subsequent years, the nomination of Bento Gonçalves for the MAKCi award, which recognizes cities around the world for their successful KBUD strategies. The award in 2019, therefore, was a confirmation of Bento Gonçalves' progress.

In fact, common to all contributions presented is the university's leadership and action, whether developing research and delivering results to society, connecting groups, or producing innovation. Strong leadership, with meaningful networks and partnerships to support it, is one essential component for a prosperous KBUD [106]. In this sense, as relevant as UCS's institutional leadership is the leadership exercised by UCS professors and researchers, who often took the lead in the city's development and planning initiatives and projects. These professors and researchers put their networks, research, and knowledge as resources for the advancement of the community. This adds to the university's role as a protagonist in Bento Gonçalves' development as well. Highlighting UCS in this role was one of the main contributions of the fourth KCWS to Bento Gonçalves' development. It is necessary to recognize, though, that the interviewees in Bento Gonçalves were very much inserted in the university context. Nonetheless, the document analysis found evidence of UCS's constant activity in matters of local development. University representatives participate in the main local planning bodies; UCS professors direct or integrate relevant sectoral organizations; the university even provides the physical structure for, for example, COREDE Serra meetings held periodically. Of course, UCS effectively acts as an innovation and technology development agent by furthering projects such as the graphene production plant.

Overall, as could be grasped by this study, the KCWS contributions compose a web of enablers and potentials that can unfold in different outcomes and achievements. They depend on stakeholder's engagement and leadership to continue and unfold into plans, actions, and results. We have found that contributions may occur through knowledge transfer on the trending topics addressed at the event, which also increases the KBUD awareness among the audience and those involved in the event. Knowledge exchanged is leveraged by the relationships and networks to which the event connects, from which comes learning, collaboration, and partnerships in initiatives that can range from academic research to the implementation of KBUD plans and projects. Moreover, because it is a highly qualified international networking platform, the event also promotes the host city to a very qualified group of professionals and thinkers worldwide, contributing to the city's image. All these contributions' potential is largely linked to local leaders' capacity to enrich and nurture these networks over time, making them evolve and grow.

The event can also contribute to making these local leaders emerge and elevating them to the role of protagonists of local KBUD. The involvement of the various local actors, especially the university, in the development processes is one of KBUD's premises. KCWS favors the interaction of these actors and helps their positioning within this process. This may be the most relevant contribution of an international event for the local KBUD.

In the interviews, continuity is pointed out as one of the main challenges of holding an event such as KCWS. It appears in the speeches of WCI executives and in those of Bento Gonçalves' representatives when they talked about untapped expectations of having development agreements and projects been launched during the event. An event is a definite point in space and time—a spark. Nevertheless, in general, whoever promotes, organizes, or participates in an event focused on city development may hope that the flame will remain lit, that someone will transform the ideas and aspirations discussed at the event into plans, actions, results, and hopefully impacts. The literature review demonstrates that events can go beyond contributions and generate effective impacts in the economic, social, environmental, and institutional fields. For that, it must be part of a structured development program instead of a dispersed and fragmented enterprise [53]. Inserting and working upon trajectories, taking the event as part of a more integrated development approach can maximize its benefits for a range of different stakeholders and enable the achievement of more meaningful goals [49,60]. The experience of other cities that tried to leverage their development process with events showed that a strategy for the post-event period, designed and planned with local actors' support and commitment, is essential [57].

Therefore, continuity depends largely on the engagement and empowerment of the local actors involved with the event. This is why raising a local actor to the role of KBUD's protagonist is such a significant contribution of KCWS. In the case of local hosts, governments often change, people move, priorities shift. However, when the development process's leadership is shared with perennial institutions, such as the university, planning is more likely to be realized.

In Bento Gonçalves, history and the Italian ancestor cultural traditions were knowledge resources used to promote economic activities and development. Throughout the years, the city transformed intangible aspects of knowledge such as cultural identity and traditional values into economic development [105] in different sectors, such as the furniture, wine, and tourism segments. In October 2020, the Bento+20 Masterplan was launched with guidelines, goals and actions to make Bento Gonçalves a smart and sustainable city by 2040. In order to fully realize the plan, Bento Gonçalves will need a systemic approach that includes a governance model capable of integrating all city actors and all development dimensions. KBUD provides such strategic and integrated approach for the transition of smart and sustainable cities [21–23] and places the university in a central position as a critical asset for development and innovation [35]. It also highlights the importance of creating institutional arrangements in a combination that favors the collaborative action of university, government, the private sector and civil society to produce innovation and economic development [39,40]. Many of these aspects were brought to Bento Gonçalves by the KCWS, and to become truly smart and sustainable, not losing sight of KBUD's strategic approach is a good practice.

5. Conclusions

When this study started in 2019, the world was not yet alert to the COVID-19 pandemic. The situation changed in 2020, drastically affecting cities' relationship with events and the way they carried them out. Nevertheless, the pandemic also invited us to reflect on the importance of gathering together and how events have an impact on city's economic and sociocultural functions. As events become viable again, the findings of this study are a powerful signal for the various cities in the context of emerging countries and those interested in building or improving their local development strategies with events.

Furthermore, it reinforces how KBUD is a tangible and viable local development strategy, even for small cities or cities located outside the central urban axes. Hosting an event focused on KBUD, such as KCWS, can not only promote the city's intent, but also help it bring together the various actors relevant to the development process and encourage them to build shared understandings enabling engagement in joint projects. Furthermore, Bento Gonçalves' case reveals some fruitful initiatives that cities can employ under the flag of a KBUD strategy. One of them is prioritizing endogenous economic activities based on local knowledge resources and intangible assets, such as cultural identity and traditional values. Placing the local university as a key agent of the development process, establishing with it strategic partnerships for knowledge transfer and innovation development, is another fundamental aspect. Creating institutional conditions for the engagement and participation of the business community and civil society in the city's development and management processes, with strong leadership and meaningful networks and partnerships to support it, completes this story.

Lastly, Bento Gonçalves' successful experience with KCWS may also serve as a reference for city managers, policymakers, and executives inserted in similar contexts and wishing to follow the KBUD path. In addition, this conclusion provides the following list of lessons learned from Bento Gonçalves, which may also be of relevance to other cities with similar goals:

- The local university has a relevant role in carrying out events of this type, as it has the means to seek and establish collaborative institutional partnerships with other universities and institutes focused on promoting events such as KCWS.
- Local leaders play an essential part in activating local networks and anchoring the ideas in the community. This is why it is crucial to have some of these leaders on the event's organizing committee, acting directly from the initial preparations to the post-event.
- In particular, Bento Gonçalves demonstrated that university professors and researchers are relevant leaders in knowledge-based development contexts and can contribute to bringing events such as KCWS to the city. Furthermore, their research, knowledge, and networks may also facilitate the unfolding of the event's topics into actions for city's development.
- Funding the event can be a challenge. Bento Gonçalves overcame this issue by seeking sponsorship with several organizations from different sectors. This arrangement also favored these organizations' engagement, as they not only sponsored, but also took part in the event;
- Engaging local actors around the event is essential in order to obtain results in city development. The involvement of the four sectors of the quadruple helix, i.e., government, companies, university, and civil society, provides representation; facilitates knowledge exchange and collaboration; and expands the reach of the ideas, concepts, and experiences addressed during the event, creating awareness city-wide.
- Involving a broader audience, such as municipalities across the country and national and international institutions, may be a way of bringing visibility to the host city and establishing strategic partnerships. This was an untapped opportunity in the Bento Gonçalves case.
- Using the event to induce local actors' commitment to development projects is a path towards the continuity of ideas, plans, and initiatives. Events such as KCWS can create the ideal environment for signing cooperation agreements or launching action plans.
- Therefore, having a strategy for the post-event period, designed and planned with local actors' support and commitment, can generate even more effective contributions to the city. This was an untapped opportunity in the Bento Gonçalves case.
- The implementation and continuity of the ideas discussed at the event depend a lot on the local actors' level of engagement and leadership over time. For this reason, sharing this leadership with perennial institutions, such as business, civil society

associations, and the university, is a strategy that contributes to the realization of the development vision.

- International events do contribute to the knowledge-based development of cities. The obtained results, however, are just a glimpse into the relationship between events and KBUD. Future studies should explore the other KCWS host cities' cases and analyze the relationship between the event and local development. Moreover, in line with Eisenhardt's [107] approach to multiple case studies, new research could draw on the replications, contrasts, and extensions to obtain theoretical generalizations. In addition, a line of investigation on how the contents presented in KCWS unfold into plans and programs in the host cities can broaden their application in public policies. A relevant aspect that can be further explored concerns the spatial dimension of KBUD and how events can affect it. Furthermore, the contributions identified in this study also pointed to the investigation of the impacts generated by the event. From the trajectories of the contributions and focusing on KBUD, one could seek to measure and analyze the direct and indirect impacts connected to the event.

Lastly, the following limitations should be noted when interpreting the research results. The interviews conducted in this study captured the perspective of actors deeply involved with the event under investigation and, therefore, may not represent all local stakeholders' views. While we contend that some unintended bias may be extant, our use of the NVivo coding platform aimed to mitigate much of the biases that seep in when conducting qualitative interviews. Concerning the analysis, Bento Gonçalves' characterization and indicator analysis were done qualitatively. The use of a consolidated assessment model, such as KBUD/AM, the quantitative performance assessment model to evaluate the KBUD achievements of cities [32], would have added objectivity and comparability to the analysis. In addition, a single case study allows for a limited generalization of the obtained results. Analyzing other host city cases could provide a gain in empirical generalization.

Author Contributions: T.T.P.C. and T.Y. designed and supervised the study and contributed to the write-up. L.D.M. collected and curated data, conducted the analysis, and prepared the first draft of the manuscript. A.C.F., L.V., and W.L. expanded the manuscript and helped to finalize it by improving its rigor, relevance, and caliber. All authors have read and agreed to the published version of the manuscript.

Funding: The article processing charges (APC) were funded by University of Caxias do Sul.

Institutional Review Board Statement: Not applicable.

Informed Consent Statement: Not applicable.

Data Availability Statement: No new data were created or analyzed in this study. Data sharing is not applicable to this article.

Acknowledgments: This manuscript is produced from the Master thesis of the first author entitled: contributions of international events on local development: the case of the knowledge cities world summit in Bento Gonçalves. A copy of the thesis is available online: https://bibliotecatede.uninove.br/handle/tede/2485. The authors thank the anonymous referees for their invaluable comments on an earlier version of the manuscript.

Conflicts of Interest: The authors declare no conflict of interest.

References

1. Alberti, A.; Senese, M. Developing capacities for inclusive and innovative urban governance. In *Governance for Urban Services. Advances in 21st Century Human Settlements*; Cheema, S., Ed.; Springer: Singapore, 2020; pp. 127–152. [CrossRef]
2. Yigitcanlar, T.; Han, H.; Kamruzzaman, M.; Ioppolo, G.; Sabatini-Marques, J. The making of smart cities: Are Songdo, Masdar, Amsterdam, San Francisco and Brisbane the best we could build? *Land Use Policy* **2019**, *88*, 104187. [CrossRef]
3. Reckien, D.; Creutzig, F.; Fernandez, B.; Lwasa, S.; Tovar-Restrepo, M.; McEvoy, D.; Satterthwaite, D. Climate change, equity and the sustainable development goals: An urban perspective. *Environ. Urban.* **2017**, *29*, 159–182. [CrossRef]
4. Sarimin, M.; Yigitcanlar, T. Knowledge-based urban development of multimedia super corridor, Malaysia: An overview. *Int. J. Knowl. Based Dev.* **2011**, *2*, 34–48. [CrossRef]

5. Shin, H.B. Urban spatial restructuring, event-led development and scalar politics. *Urban Stud.* **2014**, *51*, 2961–2978. [CrossRef]

6. Bryan, G.; Glaeser, E.; Tsivanidis, N. Cities in the developing world. *Annu. Rev. Econ.* **2020**, *12*, 273–297. [CrossRef]

7. Sarimin, M.; Yigitcanlar, T. Towards a comprehensive and integrated knowledge-based urban development model: Status quo and directions. *Int. J. Knowl. Based Dev.* **2012**, *3*, 175–192. [CrossRef]

8. Alcaide Muñoz, L.; Rodríguez Bolívar, M.P. Different levels of smart and sustainable cities construction using e-participation tools in European and Central Asian countries. *Sustainability* **2021**, *13*, 3561. [CrossRef]

9. Vojnovic, I. Urban sustainability: Research, politics, policy and practice. *Cities* **2014**, *41*, S30–S44. [CrossRef]

10. Bugliarello, G. Urban sustainability: Dilemmas, challenges and paradigms. *Technol. Soc.* **2006**, *28*, 19–26. [CrossRef]

11. UN. Sustainable Development Goals: Knowledge Platform. Available online: https://sdgs.un.org/goals (accessed on 20 February 2021).

12. Ultramari, C.; Saldiva, P.; Levy, W. A COVID-19 e as Cidades Inteligentes. Folha de S. Paulo. 12 June 2020. Available online: https://www1.folha.uol.com.br/opiniao/2020/06/a-covid-19-e-as-cidades-inteligentes.shtml (accessed on 20 June 2021).

13. Sohoo, I.; Ritzkowski, M.; Kuchta, K.; Cinar, S.Ö. Environmental sustainability enhancement of waste disposal sites in developing countries through controlling greenhouse gas emissions. *Sustainability* **2021**, *13*, 151. [CrossRef]

14. Kookana, R.S.; Drechsel, P.; Jamwal, P.; Vanderzalm, J. Urbanisation and emerging economies: Issues and potential solutions for water and food security. *Sci. Total Environ.* **2020**, *732*, 139057. [CrossRef] [PubMed]

15. Abubakar, I.R.; Aina, Y.A. The prospects and challenges of developing more inclusive, safe, resilient and sustainable cities in Nigeria. *Land Use Policy* **2019**, *87*, 104105. [CrossRef]

16. Seto, K.C.; Parnell, S.; Elmqvist, T. A global outlook on urbanization. In *Urbanization, Biodiversity and Ecosystem Services: Challenges and Opportunities*; Elmqvist, T., Fragkias, M., Goodness, J., Güneralp, B., Marcotullio, P.J., McDonald, R.I., Parnell, S., Schewenius, M., Sendstad, M., Seto, K.C., et al., Eds.; Springer: Dordrecht, The Netherlands, 2013; pp. 1–12. [CrossRef]

17. Yang, Z.; Yang, H.; Wang, H. Evaluating urban sustainability under different development pathways: A case study of the Beijing-Tianjin-Hebei region. *Sustain. Cities Soc.* **2020**, *61*, 102226. [CrossRef]

18. Cortese, T.T.P.; Coutinho, S.V.; Vasconcellos, M.D.P.; Buckeridge, M.S. Tecnologias e sustentabilidade nas cidades. *Estud. Avançados* **2019**, *33*, 137–150. [CrossRef]

19. Yigitcanlar, T.; Kamruzzaman, M. Smart cities and mobility: Does the smartness of Australian cities lead to sustainable commuting patterns? *J. Urban Technol.* **2019**, *26*, 21–46. [CrossRef]

20. Martin, C.J.; Evans, J.; Karvonen, A. Smart and sustainable? Five tensions in the visions and practices of the smart-sustainable city in Europe and North America. *Technol. Forecast. Soc. Chang.* **2018**, *133*, 269–278. [CrossRef]

21. Ahvenniemi, H.; Huovila, A.; Pinto-Seppä, I.; Airaksinen, M. What are the differences between sustainable and smart cities? *Cities* **2017**, *60*, 234–245. [CrossRef]

22. Penco, L.; Ivaldi, E.; Bruzzi, C.; Musso, E. Knowledge-based urban environments and entrepreneurship: Inside EU cities. *Cities* **2020**, *96*, 102443. [CrossRef]

23. Hara, M.; Nagao, T.; Hannoe, S.; Nakamura, J. New key performance indicators for a smart sustainable city. *Sustainability* **2016**, *8*, 206. [CrossRef]

24. Farhangi, M. Moving Esfahan toward knowledge-based urban development: The role of knowledge workers' needs. *Int. J. Knowl. Based Dev.* **2013**, *4*, 19–33. [CrossRef]

25. Knight, R. Knowledge-based development: Policy and planning implications for cities. *Urban Stud.* **1995**, *32*, 225–260. [CrossRef]

26. Yigitcanlar, T.; Dur, F. Making space and place for knowledge communities: Lessons for Australian practice. *Australas. J. Reg. Stud.* **2013**, *19*, 36–63.

27. Cabrita, M.D.R.; Cruz-Machado, V.; Cabrita, C. Managing creative industries in the context of knowledge-based urban development. *Int. J. Knowl. Based Dev.* **2013**, *2*, 318–337. [CrossRef]

28. Carrillo, F.J. Knowledge-based development as a new economic culture. *J. Open Innov. Technol. Mark. Complex* **2015**, *1*, 15. [CrossRef]

29. Yigitcanlar, T.; Sabatini-Marques, J.; Lorenzi, C.; Bernardinetti, N.; Schreiner, T.; Fachinelli, A.; Wittmann, T. Towards smart Florianópolis: What does it take to transform a tourist island into an innovation capital? *Energies* **2018**, *11*, 3265. [CrossRef]

30. Ergazakis, E.; Ergazakis, K.; Metaxiotis, K. Building successful knowledge cities in the context of the knowledge-based economy: A modern strategic framework. In *Knowledge-Based Development for Cities and Societies: Integrated Multi-Level Approaches*; IGI Global: Hershey, PA, USA, 2010; pp. 17–41.

31. Rodrigues, M.; Franco, M. Measuring the performance in creative cities: Proposal of a multidimensional model. *Sustainability* **2018**, *10*, 4023. [CrossRef]

32. Esmaeilpoorarabi, N.; Yigitcanlar, T.; Guaralda, M. Place quality in innovation clusters: An empirical analysis of global best practices from Singapore, Helsinki, New York, and Sydney. *Cities* **2018**, *74*, 156–168. [CrossRef]

33. Zhao, P. Building knowledge city in transformation era: Knowledge-based urban development in Beijing in the context of globalisation and decentralisation. *Asia Pac. Viewp.* **2010**, *51*, 73–90. [CrossRef]

34. Ponto, H.; Inkinen, T. Knowledge-based environments in the city: Design and urban form in the Helsinki metropolitan area. *Int. J. Knowl. Based Dev.* **2019**, *10*, 155–175. [CrossRef]

35. Tabibi, S.H.; Rafieian, M.; Majedi, H.; Ziari, Y.A. The role of knowledge-based and innovative cities in urban and regional development. *Urban Plan. Knowl.* **2020**, *4*, 19–32.

36. Makkonen, T.; Weidenfeld, A. Knowledge-based urban development of cross-border twin cities. *Int. J. Knowl. Based Dev.* **2016**, *7*, 389–406. [CrossRef]
37. Isaksen, A. Knowledge-based clusters and urban location: The clustering of software consultancy in Oslo. *Urban Stud.* **2004**, *41*, 1157–1174. [CrossRef]
38. Johnston, A. The roles of universities in knowledge-based urban development: A critical review. *Int. J. Knowl. Based Dev.* **2019**, *10*, 213–231. [CrossRef]
39. Etzkowitz, H.; Zhou, C. Hélice Tríplice: Inovação e empreendedorismo universidade-indústria-governo. *Estud. Avançados* **2017**, *31*, 23–48. [CrossRef]
40. Carrillo, F.J.; Batra, S. Understanding and measurement: Perspectives on the evolution of knowledge-based development. *Int. J. Knowl. Based Dev.* **2012**, *3*, 1–16. [CrossRef]
41. Alizadeh, T. The interaction between local and regional knowledge-based development: Towards a quadruple helix model. In *Knowledge-Based Development for Cities and Societies: Integrated Multi-Level Approaches*; IGI Global: Hershey, PA, USA, 2010; pp. 81–98.
42. Hortz, T. The smart state test: A critical review of the smart state strategy 2005–2015's knowledge-based urban development. *Int. J. Knowl. Based Dev.* **2016**, *7*, 75–101. [CrossRef]
43. Laasonen, V.; Kolehmainen, J. Capabilities in knowledge-based regional development–towards a dynamic framework. *Eur. Plan. Stud.* **2017**, *25*, 1673–1692. [CrossRef]
44. Getz, D. Event studies: Discourses and future directions. *Event Manag.* **2012**, *16*, 171–187. [CrossRef]
45. Liu, Y.D. Event and quality of life: A case study of Liverpool as the 2008 European capital of culture. *Appl. Res. Qual. Life* **2016**, *11*, 707–721. [CrossRef]
46. Campbell, L.M.; Corson, C.; Gray, N.J.; MacDonald, K.I.; Brosius, J.P. Studying global environmental meetings to understand global environmental governance: Collaborative event ethnography at the tenth conference of the parties to the convention on biological diversity. *Glob. Environ. Politics* **2014**, *14*, 1–20. [CrossRef]
47. Hardy, C.; Maguire, S. Discourse, field-configuring events, and change in organisations and institutional fields: Narratives of DDT and the Stockholm Convention. *Acad. Manag. J.* **2010**, *53*, 1365–1392. [CrossRef]
48. Schüssler, E.; Rüling, C.C.; Wittneben, B.B. On melting summits: The limitations of field-configuring events as catalysts of change in transnational climate policy. *Acad. Manag. J.* **2014**, *57*, 140–171. [CrossRef]
49. Richards, G. From place branding to placemaking: The role of events. *Int. J. Event Festiv. Manag.* **2017**, *8*, 8–23. [CrossRef]
50. Azzali, S. Mega-events and urban planning: Doha as a case study. *Urban Des. Int.* **2017**, *22*, 3–12. [CrossRef]
51. Gelders, D.; van Zuilen, B. City events: Short and serial reproduction effects on the city's image? *Corp. Commun. Int. J.* **2013**, *18*, 110–118. [CrossRef]
52. Hall, C.M.; Amore, A. The 2015 cricket world cup in Christchurch. *J. Place Manag. Dev.* **2019**, *13*, 4–17. [CrossRef]
53. Richards, G.; Wilson, J. The impact of cultural events on city image: Rotterdam, cultural capital of Europe 2001. *Urban Stud.* **2004**, *41*, 1931–1951. [CrossRef]
54. Waitt, G. Playing games with Sydney: Marketing Sydney for the 2000 olympics. *Urban Stud.* **1999**, *36*, 1055–1077. [CrossRef]
55. Hudec, O.; Remoaldo, P.C.; Urbančíková, N.; Cadima Ribeiro, J.A. Stepping out of the shadows: Legacy of the European capitals of culture, Guimarães 2012 and Košice 2013. *Sustainability* **2019**, *11*, 1469. [CrossRef]
56. Liang, Y.-W.; Wang, C.-H.; Tsaur, S.-H.; Yen, C.-H.; Tu, J.-H. Mega-event and urban sustainable development. *Int. J. Event Festiv. Manag.* **2016**, *7*, 152–171. [CrossRef]
57. Bernardino, S.; Freitas Santos, J.; Cadima Ribeiro, J. The legacy of European capitals of culture to the "smartness" of cities: The case of Guimarães 2012. *J. Conv. Event Tour* **2017**, *19*, 138–166. [CrossRef]
58. Burbank, M.J.; Andranovich, G.; Heying, C.H. Mega-events, urban development, and public policy. *Rev. Policy Res.* **2020**, *19*, 179–202. [CrossRef]
59. Eizenberg, E.; Cohen, N. Reconstructing urban image through cultural flagship events: The case of Bat-Yam. *Cities* **2015**, *42*, 54–62. [CrossRef]
60. Kassens-Noor, E.; Wilson, M.; Müller, S.; Maharaj, B.; Huntoon, L. Towards a mega-event legacy framework. *Leis. Stud.* **2015**, *34*, 665–671. [CrossRef]
61. Liu, Y.D. Event and sustainable culture-led regeneration: Lessons from the 2008 European capital of culture, Liverpool. *Sustainability* **2019**, *11*, 1869. [CrossRef]
62. Levy, W.; Leite, C. Interação, democracia e governança urbanas. Um ensaio sobre o conceito de cidades inteligentes. *Arquitextos* **2020**, *20*. Available online: https://vitruvius.com.br/revistas/read/arquitextos/20.237/7640 (accessed on 15 June 2021).
63. Henn, S.; Bathelt, H. Knowledge generation and field reproduction in temporary clusters and the role of business conferences. *Geoforum* **2015**, *58*, 104–113. [CrossRef]
64. Yigitcanlar, T.; Metaxiotis, K.; Carrillo, F.J. *Building Prosperous Knowledge Cities: Policies, Plans and Metrics*; Edward Elgar Publishing: Cheltenham, UK, 2012.
65. Durmaz, B.; Platt, S.; Yigitcanlar, T. Creativity, culture tourism and place-making: Istanbul and London film industries. *Int. J. Cult. Tour. Hosp. Res.* **2010**, *4*, 198–213. [CrossRef]
66. Batra, S.; Payal, R.; Carrillo, F.J. Knowledge village capital framework in the Indian context. *Int. J. Knowl. Based Dev.* **2013**, *4*, 222–244. [CrossRef]

67. Pancholi, S.; Yigitcanlar, T.; Guaralda, M. Public space design of knowledge and innovation spaces: Learnings from Kelvin Grove Urban Village, Brisbane. *J. Open Innov. Technol. Mark. Complex* **2015**, *1*, 13. [CrossRef]
68. Olcay, G.A.; Bulu, M. Technoparks and technology transfer offices as drivers of an innovation economy: Lessons from İstanbul's innovation spaces. *J. Urban Technol.* **2016**, *23*, 71–93. [CrossRef]
69. World Capital Institute. Available online: https://www.worldcapitalinstitute.org/who-we-are/ (accessed on 2 February 2021).
70. Schreier, M. Qualitative content analysis. In *The SAGE Handbook of Qualitative Data Analysis*; Flick, U., Ed.; SAGE Publications: Thousand Oaks, CA, USA, 2014; pp. 170–183.
71. Krippendorff, K. *Content Analysis: An Introduction to Its Methodology*, 2nd ed.; SAGE Publications: Thousand Oaks, CA, USA, 2004.
72. Saldaña, J. *The Coding Manual for Qualitative Researchers*; SAGE Publications: Thousand Oaks, CA, USA, 2009.
73. Yin, R.K. *Case Study Research: Design and Methods*, 3rd ed.; Sage Publications: Thousand Oaks, CA, USA, 2003.
74. Garcia, B.C. Making MAKCi: An emerging knowledge-generative network of practice in the Web 2.0. *VINE* **2010**, *40*, 39–61. [CrossRef]
75. Fachinelli, A.C.; Carrillo, F.J.; D'Arisbo, A. Capital system, creative economy and knowledge city transformation: Insights from Bento Gonçalves, Brazil. *Expert Syst. Appl.* **2014**, *41*, 5614–5624. [CrossRef]
76. Fachinelli, A.C.; D'Arrigo, F.P.; Breunig, K.J. The value context in knowledge-based development: Revealing the context factors in the development of Southern Brazils Vale dos Vinhedos region. *Knowl. Manag. Res. Pract.* **2018**, *16*, 32–41. [CrossRef]
77. Bowen, G.A. Document analysis as a qualitative research method. *Qual. Res. J.* **2009**, *9*, 27–40. [CrossRef]
78. Bertoco, C.B.; Medeiros, A.E. Sustentabilidade, planejamento urbano e instrumentos de gestão do patrimônio e da paisagem cultural em Bento Gonçalves/RS. *Paranoá Cad. Arquitetura Urban* **2015**, *14*, 105–114. [CrossRef]
79. Instituto Brasileiro de Geografia e Estatística. Estimativas da População Residente para os Municípios e para as Unidades da Federação Brasileiros com Data de Referência em 1° de Julho de 2020. 2020. Available online: https://www.ibge.gov.br/estatisticas/sociais/populacao/9103-estimativas-de-populacao.html?=&t=o-que-e (accessed on 20 December 2020).
80. Google Maps. Available online: https://goo.gl/maps/Qu316sqyAHQC4rBu5 (accessed on 20 June 2021).
81. Atlas do Desenvolvimento Humano no Brasil. Perfil Bento Gonçalves. Available online: http://www.atlasbrasil.org.br/perfil/municipio/430210 (accessed on 5 January 2021).
82. Departamento de Economia e Estatística, Secretaria de Planejamento, Governança e Gestão do Estado do Rio Grande do Sul. PIB Municipal. 2020. Available online: https://dee.rs.gov.br/pib-municipal (accessed on 21 December 2020).
83. Melhores Cidades para Fazer Negócio: Edições Anteriores. Available online: https://www.urbansystems.com.br/melhorescidadesparanegocios (accessed on 17 September 2020).
84. Panorama Socioeconômico Bento Gonçalves. Centro da Indústria, Comércio e Serviços de Bento Gonçalves: Bento Gonçalves, Brazil. 2019. Available online: http://cicbg.com.br/uploads/revista_panorama_cic_2019.pdf (accessed on 8 January 2021).
85. Sperotto, F.Q. Arranjo produtivo local móveis da Serra Gaúcha. In *Aglomerações e Arranjos Produtivos Locais no Rio Grande do Sul*; de Macadar, B.M., de Costa, R.M., Eds.; FEE: Porto Alegre, Brazil, 2016; pp. 405–443.
86. Sperotto, F.Q. Setor moveleiro brasileiro e gaúcho: Características, configuração e perspective. *Indic. Econômicos FEE* **2018**, *45*, 43–60.
87. Município de Bento Gonçalves, Conheça a Cidade. Available online: https://bentogoncalves.atende.net/cidadao/pagina/perfil-da-cidade (accessed on 12 January 2021).
88. Departamento de Economia e Estatística, Secretaria de Planejamento, Governança e Gestão do Estado do Rio Grande do Sul. Índice de Desenvolvimento Socioeconômico Estado do Rio Grande do Sul (Revisão 2020). 2020. Available online: http://visualiza.dee.planejamento.rs.gov.br/idese/ (accessed on 4 January 2021).
89. Tribunal de Contas do Estado do Rio Grande do Sul. 2010–2018. Despesa Educação MDE. Dados Abertos TCE. Available online: http://dados.tce.rs.gov.br/dataset?q=%22Despesa+Educa%C3%A7%C3%A3o+MDE%22&tags=Consolidado&sort=score+desc%2C+metadata_modified+desc (accessed on 5 January 2021).
90. Observatório da Educação Pública no Rio Grande do Sul. Assembleia Legislativa do Rio Grande do Sul: Porto Alegre, Brazil. 2019. Available online: http://www.al.rs.gov.br/FileRepository/repdcp_m505/CECDCT/CECDCT%20-%20observatorio_educacao%202.pdf (accessed on 17 October 2020).
91. Folha de S. Paulo. Ranking Universitário Folha. 2019. Available online: https://ruf.folha.uol.com.br/2019/ (accessed on 18 November 2020).
92. UCSGRAPHENE. Available online: https://www.ucsgraphene.com.br/en/ (accessed on 19 January 2021).
93. COREDE Serra. Plano Estratégico Participativo de Desenvolvimento Regional do COREDE Serra: Diagnóstico Regional. Available online: https://xadmin.s3.us-east-2.amazonaws.com/58/PhotoAssets/61240/images/original/DIAGN%C3%93STICO%20REGIONAL%20COREDE%20SERRA.pdf (accessed on 25 November 2020).
94. Torres, R.B.; Caprara, B.S. Os papéis dos agentes econômicos, políticos e religiosos na evolução urbana de Bento Gonçalves. *Bol. Gaúcho Geogr.* **2010**, *37*, 85–101.
95. Bento+20. Masterplan Bento Gonçalves. Available online: http://bentomais20.com.br/masterplan (accessed on 18 January 2021).
96. Instituto Brasileiro de Geografia e Estatística. Censo Demográfico. 2010. Available online: https://censo2010.ibge.gov.br/ (accessed on 28 December 2020).
97. Instituto Brasileiro de Geografia e Estatística. Aglomerados Subnormais. 2010. Available online: https://www.ibge.gov.br/geociencias/organizacao-do-territorio/15788-aglomerados-subnormais.html?edicao=16119&t=acesso-ao-produto (accessed on 14 February 2021).

98. Instituto Brasileiro de Geografia e Estatística. Aglomerados Subnormais. 2019. Available online: https://www.ibge.gov.br/geociencias/organizacao-do-territorio/15788-aglomerados-subnormais.html?edicao=27720&t=acesso-ao-produto (accessed on 14 February 2021).
99. Prefeitura Municipal de Bento Gonçalves. City Master Plan (Municipal Complementary Law No. 200, 2018). Available online: http://ipurb.bentogoncalves.rs.gov.br/paginas/legislacao-ipurb (accessed on 28 October 2020).
100. COREDE Serra. Plano Estratégico Participativo de Desenvolvimento Regional do COREDE Serra: 2015–2030. Available online: https://xadmin.s3.us-east-2.amazonaws.com/58/PhotoAssets/60267/images/original/Plano%20de%20Desenvolvimento%20Regional%202015-2013.pdf (accessed on 25 November 2020).
101. Carrillo, F.J. Capital cities: A taxonomy of capital accounts for knowledge cities. *J. Knowl. Manag.* **2004**, *8*, 28–46. [CrossRef]
102. Fachinelli, A.C.; D'Arrigo, F.P.; Giacomello, C.P. Open data for sustainability performance assessment in Brazilian cities. *Aust. J. Basic Appl. Sci* **2015**, *9*, 32–38.
103. Fachinelli, A.C.; Giacomello, C.P.; Larentis, F.; D'Arrigo, F. Measuring the capital systems categories: The perspective of an integrated value system of social life as perceived by young citizens. *Int. J. Knowl. Based Dev.* **2017**, *8*, 334–345. [CrossRef]
104. Silva, M.B.C.D.; Bebber, S.; Fachinelli, A.C.; Moschen, S.D.A.; Perini, R.D.L. City life satisfaction: A measurement for smart and sustainable cities from the citizens' perspective. *Int. J. Knowl. Based Dev.* **2019**, *10*, 338–383. [CrossRef]
105. Millar, C.C.J.M.; Ju Choi, C. Development and knowledge resources: A conceptual analysis. *J. Knowl. Manag.* **2010**, *14*, 759–776. [CrossRef]
106. Reffat, R.M. Essentials for developing a prosperous knowledge city. In *Knowledge-Based Development for Cities and Societies: Integrated Multi-Level Approaches*; IGI Global: Hershey, PA, USA, 2010; pp. 118–129.
107. Eisenhardt, K.M. Building theories from case study research. *Acad. Manag. Rev.* **1989**, *14*, 532–550. [CrossRef]

 *sustainability*

Article

Sustainable Smart Cities and Industrial Ecosystem: Structural and Relational Changes of the Smart City Industries in Korea

Sung-Su Jo [1], Hoon Han [2], Yountaik Leem [1] and Sang-Ho Lee [1,*]

[1] Department of Urban Engineering, Hanbat National University, Daejeon 34158, Korea; josungsu85@daum.net (S.-S.J.); ytleem@hanbat.ac.kr (Y.L.)
[2] School of Built Environment, Faculty of Arts, Design & Architecture, University of New South Wales, Sydney, NSW 2052, Australia; h.han@unsw.edu.au
* Correspondence: Lshsw@hanbat.ac.kr

Abstract: This paper examines the changing industrial ecosystem of smart cities in Korea using both input–output and structural path analysis from 1960 to 2015. The industry type of the input–output tables used in the Bank of Korea was reclassified into nine categories: Agriculture and Mining, Traditional Manufacturing, IT Manufacturing, Construction, Energy, IT Services, Knowledge Services, Traditional Services and other unclassified. The paper identified the changing patterns of an industrial ecosystem of smart cities in Korea. The study found that smart industries such as smart buildings and smart vehicles are anchor industries in Korean smart cities, and they are positively correlated with three other industries: IT Manufacturing, IT Services and Knowledge Services. The results of the input–output and structural path analysis show that the conventional industrial structure of labor-intensive manufacturing and diesel and petroleum cars has been transformed to the emerging high-tech industries and services in smart cities. Smart industries such as IT Manufacturing, IT Services and Knowledge Services have led to sustainable national economic growth, with greater value-added than other industries. The underlying demand for smart industries in Korea is rapidly growing, suggesting that other industries will seek further informatization, automatization and smartification. Consequently, smart industries are emerging as anchor industries which create value chains of new industries, serving as accelerators or incubators, for the development of other industries.

Keywords: smart city; smart city industry; industrial ecosystem; input–output analysis; structural path analysis

Citation: Jo, S.-S.; Han, H.; Leem, Y.; Lee, S.-H. Sustainable Smart Cities and Industrial Ecosystem: Structural and Relational Changes of the Smart City Industries in Korea. *Sustainability* **2021**, *13*, 9917. https://doi.org/10.3390/su13179917

Academic Editor: João J. Ferreira

Received: 28 July 2021
Accepted: 31 August 2021
Published: 3 September 2021

1. Introduction

The technological innovation of the steam and internal combustion engines from the first and second industrial revolutions significantly influenced mass production, urbanization and economic agglomeration, and the third industrial revolution applied information and communication technologies (ICTs) to manufacturing, while also leading to the emergence of virtual space [1–3].

The current fourth industrial revolution with artificial intelligence (AI) and the Internet of Things (IoT) is leading the global economy and accelerating the convergence of business, industries and IT to create new business models, including a hyper-connected society [4]. In particular, smart cities are leading industrial innovation in the fourth industrial era, instigating a knowledge industries' boom.

Korea has developed ICT-driven smart cities to reinforce the national competitiveness and enhance industry value chains and production path chains through industrial ecosystems [5]. The development of smart cities has received public attention as a global city model to foster new value creation, technological innovation and sustainable development. Smart cities perform an increasingly important role in physical infrastructure management such as transport, security and safety, power supply, sewage treatment and water supply and management in cities [6–8]. Smart cities provide a new industrial paradigm based on the convergence of the built environment and ICTs [9].

Recently, smart cities have significantly affected changes in the industrial ecosystem with new forms of living and working environments such as smart homes, smart offices, smart mobility and living lab facilities in convergence with disruptive technologies and knowledge-based industries [10–12]. Furthermore, a smart city strategy focuses on urban sustainability in response to the recent fourth industrial revolution, climate change and economic recession [13–15]. Countries that did not respond to the needs of the industrial revolution will meet challenges to sustainable development [16].

Recently, researchers have tried to understand the industrial ecosystem within smart cities such as smart city industry ecosystems [5] and smart city governance/service/data ecosystems [17–19]. These studies investigated how to measure economic efficiency rather than empirical research focusing on the mechanism of the smart city industry. Despite many studies on smart cities, there is a lack of research on the evolving industrial ecosystem within smart cities.

Thus, this study aims to examine the structural changes in the industrial ecosystem in Korea's smart cities over 60 years. Moreover, this study performed the following procedures to confirm the sustainable smart city industry. First, we defined smart city industries based on an international literature review. Second, we quantitatively measured structural changes in smart city industries. Input–output analysis is used to quantify structural changes through technical coefficients and Leontief inverse coefficients. Finally, we analyzed the relational changes in the production path of smart industries and industrial convergence using a structural path analysis model.

2. Literature Review

2.1. Concept of Smart City

A smart city is a concept of the city of the future that applies ICTs to urban services, infrastructure and governance, providing a range of ubiquitous, affordable and smart services to enhance citizens' quality of life [14,20–22]. The smart city was first introduced as a concept of ubiquitous computing, which aims to create a built environment in which computers are embedded in physical objects so that users cannot recognize the computers, known as the Internet of Things, and yet at the same time can use the objects [23]. Smart city researchers have tried to translate the ubiquitous computing environment to the city level [8]. Spin-off research on the ubiquitous city has been conducted in Korea with studies on cutting-edge technologies or successful technology implications for smart cities and communities. Most researchers agreed that smart cities should aim to improve the quality of life by using ICTs and provide integrated urban services across various fields such as local economy, health, security, education, culture and society [24–27].

Albino et al. [28] summarized the smart city into four dimensions: a city's networked infrastructure that enables social and cultural development; an emphasis on business-led urban development and creative activities for the promotion of urban growth; social inclusion of various urban residents and social capital in urban development; and preservation of the natural environment as a strategic component for the future. Neirotti et al. [24] classified the smart city concept into hardware and software domains. Smart cities focused on hardware use sensors and wireless technology to collect, store and analyze big data. In contrast, smart cities focused on software include education, culture, innovation and administration as well as communication with citizen participation. This study defines a smart city as one in which ICTs and eco-technologies (EcoTs) such as sensors, devices, artifacts and algorithms are linked, integrated and embedded with the traditional, physical city. ICTs and EcoTs are technologies such as sensors, devices, artifacts and algorithms, and a traditional city indicates the physical city. In other words, smart cities work only when there is a convergence of physically traditional cities with IT-related manufacturing business such as sensors, devices and artifacts, as well as knowledge services such as knowledge algorithms [5].

A smart city considers not only the technological aspect but also the sustainable development aspect. The United Nation's Sustainable Development Goals (SDGs) also point out the direction of achieving sustainability in every aspect of life [29]. Smart cities provide economic, cultural and social environments for residents to improve their quality of life. The primary aim of the sustainable smart city is to provide a mechanism for fulfilling the requirements of the present as well as the future generation inhabitants [30]. Nevertheless, it is debatable whether a simple implementation of the smart city could lead to a sustainable city for an improved quality of life [31,32]. Some researchers point out the negative aspects of smart cities as follows: territorial colonization in the digital age [33], the widening of inequality in technology (digital divide) [34], smart city plans that focus on corporate interests rather than citizens [35] and the negative impact within cities of new technologies, networks and infrastructure [36].

As such, various discussions in terms of smart cities are being conducted. Among these various discussions, this study focuses on a sustainable smart city from technical, industrial and economic perspectives.

2.2. Smart City Industries

Previous literature was reviewed to define smart industries. There are no clear definitions and classifications for smart industries, and they are diversely classified depending on the research objectives and subjective views of researchers. Early studies on smart city industries include Cho et al. [37], Kim et al. [38] and Jeong [39].

Cho et al. [37] classified smart industries into 15 industries, including personal life (5), equipment (7) and public administration and services (3), to analyze how the ripple effects of adopting smart cities impact the Korean economy. Kim et al. [38] analyzed the ripple effects of Hwaseong and Dongtan smart cities in Korea on the regional economy through regional input–output tables. Smart city industries were sorted into 13 main categories of the input–output table, classified and defined as personal life (social and other services and 3 others), industry and economy (electrical and electronic equipment and 6 others) and public administration (electricity, gas and water supply and 1 other).

Jeong [39] focused on services in Asan cities in Korea to analyze the economic ripple effect of smart city development and identified four smart industries: electrical and electronic equipment; construction; communications and broadcasting; and others. Early studies on smart city industries defined them based on cities with smart services and defined construction and communication as key smart city industries.

As the smart city concept was further developed, studies defined smart city industries focusing on a new framework. Since 2010, smart city industries in Korea have been defined by focusing on operation and management as well as supply and demand, reflecting the views of experts.

Lim et al. [40] examined Seoul and divided smart industries into smart city infrastructure and utilization sectors to set the policy direction for smart cities. Industries were classified into eight main categories such as electrical and electronic equipment, construction, real estate and business services. Based on an expert survey to examine the characteristics of smart city industries, Lim et al. selected six industries such as construction and transport as smart city industries. Kim et al. [41] used a Delphi survey of experts to analyze how IoT sensors are related to smart city industries, deriving 30 subcategories of smart city industries.

Recent studies include Jo and Lee [5]. This study defined smart city industries as construction businesses based on IT manufacturing (precision instruments, electrical and electronic equipment), IT services (communications and broadcasting) and knowledge services (six fields such as finance and insurance, real estate and lease, professional, scientific and technical services). Unlike other studies, Jo and Lee focused on interpreting the relationship between smart city industries and other industries. IT manufacturing is hardware such as electrical and electronic equipment (sensors), and knowledge services are software such as specialized algorithms. IT services are defined as the communications and broadcasting industry, provided for people with the convergence and integration of IT manufacturing with knowledge services and construction.

Overall, studies on classifying smart industries have been conducted by consulting with smart city experts, meaning smart industries were classified differently depending on the smart city concept defined at the time of the research. Based on the main categories, smart industries were classified into IT manufacturing such as computers, electronic and optical devices and electrical equipment, IT services such as communications and broadcasting services, and knowledge services such as finance and insurance, professional, scientific and technical services, education, health and welfare and culture and sports.

Based on previous studies that defined the smart city concept and classified smart industries, this study investigated the cases related to any major smart industries, denoted by smart-X, to determine how the technologies and industries applied to actual smart cities are connected. This study selected three smart industries, smart cars (e.g., autonomous vehicle) [42] and buildings (e.g., zero-energy building) [43]. These smart-X industries anticipate an exponential growth in production by 2025 [44–46].

The following procedures were carried out to map the technologies and services of smart-X cases with the industries. First, elements that comprise smart-X cases such as services, technologies and infrastructures were identified. Second, the elements were mapped again based on the Harmonized Classification System of ICTs developed by the Telecommunications Technology Association in Korea. Third, the mapped industries were reclassified according to the Korea Standard Industrial Classification of Statistics Korea and were finally applied to the input–output tables from the Bank of Korea [47].

By the industry analysis of smart cars, buildings and factories, it is possible to determine the detailed structure of the industries that form smart-X. Based on the subcategories of the input–output tables, smart-X industries had 20 common industries classified into IT manufacturing, IT services and knowledge services (Table 1). Based on the definitions of smart cities in the previous studies, this study defines smart city industries as IT manufacturing such as electrical and electronic equipment, precision instruments, IT services such as communication services and knowledge services such as professional, scientific and technical services.

Table 1. Components of the smart-X industry.

Industries of Input–Output Table	Smart Car	Smart Building	Classification
Semiconductor Manufacturing	•	•	
Electronic Display Manufacturing	•	•	
Printed Circuit Board Manufacturing	•	•	
Other Electronic Components Manufacturing	•	•	
Computers and Peripherals Manufacturing	•	•	
Communications and Broadcasting Equipment Manufacturing	•	•	
Medical and Measuring Devices Manufacturing		•	IT Manufacturing
Generator and Motor Manufacturing	•	•	
Electrical Conversion and Supply Control Unit Manufacturing	•	•	
Battery Manufacturing	•	•	
Wire and Cable Manufacturing	•		
Other Precision Instruments Manufacturing	•		
Wired, Wireless and Satellite Communications Services	•	•	
Other Telecommunications Services	•	•	IT Services
Information Service	•	•	

Table 1. *Cont.*

Industries of Input–Output Table	Smart Car	Smart Building	Classification
Software Development Supply Services	•	•	IT Services
Other IT Services	•	•	
Research & Development	•	•	Knowledge Services
Building and Civil Engineering Services	•		
Scientific and Technical Services	•	•	
Other Professional Service	•	•	

2.3. Industrial Ecosystem Analysis

The literature on the industrial ecosystem was reviewed, focusing on studies using structural path analysis. Defourny and Thorbecke [48] used a structural path analysis model to analyze the path spread of economic activities (the industrial ecosystem structure) using Leontief inverse coefficients apart from the economic ripple effect that can be analyzed using input–output tables. This study used the Korean Social Accounting Matrix (SAM) data for 1968 and classified and analyzed three ecosystems: the effects of production activities on factor income, the effects of production activities on households with different social and economic characteristics and the effects of households with different social and economic characteristics on production activities.

Oh and Lee [49] used structural path analysis to analyze the ripple effect of the communications sector on other industries as well as the industrial ecosystem. The data used were the 1980–1985–1990 input–output tables of the Bank of Korea. A total of 405 industries in the input–output tables were reclassified into 7 industries: communications sectors such as communication devices, communication facilities and communication services and non-communications sectors such as agriculture, forestry and fisheries and manufacturing, construction and services. Up to four structural paths (ecosystems) of the industries were analyzed. The results indicated that the ripple effect of the communications sector was increasing. The ecosystem of the communications sector was not only expanding in terms of demand, but also showing a greater change due to increased output rather than structural change.

Basu and Johnson [50] used a structural path analysis model to analyze the intersectoral connectivity of Virginia, US, and created an index. The intersectoral industrial ecosystem was analyzed to test the hypothesis that "urban connectivity becomes more complex with economic development" [51]. Intersectoral connectivity was analyzed with a focus on the number of paths, the number of arcs that form the paths and path multipliers. The results revealed that the growing influence of an industry does not necessarily lead to an increase in its paths and arcs.

Lee and Leem [11] examined the effects of ICTs on other industries. The results of the structural path analysis model showed that knowledge-based cities are most related to ICTs and are creating new production paths and value chains. Using the 2000 and 2010 input–output tables as the research data, 28 main categories were reclassified into 5 industries, and up to 4 structural paths to estimate the ecosystem were presented. The results revealed that the demand for the ICT industry continued to grow and that the industry was an intermediary for other industries.

Jo and Lee [5] analyzed how the ecosystem of smart city industries in Korea was being merged. Smart city industries were defined as construction based on IT manufacturing, IT services and knowledge services. Using the 1980 and 2014 input–output tables as the research data, 403 industries in the basic sector were reclassified into 9 industries. This study analyzed the change in the ecosystem, focusing on the structural and convergent change of smart city industries. The results showed that smart city industries exhibited remarkable growth. The industry convergence was led by smart city industries, and the ecosystem was becoming stronger around these industries. However, in the industrial aspect, traditional industries such as agriculture, mining and manufacturing held a greater position than smart city industries, showing that smart cities were yet to emerge.

Industrial ecosystem studies analyzed the ecosystem by considering both the impact exchanged among industries and also micro aspects such as the industrial structure. Previous studies focused on production, technical coefficients and multipliers in the input–output analysis and on distinct and new paths in the structural path analysis. As well as the analysis of the ecosystem, our study also analyzed catchment coverage and inter-industry convergence.

3. Model and Data

3.1. Model

3.1.1. Input–Output Model for Structural Changes in Industrial Ecosystem

The analytical model used here is the input–output model and structural path analysis. Input–output analysis is a methodology presented by Leontief to analyze the flows of all goods and services in the economy [52]. The basic structure of input–output tables is in the form of a matrix, divided into endogenous and exogenous sectors.

Endogenous sectors represent transactions of goods and services, shown in intermediate demand and intermediate input. Exogenous sectors represent final demand and value-added. A column of an input–output matrix depicts the cost of purchasing the product from industry i to manufacture a single product in industry j. The purchase structure of industry j is divided into intermediate inputs that represent the purchase of raw materials and value-added, and the total is represented as total input.

A row of an input–output table shows the product in industry i sold to industry j. The sales structure is divided into intermediate and final demand. The total demand is the sum of intermediate and final demand, and the total output is income deducted from the total amount of demand (Figure 1).

		Intermediate Demand						Final Demand	Total Supply
		1	2	...	j	...	n		
	1	X_{11}	X_{12}	...	X_{1j}	...	X_{1n}	y_1	X_1
	2	X_{21}	X_{22}	...	X_{2j}	...	X_{2n}	y_2	X_2
Intermediate		$\vdots$	$\vdots$		$\vdots$		$\vdots$	$\vdots$	$\vdots$
Input	i	X_{i1}	X_{i2}	...	X_{ij}	...	X_{in}	y_i	X_i
		$\vdots$	$\vdots$		$\vdots$		$\vdots$	$\vdots$	$\vdots$
	n	X_{n1}	X_{n2}	...	X_{nj}	...	X_{nn}	y_n	X_n
Value Added		V_1	V_2	...	V_j	...	V_n		
Total Inputs		X_1	X_2	...	X_j	...	X_n		

Figure 1. Structure of the input–output table. Source: Bank of Korea (2014), Revision.

Input–output analysis uses input coefficients calculated from the input–output tables. A technical coefficient is the intermediate input in each sector divided by total input and can be depicted as shown in Equation (1). The technical coefficient a_{ij} represents the input of industry i necessary for industry j to produce a single unit. Equation (1) is expressed as follows;

$$a_{ij} = x_{ij}/x_j \tag{1}$$

where,

a_{ij} is technical coefficient of (i, j);
x_{ij} is intermediate input of j industry;
x_j is total input of j industry.

Production inducement coefficients can be calculated through input coefficients. A Leontief inverse coefficient represents the effect on industry i when the final demand of industry j increases by a unit. The production inducement coefficient is as shown in Equations (1) and (2):

$$b_{ij} = [I - A]^{-1} \tag{2}$$

where,

b_{ij} is Leontief inverse coefficient of (i, j);
I is identity matrix;
A is matrix of technical coefficient (a_{ij}).

3.1.2. Structural Path Analysis for Relational Changes in the Industrial Ecosystem

While inter-industry linkage is explained by simple production inducement coefficients in input–output analysis, structural path analysis can decompose multipliers of input–output tables in detail [53]. This analytical model can backtrack the entire process in which the multiplier effect is calculated and analyze the path of inter-industry linkages. The ripple effects of structural path analysis are classified into direct and indirect effects depending on whether there is a multiplier in each path and transfer path (Table 2).

Table 2. Calculation in structural path analysis.

Effects	Path Type	Calculation
Direct Effect		$I_{(i\text{-}x\text{-}y\text{-}j)} = a_{xi} \cdot a_{yx} \cdot a_{jy}$
Indirect Effect(red)		$I_{(x\text{-}y\text{-}z\text{-}x)} = [1 - a_{yx}\,(a_{xy} + a_{zy} \cdot a_{xz})]^{-1}$
Total Effect		$I_{(i\text{-}x\text{-}y\text{-}j)\,(x\text{-}y\text{-}z\text{-}x)} = (a_{xi} \cdot a_{yx} \cdot a_{jy}) \cdot [1 - a_{yx}\,(a_{xy} + a_{xz} \cdot a_{zz})]^{-1}$
Global Effect		$I_{(i\text{-}x\text{-}y\text{-}j)\,(x\text{-}y\text{-}z\text{-}x)\,(i\text{-}s\text{-}j)\,(i\text{-}v\text{-}j)} = (a_{xi} \cdot a_{yx} \cdot a_{jy}) \cdot [1 - a_{yx}\,(a_{xy} + a_{xz} \cdot a_{zz})]^{-1} + a_{si} \cdot a_{js} + a_{vi} \cdot a_{jv}\,(1 - a_{vv})^{-1}$

Source: Defourny and Thorbecke (1984), Revision.

Direct effects indicate the variance of industry j that changes along the basic path when industry i has changed by a unit and can be depicted as $I_{(i\text{-}x\text{-}y\text{-}j)} = a_{xi} \cdot a_{yx} \cdot a_{jy}$. Indirect effects represent the effects from being included in the direct path and going through the intermediate stage and the effects of feedback. A related equation can be depicted as $I_{(i\text{-}x\text{-}y\text{-}j)\,(x\text{-}y\text{-}z\text{-}x)} = (a_{xi} \cdot a_{yx} \cdot a_{jy}) \cdot [1 - a_{yx}\,(a_{xy} + a_{xz} \cdot a_{zz})]^{-1}$. Total effects not only indicate the basic path from industry i to industry j, but also include indirect effects within the path and can be calculated by $I_{(i\text{-}x\text{-}y\text{-}j)\,(x\text{-}y\text{-}z\text{-}x)} = (a_{xi} \cdot a_{yx} \cdot a_{jy}) \cdot [1 - a_{yx}\,(a_{xy} + a_{xz} \cdot a_{zz})]^{-1}$. $I_{(i\text{-}x\text{-}y\text{-}j)\,(x\text{-}y\text{-}z\text{-}x)\,(i\text{-}s\text{-}j)\,(i\text{-}v\text{-}j)} = (a_{xi} \cdot a_{yx} \cdot a_{jy}) \cdot [1 - a_{yx}\,(a_{xy} + a_{xz} \cdot a_{zz})]^{-1} + a_{si} \cdot a_{js} + a_{vi} \cdot a_{jv} (1 - a_{vv})^{-1}$ represent global effects, which refer to all effects spread out to industry j when the demand of industry i has changed by a unit.

3.2. Data

This study used input–output tables, based on current market prices in the relevant year, for 1960, 1975, 1995 and 2015 from the Bank of Korea [54]. To compare the 4-year input–output tables, this study eliminated the nominal increment by inflation by applying the GDP deflator using the index with 2015 as the base year provided by the Bank of Korea (Table 3). To analyze the ecosystem of smart city industries with the 4-year input–output tables applying the GDP deflator, sub-subcategories that are the minimum unit of each industry were reclassified into nine industries: Agriculture and Mining (AM), Traditional Manufacturing (TM), IT Manufacturing (ITM), Construction (C), Energy Supply (E), IT Services (ITS), Knowledge Service (KS), Traditional Services (TS) and other unclassified (ETC).

Table 3. Classification of smart city industries.

1960 Year Industries No	1975 Year Industries No	1995 Year Industries No	2015 Year Industries No	9 Industries
1–7	1–16	1–11	1–8	Agriculture and Mining (AM)
8–11	17–27	12–18	9–12	
12–43	28–69	19–56	13–35	Traditional Manufacturing (TM)
44–55	70–93	57–75	36–51	
56–68	94–110	76–93	52–67	
69–74	111–116	94–101	83–93	
78–81	124–127	115–121	94–100	
76, 82–83	117, 120, 128–130	107, 110–111, 113–114, 122–124	74, 76, 81–82, 101–103	
75, 77	118–119, 121–123	102–106, 108–109, 112	68–73, 75, 77–80	IT Manufacturing (ITM)
90, 93	136–138, 157	125–128, 161	104–110	Energy Supply (E)
85–89	131–135	129–133	111–117	Construction (C)
94	149	145	131–132, 134–136	IT Services (ITS)
-	-	146	137–138	Knowledge Services (KS)
91	150–151	147–149	139–142	
-	-	153–154	146–150	
99	155	156	157	
100	156	159–160	158–159	
104	160	162	160–161	
-	-	134	118	Traditional Services (TS)
98	139–141	135–136	129–130	
95–97	142–148	137–144	119–128	
92	152–153	150	143–145	
102–103	159	152	151–154	
101	154	155	155–156	
105	158, 161	163–165	162–164	
106–109	162–164	166–168	165	other unclassified (ETC)

Source: The Bank of Korea (1960, 1975, 1995, 2015).

Agriculture and Mining (AM) was reclassified around primary industries. Traditional Manufacturing (TM) was classified around secondary industries. IT Manufacturing was separated from Traditional Manufacturing based on the key indicators of previous studies and smart-x cases. IT Manufacturing had 11 sub-subcategories, such as semiconductors, electronic display devices, circuit boards, electronic components, computers and communications equipment.

Construction (C) was classified around building construction and civil engineering, and energy supply (E) included electricity, gas, water supply and renewable energy. IT

Services included wired and wireless communications, information services and software. Knowledge Services were classified based on previous studies and included finance and insurance, professional, scientific and technical services, education, health and culture [55,56]. Knowledge Services is an industry that includes IT Services, but these were separated to specifically examine the ecosystem of smart city industries. The industry titled other included businesses that are not clearly classified such as residuals and office supplies, but this study did not include others in the interpretation of the ecosystem of smart city industries. Traditional Service (TS) was classified into food and accommodation, real estate, business support and other services. Other unclassified industries (ETC) included businesses that are not clearly classified such as residuals and office supplies. This study did not include ETC in interpretation related to the ecosystem of smart city industries.

4. Results

4.1. Quantitative Ecosystem of Smart City Industries

4.1.1. Changes in the Industry Spectrum of Smart City

The industry spectrum refers to the relative proportion that a particular industry occupied in the entire industry at a given time. The spectrum analysis of smart city industries reveals the growth in the number of smart city industries among the total number of industries from 1960 to 2015 (see Figure 2 and Table 4). The spectrum coverages of smart city industries are growing, increasing from 112 in 1960 to 165 in 2015, while those of traditional industries are declining. Agriculture and Mining and Traditional Manufacturing showed a simultaneous decline in both catchment coverages and shares starting from 1975.

The three smart city industries showed a constant increase from 1960 to 2015 as their shares in all industries increased by 4.9% for IT Manufacturing, 2.1% for IT Services and 4.9% for Knowledge Services, meaning the spectrum coverages of the smart city industries are becoming broader. In particular, IT Manufacturing and Knowledge Services showed a rapid increase in spectrum coverages compared to other smart city industries.

The industry spectrum analysis showed that smart city industries are being increasingly subdivided. In 2015, IT Services broadened its catchment coverages. However, Traditional Manufacturing still has many spectrum coverages, which implies that the Korean ecosystem of smart city industries is still in its initial stage.

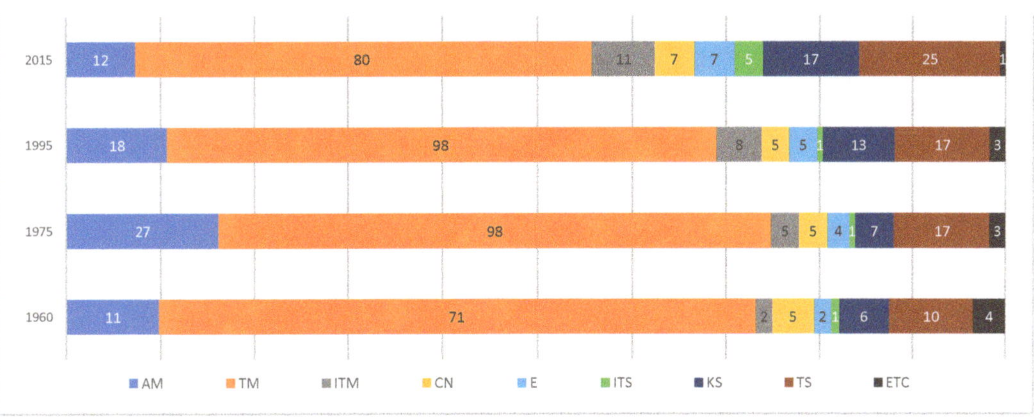

Figure 2. Result of industrial spectrum changes (unit: the number of industries).

Table 4. Result of industrial spectrum changes (unit: the number of industries, %).

Industries	1960	1975	1995	2015
AM	11 (9.8)	27 (16.2)	18 (10.7)	12 (7.3)
TM	71 (63.4)	98 (58.7)	98 (58.3)	80 (48.5)
ITM	2 (1.8)	5 (3.0)	8 (4.8)	11 (6.7)
CN	5 (4.5)	5 (3.0)	5 (3.0)	7 (4.2)
E	2 (1.8)	4 (2.4)	5 (3.0)	7 (4.2)
ITS	1 (0.9)	1 (0.6)	1 (0.6)	5 (3.0)
KS	6 (5.4)	7 (4.2)	13 (7.7)	17 (10.3)
TS	10 (8.9)	17 (10.2)	17 (10.1)	25 (15.2)
ETC	4 (3.6)	3 (1.8)	3 (1.8)	1 (0.6)
Sum.	112 (100)	167 (100)	168 (100)	165 (100)

4.1.2. Production Changes

Changes in industrial production are analyzed to examine the changes in the sum of intermediate inputs, and the results are reported in Table 5. All industrial production constantly increased from 1960 to 2015. In 1960–1975, Agriculture and Mining and Traditional Manufacturing had more production than the total average. However, Agriculture and Mining were replaced by Traditional Services in 1995. Increasing urbanization weakened agriculture and fisheries while increasing transport and real estate. The production ratios of Agriculture and Mining (−16.8%) and Construction (−1.6%) decreased from 1960 to 2015, whereas those of other industries increased. In particular, IT Manufacturing and Knowledge Services showed a remarkable increase. The average rate of increase in the amount of production was greatest for IT Manufacturing (241,311%), followed by Knowledge Services (93,082%), Traditional Manufacturing (51,829%) and IT Services (42,854%).

Table 5. Result of production changes (unit: million Korean won, %).

Industries	1960 Year	1975 Year	1995 Year	2015 Year	Avg. Growth Rate (%)
AM	268 (24.6)	188,109 (18.4)	24,320,968 (8.2)	170,481,650 (7.8)	27,840
TM	441 (40.5)	566,736 (55.3)	150,823,687 (50.7)	998,062,109 (45.4)	51,829
ITM	5 (0.5)	32,936 (3.2)	21,299,216 (7.2)	179,571,645 (8.2)	241,311
CN	22 (2.1)	7679 (0.7)	4,424,448 (1.5)	11,441,286 (0.5)	30,827
E	30 (2.8)	23,183 (2.3%)	7,917,369 (2.7)	89,307,901 (4.1)	37,419
ITS	10 (0.9)	7321 (0.7)	3,970,270 (1.3)	56,467,957 (2.6)	42,854
KS	17 (1.6)	25,492 (2.5)	32,842,130 (11.0)	249,525,295 (11.4)	93,082
TS	197 (18.1)	133,394 (13.0)	39,973,548 (13.4)	437,109,558 (19.9)	32,824
ETC	98 (9.0)	39,782 (3.9)	11,988,052 (4.0)	4,144,011 (0.2)	23,487
Total	1088 (100)	1,024,631 (100)	297,559,688 (100)	2,196,111,412 (100)	-
Avg.	121	113,848	33,062,188	244,012,379	-

4.2. Qualitative Ecosystem of Smart City Industries

4.2.1. Changes in Technical Coefficient

Structural changes in the input–output coefficients were analyzed by the sum of the input–output coefficients of intermediate input in each year ($\sum_{j=1}^{9} a_{ij}$). The input–output coefficient refers to the amount of goods and services that must be input by each industry when all industries produce a single-unit product. The results not only reveal that smart city industries are growing continuously but also how much value-added is produced at the same time.

Table 6 shows that Traditional Manufacturing had a remarkably higher input than other industries. Although its ratio is decreasing, it is still a key industry in Korea. The smart city industries of IT Manufacturing (146%), IT Services (114%) and Knowledge Services (171%) had more growth, showing that smart city industries are important for other industries.

Table 6. Result of changes in technical coefficient (unit: %).

Industries	1960 Year	1975 Year	1995 Year	2015 Year	Avg. Growth Rate
AM	0.5261 (13.3)	0.4002 (8.5)	0.3264 (7.0)	0.4477 (8.4)	−2
TM	1.5184 (38.5)	2.3823 (50.7)	1.9877 (42.9)	1.7936 (33.7)	10
ITM	0.0943 (2.4)	0.5040 (10.7)	0.4534 (9.8)	0.5195 (9.8)	146
CN	0.0704 (1.8)	0.0419 (0.9)	0.0813 (1.8)	0.0271 (0.5)	−4
E	0.1087 (2.8)	0.0899 (1.9)	0.1916 (4.1)	0.2699 (5.1)	46
ITS	0.0220 (0.6)	0.0326 (0.7)	0.0776 (1.7)	0.1994 (3.7)	114
KS	0.0472 (1.2)	0.1269 (2.7)	0.5681 (12.3)	0.5476 (10.3)	171
TS	0.9726 (24.7)	0.8719 (18.6)	0.7687 (16.6)	1.4858 (27.9)	24
ETC	0.5853 (14.8)	0.2472 (5.3)	0.1767 (3.8)	0.0315 (0.6)	−56
Total	3.9452 (100.0)	4.6970 (100.0)	4.6315 (100.0)	5.3221 (100.0)	-

Other industries are in constant need of smart city industries, and the value-added of these smart city industries is increasing. From the perspective of input (considering the ratio of all), the three smart city industries were growing; however, they still had a lower ratio than traditional industries of Agriculture and Mining, Traditional Manufacturing and Traditional Services. Energy showed a constant increase in input to other related industries such as smart energy, smart grid and smart energy monitoring.

4.2.2. Changes in Leontief Inverse Coefficient

Changes in the production inducement coefficients were analyzed by the sum of the production inducement coefficients of intermediate demand by year ($\sum_{i=1}^{9} b_{ij}$) and the results are reported in Table 7. The results represent the direct and indirect effects on all industries when each industry increases by one production unit. The industries with the largest ripple effect were Traditional Manufacturing, IT Manufacturing, Construction and

Energy. Manufacturing industries showed higher production inducement coefficients than service industries because they have more input of domestically produced raw materials.

Table 7. Changes in Leontief inverse coefficient (unit: %).

Industries	1960 Year	1975 Year	1995 Year	2015 Year	Avg. Growth Rate	Avg. of b_{ij}
AM	1.2971 (8.0)	1.5872 (7.8)	1.8057 (9.1)	2.1990 (10.0)	19	1.7223
TM	2.1481 (13.2)	2.7158 (13.3)	2.7067 (13.7%)	2.9295 (13.4)	11	2.6250
ITM	2.5753 (15.8)	2.9985 (14.7)	2.5779 (13.0)	2.7922 (12.7)	4	2.7360
CN	2.2264 (13.7)	2.6165 (12.8)	2.4182 (12.2)	2.5255 (11.5)	5	2.4466
E	1.7965 (11.0)	2.5853 (12.7)	2.0790 (10.5)	2.4100 (11.0)	13	2.2177
ITS	1.3172 (8.1)	1.4941 (7.3)	1.3949 (7.0)	1.8897 (8.6)	14	1.5240
KS	1.3587 (8.4)	1.6591 (8.1)	1.7110 (8.6)	1.9938 (9.1)	14	1.6806
TS	1.4372 (8.8)	1.6241 (8.0)	1.8246 (9.2)	1.9892 (9.1)	11	1.7188
ETC	2.1070 (13.0)	3.1260 (15.3)	3.2772 (16.6)	3.2045 (14.6)	17	2.9287
Total	16.2635 (100)	20.4067 (100)	19.7952 (100)	21.9334 (100)	-	-

IT Manufacturing including sensors, computers and networks can maximize the value-added in convergence with other industries (e.g., smart building), thereby showing high production inducement coefficients. The smart city industries of IT Services (14%) and Knowledge Services (14%) had a growing influence on all industries. The multiplier effect of 30.4% of smart city industries in 2015 was one percentage point more than 29.7% in 1960, representing approximately one-third of all industries.

4.3. Ecosystem of Smart City Industries

4.3.1. Relational changes in Industrial Path

Input–output analysis can be used to calculate the effect of the final demand change on each industry. However, input–output analysis has limitations in detailed correlation analysis. Structural path analysis can decompose and analyze the multiplier effect in detail [52]. The increase in the number of structural paths is the increase in the production inducement coefficient, which indicates that an industry's demand is growing and its scope of influence is expanding.

This study shows that the number of paths decreased from 141 in 1960 to 117 in 2015 because the ripple effect of direct paths became greater than that of indirect paths. Inter-industry linkage showed a greater ripple effect in simplified paths than complicated paths [55]. The number of paths through which IT Manufacturing affects other industries decreased in 2015 compared to 1960. There was a decrease in direct paths, paths that go through one industry and paths that go through three industries.

This also implies that other industries in need of IT Manufacturing also decreased. IT Services showed an increase in the number of paths through which it affects other industries from 1960 to 2015. The path type with the largest increase was "via two paths" (from 13 in 1960 to 27 in 2015), which suggests that the linkage between IT Services and other industries is being reinforced.

This result shows that IT Services play an accelerator role for other industries. In particular, the paths from IT Services to other industries and Knowledge Services to other industries have increased via three paths from 1995 to 2015 (Table 8), which means that the value chain of new industries increased. Knowledge Services showed an increase of 18 paths in 2015 compared to 1960. The number increased in all path types, which proves that the increasing demand of Knowledge Services results in distribution to other industries. The extinct paths were identified most in "via two paths". This means that the link of industrial paths in "via two paths" is weak. Therefore, the smart city industry influences other industries in a greater scale in "direct paths" and "via one path" than in "via two paths".

Table 8. Relational changes in industrial path (unit: no).

		1960 Path (NP, EP)	1975 Path (NP, EP)	1995 Path (NP, EP)	2015 Path (NP, EP)
	Direct path	8 (-, -)	8 (0, 0)	8 (0, 0)	8 (0, 0)
	via 1 path	25 (-, -)	22 (5, 8)	26 (6, 2)	23 (3, 6)
IT Manufacturing → Other Industry	via 2 paths	28 (-, -)	22 (6, 12)	29 (14, 7)	28 (11, 12)
	via 3 paths	3 (-, -)	1 (1, 3)	2 (2, 1)	0 (0, 2)
	Total	64 (-, -)	53 (12, 23)	65 (22, 10)	59 (14, 20)
	Direct path	7 (-, -)	7 (1, 1)	7 (0, 0)	7 (0, 0)
	via 1 path	17 (-, -)	26 (10, 1)	26 (8, 8)	26 (5, 5)
IT Services → Other Industry	via 2 paths	13 (-, -)	19 (10, 3)	18 (9, 10)	27 (20, 11)
	via 3 paths	1 (-, -)	1 (0, 0)	0 (0, 1)	2 (2, 0)
	Total	38(-, -)	53(21, 5)	51 (17, 19)	62 (27, 16)
	Direct path	8 (-, -)	8 (0, 0)	8 (0, 0)	8 (0, 0)
	via 1 path	20 (-, -)	27 (9, 2)	22 (4, 7)	25 (9, 6)
Knowledge Services → Other Industry	via 2 paths	10 (-, -)	20 (11, 1)	15 (3, 8)	22 (16, 9)
	via 3 paths	1 (-, -)	1 (0, 0)	0 (0, 1)	2 (2, 0)
	Total	39 (-, -)	56 (20, 3)	45 (6, 17)	58 (27, 14)

Note: NP (New path), EP (Extinct path), Direct path (Industry 1 → Industry 2), via 1 path (Industry 1 → Industry 3 → Industry 2), via 2 paths (Industry 1 → Industry 3 → Industry 4 → Industry 2), via 3 paths (Industry 1 → Industry 3 → Industry 4 → Industry 5 → Industry 2).

4.3.2. Changes in Industrial Convergence: Multiplier Effects

Table 9 presents how smart city industries converge in the structural paths of all industries, directly showing how many industries and value chains are created through smart city industries. For instance, the convergence of IT Services and Knowledge Industry could lead to new IT manufacturing industries such as smart factories. Analyzing structural paths in all industries shows there were 83 single paths in 1960, growing to 277 single paths in 2015 involving IT Manufacturing, IT Services and Knowledge Services. Knowledge Services showed more convergence with other industries than IT Manufacturing and IT Services.

However, in 2015, IT Services showed a rapid change in convergence, demonstrating that the industry is becoming an important element in convergence with other industries. IT Manufacturing was constantly increasing. Based on the single path average, the convergence ratio was highest in Knowledge Services (56.9%), followed by IT Manufacturing (31.6%) and IT Services (11.5%), and the year-on-year increase was highest in Knowledge Services, followed by IT Services and IT Manufacturing.

Table 9. Changes in convergence (unit: the number of industries, %).

Path Type		1960 Year	1975 Year (Ratio, Rate of Increase)	1995 Year (Ratio, Rate of Increase)	2015 Year (Ratio, Rate of Increase)	Avg. (%)
Path type of Including Single Industry	ITM	53	61 (37.7, 15.1)	64 (27.1, 4.9)	83 (30.0, 29.7)	65 (31.6)
	ITS	11	20 (12.3, 81.8)	20 (8.5, 0.0)	38 (13.7, 90.0)	22 (11.5)
	KS	19	81 (50.0, 326.3)	152 (64.4, 87.7)	156 (56.3, 2.6)	102 (56.9)
	Total	83	162 (100)	236 (100)	277 (100)	189 (100)
Path type of Including Double Industry	ITS, ITM	0	2	0	4	6
	ITM, KS	0	0	11	16	27
	ITS, KS	0	0	0	11	11
	Total	0	2	11	31	44
Path type of Including Triple Industry	ITM, ITS, KS	0	0	0	0	0
	Total	0	0	0	0	0

Analyzing double paths showed 27 paths with IT Manufacturing and Knowledge Services, 11 with IT Services and Knowledge Services and 6 with IT Services and IT Manufacturing. This result implies that IT Manufacturing and Knowledge Services are creating more value chains than other double paths. Paths with IT Services and Knowledge Services first appeared in 2015, showing that the two industries are becoming converged into new industries. Triple paths are not yet shown, indicating that the convergence of smart city industries is still at an early stage.

5. Conclusions and Discussion

This study analyzed the innovation ecosystem of smart city industries in Korea using input–output models and structural path analysis on data from 1960, 1975, 1995 and 2015 input–output tables and applying Korea's GDP deflator. The industries were classified into nine industries through minimum units of input–output tables by year: Agriculture and Mining, Traditional Manufacturing, IT Manufacturing, Energy, Construction, IT Services, Knowledge Services, Traditional Services and Other unclassified.

The spectrum of primary and secondary industries such as Agriculture and Mining and Traditional Manufacturing decreased over time, whereas those of smart city industries such as IT Manufacturing, IT Services and Knowledge Services are relatively increasing. This indicates that the external industrial structure is changing toward smart city industries. The smart city ecosystem analyzed with a focus on production showed an explosive quantity growth in IT Manufacturing, which showed approximately 10 times higher growth than the industry showing the lowest growth, and 2.5 times higher growth than the industry showing the second-highest growth. Typical examples of IT Manufacturing as a smart city element are semiconductors, computers, internet network and sensors. The rapid growth of IT Manufacturing as a smart element industry has great implications for the entire economic structure, indicating that the value-added is greater than other industries.

Growth rates for industries in which goods and services are input in terms of technical coefficients were relatively higher in smart city industries such as IT Manufacturing, IT Services and Knowledge Services than others. This indicates that the demand for smart city industries is rapidly increasing, which means that other industries took informatization and smartification with a focus on IT Manufacturing, IT Services and Knowledge Services industries and that smart city industries are replacing others. By the analysis of production inducement coefficients, manufacturing industries such as IT Manufacturing were found to have a greater ripple effect than service industries such as IT Services and Knowledge Services. The average increase rate of production inducement coefficients was highest in IT Services and Knowledge Services, which indicates that the potential for the ripple effect in these two industries was growing, and they create greater value-added than other industries.

The number of paths in structural path analysis indicates the complexity and connectivity of the entire ecosystem. The path analysis found the structure of the smart city industry ecosystem had complexity and connectivity and was evolving. Unlike analysis of production, technical coefficients and production inducement coefficients, the structural paths from IT Manufacturing to other industries decreased due to the stronger direct paths of IT Manufacturing, indicating that direct transactions between IT Manufacturing and other industries are increasing. IT Services and Knowledge Services showed an increase in all structural paths, indicating that they are emerging as key industries that create value chains of new industries and are serving as accelerators for the development of other industries.

Smart city industries are at the center of industry convergence and reinforce transactions among other industries. The number of paths needing the intermediary role of smart city industries is increasing. Paths including two or more of the three smart industries are also increasing in a highly significant result. In particular, paths with IT Services and IT Manufacturing or paths with IT Services and Knowledge Services, which had not existed before 1995, first appeared in 2015, consistent with industry convergence. IT Services creates value chains of new industries. Overall, the ecosystem of smart city industries showed convergence and evolution, creating value chains of new industries. IT Manufacturing, IT Services and Knowledge Services are growing as important industries. The qualitative, quantitative and convergence path analysis results presented in this study show sustained growth in the smart city industry.

This paper makes several contributions. First, we have defined smart city industries based on actual smart cities such as smart cars and smart buildings. Our study classified industries with a focus on smart-X cases and technologies directly related to the fourth industrial revolution. Second, we provide a rigorous set of evidence about the link between smart cities and the industrial ecosystem and found that smart city industries of IT Manufacturing, IT Services and Knowledge Services have transformed Korean smart cities over a period of 60 years. Smart city industries in Korea have sustained growth through the interaction of these three industries. In addition, the results of this study further suggest policy implications and industrial development direction for smart cities as a new growth engine of the country in the future.

There is no international standard for smart industries. Existing studies arbitrarily classified the smart city industries based on their industrial code. This study actively reviewed the classification of existing studies and classified the smart city industry through smart-X cases such as smart cars and smart buildings. Nevertheless, this study has limitations in the classification of the smart city industry. Furthermore, the input–output model and structural path analysis methods that include specific characteristics of the input–output table have limitations. There are restrictions in the particular analysis of the rapid changes in economic conditions because of the analysis on the macro aspects of the smart city industry.

For a future study, it is necessary to analyze the industrial ecosystem of the smart city by reclassifying the nine industries into sub-categories and to analyze its convergence impact in a consistent manner. It is necessary to examine how much smartization each industry (e.g., agriculture, manufacturing, automobile and finance, etc.) are progressing through the smart industry.

Author Contributions: Conceptualization, S.-H.L. and S.-S.J.; methodology, S.-H.L. and S.-S.J.; data curation, S.-S.J.; formal analysis, S.-S.J.; writing—original draft, S.-S.J.; writing—review and editing, H.H., S.-H.L. and Y.L.; supervision, S.-H.L.; validation, S.-H.L., H.H. and Y.L. All authors have read and agreed to the published version of the manuscript.

Funding: This work was supported by the National Research Foundation of Korea grant funded by the Korea government (Ministry of Science and ICT) (No. 2019R1F1A1062708 and 2021R1F1A1049301).

Institutional Review Board Statement: Not applicable.

Informed Consent Statement: Not applicable.

Data Availability Statement: Not applicable.

Conflicts of Interest: The authors declare no conflict of interest.

References

1. Lucas, R.E. *Lectures on Economic Growth*; Harvard University Press: Cambridge, MA, USA, 2002.
2. Mokyr, J. The second industrial revolution, 1870–1914. In *Storia dell'Economia Mondiale*; Citeseer: Princeton, NJ, USA, 1998; pp. 219–245. Available online: http://citeseerx.ist.psu.edu/viewdoc/download?doi=10.1.1.481.2996&rep=rep1&type=pdf (accessed on 19 August 2021).
3. Rifkin, J. *The Third Industrial Revolution: How Lateral Power is Transforming Energy, the Economy, and the World*; Macmillan: New York, NY, USA, 2011.
4. Park, S.J.; Kim, B.W. 4 th Industrial Revolution and Open Access Network for Smart City. In Proceedings of the 2018 Portland International Conference on Management of Engineering and Technology (PICMET), Honolulu, HI, USA, 19–23 August 2018; pp. 1–10.
5. Jo, S.S.; Lee, S.H. An Analysis on the Change of Convergence in Smart City from Industrial Perspectives. *J. Korean Reg. Sci. Assoc.* **2018**, *34*, 61–74.
6. Lee, S.H.; Yigitcanlar, T.; Han, J.H.; Leem, Y.T. Ubiquitous urban infrastructure: Infrastructure planning and development in Korea. *Innovation* **2008**, *10*, 282–292. [CrossRef]
7. Yigitcanlar, T.; Han, H.; Kamruzzaman, M.; Ioppolo, G.; Sabatini-Marques, J. The making of smart cities: Are Songdo, Masdar, Amsterdam, San Francisco and Brisbane the best we could build? *Land Use Policy* **2019**, *88*, 104187. [CrossRef]
8. Leem, Y.; Han, H.; Lee, S.H. Sejong Smart City: On the Road to Be a City of the Future. In *International Conference on Computers in Urban Planning and Urban Management*; Springer: Cham, Germany, 2019; pp. 17–33.
9. Lee, S.H.; Han, J.H.; Leem, Y.T.; Yigitcanlar, T. Towards ubiquitous city: Concept, planning, and experiences in the Republic of Korea. In *Knowledge-Based Urban Development: Planning and Applications in the Information Era*; Igi Global: Hershey, PA, USA, 2008; pp. 148–170.
10. Lee, S.H.; Leem, Y. Identification of Knowledge Driven Production Path through ICTs Industry as a Tool of Knowledge Sharing and Knowledge Management in Knowledge City. *J. Korean Urban Manag. Assoc.* **2015**, *28*, 409–434.
11. Jo, S.S.; Baek, H.J.; Han, H.; Lee, S.H. An Analysis on the Expert Opinions of Future City Scenarios. *J. Korean Reg. Sci. Assoc.* **2019**, *35*, 59–76.
12. Hawken, S.; Hoon Han, J. Innovation districts and urban heterogeneity: 3D mapping of industry mix in downtown Sydney. *J. Urban Des.* **2017**, *22*, 568–590. [CrossRef]
13. Cretu, L.G. Smart cities design using event-driven paradigm and semantic web. *Inform. Econ.* **2012**, *16*, 57.
14. Guan, L. Smart steps too a better city. *Gov. News* **2012**, *32*, 24–27.
15. Jo, S.S.; Lee, S.H.; Leem, Y. Temporal Changes in Air Quality According to Land-Use Using Real Time Big Data from Smart Sensors in Korea. *Sensors* **2020**, *20*, 6374. [CrossRef]
16. Pomeranz, K. *The Great Divergence*; Princeton University Press: Princeton, NJ, USA, 2021.
17. Gupta, A.; Panagiotopoulos, P.; Bowen, F. An orchestration approach to smart city data ecosystems. *Technol. Forecast. Soc. Chang.* **2020**, *153*, 119929. [CrossRef]
18. Rotună, C.; Gheorghiță, A.; Zamfiroiu, A.; Anagrama, D.S. Smart City Ecosystem Using Blockchain Technology. *Inform. Econ.* **2019**, *23*, 41–50. [CrossRef]
19. Pellicano, M.; Calabrese, M.; Loia, F.; Maione, G. Value co-creation practices in smart city ecosystem. *J. Serv. Sci. Manag.* **2018**, *12*, 34–57. [CrossRef]
20. Bakıcı, T.; Almirall, E.; Wareham, J. A smart city initiative: The case of Barcelona. *J. Knowl. Econ.* **2013**, *4*, 135–148. [CrossRef]
21. Caragliu, A.; Del Bo, C.; Nijkamp, P. Smart cities in Europe. *J. Urban Technol.* **2011**, *18*, 65–82. [CrossRef]
22. Chen, T.M. Smart grids, smart cities need better networks [Editor's Note]. *IEEE Netw.* **2010**, *24*, 2–3. [CrossRef]
23. Weiser, M. Some computer science issues in ubiquitous computing. *Commun. ACM* **1993**, *36*, 75–84. [CrossRef]
24. Eger, J.M. Smart growth, smart cities, and the crisis at the pump a worldwide phenomenon. *I-WAYS-J. E-Gov. Policy Regul.* **2009**, *32*, 47–53. [CrossRef]
25. Chia, J.; Lee, J.B.; Han, H. How Does the Location of Transfer Affect Travellers and Their Choice of Travel Mode?—A Smart Spatial Analysis Approach. *Sensors* **2020**, *20*, 4418. [CrossRef] [PubMed]
26. Han, H.; Lee, S.H.; Leem, Y. Modelling Interaction Decisions in Smart Cities: Why Do We Interact with Smart Media Displays? *Energies* **2019**, *12*, 2840. [CrossRef]
27. Han, J.H.; Hawken, S.; Williams, A. Smart CCTV and the management of urban space. In *Smart Technologies: Breakthroughs in Research and Practice*; IGI Global: Hershey, PA, USA, 2018; pp. 508–526.

28. Albino, V.; Berardi, U.; Dangelico, R.M. Smart cities: Definitions, dimensions, performance, and initiatives. *J. Urban Technol.* **2015**, *22*, 3–21. [CrossRef]
29. Pedersen, C.S. The UN sustainable development goals (SDGs) are a great gift to business! *Procedia Cirp* **2018**, *69*, 21–24. [CrossRef]
30. Ahad, M.A.; Paiva, S.; Tripathi, G.; Feroz, N. Enabling technologies and sustainable smart cities. *Sustain. Cities Soc.* **2020**, *61*, 102301. [CrossRef]
31. Höjer, M.; Wangel, J. Smart Sustainable Cities: Definition and Challenges. In *ICT Innovations for Sustainability*; Advances in Intelligent Systems and Computing; Hilty, L., Aebischer, B., Eds.; Springer: Cham, Germany, 2015; Volume 310.
32. Hollands, R.G. Will the real smart city please stand up? Intelligent, progressive or entrepreneurial? *City* **2008**, *12*, 303–320. [CrossRef]
33. Datta, A.; Odendaal, N. Smart cities and the banality of power. *Environ. Plan. D Soc. Space* **2019**, *37*, 387–392. [CrossRef]
34. Woyke, E. Smart cities could be lousy to live in if you have a disability. *MIT Technol. Rev.* 9 January 2019.
35. Greenfield, A. *Against the Smart City: A Pamphlet. This is Part I of "The City is Here to Use"*; Do Projects: New York, NY, USA, 2013.
36. Sennett, R. No one likes a city that's too smart. *Guardian*, 4 December 2012.
37. Jeong, S.Y. Economic Impact Analysis on the u-City Development. Master's Thesis, University of Seoul, Seoul, Korea, 2008.
38. Cho, B.S.; Jeong, W.S.; Kim, P.R. An Analysis on the Economic Effects for launching the ubiquitous City. In Proceedings of the Korea Technology Innovation Society Conference, Portland, OR, USA, 5–9 August 2006; pp. 273–286.
39. Kim, P.R.; Cho, B.S.; Jeong, W.S. The propagation effects on the regional economy induced by u-City construction in Wha-sung and Dong-tan city. *J. Korean Inst. Commun. Inf. Sci.* **2006**, *31*, 1087–1098.
40. Lim, S.Y.; Lim, Y.M.; Hwang, B.J.; Lee, J.Y. A study on the characteristics of the U-CITY industry using the IO tables. *Spat. Inf. Res.* **2013**, *21*, 37–44.
41. Sun, J.; Zhang, Y.; He, K. Providing context-awareness in the smart car environment. In Proceedings of the 2010 10th IEEE International Conference on Computer and Information Technology, Bradford, UK, 29 June–1 July 2010; pp. 13–19.
42. Kim, K.; Jung, J.K.; Choi, J.Y. Impact of the smart city industry on the Korean national economy: Input-output analysis. *Sustainability* **2016**, *8*, 649. [CrossRef]
43. Thomas, W.D.; Duffy, J.J. Energy performance of net-zero and near net-zero energy homes in New England. *Energy Build.* **2013**, *67*, 551–558. [CrossRef]
44. Markets and Markets. Market Research Report. 2019. Available online: https://www.marketsandmarkets.com/PressReleases/smart-building.asp (accessed on 19 August 2021).
45. Markets and Markets. Market Research Report. 2019. Available online: https://www.marketsandmarkets.com/Market-Reports/smart-factory-market-1227.html?gclid=CjwKCAjwY8BRBiEiwA5MCBJupK5L5DQ-7j0mCUsqztan-WT3a4KOL6-vFLDuLoEN8qU17HuG9hoCjlkQAvD_BwE (accessed on 19 August 2021).
46. Markets and Markets. Market Research Report. 2020. Avaliable online: https://www.marketsandmarkets.com/Market-Reports/connected-car-market-102580117.html?clid=CjwKCAjwY8BRBiEiwA5MCBJjpuNcMnddDSRB2lsq6I2EInPWcrc1XEj2JpSU226ph-WWarPNTuBoCiFIQAvDBwE (accessed on 19 August 2021).
47. The Bank of Korea. Input-Output Statistics. In *The Executive Summary of the 2015 Input-Output Tables*; The Bank of Korea Press: Seoul, Korea, 2019. Available online: https://www.bok.or.kr/portal/bbs/P0000559/view.do?nttId=10050567&menuNo=200690 (accessed on 19 August 2021).
48. Defourny, J.; Thorbecke, E. Structural path analysis and multiplier decomposition within a social accounting matrix framework. *Econ. J.* **1984**, *94*, 111–136. [CrossRef]
49. Oh, B.H.; Lee, S.H. The Estimation of Telecommunication Demand and Analysis of the Structure Changes. *J. Korea Plan. Assoc.* **1995**, *30*, 123–140.
50. Basu, R.; Johnson, T.G. The development of a measure of intersectoral connectedness by using structural path analysis. *Environ. Plan. A* **1996**, *28*, 709–730. [CrossRef]
51. Hirschman, A.O. *The Strategy of Economic Development*; Yale University Press: New Haven, CT, USA, 1958.
52. Leontief, W. *Input-Output Economics*; Oxford University Press: New York, NY, USA, 1986.
53. Lewis, B.D.; Thorbecke, E. District-level economic linkages in Kenya: Evidence based on a small regional social accounting matrix. *World Dev.* **1992**, *20*, 881–897. [CrossRef]
54. The Bank of Korea. 1960, 1975, 1995 and 2015 Input-Output Tables, The Bank of Korea: Seoul, South Korea. Available online: https://ecos.bok.or.kr/ (accessed on 19 August 2021).
55. Organisation for Economic Co-Operation and Development. *The Knowledge-Based Economy: A Set of Facts and Figures*; Organisation for Economic Co-Operation and Development: Paris, France, 1999.
56. Ronnie, J.; Neto, J.V.; Quelhas, O.L.G.; de Matos Ferreira, J.J. Knowledge Intensive Business Services (KIBS): Bibliometric analysis and their different behaviors in the scientific literature: Topic 16–Innovation and services. *RAI Rev. De Adm. E Inovação* **2017**, *14*, 216–225.

sustainability

MDPI

Article

Redesigning the Municipal Solid Waste Supply Chain Considering the Classified Collection and Disposal: A Case Study of Incinerable Waste in Beijing

Xiaoyu Yang [1], Xiaopeng Guo [1,2,*] and Kun Yang [1]

[1] School of Economics and Management, North China Electric Power University, Beijing 102206, China; yxyxiaoyu2016@163.com (X.Y.); yangkun@ncepu.edu.cn (K.Y.)
[2] Beijing Key Laboratory of New Energy and Low-Carbon Development, North China Electric Power University, Beijing 102206, China
* Correspondence: 13520328997@163.com

Abstract: The output of municipal solid waste is growing rapidly, which has brought tremendous pressure to urban development. The supply chain of municipal solid waste (MSW) in China mainly contains three processes: collection, transportation, and disposal. The waste is sorted at the collection and disposed of according to the classification. However, it is mixed at the transportation stage. Mixed transportation remixes the separately collected waste, which seriously affects the disposal effect. The supply chain of MSW urgently needs to be redesigned to improve the MSW disposal effect. First of all, on the ground of the waste treatment situation, we redesigned the supply chain of MSW in China. Secondly, combined with the redesign of the MSW supply chain, this paper established the function allocation model for collection stations, making a collection station only gather one type of waste, and built the transportation path planning model for vehicles, reducing the impact of waste storage on residents. Finally, based on the data of Xuanwu District in Beijing, the supply chain redesigning practical example of incinerable waste was given. The supply chain redesigning model in this paper not only makes full use of the existing infrastructure but also improves the disposal effect of waste. The supply chain redesigning model has practical application value.

Keywords: waste sorting; supply chain redesigning; function allocation; path planning; incinerable waste

Citation: Yang, X.; Guo, X.; Yang, K. Redesigning the Municipal Solid Waste Supply Chain Considering the Classified Collection and Disposal: A Case Study of Incinerable Waste in Beijing. *Sustainability* **2021**, *13*, 9855. https://doi.org/10.3390/su13179855

Academic Editor: Tan Yigitcanlar

Received: 19 July 2021
Accepted: 20 August 2021
Published: 2 September 2021

Publisher's Note: MDPI stays neutral with regard to jurisdictional claims in published maps and institutional affiliations.

1. Introduction

The acceleration of urbanization has led to a sharp increase in MSW output. In China, the output of waste has increased at a rate of 8–10% every year, and disposed MSW capacity in 2019 reached 242 million tons (Data from National Bureau of Statistics of China). Effectively dealing with MSW has become a major problem that must be solved in urban development. Waste incineration power generation is the best method of waste disposal under the principle of "reduction, harmlessness, and resource reuse", which has received increased attention. After years of development, waste incineration has achieved significant economic benefits in some developed countries. However, in China, the economic efficiency and emissions of incinerating waste for power are unsatisfactory. Even in large developed cities, such as Beijing and Shanghai, the problem is also serious.

The costs of incinerating waste for power in Beijing Gaoantun Waste to Energy co., Ltd., Beijing Shunyi District Municipal Waste Treatment Plant, Beijing Shougang Bioenergy Science & Technology Co., Ltd., have been analyzed. The results show that the social cost of burning 1 ton of MSW is about 1088.49 yuan, of which 70% comes from the loss of health in society caused by the dioxin generated by waste incineration [1]. It is common knowledge that incomplete waste classification and excessive impurities are the roots of dioxins in waste incineration. Improving the waste incineration effect is an urgent problem for the waste treatment industry.

In China, from generation to disposal, the MSW supply chain mainly contains collection, transportation, and disposal processes. Many cities, such as Beijing and Shanghai, have already realized the sorting collection and disposal of MSW. The MSW is divided into recyclables, hazardous waste, kitchen waste (organic waste), and other waste. In order to improve waste recycling and reduction, since 2019, many local governments have issued strict waste classification policies, such as the "Management Regulations of Shanghai Municipal Solid Waste Classification" and the "Management Regulations of Beijing Municipal Solid Waste". However, in transportation, MSW is mostly mixed transported, which mixes the sorting waste together, seriously destroying the waste classification results and affecting the waste disposal effect. It is urgent to change the supply chain mode. Therefore, combined with the waste treatment statute, we redesign the MSW supply chain and replan the waste collection and transportation under the existing waste treatment facilities. The specific structure of this paper is shown in Figure 1.

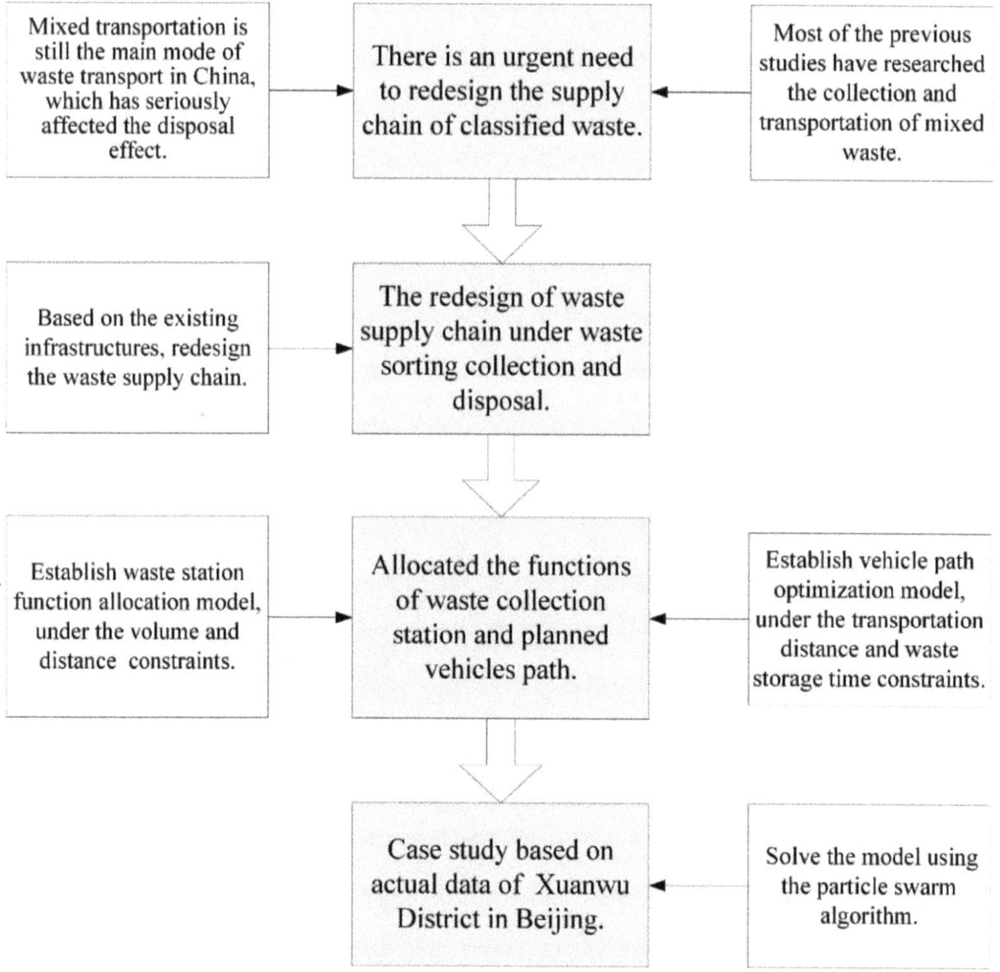

Figure 1. The specific structure of the paper.

2. Literature Review

The redesign of the MSW supply chain should consider the situation of waste collection and transportation [2]. Therefore, we summarize the research about waste supply chain design, the waste collection facilities layout, and the transportation route optimization.

2.1. The Design of MSW Supply Chain

MSW supply chain design and management can effectively resolve energy supply, waste management, and greenhouse gas issues [3]. In order to promote waste recovery, treatment, and recycling, previous research has studied the methods to improve the MSW collection effect [4,5], the waste governance modes [6], and the distribution of profits among all parties in the MSW supply chain [7,8], which provide the basis for MSW supply chain design. In MSW supply chain research, several studies using multi-objective programming methods design or optimize the MSW supply chain with economic, environmental, and social impact [9–11]. Additionally, with the deepening of MSW supply chain research, the effect of the uncertainty of waste collection levels [12] and sustainability issues such as land-use and public health impacts [13] on waste supply chain design have been focused on.

The mixed-integer linear programming (MILP) method has also been applied in the design of the waste supply chain. Previous studies mainly use the MILP method to explore the environmental impact [14] and the conflicts of waste collection and supply [15] on supply chain design. After that, a two-stage stochastic MILP model is formulated to examine the effects of the power price uncertainties on the MSW supply chain design [16] [15]. Additionally, an MILP model for waste supply chain networks containing the logistics, production, and distribution is presented with the aim of maximum profit of the entire supply chain [17].

Generally, the design of the waste supply chain mostly uses multi-objective planning methods, considering the economic and environmental impacts. However, previous studies have mostly neglected the waste classified and the existing infrastructure, which is not suitable for the redesign of the MSW supply chain. Therefore, combined with the current status of waste collection, transportation, and facility layout, this paper redesigns the waste supply chain, which effectively improves the waste supply chain in the case of waste sorting clearance and treatment.

2.2. The Layout of Waste Treatment Facility

The layout of waste treatment facilities is closely related to environmental protection and residents' lives. In different processing stages, treatment facilities are different. They mainly contain various waste treatment plants and waste collection stations that collect waste generated by residents. The location of waste collection stations is mostly close to the residential quarter. When waste is stored and removed, collection stations generate irritating gas, affecting nearby residents' lives [18]. There is a direct relationship between the surrounding environment of the waste collection station, the storage time of waste, and the times of vehicle clearance operations [19]. Additionally, when the waste classification level improves, the energy consumption and emission of waste collection stations is reduced [20].

Multi-objective programming and the geographic information system (GIS) are the most common methods used to research the layout of waste treatment facilities. Yadav et al. combined GIS with multi-objective programming to research the layout of waste transfer centers [21]. Based on the same method, Yadav et al. further studied the layout of waste transfer centers with the optimal economic cost when the waste output is uncertain [22]. Liu et al. analyzed the natural environment and human environment of Beijing by GIS and, using the multi-objective programming method, planned the layout of the waste collection station [23]. Furthermore, the disposal of waste has the most serious impact on the environment, so the location problem of waste disposal plants is mainly discussed, and the layout of landfill is mostly focused on plants' location selection [24–28]. The energy conversion rate also is one of the factors that affect the location of waste disposal plants. Kyriakis et al. selected a site for a waste-to-energy facility, under the constraints of maximum output energy and minimization of the gate fee [29]. Summarizing the previous studies, we found that few studies have studied the layout of facilities in waste classification.

It is found that the waste collection station has a great impact on residents' lives, and the impact of waste collection stations on the surrounding environment is directly proportional to the waste storage time and the clearance times of vehicles. Therefore, the redesign of the MSW supply chain should fully consider the impact on the surrounding residents, reducing the storage time and clearance times of the waste collection station.

2.3. The Optimization of Waste Transportation Routes

Transportation is an important process in the MSW supply chain. The optimization of waste transportation mostly has the goal of minimum economic cost [30] or transportation distance [31–33]. However, the above studies only optimized waste transportation under a single objective. The single-objective method has significant shortcomings in waste transportation optimization [34]. The multi-objective optimization model was established considering both transportation costs and social environmental impact. With the constraints of minimum transportation distances and emissions, Zdena et al., using GIS, optimized the path and schedule of waste transportation vehicles [35]. Dirk et al. studied the feasibility of using multimodal trucks and waterway transportation instead of waste highway transportation in the waste supply chain with the restrains of transportation costs and social impact [36].

GIS also is one of the common methods in transportation route optimization [37]. Kinobe et al. used GIS to select the site of the waste landfill plant in Kampala under the constraint of waste transportation routes and time [38]. Additionally, the simulation method has also been applied in waste transportation optimization. Khanh et al. established a multi-intelligent waste transportation simulation model for Hagiang city in Vietnam [39]. Xue et al. reviewed the problems of waste collection and disposal in Singapore and utilized the space allocation model, studying the resource allocation of waste incineration [40].

Generally, in transportation route optimization, previous studies mostly use GIS or mathematical algorithms to optimize vehicle routes under the constraints of the economy. Moreover, most studies were conducted under waste mixed collection and transportation. Few scholars have researched the collection and transportation of waste in classification. Additionally, very little research has considered the perishability of waste, combining waste storage and transportation together, exploring waste transportation paths with the minimum impact on the surrounding environment.

Based on the current situation of waste collection, transportation, and disposal, this paper redesigned the supply chain of waste sorting. First of all, on the basis of the existing layout of the waste collection stations, we distributed the function of waste collection stations, changing the mode of one collection station collecting various types of waste to one type of waste. Secondly, in order to reduce the impact of waste collection and transportation on surrounding residents, with the goal of minimum storage time and transportation distance, this paper optimized the path of waste transportation vehicles. Finally, the case study about Xuanwu District in Beijing was given. It is hoped that the research in this paper can help to improve the effect of the waste disposed and provide a reference for the supply chain mode redesign in waste sorting.

3. The Methodology for Supply Chain Redesign

3.1. The Supply Chain Redesign

In the actual treatment of waste, from generation to disposal, the waste mainly passes the three processes: collection, transportation, and disposal. Residents transfer the waste generated by the family to the waste cans in the community collection point; then, the property in the community mixes the separately collected waste and transports it to the nearby waste collection station. Additionally, after a short period of storage, the waste is transported to the waste treatment plant by truck. When the distance between the waste collection station and the treatment plant is far, it is necessary to set up a waste transfer center. To improve the treatment effect, the transfer center compresses and sorts the waste. To vividly express the waste treatment processing, Figure 2, the current waste collection–transportation–disposal process in China, has been given.

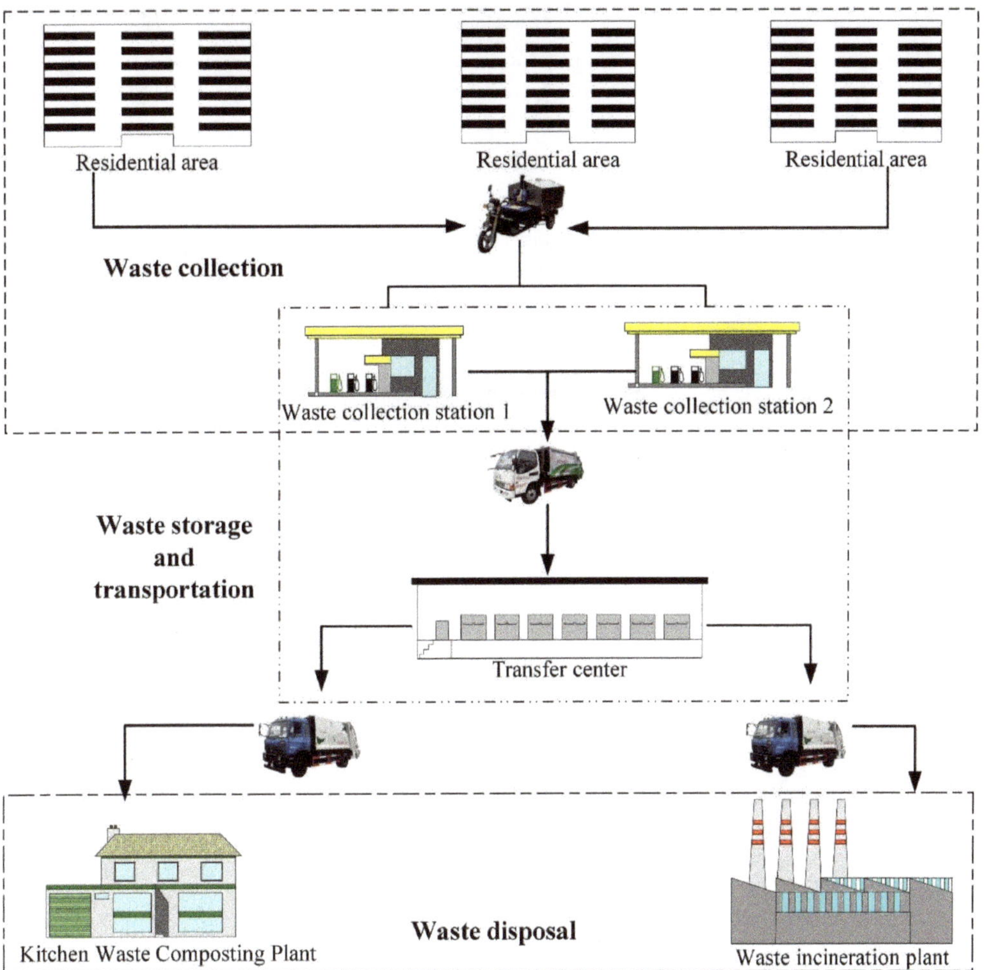

Figure 2. The current waste collection–transportation–disposal process in China.

Through investigation, it was found that when waste was transported from the residential area to the collection station, the sorted waste began to be mixed. It is necessary to prevent the mixed transportation of waste in this process. In order to restrict the mixed transportation, this paper redesigned the MSW supply chain, in which the waste collection station only gathers and transports one type of waste. The supply chain of sorting waste is shown in Figure 3.

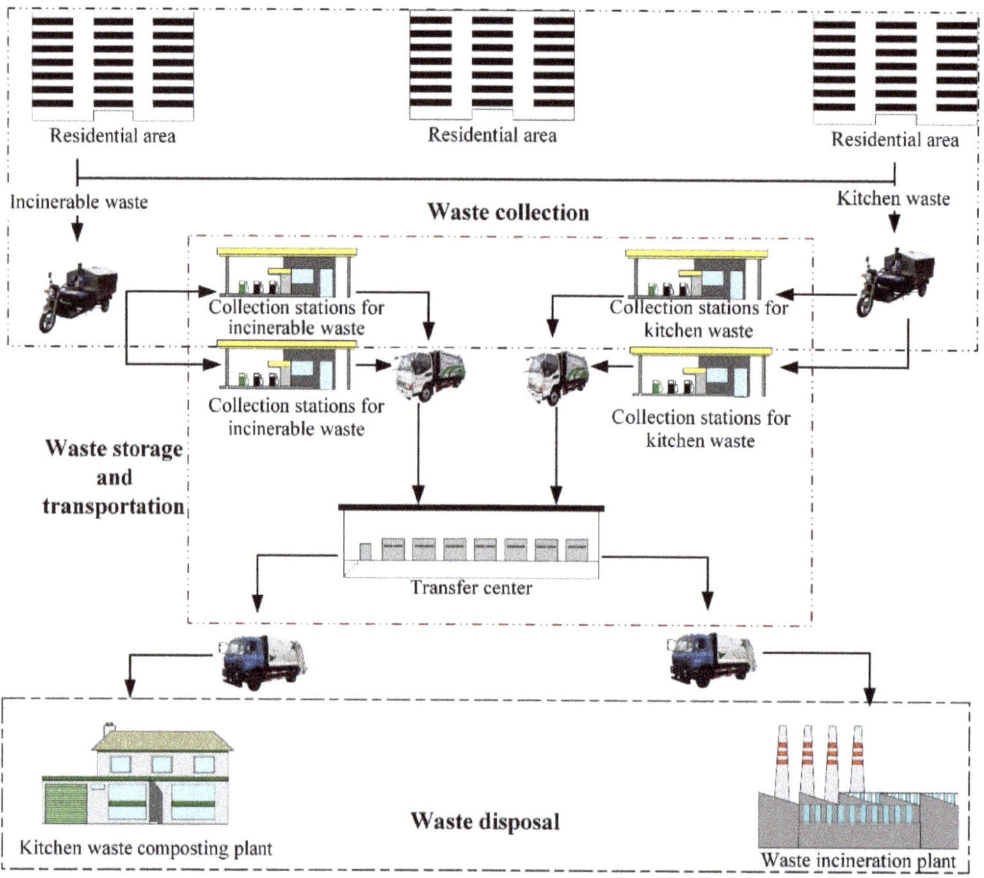

Figure 3. The redesigned collection–transportation–disposal model under waste classification.

3.2. The Function Allocation Model for Waste Collection Station

The essential basis of the redesigned MSW supply chain is that each waste collection station only collects and transports one type of waste. In order to realize the design, the function of collection stations needs to be redistributed. Additionally, the number and capacity of every type of waste collection station should fulfill the waste output and the proportion in total waste output. The proportions of kitchen waste, recyclable waste, disposable packaging, and other waste within the total amount of MSW, respectively, are 52.2%, 17.2%, 9.9%, and 20.7% [41]. Disposable packaging and other waste both can be incinerated, so the proportions of kitchen waste, incinerable waste, and recyclable waste in MSW, respectively, are 52.2%, 30.6%, and 17.2%. Therefore, kitchen waste collection stations account for 52.2% of the total collection stations, incinerable waste collection stations account for 30.6%, and the rest are recyclable waste collection stations. Affected by local characteristics and urban functions, the composition of MSW is various in different cities [42] and locations [43]. In actual applications, the proportion of collection stations needs to be adjusted according to the local conditions. The amount of renewable waste is relatively small, and most of it is collected by waste pickers; this paper focuses on kitchen waste and incinerable waste.

3.2.1. Modeling Assumptions

Since the function allocation model is built on the existing collection stations, the construction cost of the collection station is no longer considered in the total cost. In order to make the model reasonable, this paper makes assumptions about the distribution of waste generation sources, waste transportation costs, and other factors.

(1) The waste production sources are evenly distributed around the collection station.
(2) The capacity of the collection station meets the maximum volume of waste in one day.
(3) The unit cost of waste transportation is the same between all stations.
(4) Collection stations can gather the same kind of waste from multiple residential areas.
(5) The same kind of waste in a waste production source can be transported to multiple collection stations.
(6) Straight-line transportation can be realized from the waste production source to the collection station, and the transportation distance can be calculated by the Euclidean Distance.
(7) The construction and operation of collection stations are the same, and the impact of collection stations on the surrounding environment is equal.
(8) There is no demolished risk for the existing collection stations.

3.2.2. Symbol Description

There are many variables. Additionally, we use the following symbols to represent them.

(1) Common variables:

N The aggregate of waste collection stations, $i \in N$, $i = \{1, 2, \ldots n\}$.
N_0 The aggregate of waste production source, $j \in N_0$
C_i The capacity of waste collection station, $i \in N$
g_j The volume of kitchen waste in waste production source j, $j \in N_0$
p_j The volume of incinerable waste in waste production source j, $j \in N_0$
f_{ij} The volume of kitchen waste transferred from production source j to collection station i
f_{1ij} The volume of incinerable waste transferred from production source j to collection station i
d_{ij} The distance between i and j

(2) Decision variables:

$$Z_i = \begin{cases} 1, & \text{if collection station } i \text{ gather kitchen waste} \\ \\ 0, & \text{if not} \end{cases}$$

$$Z_{ij} = \begin{cases} 1, & \text{if the kitchen waste in production source } j \text{ transported to collection station } i \\ \\ 0, & \text{if not} \end{cases}$$

$$M_i = \begin{cases} 1, & \text{if collection station } i \text{ gather incinerated garbage} \\ \\ 0, & \text{if not} \end{cases}$$

$$M_{ij} = \begin{cases} 1, & \text{if the incinerated gargage in production source } j \text{ transported to collection station } i \\ \\ 0, & \text{if not} \end{cases}$$

3.2.3. The Function Allocation Model

With the above assumptions and symbol definitions, in waste classification, the function allocation for waste collection stations can be described as the following mathematical model.

$$\min F = \sum_{i \in N} \sum_{j \in N_0} Z_i * Z_{ij} * g_j * d_{ij} + \sum_{i \in N} \sum_{j \in N_0} M_i * M_{ij} * p_j * d_{ij} \tag{1}$$

Formula (1) is the objective function, which has the goal of minimizing transportation volume and distance of kitchen waste and incinerable waste from production sources to collection stations.

Formulas (2)–(9) are constraints.

$$M_i + Z_i \leq 1, \ \forall i \in N \tag{2}$$

Formula (2) indicates that a collection station only gathers and transports one type of waste.

$$\sum_{j \in N_0} g_j - \sum_{i \in N} Z_i * C_i \leq 0 \tag{3}$$

$$\sum_{j \in N_0} P_j - \sum_{i \in N} M_i * C_i \leq 0 \tag{4}$$

Formulas (3) and (4) indicate that the total capacity of the selected collection stations for kitchen waste or incinerable waste should meet the output of kitchen waste or incinerable waste.

$$g_j - \sum_{i \in N} Z_i * Z_{ij} * f_{ij} = 0, \ \forall j \in N_0 \tag{5}$$

$$p_j - \sum_{i \in N} M_i * M_{ij} * f_{1ij} = 0, \ \forall j \in N_0 \tag{6}$$

Formulas (5) and (6) indicate that all kitchen waste and incinerable waste in waste production sources are collected by collection situations.

$$\sum_{i \in N} Z_i = 52.2\% * n \tag{7}$$

$$\sum_{i \in N} M_i = 30.6\% * n \tag{8}$$

Formulas (7) and (8) are the quantity constraints of collection stations. The number of collection stations should satisfy the proportion of kitchen waste and incinerable waste. Because one collection station only collects one kind of garbage, it is necessary to take the integer of the number of collection stations. This paper rounds up the number of collection stations to ensure the collection station is enough to accommodate all the waste.

$$Z_i, M_i, Z_{ij}, M_{ij} \in \{0, 1\} \tag{9}$$

Formula (9) is the range of decision variables, which is 1 or 0.

3.3. The Transportation Path Planning Model for Waste Truck

Through the allocation of the waste collection station function, the design that one collection station receives and transports one type of waste has been realized. In order to prevent the mixed transportation of waste from the collection station to the transfer center, this paper plans the path of the waste truck that transfers the same kind of waste in collection stations on the basis of the function allocation. Combined with the impact of the waste collection station on the surrounding environment, under the goals of the shortest transportation path of the truck, the shortest storage time of waste in the collection station, and the minimum transported times of waste for the collection station, a waste truck path optimization model for one kind of waste was established.

3.3.1. Modeling Assumptions

In order to make the model reasonable, this paper makes the following assumptions.

(1) The amount of waste in the collection station is known and will not significantly change.
(2) When the load and time permit, the truck can go to multiple collection stations to transport waste.
(3) The truck does not affect each other in driving and operation;
(4) The opening hours of the collection station are consistent with the waste truck.
(5) The truck travels at the same speed.
(6) Straight-line transportation can be realized between collection stations, and the distance is equal to the Euclidean Distance.
(7) All trucks start from the transfer center, and after finished waste transportation, they will return to the transfer center.
(8) The unit transportation cost of trucks is the same.

3.3.2. Symbol Description

There are many variables. Additionally, we use the following symbols to represent them.

(1) Common variables:

N_1 The aggregate of waste collection stations i that gather the same kind of waste, $i \in N_1, i = \{1, 2, \ldots n\}$
N_2 The aggregate of the transfer center and the aggregate of the waste collection station N_1
q_i The volume of waste in the collection station i, $i \in N_1$.
K The aggregate of a waste truck k, $k \in K, k = \{1, 2, \ldots k\}$
C The number aggregate of a truck, $c = \{1, 2, 3 \ldots c\}, c \in C$.
W The maximum load of a truck
e The collection station starting work time
l The collection station ending work time
v The speed of a truck
S Time for waste truck loading
T_{ic}^k The time point that the collection station i was cleared by the truck k when it starts from transfer center in Cth time.
q_{ic}^k The transported volume of waste in collection stations i, if the truck k collected waste from collection station i when it starts from transfer center in Cth time.

(2) Decision variables:

$$X_{ic}^k = \begin{cases} 1, & \text{if the truck } k \text{ collected the garbage in the collection} \\ & \text{station } i \text{ when it departs from transfer center in Cth time} \\ \\ 0, & \text{if not} \end{cases}$$

$$X_{ijc}^k = \begin{cases} 1, & \text{if the trunk pass through path between } i \text{ and } j \\ & \text{when it departs from transfer center in Cth time } (i, j \in N_2) \\ \\ 0, & \text{if not} \end{cases}$$

3.3.3. The Transportation Path Planning Model

In transportation path planning, with the same unit transportation cost of trucks, the smaller the mileage of vehicles, the lower the total cost of transportation [31–33]. Therefore, the mileage of vehicles is used to represent the economic benefits of waste transportation.

The irritating gas generated by the waste in collection stations significantly affects the nearby residents' lives [18]; transportation path planning should consider the impact of irritating gas on residents. There is a direct relationship between the surrounding environment of the waste collection station, the storage time of waste, and the times of vehicle clearance operations [19], so the storage time of waste and the transported times of collection stations are used to represent the impacts of the waste collection station on the surrounding environment. The path optimization model for the waste truck aims at achieving the shortest travel path for the truck, the shortest waste storage time for the collection station, and the minimum times for waste transportation in the collection station. The constraints of the model include vehicle load, waste transportation volume, and the working time of the collection station.

$$\min F = w_1 F_1 + w_2 F_2 + w_3 F_3 \tag{10}$$

$$F_1 = \sum_{k \in K} \sum_{c \in C} \sum_{i \in N_2} d_{ij} * X_{ijc}^k \tag{11}$$

$$F_2 = \sum_{i \in N_1} \max\left\{ T_{ic}^K, i = 1, 2, \dots n \right\} \tag{12}$$

$$F_3 = \sum_{k \in K} \sum_{c \in C} \sum_{i \in N_2} X_{ic}^k \tag{13}$$

Equation (10) is the overall objective function; F_1, F_2, and F_3 are the sub-objectives, whose specific explanations are Formulas (11)–(13); w_1, w_2, and w_3 are the weight of the sub-objective.

F_1 is the total travel distances of trucks, which are the sum distances of every truck. Additionally, every trucks' travel distances are affected by d_{ij}, the distance between i and j, and X_{ijc}^k, whether the truck passes through the way between i and j when the truck starts from the transfer center in Cth time.

F_2 is the total waste storage time of collection stations, which is the sum of the maximum of T_{ic}^k, $\forall i \in N_1$, the time that every collection station i was transferred by the truck k, when it starts from the transfer center in Cth time.

F_3 is the total time for collection stations transported by trucks, which is the sum of times that all collection stations are transported by all waste trucks. The specific calculation is shown in Equation (13).

The waste collection stations are close to the residential quarter. Waste with long-time storage produces a significant amount of irritating gas, which has a serious impact on the lives of surrounding residents, especially in summer. In order to achieve the optimal economic and environmental benefits, this paper assumes that economic impacts and environmental impacts are equally important when transporting waste. Therefore, w_1, the weight of vehicles mileage, is equal to the sum of w_2, the weight of waste storage time, and w_3, the weight of the waste in collection stations be transported times. Since the effect of waste storage time on the environment is more than the impact of transported times, the weight of storage time should be greater than the transported times. We assume that w_1, w_2, and w_3, respectively, are 0.5, 0.3, and 0.2.

$$\sum_{k \in K} \sum_{c \in C} X_{ic}^k \geq 1, \; \forall i \in N_1 \tag{14}$$

$$\sum_{i \in N_1} \sum_{k \in K} \sum_{c \in C} X_{ic}^k \geq n \tag{15}$$

Equations (14) and (15) are the constraints of waste transported times in the collection station. Formula (14) indicates that the waste in the collection station must be transported, and Formula (15) indicates that a collection station can be transported multiple times.

$$\sum_{i \in N_1} \sum_{k \in K} \sum_{c \in C} q_{ic}^k * X_{ic}^k \geq \sum_{i \in N_1} q_i \tag{16}$$

$$\sum_{k \in K} \sum_{c \in C} q_{ic}^k * X_{ic}^k \geq q_i, \ \forall i \in N_1 \tag{17}$$

Formulas (16) and (17) are the transportation volume constraints. Formula (16) indicates that the total transportation volume of waste trucks is equal to the storage volume of collection stations. Formula (17) indicates that the amount of waste being transported at each collection station is equal to the amount stored. Additionally, q_{ic}^k is the shipment volume of the truck when it passes through collection station i, which is greater than 0.

$$\sum_{c \in C} \sum_{i \in N_1} q_{ic}^k \leq w, \forall k \in K \tag{18}$$

Formula (18) is the truck's load constraints, which mean that for any vehicle, the load cannot exceed the vehicle maximum load.

$$T_{ic}^k = e + \sum_{c=1}^{c} X_{ijc}^K * d_{ij} \div v + \left[\sum_{i=1}^{n} \sum_{c=1}^{c} X_{ic}^k \right] * S \tag{19}$$

Formula (19) is the working time constraints, which mean that when the waste in the collection station i is transported, the time must be the working time of collection station, $e \lhd T_{ic}^k \lhd l$. Additionally, T_{ic}^k is the sum of the time collection station starting work, the time truck driving on the road, and the loading time. The driving time is calculated by the travel distance and truck speed. The loading time is equal to the product of the number of loading times and the time required to load once.

4. Data Collection and Processing

Waste disposal should follow the principles of reduction, recycling, and harmlessness. Due to the large demand for land, the landfill has no longer been used in an increasing number of countries and districts. In China, renewable waste is mostly collected by waste pickers. Therefore, we mainly discuss the supply chain of waste for incinerated, based on the actual data of Xuanwu District (It has been merged with other districts, becoming Xicheng District) in Beijing.

Xuanwu District is located in the southwest of Beijing. The shape of Xuanwu District is similar to a rectangle, and the total area is 19.04 square km. There are about 107 communities in Xuanwu District, with an approximate population of 544,000. In general, Xuanwu District has a large population, frequent personnel activities, and a large amount of waste production. There are 30 waste collection stations in Xuanwu District, and their distribution is shown in Figure 4. After collecting and temporarily storing, the waste in Xuanwu District is transported to the Majialou Sorting Transfer Station (Majialou transfer center). In the transfer station, the waste is transported to the corresponding disposal plant for processing after weighing, sorting, compressing, and removing impurities.

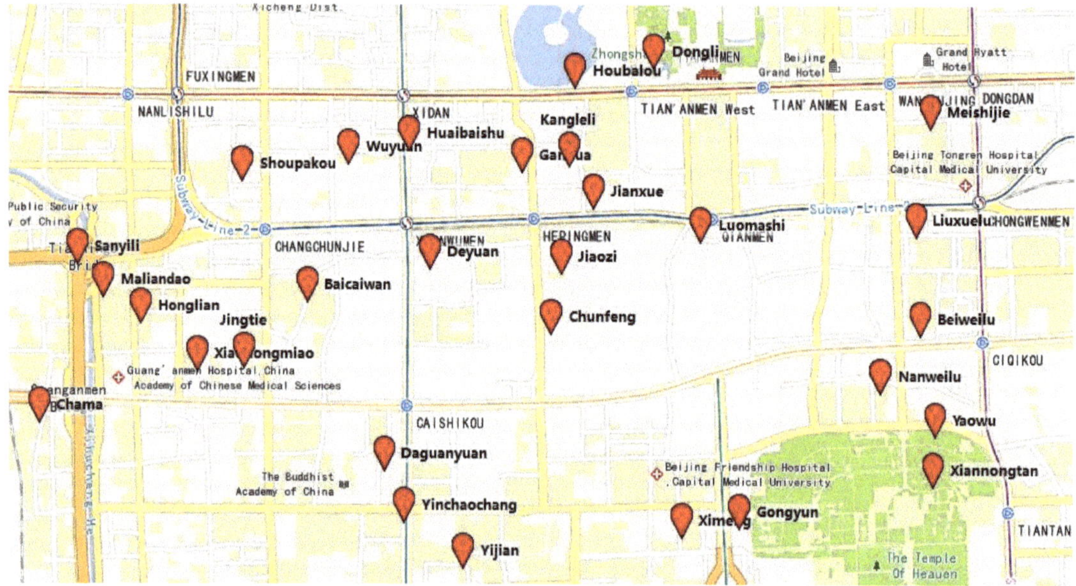

Figure 4. The distribution of waste collection stations.

(1) The basic data of waste output and collection stations

This paper quotes the basic data of collection stations and waste output in Xuanwu District from the "Research on Vehicle Dispatching and Optimization of Beijing Refuse Collection based on Municipal Solid Waste Classification" [44]. There are two working hours for waste collection and transportation. Since the waste output and the trucks are the same, we design the waste collection and transportation model using the data in the working hours from 5:30 to 13:30. We calculated the quantity of kitchen waste, incinerable waste, and recyclable waste in each collection station according to the proportion of them in total waste output [41] and established the coordinate system using the point (116.3334, 39.840239) as the origin. The coordinate value of each point is calculated by gpsCalc (2.0). The waste output and coordinate value are shown in Table 1.

The layout of the waste collection stations shows that the distribution of waste collection stations is uneven, and the waste collection stations are dense in some areas. For example, for Nanweilu, Xiannongtan, and Yaowu, the longest distance between them is only 696 m, and the waste collection volume of Yaowu and Xiannongtan is only 3 tons. This violates the construction requirements of the collection stations. Therefore, the waste collection stations in Xuanwu District have been filtrated. Additionally, 8 waste collection stations with 3 tons maximum capacity were eliminated; they are Yaowu, Xiannongtan, Gongyun, Kangleli, Jianxue, Wuyuan, Yinchaochang, and Sanyili.

(2) The basic data of vehicles

At present, there are 22 vehicles in the Majialou transfer center, including 18 vehicles with a load of 3 tons and 4 waste compression vehicles with a load of 6.1 tons. Considering the amount of waste, and the professional development of waste transportation, we assume that the 4 waste compression vehicles with the load of 6.1 tons are responsible for the transportation of incinerable waste.

(3) The basic data of working time

The waste collection stations and the vehicles both start working at 5:30 in the morning. Additionally, the vehicles need to complete the transportation of all waste in the collection stations before 13:30 in the afternoon. It is assumed that the loading time of waste compression vehicles is 20 min, 0.33 h. Additionally, the speed of the vehicles is 30 km per hour.

Table 1. The basic data of collection stations and waste output.

Number	Name	X Position (m)	Y Position (m)	The Amount of Kitchen Waste (Kg)	The Amount of Incinerable Waste (Kg)	The Amount of Recyclable Waste (Kg)	The Maximum Capacity of Collection Station (Kg)
1	Meishijie	7659	6925	12,685	7436	4180	27,000
2	Liuxuelu	7536	6233	5638	3305	1858	12,000
3	Dongli	5288	7309	2819	1652	929	6000
4	Luomashi	5688	6193	5638	3305	1858	12,000
5	Beiweilu	7581	5590	4228	2479	1393	9000
6	Nanweilu	7236	5227	7047	4131	2322	15,000
7	Yaowu	7714	4951	1409	826	464	3000
8	Xiannongtan	7692	4624	1409	826	464	3000
9	Gongyun	6022	4356	1409	826	464	3000
10	Ximeng	5533	4298	5638	3305	1858	12,000
11	Houbalou	4620	7183	7047	4131	2322	15,000
12	Kangleli	4564	6677	1409	826	464	3000
13	Jianxue	4776	6408	1409	826	464	3000
14	Jiaozi	4497	5993	4228	2479	1393	9000
15	Chunfeng	4419	5606	4228	2479	1393	9000
16	Ganhua	4163	6644	4228	2479	1393	9000
17	Deyuan	3373	6020	4228	2479	1393	9000
18	Huaibaishu	3195	6782	4228	2479	1393	9000
19	Wuyuan	2672	6700	1409	826	464	3000
20	Daguanyuan	2983	4723	8456	4957	2786	18,000
21	Yinchaochang	3162	4392	1409	826	464	3000
22	Yijian	3662	4103	4228	2479	1393	9000
23	Shoupakou	1759	6588	9866	5783	3251	21,000
24	Baicaiwan	2315	5813	7047	4131	2322	15,000
25	Jingtie	1770	5387	4228	2479	1393	9000
26	Xiaohongmiao	1369	5364	7047	4131	2322	15,000
27	Honglian	868	5677	7047	4131	2322	15,000
28	Maliandao	546	5845	7047	4131	2322	15,000
29	Sanyili	323	6058	1409	826	464	3000
30	Chama	0	5036	4228	2479	1393	9000
31	Majialou transfer center	2661	0	–	–	–	–

5. Results and Discussion

The supply chain of incinerable waste under the classification is redesigned. Based on the data of Xuanwu District, a supply chain redesign case is given, which provides a reference for supply chain redesign in other places.

5.1. The Function Allocated Result of Collection Stations

In order to sort collection and transportation waste, 22 eligible collection stations in Xuanwu District were analyzed. Due to limited data, the 30 waste collection stations are used to simulate the distribution of waste generation sources in Xuanwu District. The function allocation model with the goal of minimum waste transportation volume and transportation distance is established. Incinerable waste accounts for 30.6% of the total output of waste, so the number of incinerable waste collection stations should be 6.44. Rounding up the number of collection stations, the number of incinerable waste collection stations is set at 7. This paper uses Lingo 11 software to solve the function allocation model.

After the function allocation, the waste that can be used for incinerating power generation in Xuanwu District only needs 7 waste collection stations to gather. They are Chunfeng, Huaibaishu, Yijian, Baicaiwan, Xiaohongmiao, Nanweilu, and Maliandao collection station. The specific results are shown in Table 2. The distribution of incinerable waste collection stations is shown in Figure 5, and the collection stations with purple are incinerable waste collection stations.

Table 2. The location and inventory of incinerable waste collection stations.

Number	Name	X Position (m)	Y Position (m)	The Maximum Capacity of Collection Station (Kg)	The Volume of Incinerable Waste (Kg)
1	Chunfeng	4419	5606	9000	8689
2	Huaibaishu	3195	6782	9000	8689
3	Yijian	3662	4103	9000	8689
4	Baicaiwan	2315	5813	15,000	14,482
5	Xiaohongmiao	1369	5364	15,000	14,482
6	Nanweilu	7236	5227	15,000	14,482
7	Maliandao	546	5845	15,000	14,482
0	Majialou Transfer Center	2661	0	–	–

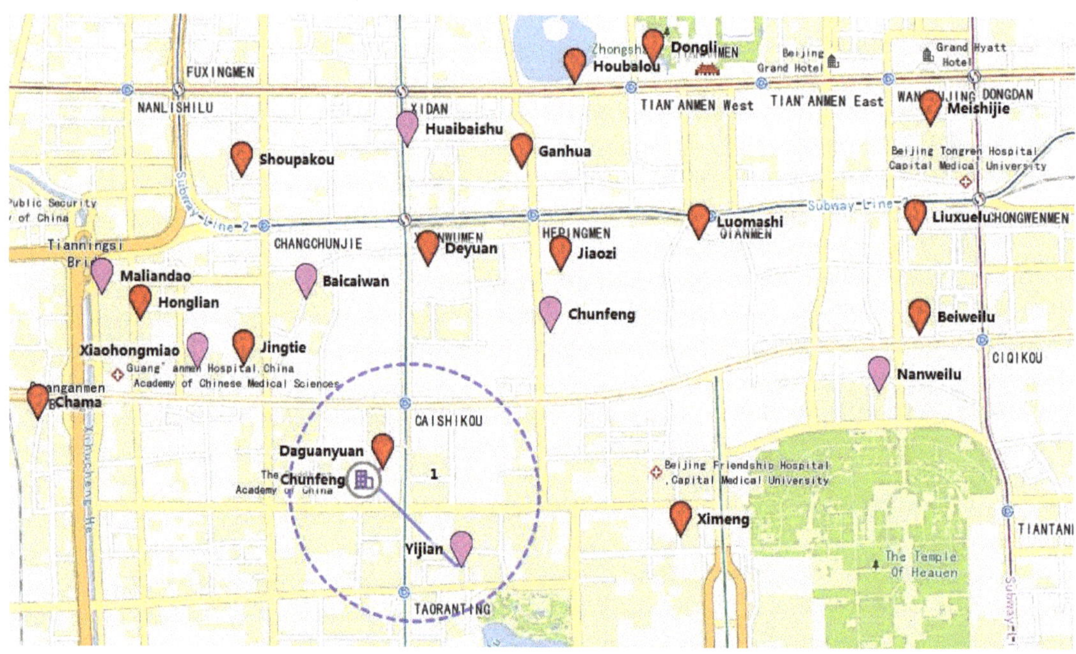

Figure 5. The distribution of incinerable waste collection stations.

It should be noted that due to the function allocation of collection stations, the transportation of waste in residential areas to collection stations has changed. As shown in the dotted box in Figure 5, in the original MSW supply chain mode, the community of Chunfeng only needs to transport all the waste to the nearest collection station Daguanyuan, but in the new mode, it should transport the kitchen waste to the Daguanyuan collection station and transport the incinerable garbage to Yijian collection station, which significantly increased the waste transportation distance and workload from residential areas to the collection station. Owing to shortages of waste production and transportation data in residential areas, this paper cannot measure the change of waste transportation workload from residential areas to collection stations.

The optimization model of the waste transportation path is carried out on the basis of the distribution of incinerable waste collection stations. The Majialou transportation center needs to complete the waste transportation in the seven collection stations.

5.2. The Results of Transportation Path Planning

To prevent mixed transportation and reduce the impact of collection stations on residents' lives, the transportation path of incinerable waste has been studied. The destination of waste transportation is the Majialou transfer center, and the transportation vehicle is the waste compression truck with a load of 6.1 tons. The start and end of the waste truck are both the Majialou transfer center. With the help of Matlab R2014b software, this paper uses a particle swarm algorithm to solve the path planning model. The specific solution processes are as follows.

(1)   The dimensionless treatment of objective functions

Due to the different measurement units, the objective functions need to be processed. In this paper, the extremum method was utilized to perform dimensionless processing. First, under a single objective, the optimal value was calculated and used as a magnitude standard. Then, using the objective function values under multiple objectives, the corresponding magnitude standard as the dimensionless values was divided.

(2)   Model solution using particle swarm optimization

The particle swarm algorithm is a commonly used intelligent solution algorithm, which has the advantages of fewer parameters and easier implementation. Combined with previous research [45], the initialization parameters are set. This paper sets the range of particle position as [0, 1] and the speed range as [−0.1, 0.1]; the population size is 200, the number of iterations is 200, and the acceleration coefficient C_1 and C_2 both are 2; the parameter inertia weight is 0.8 and 0.4. In order to improve the ability to seek optimization, during the calculation process, the parameter inertia weight is adjusted from large to small.

The particle swarm algorithm relies on iterative learning to find the optimal solution of the model. In iteration, the optimal solution of the model was determined by the optimal value of the particle and the average value of the population. After solving the waste truck transportation path model by particle swarm optimization, we obtained the value of the average target and optimal target (Figure 6). The results show that after 20 times iteration, the optimal transportation path scheme was found, and the vehicle's transportation path is shown in Table 3.

Since the amount of waste in one collection station is greater than the load of the waste truck, the waste truck needs to go to the collection station several times. At the beginning of waste transportation, waste trucks will be fully loaded by transporting waste in one collection station. Additionally, then, waste trucks should collect waste in several stations to achieve a full load. The waste trucks 1 and 4 both completed transportation four times, and the waste trucks 2 and 3 completed transportation three times. The waste in collection station 7 (Maliandao, the number of collection station is shown in Table 2) was finished being first transported, and the waste storage time in collection station 7 is the shortest; the waste in collection station 2 (Huaibaishu) was finished being transported last, and the waste storage time in collection station 2 is the longest.

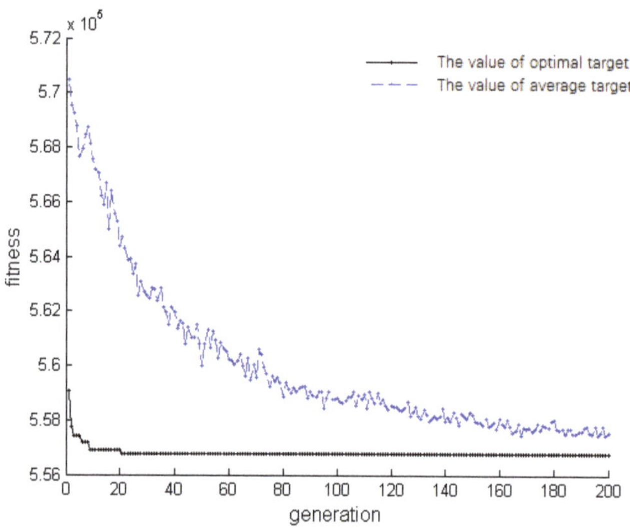

Figure 6. The change of average target and optimal target.

Table 3. The path and time of vehicles.

Vehicles	Name	Path and Time										
Truck 1	The number of collection station	0	5	0	5	0	3	0	3	1	2	0
	Arrival time	5.5	6.02	6.83	7.35	8.17	8.49	9.11	9.43	9.8	10.12	10.79
	Departure time	5.5	6.32	6.83	7.65	8.17	8.79	9.11	9.73	10.1	10.42	11.09
Truck 2	The number of collection station	0	4	0	6	0	7	0	–	–	–	–
	Arrival time	5.5	6.02	6.84	7.36	8.18	8.71	9.54	–	–	–	–
	Departure time	5.5	6.32	6.84	7.66	8.18	9.01	9.54	–	–	–	–
Truck 3	The number of collection station	0	7	0	6	0	7	6	5	0	–	–
	Arrival time	5.5	6.03	6.85	7.37	8.19	8.72	9.03	9.34	10.16	–	–
	Departure time	5.5	6.33	6.85	7.67	8.19	9.02	9.33	9.64	10.46	–	–
Truck 4	The number of collection station	0	4	0	2	0	1	0	5	4	3	0
	Arrival time	5.5	6.02	6.84	7.21	7.88	8.23	8.88	9.4	9.73	10.07	10.69
	Departure time	5.5	6.32	6.84	7.51	7.88	8.53	8.88	9.7	10.03	10.37	10.99

5.3. Discussion of the Supply Chain Redesign Results

There is an inevitable requirement for sustainable development of cities that decrement harmless and resources for waste treatment. In China, waste treatment has become a major issue in the sustainable development of cities, either large cities or small towns. Based on the present situation that mixed transportation seriously affects the waste classified disposal and sorted collection effect, this paper redesigns the waste supply chain to improve the waste treatment effect. In collection processing, the redesigned waste supply chain can supervise the waste classification effect, which is conducive to evaluating the waste classification effect of each enterprise, public institution, and community. In transportation processing, it can effectively avoid waste mixed transportation, reduce mutual pollution between waste, and solve the problem of false waste classification. Additionally, in disposal processing, it is helpful to improve the final disposal effect of waste and solve the problem caused by excessive impurities [46].

To achieve sustainable development, Beijing, Shanghai, and many other cities have issued strict waste classification policies, hoping to improve the level of classification of waste collected. However, mixed transportation mixes the classified collected waste together, which makes the front-end of the waste supply chain classified collection invalid and seriously affects the waste disposal effect of the supply chain back-end. Based on the existing infrastructure, this paper redesigns the sorting waste collection–transportation–disposal supply chain and allocates the function of the waste collection station, making the waste collection station only collect and transport one type of waste. Compared with the original waste supply chain mode, the new waste supply chain mode proposed in this paper can provide corresponding supporting facilities for the waste classification policy. The improvement of the waste supply chain can ensure the effect of the waste classification policy and promote the implementation of the waste classification policy.

In addition, the waste supply chain proposed in this paper is contributed to promote the waste transportation infrastructure upgrades. The waste classified collection and transportation is conducive to maximizing the advantages of professional waste removal facilities, improving the waste removal efficiency. In waste transportation, with the goal of optimal transportation distance and impact of the collection station on surrounding residents, this paper studies the path planning under the waste classification. Affected by the waste transportation facilities situation, the trucks actually used in transportation fail to fully meet the professional standards of trucks used in this paper. Therefore, it is impossible to quantitatively calculate the improved benefits of waste transportation. Combined with the specialization requirements of waste removal and treatment, the waste trucks will be more specialized in the future.

It should be noted that the waste supply chain mode proposed in this paper may increase the workload of transporting waste from residential areas to collection stations and augment the specialized equipment for waste transportation. In the original model, the waste in one community only needs to be transported to one collection station. However, in the redesigned waste supply chain, the waste in one community needs to be transported to the corresponding sorting collection stations, which will increase the transport distance. Due to the lack of residential areas' waste production data, it is impossible to specifically measure the increased workload of waste transportation from residential areas to collection stations. Additionally, because the redesigned supply chain has not yet been applied, it is impossible to concretely measure the improved ecological footprint or carbon emission in waste disposal. However, considering the situation that the social health cost of waste incineration is 764 yuan/ton, accounting for 70% of the waste incineration cost, which is close to the total cost of waste collection cost in residential areas and waste transportation costs from residential areas to the collection station [1], it is feasible to improve the effectiveness of waste incineration by optimizing the collection and transportation mode since, in long-term operational effects, the redesigned supply chain will play a positive role in the sustainable development of cities.

6. Conclusions

In China, mixed transportation is still the main mode of waste transportation, which mixes the sorted waste, seriously affecting the waste disposal effect. It is the general trend for the waste treatment industry to replace waste mixed transportation with classified transportation. Combined with the impact of waste collection stations on the surrounding environment, this paper, with the goal of optimal economic benefits and minimum environmental impact, has redesigned the supply chain of sorted waste. Based on the data of Xuanwu District in Beijing, an application case of the design model is given.

In general, the method proposed in this paper fully considered the existing waste treatment facilities, which are applicable to most cities in China. However, the model in this paper still has deficiencies, and the following questions are proposed for subsequent related research.

(1)   When allocating the functions of waste collection stations, this paper only considers the transportation volume and collection station capacity and does not consider the environment around the collection station and the residents' attitude. Because of the limited data, in the function allocation model, we only use 30 waste collection stations

simulating the waste generation sources. The actual waste generation sources are more complicated.

(2)   The production and composition of waste will vary with seasons and positions. The composition of waste in residential areas and commercial areas is different. In this paper, neither the function allocation model for waste collection stations nor the transportation path planning model for waste trucks takes the changes of waste production and composition into account. In the model actual application, the changes in waste production and composition should be considered.

Author Contributions: X.Y.: investigation, methodology, model analysis, writing—original draft, and writing and editing. X.G.: methodology, investigation, writing—review and editing, and funding acquisition. K.Y.: supervision and writing review—editing. All authors have read and agreed to the published version of the manuscript.

Funding: Project supported by the National Key R&D Program of China (2020YFB1707802) and the Fundamental Research Funds for the Central Universities (2019FR002).

Institutional Review Board Statement: Not applicable.

Informed Consent Statement: Not applicable.

Data Availability Statement: The datasets used and/or analyzed during the current study are available from the corresponding author on reasonable request.

Conflicts of Interest: The authors declare that they agree with the publication of this paper in this journal. The authors declare that they have no known competing financial interests or personal relationships that could have appeared to influence the work reported in this paper.

References

1.   Guojun, S.; Yueyang, S.; Chang, Z.; Shuai, L.; Ying, W. Social cost accounting for municipal solid waste incineration in Beijing. *China Popul. Resour. Environ.* **2017**, *27*, 17–27.
2.   Mota, B.; Gomes, M.I.; Carvalho, A.; Barbosa-Povoa, A.P. Sustainable supply chains: An integrated modeling approach under uncertainty. *Omega* **2018**, *77*, 32–57. [CrossRef]
3.   Pan, S.Y.; Du, M.A.; Huang, I.T.; Liu, I.H.; Chang, E.E.; Chiang, P.-C. Strategies on implementation of waste-to-energy (WTE) supply chain for circular economy system: A review. *J. Clean. Prod.* **2015**, *108*, 409–421. [CrossRef]
4.   Rai, R.K.; Nepal, M.; Khadayat, M.S.; Bhardwaj, B. Improving Municipal Solid Waste Collection Services in Developing Countries: A Case of Bharatpur Metropolitan City, Nepal. *Sustainability* **2019**, *11*, 3010. [CrossRef]
5.   Hannan, M.A.; Hossain Lipu, M.S.; Akhtar, M.; Begum, R.A.; Al Mamun, M.A.; Hussain, A.; Mia, M.S.; Basri, H. Solid waste collection optimization objectives, constraints, modeling approaches, and their challenges toward achieving sustainable development goals. *J. Clean. Prod.* **2020**, *277*. [CrossRef]
6.   Peng, L.; Gu, M.; Peng, Z. Study on the Optimized Mode of Waste Governance with Sustainable Urban Development Case from China's Urban Waste Classified Collection. *Sustainability* **2020**, *12*, 3706. [CrossRef]
7.   Jafari, H.; Hejazi, S.R.; Rasti-Barzoki, M. Sustainable development by waste recycling under a three-echelon supply chain: A game-theoretic approach. *J. Clean. Prod.* **2017**, *142*, 2252–2261. [CrossRef]
8.   Ghalehkhondabi, I.; Maihami, R. Sustainable municipal solid waste disposal supply chain analysis under price-sensitive demand: A game theory approach. *Waste Manag. Res.* **2020**, *38*, 300–311. [CrossRef]
9.   Van Engeland, J.; Beliën, J.; De Boeck, L.; De Jaeger, S. Literature review: Strategic network optimization models in waste reverse supply chains. *Omega* **2020**, *91*, 102012. [CrossRef]
10.   Mamashli, Z.; Javadian, N. Sustainable design modifications municipal solid waste management network and better optimization for risk reduction analyses. *J. Clean. Prod.* **2021**, *279*, 123824. [CrossRef]
11.   Yılmaz Balaman, Ş.; Wright, D.G.; Scott, J.; Matopoulos, A. Network design and technology management for waste to energy production: An integrated optimization framework under the principles of circular economy. *Energy* **2018**, *143*, 911–933. [CrossRef]
12.   Xu, Z.; Elomri, A.; Pokharel, S.; Zhang, Q.; Ming, X.G.; Liu, W. Global reverse supply chain design for solid waste recycling under uncertainties and carbon emission constraint. *Waste Manag.* **2017**, *64*, 358–370. [CrossRef] [PubMed]
13.   Olapiriyakul, S.; Pannakkong, W.; Kachapanya, W.; Starita, S. Multiobjective Optimization Model for Sustainable Waste Management Network Design. *J. Adv. Transp.* **2019**, *2019*, 3612809. [CrossRef]
14.   Mohammadi, M.; Jämsä-Jounela, S.-L.; Harjunkoski, I. Optimal planning of municipal solid waste management systems in an integrated supply chain network. *Comput. Chem. Eng.* **2019**, *123*, 155–169. [CrossRef]

15. Mohammadi, M.; Harjunkoski, I. Performance analysis of waste-to-energy technologies for sustainable energy generation in integrated supply chains. *Comput. Chem. Eng.* **2020**, *140*, 106905. [CrossRef]

16. Saif, Y.; Rizwan, M.; Almansoori, A.; Elkamel, A. Municipality solid waste supply chain optimization to power production under uncertainty. *Comput. Chem. Eng.* **2019**, *121*, 338–353. [CrossRef]

17. Mohammadi, M.; Jämsä-Jounela, S.L.; Harjunkoski, I. Sustainable supply chain network design for the optimal utilization of municipal solid waste. *AIChE J.* **2018**, *65*, e16464. [CrossRef]

18. Zimin, Z. Status and Countermeasures of Waste Collection and Transportation in Xicheng District of Beijing. *Environ. Sanit. Eng.* **2017**, *6*, 7–9.

19. Ying, L.; Shaohua, X.; Jing, Z. Study of Collection and Transportion System of Beijing Municipal Solid Wastes. *China Popul. Resour. Environ.* **2011**, *21*, 136–139.

20. Zhao, W.; Zhenshan, L.; Yabin, F.; Anying, J.; An, X. Characteristics and Influence Factors of the Energy Consumption and Pollutant Discharge of Municipal Solid Waste Transfer Stations in Beijing. *Environ. Sci.* **2013**, *4*, 2456–2463.

21. Yadav, V.; Karmakar, S.; Dikshit, A.K.; Vanjari, S. A feasibility study for the locations of waste transfer stations in urban centers: A case study on the city of Nashik, India. *J. Clean. Prod.* **2016**, *126*, 191–205. [CrossRef]

22. Yadav, V.; Bhurjee, A.K.; Karmakar, S.; Dikshit, A.K. A facility location model for municipal solid waste management system under uncertain environment. *Sci. Total. Env.* **2017**, *603–604*, 760–771. [CrossRef]

23. Sijun, L.; Changfeng, J.; Ming, D.; Yanli, F. Spatial layout optimization and location of urban waste buildings based on GIS-multicria. *Sci. Surv. Mapp.* **2018**, *43*, 45–49.

24. Wang, Y.; Li, J.; An, D.; Xi, B.; Tang, J.; Wang, Y.; Yang, Y. Site selection for municipal solid waste landfill considering environmental health risks. *Resour. Conserv. Recycl.* **2018**, *138*, 40–46. [CrossRef]

25. Demesouka, O.E.; Anagnostopoulos, K.P.; Siskos, E. Spatial multicriteria decision support for robust land-use suitability: The case of landfill site selection in Northeastern Greece. *Eur. J. Oper. Res.* **2019**, *272*, 574–586. [CrossRef]

26. Soroudi, M.; Omrani, G.; Moataar, F.; Jozi, S.A. A comprehensive multi-criteria decision making-based land capability assessment for municipal solid waste landfill sitting. *Env. Sci. Pollut. Res. Int.* **2018**, *25*, 27877–27889. [CrossRef]

27. Spigolon, L.M.; Giannotti, M.; Larocca, A.P.; Russo, M.A.; Souza, N.D.C. Landfill siting based on optimisation, multiple decision analysis, and geographic information system analyses. *Waste Manag. Res.* **2018**, *36*, 606–615. [CrossRef]

28. Liu, K.M.; Lin, S.H.; Hsieh, J.C.; Tzeng, G.H. Improving the food waste composting facilities site selection for sustainable development using a hybrid modified MADM model. *Waste Manag.* **2018**, *75*, 44–59. [CrossRef] [PubMed]

29. Kyriakis, E.; Psomopoulos, C.; Kokkotis, P.; Bourtsalas, A.; Themelis, N. A step by step selection method for the location and the size of a waste-to-energy facility targeting the maximum output energy and minimization of gate fee. *Env. Sci. Pollut. Res. Int.* **2018**, *25*, 26715–26724. [CrossRef]

30. Vecchi, T.P.B.; Surco, D.F.; Constantino, A.A.; Steiner, M.T.A.; Jorge, L.M.M.; Ravagnani, M.A.S.S.; Paraíso, P.R. A sequential approach for the optimization of truck routes for solid waste collection. *Process. Saf. Environ. Prot.* **2016**, *102*, 238–250. [CrossRef]

31. Carlos, M.; Gallardo, A.; Edo-Alcon, N.; Abaso, J.R. Influence of the Municipal Solid Waste Collection System on the Time Spent at a Collection Point: A Case Study. *Sustainability* **2019**, *11*, 6481. [CrossRef]

32. Yonggang, W.; Xiangxin, X.; Mei, L.; Long, Z. Research on City Waste Removal Routes Based on Genetic Algorithm. *Comput. Simul.* **2012**, *4*, 259–262.

33. Son, L.H.; Louati, A. Modeling municipal solid waste collection: A generalized vehicle routing model with multiple transfer stations, gather sites and inhomogeneous vehicles in time windows. *Waste Manag.* **2016**, *52*, 34–49. [CrossRef]

34. Chunping, L. Atmosphere Monitoring and Safe Protective Distance inside and outside Municipal Solid Waste Collecting Station in Beijing. *Urban Environ. Urban Ecol.* **2017**, *24*, 39–42.

35. Zsigraiova, Z.; Semiao, V.; Beijoco, F. Operation costs and pollutant emissions reduction by definition of new collection scheduling and optimization of MSW collection routes using GIS. The case study of Barreiro, Portugal. *Waste Manag.* **2013**, *33*, 793–806. [CrossRef]

36. Inghels, D.; Dullaert, W.; Vigo, D. A service network design model for multimodal municipal solid waste transport. *Eur. J. Oper. Res.* **2016**, *254*, 68–79. [CrossRef]

37. Chatzouridis, C.; Komilis, D. A methodology to optimally site and design municipal solid waste transfer stations using binary programming. *Resour. Conserv. Recycl.* **2012**, *60*, 89–98. [CrossRef]

38. Kinobe, J.R.; Bosona, T.; Gebresenbet, G.; Niwagaba, C.B.; Vinnerås, B. Optimization of waste collection and disposal in Kampala city. *Habitat Int.* **2015**, *49*, 126–137. [CrossRef]

39. Nguyen-Trong, K.; Nguyen-Thi-Ngoc, A.; Nguyen-Ngoc, D.; Dinh-Thi-Hai, V. Optimization of municipal solid waste transportation by integrating GIS analysis, equation-based, and agent-based model. *Waste Manag.* **2017**, *59*, 14–22. [CrossRef]

40. Xue, W.; Cao, K.; Li, W. Municipal solid waste collection optimization in Singapore. *Appl. Geogr.* **2015**, *62*, 182–190. [CrossRef]

41. Li, Z.; Guiqin, W.; Dian, W. Investigation and Analysis of Domestic Waste Output of Residential Area in Central Six Districts of Beijing. *Environ. Sanit. Eng.* **2018**, *26*, 59–62.

42. Benis, K.; Safaiyan, A.; Farajzadeh, D.; Nadji, F.; Shakerkhatibi, M.; Harati, H.; Safari, G.; Sarbazan, M. Municipal solid waste characterization and household wast behaviors in a megacity in the northwest of Iran. *Int. J. Environ. Sci. Technol.* **2019**, *16*, 4863–4872. [CrossRef]

43. Zakarya, I.A.; Fazhil, N.S.A.; Izhar, T.N.T.; Zaaba, S.K.; Jamaluddin, M.N.F. Municipal Solid Waste Characterization and Quantification as A Measure Towards Effective Solid Waste Management in UniMAP. *IOP Conf. Ser. Earth Environ. Sci.* **2020**, *616*, 12047. [CrossRef]
44. Chendi, W. *Research on Vehicle Dispatching and Optimization of Beijing Refuse Collection Based On Municipal Solid Waste Classification*; Beijing Jiaotong University: Beijing, China, 2018.
45. Feng, P.; Weixng, L.; Qi, G.; Al, E. *Particle Swarm Optimizer and Multi-Object Optimization*; Beijing Institute of Technology Press: Beijing, China, 2013; pp. 89–92.
46. Guo, X.; Yang, X. The economic and environmental benefits analysis for food waste anaerobic treatment: A case study in Beijing. *Env. Sci. Pollut. Res. Int.* **2019**, *26*, 10374–10386. [CrossRef] [PubMed]

 *sustainability*

Article

Empowering a Sustainable City Using Self-Assessment of Environmental Performance on EcoCitOpia Platform

Ratchayuda Kongboon [1], Shabbir H. Gheewala [2,3] and Sate Sampattagul [4,5,*]

1   Research Unit for Energy, Economic and Ecological Management, Science and Technology Research Institute, Chiang Mai University, Chiang Mai 50200, Thailand; ratchayuda.k@cmu.ac.th
2   The Joint Graduate School of Energy and Environment (JGSEE), King Mongkut's University of Technology Thonburi, Bangkok 10140, Thailand; shabbir_g@jgsee.kmutt.ac.th
3   Center of Excellence on Energy Technology and Environment, PERDO, Ministry of Higher Education, Science, Research and Innovation, Bangkok 10140, Thailand
4   Department of Industrial Engineering, Faculty of Engineering, Chiang Mai University, Chiang Mai 50200, Thailand
5   Excellence Center in Logistics and Supply Chain Management, Chiang Mai University, Chiang Mai 50200, Thailand
*   Correspondence: sate@eng.cmu.ac.th

Abstract: In Thailand, many municipalities lack the information to guide decision-making for improving environmental performance. They need tools to systematize the collection and analysis of data, and then to self-assess environmental performance to increase efficiency in environmental management toward a sustainable city. The aim of this study is to develop a platform for self-assessment of an environmental performance index. Nonthaburi municipality, Hat Yai municipality, and Yasothon municipality were selected to study the work context for six indicators, viz., energy, greenhouse gas, water, air, waste, and green area, which were important environmental problems. The development of an online system called "EcoCitOpia" divides municipality assessment into four parts: data collection, database creation, data analysis, and data display. The municipality can use the system for the assessment of environmental performance and the creation of a separate database based on indicators. The system can analyze the results and display them in the form of radar graphs, line graphs, and tables for use in public communication that will lead to cooperation in solving environmental problems at the policy level for urban development to meet the Sustainable Development Goals.

Keywords: sustainable city; sustainable development; environmental performance; online platform; municipalities

Citation: Kongboon, R.; Gheewala, S.H.; Sampattagul, S. Empowering a Sustainable City Using Self-Assessment of Environmental Performance on EcoCitOpia Platform. *Sustainability* **2021**, *13*, 7743. https://doi.org/10.3390/su13147743

Academic Editor: Tan Yigitcanlar

Received: 15 June 2021
Accepted: 7 July 2021
Published: 12 July 2021

Publisher's Note: MDPI stays neutral with regard to jurisdictional claims in published maps and institutional affiliations.

1. Introduction

The current environment crisis, whether it is climate change, lack of production resources, waste crisis, lack of clean water, and air pollution, affects the world population in terms of living and health. The impacts are not limited to one country but affect the whole world. One of the causes of the current environmental crisis is the relentless economic growth as a result of the exponential increase in the world population. The United Nations stated that by 2050, 68% of the earth's population is projected to be urban, which is about 14% more than in 2018 [1], leading to the increasing demand for resources while the remaining resources are on the verge of disappearing. In addition, our world has been using natural resources but, ultimately, there is a limit on economic growth that depends on the natural costs of the world. Today, cities are responsible for over 78% of the global energy consumption and over 60% of greenhouse gas (GHG) emissions, but less is known about how they drive resource use and sustainability impacts [2]. Therefore, global community cooperation in driving each country to set goals for sustainable urban development is widely discussed nowadays. Thailand is one of the countries responding to the need

137

for sustainable development; this can be seen from the 20-Year National Strategy (2018–2037) [3], the main master plan for national development and the Sustainable Development Goals (SDGs) [4]. This has led to the development of plans, requirements, and regulations, including guidelines for all sectors in order to proceed in the same direction.

Sustainable development is the long-term development plan to fulfill the next generation's needs through planning today. It is a notable notion worldwide, especially in developing and emerging countries [5]. A sustainable city is a city that is designed taking into consideration environmental impacts, inhabited by people dedicated to minimization of required inputs of energy, water, and food, and waste outputs of heat, air pollution, CO_2, methane, and water pollution [6]. The environmental indicators have become a fundamental tool in environmental assessment at detailed local, regional, and national levels [7]. They are cost-effective and powerful tools for tracking environmental progress and measuring environmental performance [8]. Various indicators have been used to create a "sustainable city" index by some institutions in Thailand and other countries. Most of the monitoring tools are based on a set of indicators, with the intention to assess the performance of local governments, identify actions for sustainable urban development, and review the challenges facing sustainable urban development. The use of indicators to emphasize the relevance of environmental data has many advantages [9]. The set of indicators on which the tools base their evaluation usually encompasses between 10 and 30 indicators. These can be either qualitative or quantitative and are inspired or derived from a number of sources [10]. For example, the Urban Ecosystem Europe (UEE) tool is the result of collaboration between DEXIA, an international banking group, and Ambiente Italia, a research consultancy and creator of the tool [6,10]. The UEE assessment is based on a questionnaire comprising 25 indicators [6,10]. European Green City Index (EGCI) is an indicator developed by *The Economist* Intelligence Unit (EIU) in collaboration with Siemens. The study indicators are divided into eight categories consisting of CO_2, energy, buildings, transport, waste and land use, water, air quality, and environmental governance [11–15]. In Thailand, the Office of Natural Resources and Environmental Policy and Planning (ONEP) proposed the implementation frameworks of the eco-city development of the indicators under the project to drive development, according to the concept of an eco-city model for sustainable urban and rural environmental management. The project defined eighteen indicators for evaluating the eco-city.

In this age of information, sustainable urban development will involve a lot of data and a variety of data formats. Previously, the data were of a certain form and were sourced only from the system within the department; but nowadays, data sources come from many different organizations and forms, including social media, which are becoming increasingly more important, with new data generated every millisecond. The current volume of data and information collected, stored, shared, and used in urban agglomerations is almost unlimited [16]. Therefore, the most important part of city management at the present time is information; the executive decisions should be based on reliable information. Currently, however, according to the municipal administration of Thailand, sufficient qualitative and quantitative data have not been collected for use in environmental management. Most municipalities collect environmental data qualitatively rather than quantitatively; therefore, troubleshooting or development may not be able to measure performance properly. For this reason, three issues were raised that led to this study. First, the leaders make environmental management decisions by using the information or the thoughts of one person. In fact, this choice is based on the fact that key strategic decisions on sustainability-related issues are mostly made on an upper management level [17]; second, in the age when information is vital to environmental management, how many municipalities are using the information to resolve the problem; and last, if the municipality does not have information now, how can they manage the environment effectively?

In Thailand, the municipality is an organization responsible for various functional contexts. The data are stored separately by the department in the form of hard or soft copy. In terms of operation, this often causes data problems such as lost data, missing data, retrospective data collection taking a long time, and no information management system supporting the use on demand. With a systematic storage database, users can share work-related information without duplication of information and avoid data conflicts, including data that are accurate, reliable and with a standardized collection system. It is important to develop the data to be appropriate and consistent with the municipality's use. Part of the problem that arises is that the ministries, departments, or offices have duplicated data requests or duplicated assessing of the municipality operational efficiency. This may cause the municipality to gather the same data several times, and if there are new indicators, there may be problems such as having no data for reporting or assessing. Based on the municipality's obligations, the data are used in two forms of assessment: (1) a mandatory assessment based on the obligations, such as a local performance assessment (LPA) with the Department of Local Administration (DLA) [18], and (2) a voluntary assessment of the project participation, such as carbon footprint assessment of the organization with the Thailand Greenhouse Gas Management Organization (TGO).

The challenge of this study is when the municipality is an organization that must be assessed with environmental indicators. Many indicators share the same data set but are displayed in different units. The current reality in gathering the municipal data is to collect on demand, and sometimes the work is repetitive, which wastes time to gather the same data while using reports from different departments. Therefore, creating a municipal information management system with a data platform can support both the organization's internal and external data and create a central data storage system that can be used in the formulation of policies, plans, and activities for urban development and assessment of the efficiency of the municipality's environmental performance.

According to the municipality's mission, it is important to use information or analysis results to communicate with the public in order to achieve understanding, awareness, warning, and prevention, resulting in cooperation driven by the same goal. If every municipality in Thailand has an established database, a national database can be created that will be greatly useful in the development of the country. Therefore, the objective of this study was to study a platform for a self-assessment environmental performance index called EcoCitOpia. The platform is a tool to determine a city's progress toward sustainability and to indicate how the decision maker can improve the state of sustainability in municipalities.

2. Materials and Methods

The municipal environmental performance assessment allows the municipality to measure its own performance. Moreover, the development of the environmental data collection platform will enable municipalities to manage the information more efficiently. Therefore, the efficiency assessment and data collection platform needs to be flexible and suitable for the municipal work system in Thailand, since the aim is to develop a city based on accurate and reliable information. If the municipality has good information on the management, it may be beneficial in many areas, such as cost-effective budgeting, increasing the productivity of employees in the organization, being able to measuring before and after performance, and providing information for decision-making. The concept of this study is presented in Figure 1.

2.1. Study Municipality

In this study, a few municipalities were selected to study the municipal work system, and to develop a performance evaluation model and a data collection platform that can be applied to all municipalities in Thailand. The selected municipalities were chosen from those that volunteered to participate, because the data collection must be authorized by the mayor and the team set for providing information. Currently, Thailand has a total of 7850 municipalities [19], but in this study, three municipalities with different topographies, policies, and working processes were selected, as shown in Figure 2: Nonthaburi municipality, Hat Yai municipality, and Yasothon municipality.

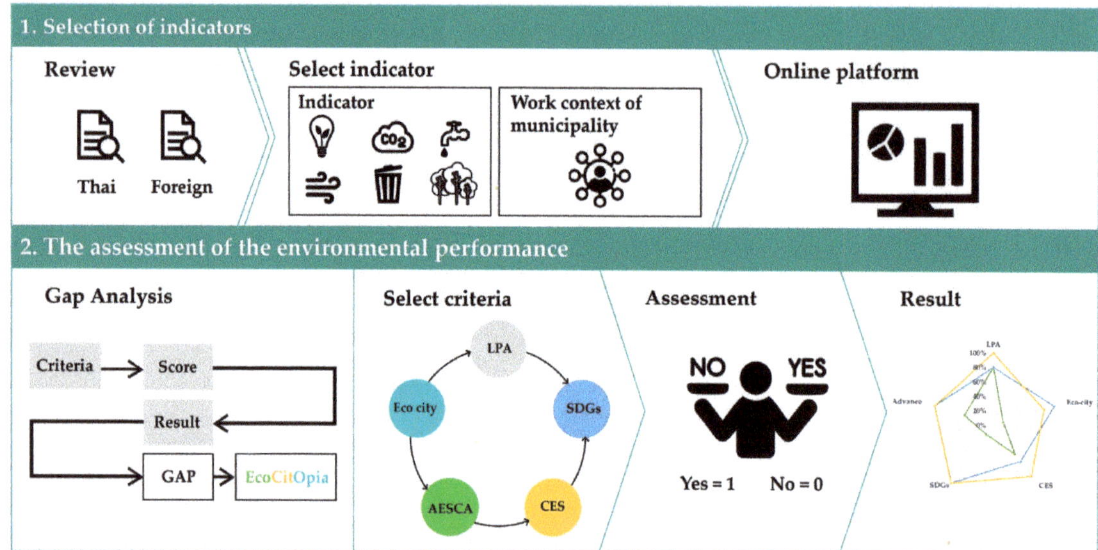

Figure 1. Concepts for the selection of indicators and assessment of environmental performance.

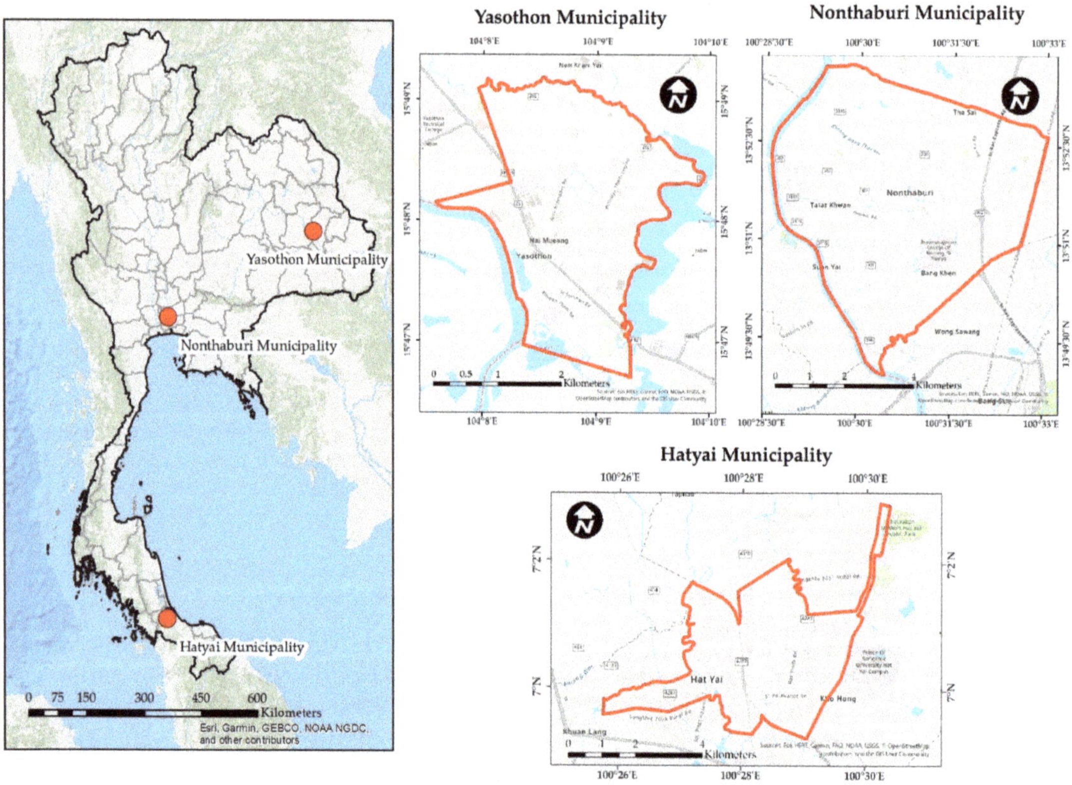

Figure 2. The three municipalities in this study.

Nonthaburi is a large municipality located in the district of Mueang Nonthaburi, Nonthaburi Province. There are over 250,000 people within the municipality, making it the most populous municipality in Thailand because it is located in a suburb that has a border with Bangkok and is part of the Bangkok metropolitan region. It has an administrative boundary of 38.9 square kilometers [20]. There are four important aspects of operating policies for sustainable urban development on the environment: (1) solid waste management, (2) sewage management, (3) wastewater management, and (4) environmental quality monitoring with the goal to promote and support the environment, health and safety of the community by acting regarding the safety, occupational health, and environment for a sustainable and balanced development. The important environmental problems are (1) the solid waste management problem, because the amount of solid waste tends to be higher; (2) problems in wastewater management, where the quality of water in the municipality's canal and river are currently deteriorated, and (3) air pollution management problems where the amount of dust particles is up to 2.5 microns, which is higher than the standard in winter. Nonthaburi is a municipality that is ready for information requests and personnel. There are both public and private departments to provide information such as electricity consumption data from the Metropolitan Electricity Authority, water usage data from the Metropolitan Waterworks Authority, and data on the amount of solid waste from the Nonthaburi Provincial Administrative Organization.

Hat Yai, regarded as the largest city in the south, is a municipality located in Hat Yai district, Songkhla province. It is an important economic, commercial, and transportation center of the south and is the third most populous municipality in Thailand, after Nonthaburi and Pakkred municipalities. With an administrative area of 21 square kilometers covering the entire Hat Yai sub-district area, it has a population of more than 150,000 people [21]. There is an action plan called "Green City" for 2017–2027 proposing (1) a garden city aiming to explore, maintain, and develop green spaces; (2) a city of waste minimization focusing on solid waste management, wastewater management, and air pollution management; (3) a city of energy efficiency consisting of energy for transport by supporting monorail construction, promoting the use of bicycles, and, on the electric power side, promoting the use of renewable energy and increasing energy efficiency use; (4) a city with sustainable consumption and tourism by encouraging households to implement the sufficiency economy guideline and establishing the environmental division; and (5) a city with flood resilience. In terms of environmental problems, these include (1) the amount of solid waste that tends to be higher, (2) at least 20% of wastewater that has not been treated in the treatment system because the collection system is clogged and the wastewater smuggling of many establishments affects water quality in the canals, (3) air quality that was found to exceed the standard, caused by the denser traffic within the city, (4) urban green areas and relaxation areas which are inadequate for the population.

At present, Hat Yai has an ongoing environment policy. Therefore, it is well prepared in terms of executives, teams, and information, including having a good understanding of how to utilize the environmental database, since it is a municipality that focuses on measuring performance through project participation. Hat Yai is also aware of the limitations of data storage in traditional working models, being a municipality with a clear goal to become a city with environmental operations under sustainable urban development guidelines.

Yasothon is a municipality in Mueang Yasothon District, Yasothon Province, with an area of 9.7 square kilometers, covering the entire Nai-Mueang sub-district, and with a population of more than 20,000 people [22]. A sustainable urban development policy and plan were established in the development strategy and approach to Yasothon city development outlined in the 4-Year Local Development Plan (2018–2021). It has been implemented through projects such as the carbon footprint project in Yasothon, the beautiful front yard project, the garbage bank project, the green road project, the global warming problem-solving project, and the green office project. The most important environmental problems are related to solid waste management, energy management in the road transport sector that primarily uses fossil fuels, the public transportation system that does not cover the administrative area, and the use of electric power that has increased.

For Yasothon, there is environmental action and the results are measured through participation in activities such as Thai municipalities caring for global warming, and the ASEAN awards for sustainable environment cities. The executives and staff are also well prepared and experienced in establishing an environmental database and recognizing problems in gathering the data; the comments and suggestions from work can be excellently applied for developing online databases to carry out this research.

Therefore, all three municipalities have the potential of executive, personnel, and cooperation from data owners. The experience of these three municipalities can be transformed into establishing an online database that is suitable and flexible to use, and there are concrete benefits for the municipality in the management of the information of the city. It can also be a municipality model that will establish the best practice for other municipalities to learn from.

2.2. Selection of Indicators

Sustainability indicators are increasingly recognized as a useful tool to provide policy information in support of the environment [23]. Thus, several approaches have been developed to assess environmental performance based on indicators. For example, the UEE tool, which is based on a questionnaire comprising 25 indicators [6]; the EGCI comprising 17 quantitative indicators; the European Green Capital Award (EGCA) with a set of 12 environmental indicators [6,14]; the Sustainable Cities Index (SCI) exploring the three demands of People, Planet and Profit to develop an indicative ranking of 50 of the world's leading cities [24]; the Sustainable Development Goals (SDGs), which are a set of 17 goals designed to lead global development [25]; the ASEAN Environmentally Sustainable City Award (AESCA) program focused on indicators for clean air, clean land, and clean water [26]; the environmentally sustainable city assessment (ESC) guide developed for urban development according to the livable city [27]; and the Eco-city indicator for sustainable development assessment consisting of 18 indicators [28].

The selection of indicators depends on the characteristics of the area under study [29]. The indicators should fulfill the following criteria: (1) they have either a direct or an indirect linkage to city development; (2) they can manifest the diverse facets of sustainability performance; and (3) the raw data for the indicator system is accessible and measurable [30].

Therefore, the selection of indicators in this study consisted of three steps. The first step was the review of the environmental indicators of both Thai and foreign assessment criteria to group the indicators and categorize the types of data that need to be collected for use in assessing the results of each indicator, as shown in Table 1. Table 1 gives a review of the nine criteria that were included in this study. For each included criterion, the report and methodological background documents that were found on the website were studied. For the next step, the indicators were selected by cluster analysis. This is a popular technique in unsupervised learning, where the objective is to group the studied data in different clusters according to the similarity of data characteristics [31]. The indicator must be classified as a data type and the data must be associated with the contextual work of the municipality to establish which department is the owner of the data, and what the format is for collecting the data. The last step was the study of the quantitative data that need to be analyzed in order to measure results in each of the indicators compared to each evaluation criterion and consider sources of the data that are accessible for collection. However, the focus of the selection of indicators in this study was data acquisition, which was the data that the municipality had already collected or was able to collect according to the work context.

Table 1. Review of environmental indicators of the nine criteria included in this study.

Criteria	Initiator	Spatial Scope	Indicator	Index Score	Result
1. UEE [6]	Ambiente Italia	Europe/32	Air quality, Acoustic environment, Water, Energy, Waste, Transport, Green areas and land use, Building, CO_2, Health, Equity, Education, Participation	0–100	Spider-web diagram
2. EGCI [6,14]	EIU	Europe/30	CO_2, Energy, Building, Transport, Water, Waste and land use, Air quality	0–10	Spider-web diagram
3. EGCA [6,14]	European Commission	Europe	Climate change, Local transport, Green urban areas, Nature and biodiversity, Air quality, Quality of acoustic environment, Waste, Water, Wastewater, Eco-innovation, Energy, Environmental management	0–100	Spider-web diagram
4. SCI [25]	Arcadis	World	Green spaces, Energy, Air pollution, Greenhouse gas emissions, Waste management, Drinking water	0–100	Spider-web diagram
5. SDGs (6, 7, 11, 12, 13, 16) [26,27]	United Nations	World	Water, Wastewater, Energy, Building, Transport, Air quality, Waste, Green areas and land use, Education, Equity, Safety, Health, Participation	0–100	Spider-web diagram
6. AESCA [28]	ASEAN	ASEAN	Clean air, Clean land, and Clean water	0–100	N/A
7. LPA (Section 4) [29]	DLA	Thailand	CO_2, Water, Waste, Air, Energy, Green areas	0, 1, 3, 5	N/A
8. ESC [30]	DEQP	Thailand	Water, Air, Energy, Wastewater, Waste, Green areas	0, 1, 3, 5	N/A
9. Eco-City [31]	ONEP	Thailand	Water, Energy, Wastewater, Waste, Air, Green areas	0, 1, 3, 5	N/A

According to the study of environmental indicators, the same and different indicators were available depending on the principle of each criterion. In this regard, if the criteria were related to municipal operations, each assessment criterion could be classified in order to classify the groups of indicators, such as greenhouse gas, energy, water and sanitation, solid waste, air quality, transportation, green space, and land use. The results from the sub-criteria in the assessment to group the indicators showed the details of the measurement results of each criterion. This made it possible to separate the types of data in terms of quantities that needed to be collected for the calculation of the indicator index, including classifying the sources of information. The criteria used in this study were LPA, Eco-city, sustainable environmental cities, and SDGs, which were related to municipalities based on their obligations.

2.3. The Assassessment of the Environmental Performance

Environmental performance assessment started from reviewing the criteria for assessing environmental performance both in Thailand and abroad that were relevant to the operation of the municipality in order to learn the details, such as assessment form, criteria used in the assessment, scoring criteria, and the demonstration of the result. The results of the review were based on the principle of gap analysis to analyze the existing evaluation criteria, whether not the environmental dimensions linked to the working context of the municipality should be considered. The goal of the performance assessment was to let the municipality know its own working status and raise awareness of the importance of developing an environmental database with relevance, completeness, consistency, accuracy, and transparency.

2.4. The EcoCitOpia Platform

The development covered performance assessment, establishing a database, data analysis, and demonstrating the result to be utilized, which the system design considered to be the type of user, such as the user group, each group pattern of entry, and the level of access to the user's information, as well as the security of the information. In this study, we developed an online platform by considering the connections to existing data storage formats that were already in the municipal works, and achieving a modification of the traditional data storage format used for paper or computer data storage. This process involved changing to a new form of data storage by storing in online databases to order to fix data loss problems or difficulty in searching for data when there was a need. The details of each part are as follows.

2.4.1. The Collection of the Data

The study found that primary data and secondary data used in the evaluation of environmental information were either from within the organization or outside the organization, so the data collection model concept was a flexible online platform compatible with municipal data collection formats, which may have different follow-up data collection formats. The information collected should be comprehensive for use in various dimensions, such as the management of the city, the measurement of results of projects or activities, and the use of information for funding requests from the different funding sources. Therefore, data types, analysis method, and the method of each selected indicator result demonstration were examined in establishing an online storage platform

2.4.2. Database

The data collected through the online platforms were stored in a separate database of each indicator to facilitate the use or analysis of the data. The data stored in the database were stored according to the calendar year, separately by month.

2.4.3. Data Analysis

Most indicator data were queried in terms of ratio, percentage, increase or decrease value. Therefore, the data analysis model utilized a format that could be used to create equations, ratios, and percentages.

2.4.4. The Demonstration of the Result for Utilization

Currently, in the age of information, the focus is to utilize information. In this study, the result demonstration model supported municipal uses such as radar graphs, was easy to understand, and could be summarized in a single image, such as a linear graph for viewing data trends useful in policy management to visualize the trends of the tracked data and to show the results in the form of a table.

3. Results

3.1. Environmental Indicators

The results of classifying the categories of environment indicators are shown in Table 2. These indicators allowed assessing the environmental performance of municipalities in six categories: Greenhouse gases, Energy, Water, Air quality, Waste, and Green area, which were analyzed in the criteria in both Thailand and abroad. The municipalities can effectively collect data according to their work context to be utilized in the city's environmental management.

Table 2. The result of classifying categories of environment indicators for municipalities.

Criteria	Category [1]					
	C	**E**	**WSS**	**SW**	**AQ**	**GL**
LPA	✓	✓	✓	✓	✓	✓
ESC	✓	✓	✓	✓	✓	✓
AESCA			✓	✓	✓	✓
Eco-city	✓	✓	✓	✓	✓	✓
SDGs		✓		✓	✓	✓

[1] C = Greenhouse gases, E = Energy, WSS = Water supply and sanitation, SW = Solid waste, A = Air quality, GL = Green area and land use.

However, when studying the details of indicators in each category, it was found that even though they were the same indicators, they had different descriptions of the data used to measure the results. For example, for the indicators of air quality, LPA was measured as the amount of activity that the municipality undertakes to manage the air quality, while the Eco-city considered the number of days in one year with pollutants in the air that exceeded the standard. According to the foreign criteria, EGCI tracked air quality measurements of nitrogen dioxide (NO_2), ozone (O_3), and sulfur dioxide (SO_2) in micrograms per cubic meter compared to the European standard. The EGCA criteria considered particulate matter size up to 2.5 and 10 microns ($PM_{2.5}/PM_{10}$) and nitrogen dioxide (NO_2). Therefore, a data acquisition platform needed to be flexible for municipalities to use in assessing environmental performance according the indicators of each criterion. Table 3 lists the categories and indicators of the data to be collected that were relevant to a data acquisition platform.

Table 3. List of categories and indicators.

Category	Indicator	Description
Energy	Electricity consumption	Total annual electricity consumption, in kilowatt-hour (kWh) per capita.
CO_2	CO_2 emissions	Total annual greenhouse gas emissions, in tonnes per capita.
Water	Water consumption	Total annual water consumption, in cubic meters (m^3) per capita.
	Wastewater treatment	Percentage of households connected to the sewage system.
	Wastewater management efficiency	Percentage of the total annual amount of wastewater entering the treatment system.
	Wastewater treatment system efficiency	Quality of treated wastewater in milligram per liter (mg/L).
Waste	Municipal waste production	Total annual municipal waste collected, in kilogram (kg) per capita.
	Municipal waste recycling	Percentage of municipal waste recycled.
Air	Air quality	Annual daily mean of CO, NO_2, O_3, SO_2, Pb, PM_{10}, and $PM_{2.5}$ emissions in microgram per cubic meters ($\mu g/m^3$).
Green area	Green area	Green area, in square meters per capita.

3.2. The Environmental Performance

The results from gap analysis showed that municipal work had various criteria to measure results in environmental dimensions, either in the form of duty criteria, such as the LPA, where municipalities were assessed annually, or the criteria that were based on voluntary actions to challenge the municipality's performance, such as the sustainable environmental city assessment or the ASEAN Environmentally Sustainable City Award. At present, the evaluation criteria have started to use more quantitative measurements, and this makes it difficult for municipalities to gather information to be used in measuring results. Therefore, the study found that, currently, there is a lack of performance assessment criteria for municipalities that let the municipality know what their state of operations is at the moment, when compared to each of the criteria. It should also be a simple form of assessment that does not require a large amount of data to be assessed. This study used "Yes or No" questions to assess the performance of each indicator. The six criteria used to reference were LPA, CES, AESCA, SDGs and Advance. The Advance criterion was set specifically for this study, which assessed the municipality data collection continuously and classified the data according to the sources. The basis of this study was to evaluate the efficacy of quantitative data for evaluation in Advance, such as using data for various simulations or predicting trends. The scoring method 'Yes' was equal to 1 point, 'No' was equal to 0 points. The answers were calculated and shown in the percentage of answers compared to total answers. The score criteria used in assessing performance are shown in the Table 4.

Table 4. Scoring criteria for assessing performance in each of the indicators.

Indicators	Score						Total
	LPA	**Eco-city**	**CES**	**AESCA**	**SDGs**	**Advance**	
Energy	6	5	13	-	6	6	36
CO$_2$	5	7	16	-	6	7	41
Waste	8	9	12	6	9	12	56
Water	7	6	10	5	6	8	42
Air	3	2	11	-	3	10	29
Green space	2	4	3	2	3	2	16

3.3. Self-Assessment Environmental Performance Index on EcoCitOpia Platform

The EcoCitOpia platform is an online form available at "http://www.ecocitopia.org (accessed on 19 April 2021)". A working system is shown in Figure 3. As a result of the study of the three model municipalities, it was adapted to be a flexible platform suitable for municipalities of Thailand. The EcoCitOpia platform divides a system into two parts: the first part is the environmental performance assessment beginning with the municipalities answering questions on each of the indicators to assess the score. The system processes the answers and compares them with the criteria from Table 4 to show the results of the performance assessment for each of the indicators. The results of the municipal assessment can be utilized in municipal work and the information can be released to the public, aiming to create cooperation in solving environmental problems. When the municipalities know their state of work, 'Yes' means to continue and make it better, but 'No' means it should be improved to 'Yes', which would allow municipalities to develop procedures to make their work more efficient.

Figure 4 shows the results of the performance assessment compared to the assessment criteria of the three municipalities for each of the indicators in the base year 2019. LPA, which is a fundamental criterion that all municipalities in Thailand must assess annually, is based on the assessment results in terms of energy, water, air and green space, and the three municipalities had an operational performance equal to 100%. Other criteria, assessed by the municipalities voluntarily, found that the reference performance in each of the three municipalities had a value greater than 50%. However, if the results were taken separately for each indicator, they would have the lowest GHG performance compared to Hat Yai and Nonthaburi, except for LPA, with an 80% rating equal to Hat Yai. The air problem was a major one for Hat Yai, especially for PM$_{2.5}$. As a result, Hat Yai is equipped with real time monitoring and measurement sensors, causing it to have a more operational performance than the other two municipalities.

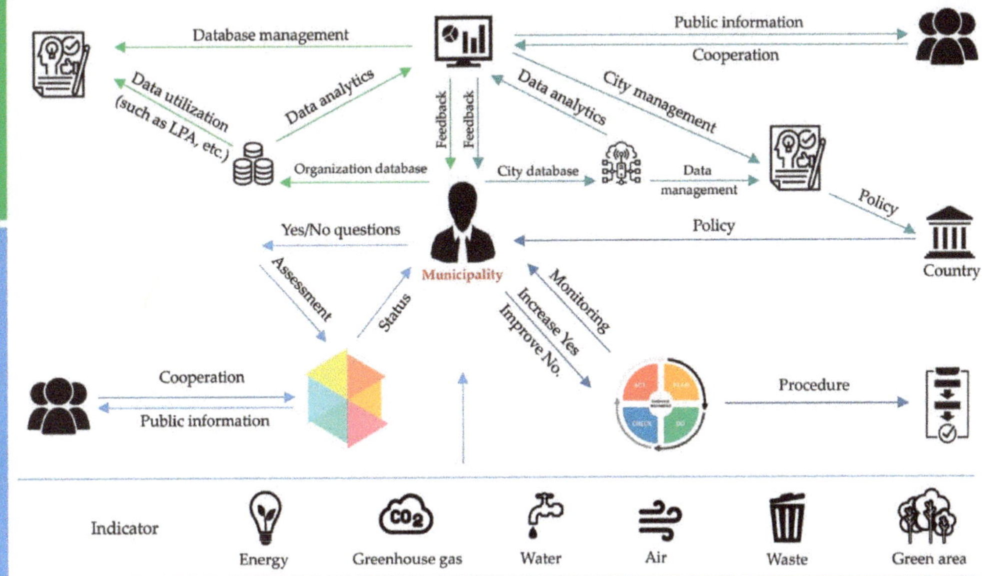

Figure 3. The operating system of EcoCitOpia platform.

In terms of the waste indicator, Yasothon was the municipality with the best performance with an assessment of 100% in all criteria, except for Advance, which was 91.67%, because they did not have plans to gather waste data based on their sources of waste such as residential, commercial business, and industrial sources. As for the water indicators, Hat Yai and Yasothon had the same performance assessments across all criteria; Nonthaburi had the same performance assessments as Hat Yai and Yasothon, except for AESCA, which was 80%, less than the other two for which the assessment results were 100%. The Hat Yai and Nonthaburi green space indicators had an operational performance of 100% across all evaluation criteria. For Yasothon, the assessment results were the least because it collected fewer data on green spaces, resulting in it having less performance than the other two municipalities.

Considering the overall performance assessment shown in Figure 5, it was found that the three municipalities had policies and operations that emphasized the environment continuously in accordance with their own context and duties. The municipality with the best performance was Hat Yai. The result of Nonthaburi's greenhouse gas assessment was 41.46%, less than Yasothon and Nonthaburi, which were 87.80 and 78.05%, respectively. Nonthaburi started its greenhouse gas action policy in 2017, while Hat Yai and Yasothon have been operating since 2012 and 2013, respectively, so its readiness in policies, setting goals, and operating activities to reduce greenhouse gas emissions, including the preparation of the greenhouse gas database, was only one year old.

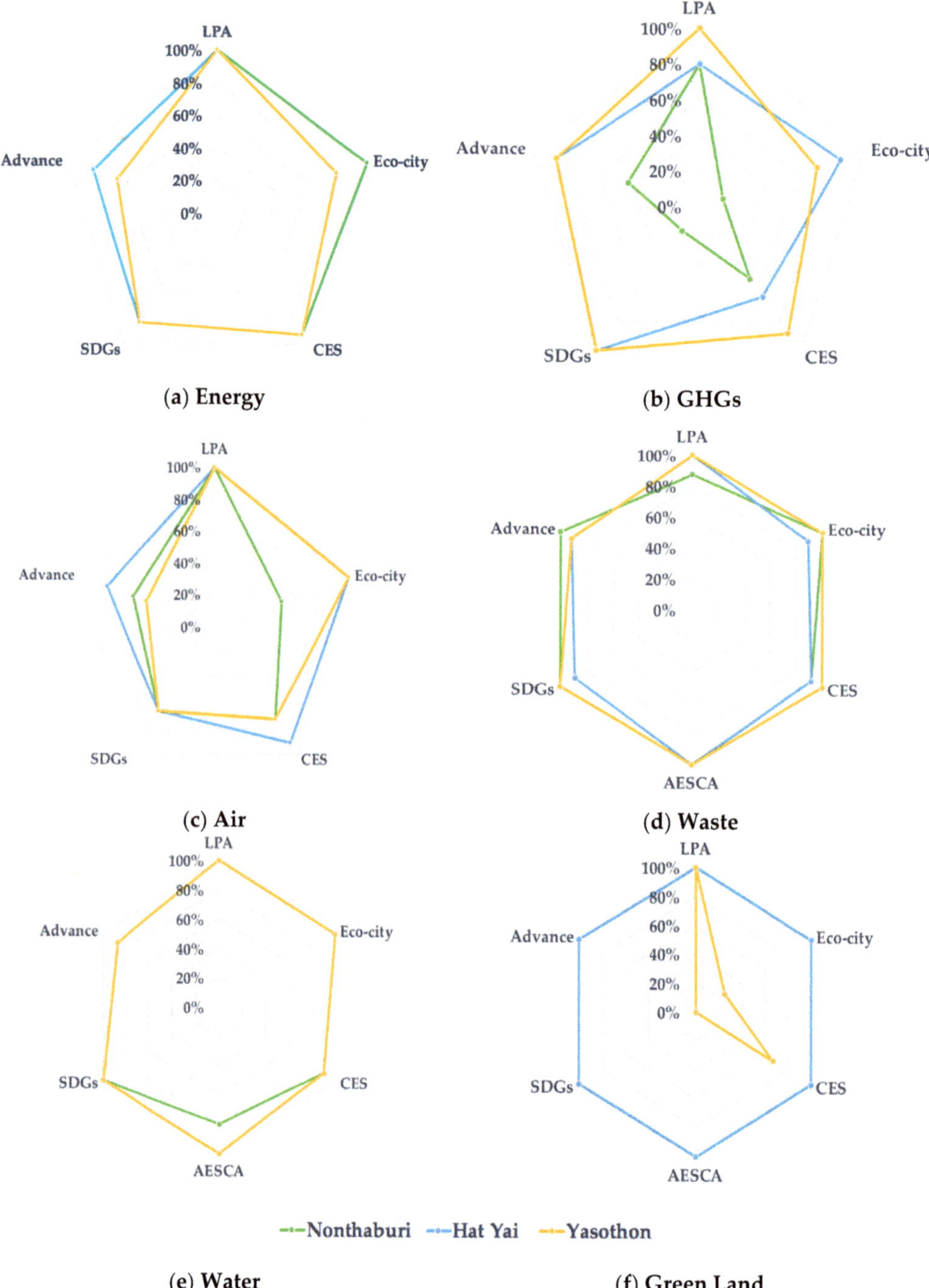

Figure 4. Results of all 6 indicators of environmental performance assessment for the 3 municipalities.

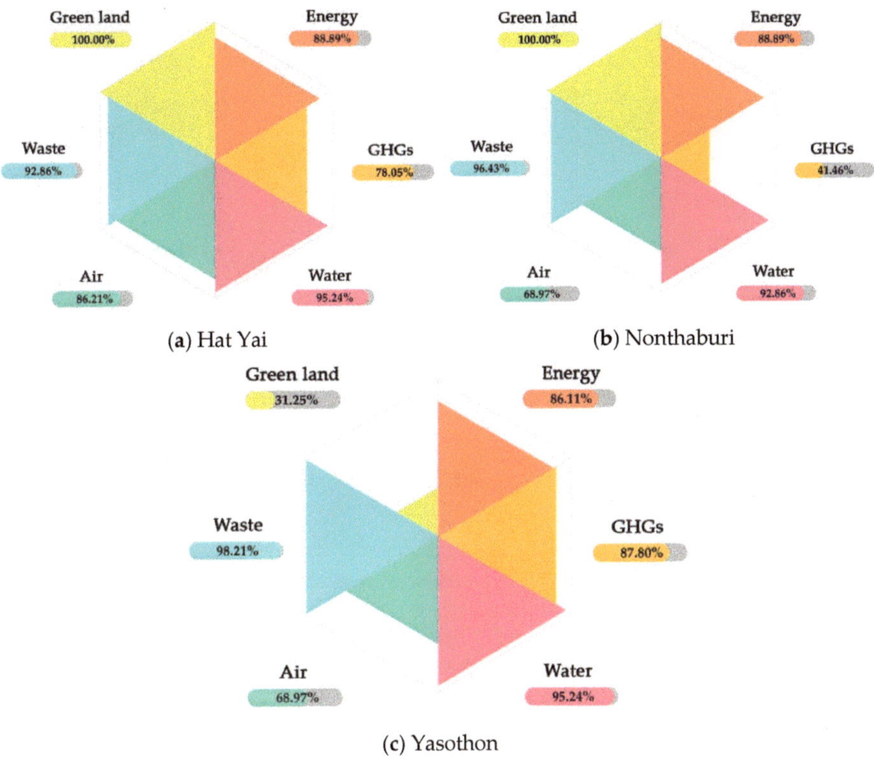

(**a**) Hat Yai

(**b**) Nonthaburi

(**c**) Yasothon

Figure 5. The results of the environmental performance assessment of 3 municipalities.

As for the air indicators, Hat Yai had the highest score at 86.21%, while Nonthaburi and Yasothon were equal at 68.97%. The green area score was the lowest at 31.25% for Yasothon, as it had less operational performance than the other two because there was no continuous database to be used effectively in a policy or plan.

The results of the assessment of performance from this study helped the municipalities to know their current state of performance, which could lead to consistent policies and plans and increase their operational efficiency in every indicator. This may lead to improvements in the municipality's procedures to be more efficient, and it reflects the importance of establishing a quantitative environmental database that can measure the performance.

For creating a database, data were divided into organizational level and city level. The organizational-level data were intended for the municipalities to use to manage data within the organization. The city-level data were intended for use in policy-level data management for further development of the city. The working system shown in Figure 6 was divided into four parts. The first part was the data collection platform, which included two parts: the first was the basic municipal information, such as population, number of households, and the municipal administrative area, and the second part to collect data on each indicator, such as the consumption of electricity, water consumption, and amount of waste. The data were stored in a database classified by indicators, to be used for data analysis such as the rate of electricity consumption per capita or the rate of water consumption per capita. The data obtained from the analysis can be displayed in the form of a line graph and a table, separated by yearly data.

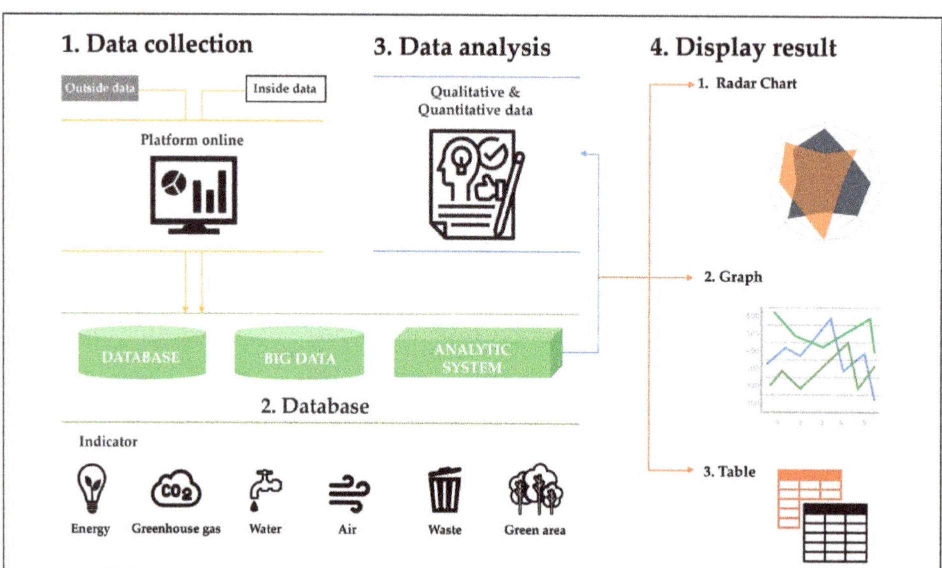

Figure 6. The database system for the EcoCitOpia platform.

4. Discussion

Environmental databases can be used to solve environmental problems if the municipalities have the information available according to the situation. The information should be updated and have quality in both data collection and verification before it can be used for policy planning, work plans, and action plans in various activities that will be limited by people, budget, and duration. Good information allows the municipalities to make precise decisions about the operation, including the use of information to request budget funds for the development of the city in the work plan. The key to all these is having available information. In the past, municipalities had several environmental operations based on their work context, mostly qualitative measures. However, they may not have been able to collect quantitative data that could lead to numerical measurements.

Starting from the energy indicator, electric consumption data from the database allows the city to know the overall electricity consumption trends and classifications such as housing, and small, medium, and large commercial businesses. If cities aim to reduce electricity consumption, they may focus on increasing energy efficiency by using various technologies and increasing the use of renewable energy. For greenhouse gas, the emissions database provides the municipality with the knowledge of significant sources of greenhouse gas emissions to define the policies, work plans, or activities that can be used to reduce the amount of greenhouse gas emissions appropriately. In terms of water, the management of wastewater is mainly focused on. However, when assessing the environmental performance, it is concerned with water for consumption, including not only the water supply system but also the natural water sources. Therefore, having a comprehensive database will enable the city to manage all dimensions of water problems. On the air side, tracking air quality data at different points in the city can be used for public communication. This includes knowing activities or sources that cause air quality problems, which will enable the city to monitor and resolve them promptly. For Hat Yai and Nonthaburi municipalities, which have problems with $PM_{2.5}$, continuous measurement data are used in planning action guidelines to prepare for the crisis that will occur each year. Whether or not it is a form of alarm to the public, if the sector information is known, there may be the creation of safe zones in office buildings, schools, and various sectors.

The waste indicator is one that all municipalities are continually monitoring, but the previous problem was that the information was scattered and the data reporting in each section did not match. The database from this study allows the municipalities to verify their own waste data to be accurate, usable, referenced, and corresponding. As a result of the activities that the municipalities operate, if the data are collected continuously, they can be used to measure results. For indicators of green spaces, most municipalities only have information on the areas that are in the municipalities, such as parks, roads, and road islands, but do not have information on the privately owned areas that can be used for management. Therefore, establishing a comprehensive database by applying modern technology such as GIS to help collect data can assist municipalities to increase the work performance.

5. Conclusions

Local government organizations at the municipal level are essential for driving urban development in order to support a policy in accordance with the concept of sustainable urban development. It is important to develop a balance between economic, social, and environmental aspects. In this study, emphasis is placed on the dimensions of the environment and the working context of the municipality with the aim of the municipality to be able to manage the environment well. Nowadays, however, most municipalities are inefficient in collecting environmental data; therefore, solutions to environmental problems sometimes lack information to make decisions. Although the municipalities have a wide variety of local characteristics and limitations, tools or approaches are needed to develop an efficient storage system in a flexible online platform. In the past, the format for storing data would be paper-based or on computers that may have a data loss problem, including scattered information that will be searched only when needed. As a result, the management of the previous system may not use the information efficiently.

Therefore, this study aimed to (1) study a model for assessing the environmental performance in accordance with the context and duty of municipalities, which is a simple assessment system, and (2) create a platform to establish an environmental database for municipalities to benefit from information management for sustainable urban development. Pilot municipalities with different city characteristics were selected to study the environmental dimensions according to the context and duties of the municipalities, current data collection models, current environmental databases, to be compared with the reference criteria that summarize the information that will be used to create the platform, whether it is a data type, data collection process, source of information, data analysis or the form of utilization. The platform will enable municipalities to collect data easily, conveniently, quickly, and flexibly. Environmental data are usually collected from internal and external sources of the municipality because the municipality is the center of information that relates to all sectors, including government, business, commercial, and the public sector. Therefore, the developed online platform can import data to be stored in a separate database sorted by indicators. This information can be used by municipalities for analysis or processing in decision-making or work.

Good or effective operations must be based on having good information that can be used in the environmental management of a city in the age of information. Currently, there are various criteria to measure the environmental performance of municipalities, whether Thai or foreign, so the development of the Self-Assessment Environmental Performance Index on EcoCitOpia platform will be a tool to improve the efficiency of information management that will benefit the drive for sustainable urban development in the future. However, experience also shows significant lags between the demand for environmental indicators, the related conceptual work, and the actual capacity for mobilizing and validating underlying data. In the field of environmental statistics, differences among countries may be considerable, and there may be a need to establish reliable and internationally comparable data that calls for continuous monitoring, analysis, treatment, and checking.

Author Contributions: Conceptualization, R.K. and S.S.; methodology, R.K. and S.S.; resources, R.K.; data curation, R.K.; writing—original draft preparation, R.K.; writing—review and editing, S.H.G.; visualization, R.K.; supervision, S.S. All authors have read and agreed to the published version of the manuscript.

Funding: This research received no external funding.

Institutional Review Board Statement: Not applicable.

Informed Consent Statement: Not applicable.

Data Availability Statement: Not applicable.

Acknowledgments: This study was supported by Thailand Science Research and Innovation (TSRI), Excellence Center in Logistics and Supply Chain Management (E-LSCM), Faculty of Engineering, Chiang Mai University, and the Research Unit for Energy Economic & Ecological Management, Science and Technology Research Institute, Chiang Mai University.

Conflicts of Interest: The authors declare no conflict of interest.

References

1. Klumbytė, E.; Bliūdžius, R.; Medineckienė, M.; Fokaides, P.A. An MCDM Model for Sustainable Decision-Making in Municipal Residential Buildings Facilities Management. *Sustainability* **2021**, *13*, 2820. [CrossRef]
2. Harris, S.; Weinzettel, J.; Levin, G. Implications of Low Carbon City Sustainability Strategies for 2050. *Sustainability* **2020**, *12*, 5417. [CrossRef]
3. National Strategy Secretariat Office. National Strategy 2018–2037 (Summary). 2018. Available online: https://www.bic.moe.go.th/images/stories/pdf/National_Strategy_Summary.pdf (accessed on 25 March 2021).
4. UNDP. Sustainable Development Goals (SDGs). Available online: https://www.th.undp.org/content/thailand/en/home/sustainable-development-goals.html (accessed on 25 March 2021).
5. Abbas, H.S.M.; Xu, X.; Sun, C.; Ullah, A.; Nabi, G.; Gillani, S.; Raza, M.A.A. Sustainable Use of Energy Resources, Regulatory Quality, and Foreign Direct Investment in Controlling GHGs Emissions among Selected Asian Economies. *Sustainability* **2021**, *13*, 1123. [CrossRef]
6. Pace, R.; Churkina, G.; Rivera, M. How green is a "Green City"? A Review of Existing Indicators and Approaches. 2016. Available online: http://doi.org/10.2312/iass.2016.035 (accessed on 25 March 2021).
7. Jackson, L.E.; Kurtz, J.C.; Fisher, W.S. Evaluation Guidelines for Ecological Indicators; Report, No. EPA/620/ R-99/005. Environmental Protection Agency: Washington, DC, USA, 2000.
8. Organisation for Economic Cooperationand Development (OECD). *Environmental Indicators: Towards Sustainable Development*; OECD: Paris, France, 2001. Available online: https://www.oecd.org/site/worldforum/33703867.pdf (accessed on 27 March 2021).
9. Rodrigo-Ilarri, J.; Romero, C.P.; Rodrigo-Clavero, M.-E. Land Use/Land Cover Assessment over Time Using a New Weighted Environmental Index (WEI) Based on an Object-Oriented Model and GIS Data. *Sustainability* **2020**, *12*, 10234. [CrossRef]
10. Joas, M. *Informed Cities: Making Research Work for Local Sustainability*; Routledge: Abingdon, Oxon, UK, 2014.
11. Economist Intelligence Unit. European Green City Index. Assessing the Environmental Impact of Europe's Major Cities. 2009. Available online: http://sg.siemens.com/city_of_the_future/_docs/greencityindex_report_en.pdf (accessed on 27 March 2021).
12. Economist Intelligence Unit (EIU). The Green City Index. A Summary of the Green City Index Research Series. 2012. Available online: https://www.siemens.com/entry/cc/features/greencityindex_international/all/en/pdf/gci_report_summary.pdf (accessed on 26 March 2021).
13. Economist Intelligence Unit (EIU). Asian Green City Index Green City Index. Assessing the Environmental Performance of Asia's Major Cities. 2012. Available online: http://sg.siemens.com/city_of_the_future/_docs/Asian-Green-City-Index.pdf (accessed on 26 March 2021).
14. Gong, W.; Lyu, H. Sustainable City Indexing: Towards the Creation of an Assessment Framework for Inclusive and Sustainable Citation Urban-Industrial Development. 2017. Available online: https://www.unido.org/sites/default/files/files/2018-02/BRIDGE%20for%20Cities_Issue%20Paper_2.pdf (accessed on 26 March 2021).
15. European Commission. Indicators for Sustainable Cities. 2015. Available online: https://ec.europa.eu/environment/integration/research/newsalert/pdf/indicators_for_sustainable_cities_IR12_en.pdf (accessed on 26 March 2021).
16. Kourtit, K.; Nijkamp, P. Big data dashboards as smart decision support tools for i-cities—An experiment on stockholm. *Land Use Policy* **2018**, *71*, 24–35. [CrossRef]
17. Angelakoglou, K.; Gaidajis, G. A Conceptual Framework to Evaluate the Environmental Sustainability Performance of Mining Industrial Facilities. *Sustainability* **2020**, *12*, 2135. [CrossRef]
18. Department of Local Administration. Local Performance Assessment: LPA. 2020. Available online: http://www.dla.go.th/upload/ebook/column/2020/5/2292_6137.pdf (accessed on 7 April 2021). (In Thai)
19. Department of Local Administration. Summary of Information of Local Government Organizations of Thailand. Available online: http://www.dla.go.th/work/abt/summarize.jsp (accessed on 29 March 2021). (In Thai)

20. Nonthaburi. Available online: https://en.wikipedia.org/wiki/Nonthaburi (accessed on 29 March 2021).
21. Hat Yai. Available online: https://en.wikipedia.org/wiki/Hat_Yai (accessed on 29 March 2021).
22. Yasothon. Available online: https://en.wikipedia.org/wiki/Yasothon (accessed on 29 March 2021).
23. Perchinunno, P.; Cazzolle, M. A clustering approach for classifying universities in a world sustainability ranking. *Environ. Impact Assess. Rev.* **2020**, *85*, 106471. [CrossRef]
24. Arcadis, Sustainable Cities Index 2015: Balancing the Economic, Social and Environmental Needs of the World's Leading Cities. Available online: https://s3.amazonaws.com/arcadis-whitepaper/arcadis-sustainable-cities-index-report.pdf (accessed on 1 April 2021).
25. UN Statistical Commission. Report of the Inter-Agency and Expert Group on Sustainable Development Goal Indicators. 2016. Available online: https://unstats.un.org/unsd/statcom/47th-session/documents/2016-2-IAEG-SDGs-Rev1-E.pdf (accessed on 4 April 2021).
26. Association of Southeast Asian Nations. ASEAN Award. 2016. Available online: https://environment.asean.org/awgesc/#:~:text=ASEAN%20Environmentally%20Sustainble%20City%20(ESC)%20Award&text=The%20ESC%20Award%20aims%20to,clean%2C%20green%2C%20and%20liveable (accessed on 4 April 2021).
27. The Sustainable City Assessment of Department of Environmental Quality Promotion. Sustainable City Assessment Guide. 2013. Available online: https://datacenter.deqp.go.th/service-portal/green-city/municipal-livable-system/download/manual/ (accessed on 7 April 2021). (In Thai)
28. Office of Natural Resources and Environmental Policy and Planning. The Project to Drive the Development of the Eco-City Concept for Sustainable Management of the Environment, City and Community. 2019. Available online: https://www.dmcr.go.th/detailLib/4733 (accessed on 7 April 2021). (In Thai)
29. Hiremath, R.B.; Balachandra, P.; Kumar, B.; Bansode, S.S.; Murali, J. Indicator-based urban sustainability—A review. *Energy Sustain. Dev.* **2013**, *17*, 555–563. [CrossRef]
30. Chen, Y.; Zhang, D. Evaluation and driving factors of city sustainability in Northeast China: An analysis based on interaction among multiple indicators. *Sustain. Cities Soc.* **2021**, *67*, 102721. [CrossRef]
31. Mitra, D.; Chu, Y.; Cetin, K. Cluster analysis of occupancy schedules in residential buildings in the United States. *Energy Build.* **2021**, *236*, 110791. [CrossRef]

 *sustainability*

Article

Sustainability Understanding and Behaviors across Urban Areas: A Case Study on Istanbul City

Hasan Fehmi Topal *, Dexter V. L. Hunt and Christopher D. F. Rogers

Department of Civil Engineering, School of Engineering, University of Birmingham, Edgbaston, Birmingham B15 2TT, UK; d.hunt@bham.ac.uk (D.V.L.H.); c.d.f.rogers@bham.ac.uk (C.D.F.R.)
* Correspondence: HFT718@bham.ac.uk

Abstract: The success of urban sustainability is very much dependent on a number of human factors. Therefore, it becomes even more important to explore how people understand urban sustainability and how they behave accordingly. Based on a formerly developed conceptual framework and on specified influencing factors, this study aimed to evaluate and elucidate the urban sustainability understanding and behavior of individuals in the city of Istanbul. This was assessed through the use of a quantitative questionnaire survey of 535 respondents. Therein, socio-psychological processes of sustainability understanding (i.e., determinants of awareness, perception, and attitude) and sustainability behaviors along with personality traits and influential factors were assessed and analyzed through the use of bivariate and multivariate methods (i.e., correlation tests, ANOVA, *t*-tests, and multiple linear regression). The results showed that sustainability awareness was more strongly correlated with attitude than perception, whereas behavior was found to be strongly correlated with both awareness and attitude and was (significantly) predicted by all determinants. The associations/influences of personality traits with determinants were found to be mostly insignificant. Conversely, for behavior, they were significant. The most influential factors found (in hierarchical ordering) were awareness of consequences, trust in society, social appraisement, world-mindedness, willingness to pay, trust in science and technology, ascription of responsibility, age and gender.

Keywords: urban sustainability; sustainable behavior; sustainability understanding; awareness; perception; attitude; pro-environmental behavior; influencing factors; Turkey; Istanbul

Citation: Topal, H.F.; Hunt, D.V.L.; Rogers, C.D.F. Sustainability Understanding and Behaviors across Urban Areas: A Case Study on Istanbul City. *Sustainability* **2021**, *13*, 7711. https://doi.org/10.3390/su13147711

Academic Editor: Tan Yigitcanlar

Received: 2 June 2021
Accepted: 25 June 2021
Published: 9 July 2021

Publisher's Note: MDPI stays neutral with regard to jurisdictional claims in published maps and institutional affiliations.

1. Introduction

Over the last 35 years, sustainability studies have featured prominently within research agendas around the world. Starting with the Brundtland Report in 1987 [1], sustainability's prominence has increased with time. In its early approaches, it was understood as the balance between nature and humans that provides continuous development [2]. Early interpretations of sustainable development were defined by the three pillars approach, which includes economic development, social equity, and environmental protection [3].

Sustainability in urban areas plays a critical role for sustainable development, since the planet is facing critical challenges that have arisen as a result of unsustainable urbanization. These problems require precautions such as using natural resources efficiently and balancing the human–nature relationship at both micro- and macro-levels. According to the United Nations, by 2050, while the world population is expected to surpass nine billion [4], urbanization rates are anticipated to reach 66% [5]. Therefore, urban living areas emerge as one of the most critical elements of sustainability studies. In line with this, the United Nations has addressed the importance of urban areas within the Sustainable Development Goals (SDGs) under the heading of "Sustainable cities and communities" [6].

In line with the Sustainable Development Goals, Istanbul, being the largest city in Turkey (with the population reaching 15.5 million by 2020) [7], aims to integrate sustainability into its urbanization and development strategies [8,9]. This rapid urbanization, together with a rapidly developing economy, has prompted Istanbul's Urban Regeneration

Law, commonly referred to as "Transformation of Areas Under the Disaster Risk" [10], making sustainability even more prominent for the case of Turkey in the city of Istanbul. Adopting development aspirations for 2023 in Istanbul [11], Istanbul's master plan has been redesigned according to 2023 goals, including new headings such as sustainable urban development, spatial quality, sustainable transportation, sustainable environment, and environmentally friendly energy [12]. Considering all of this, it is obvious that sustainability is becoming a very prominent topic for Istanbul. However, the development of sustainable behavior needs the support of local and central authorities in order that any sustainability targets are met.

The overall success of sustainability policies within cities is subject to their acceptance by the people who reside therein. Therefore, individual and social behaviors are of critical importance for sustainable living areas. Moreover, combatting the global challenges is immensely reliant on the sustainable practices of individuals. In light of this, it is of prominent importance to study the sustainability understanding and behavior of people in the urban context.

Numerous studies have attempted to explore pro-environmental or sustainable behavior [13–17] from various perspectives such as ecological economics [18], environmental policy [13], household energy conservation [19], waste management [20], and climate change [21]. However, earlier research has mainly adopted either environmental [22] or psychological approaches [23]. On the other hand, a synthesized approach that puts urban sustainability at the very core is required, and, more importantly, this approach was adopted within the current research paper. Moreover, in a Turkish context, studies within this field have, until recently, been relatively unobserved. While there are increasing numbers of studies investigating sustainability from several perspectives, such as transportation [24], climate change [25], energy [26], natural resources [27], and urbanization [28,29], very few studies have concentrated on the understanding and behaviors of urban sustainability [30]. Therefore, this research aimed to fill an important gap within the literature.

2. Methodology

2.1. Conceptual Approach

Based on the literature review conducted previously by Topal et al. [31], it was revealed that the complexity of the urban sustainability behavior context required specification of variables that influence sustainability understanding and behavior. Moreover, it was necessary to specify the relationship between socio-psychological determinants of urban behavior in order to evaluate and analyze the individual practices (a determinant is defined herein as a factor that decisively affects the nature or outcome of something).

It is obvious that methodological approaches to sustainability assessment studies require deliberate efforts. In this manner, Sharifi et al. [32] have described critical methodological shortcomings in sustainability assessment tools. Some of the critical ones are:

- Limitations and the lack of harmony about sustainability dimensions;
- Lack of lucidity and the dominance of top-down approaches;
- Insufficient consideration of context-sensitive subjects;
- Lack of flexibility in design stages;
- Lack of compliance among various methodologies; and
- The complexity of the instrument.

On the other hand, the importance of the balance of the complementary relationship between nature-based recipes and urban-wilding approaches should be kept in mind and needs to be considered in urban sustainability studies [33]. For instance, as stated by Morano et al. [34], the inherent contradiction between the precedence of real estate developers and public authorities defines the natural limits of urban sustainability approaches. While the former has mainly focused on profit maximization, the latter has prioritized the quality of and livability within the city.

Considering the above requirements, a novel conceptual framework was proposed in Topal et al. [35] based on an in-depth literature search. A simplified version of the proposed conceptual framework is shown in Figure 1, which outlines the multiple linkages between awareness, perception, attitude, and behavior. Moreover, the associations between these socio-psychological processes and a range of influencing factors and personality traits have been specified [35].

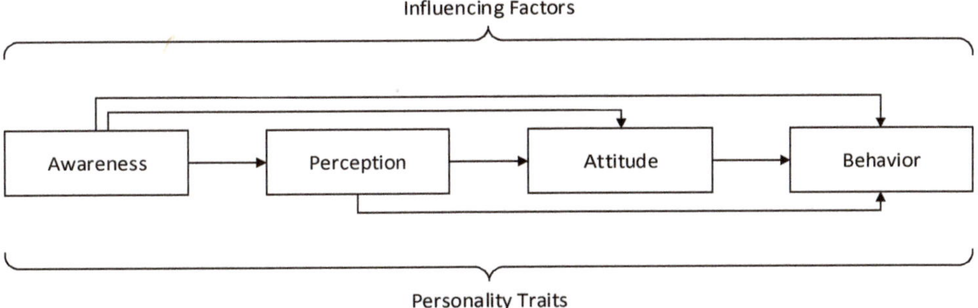

Figure 1. Conceptual framework used in the study.

The development process of the conceptual framework consists of four stages, which are presented in Figure 2. Topal et al. [31] provided a systematic review of the influencing factors of urban sustainability behavior(s), fulfilling the requirements of stage 1. Topal et al. [35] thereafter examined prominent environmental and sustainability behavior theories (stage 2), investigated socio-psychological determinants of urban sustainability understanding and behavior (stage 3), and synthesized these findings into a newly developed conceptual framework (stage 4), which has formed the basis for this current study.

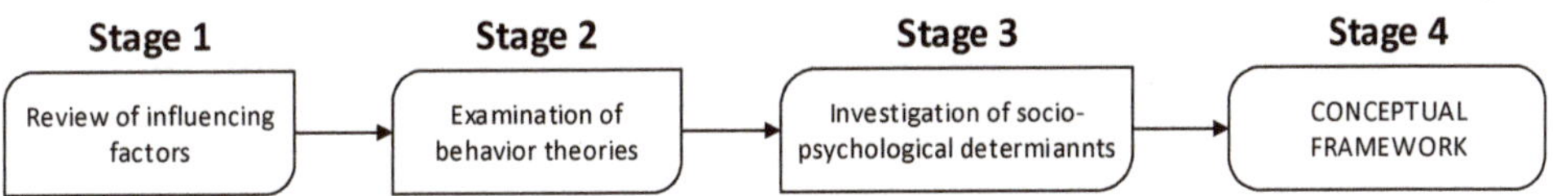

Figure 2. Methodological stages of the conceptual framework development.

2.2. Research Aim and Objectives

The research aim was to develop a holistic approach to evaluate and elucidate the urban sustainability understanding and behavior of individuals through the application of a developed conceptual framework. In so doing the formation of sustainable urban behavior through socio-psychological processes (i.e., determinants of awareness, perception, and attitude) and the effect of influential factors and personality traits on urban sustainability understanding and behavior were investigated.

In order to meet the research aim, the following objectives were defined:

- To assess the urban sustainability understanding and behavior of individuals, and determine how they are related
- To analyze the relationship between this urban sustainability understanding and behavior and both influencing factors and personality traits
- To develop policies that can improve urban sustainability understanding and the behavior of individuals

2.3. Methods

2.3.1. Study Location and Sample

The focus of the project was the sustainable behaviors of the people living in urban areas; therefore, the two main elements were the urban built environment and people. Accepting that sustainability is a highly local context-dependent subject [36], a specific urban area has to be selected. Istanbul was chosen for this study since it is the largest city in Turkey and well-known around the world.

Istanbul is located in the north-west of Turkey in the Marmara region. It had a population of 15.5 million by 2020 [7], and consists of 39 counties. Given its logistically advantageous geography, its strong history and the fact it hosts several crucial economic activities, Istanbul is expected to reach a population of 17 billion by 2030 [6].

Following ethical approval from the University of Birmingham (reference number ERN_19-0513A), 20 pilot questionnaire surveys were conducted with colleagues at The University of Birmingham and colleagues who were resident in Istanbul. Based on the results of and feedback from the pilot study, the final questionnaire was formulated and administered between 1 December 2020 and 20 January 2021. The survey was undertaken on all days of the week between early morning (7 a.m.) and late in the evening (9 p.m.), in order to minimize sampling bias. Survey respondents were randomly selected from various counties of Istanbul in order to best represent the socio-demographic distribution of the population.

A professional survey company was employed to collect the data. Trained surveyors were involved in the data collection process, armed with tablets to provide people with the option to fill web-surveys. In total, 535 initial responses were obtained.

2.3.2. Measures

Quantitative data were obtained through a comprehensive questionnaire, mainly using a five-point Likert scale [37]. The question wording was designed to be as concise as possible, while the structure was designed to be understandable by grouping questions according to the elements of the conceptual framework.

Respondents were presented with a cover letter, which gave a brief introduction about the survey, and a participant information sheet, which presented the purpose of the study, targeted audiences, confidentiality issues, and contact details. A short explanation of urban sustainability and sustainable behavior was given at the beginning of the survey in order to prevent any misunderstanding or misinterpretation of the terms. In total, the questionnaire consisted of five sections:

1. General socio-demographic information (12 questions)
2. Understanding, i.e., determinants of awareness, perception, and attitudes (14 questions)
3. Behavior (18 questions)
4. Influencing factors (19 questions)
5. Personality traits (5 questions)

As stated earlier, the questions were based on the findings of an in-depth literature review [31] and conceptual framework [35] created by the current authors. The details of the questionnaire are given in Appendix A.

2.3.3. Data Analysis

All statistical analyses for the data collected from the questionnaire survey were performed using Python 3.8 software. In all statistical analysis, $\alpha = 0.05$ was used as the significance level since it has become the most commonly used threshold by researchers [38]. Three stages of analysis have been conducted:

Firstly, in the preliminary analysis, descriptive statistics including frequency, mean, standard deviation, and scale construction have been performed on each element of the conceptual framework: determinants of understanding (i.e., awareness, perception, and attitude), behavior, personality traits, and influencing factors.

Secondly, bivariate analysis was conducted between the elements of the conceptual framework. Thereafter, correlation tests, *t*-tests, and ANOVA tests were conducted.

In the third stage, multivariate analysis was carried out on the elements of the conceptual framework. A series of multiple linear regression tests were performed on the outcome variables specified in the framework.

3. Descriptive Analyses and Results

3.1. Understanding

3.1.1. Determinants of Understanding

Determinant 1: Awareness

Awareness was assessed by five different questions (Aw1 to Aw5). The fifth question (Aw 5.1 to Aw 5.9) was about familiarity level and consisted of nine sub-questions (Figure 3).

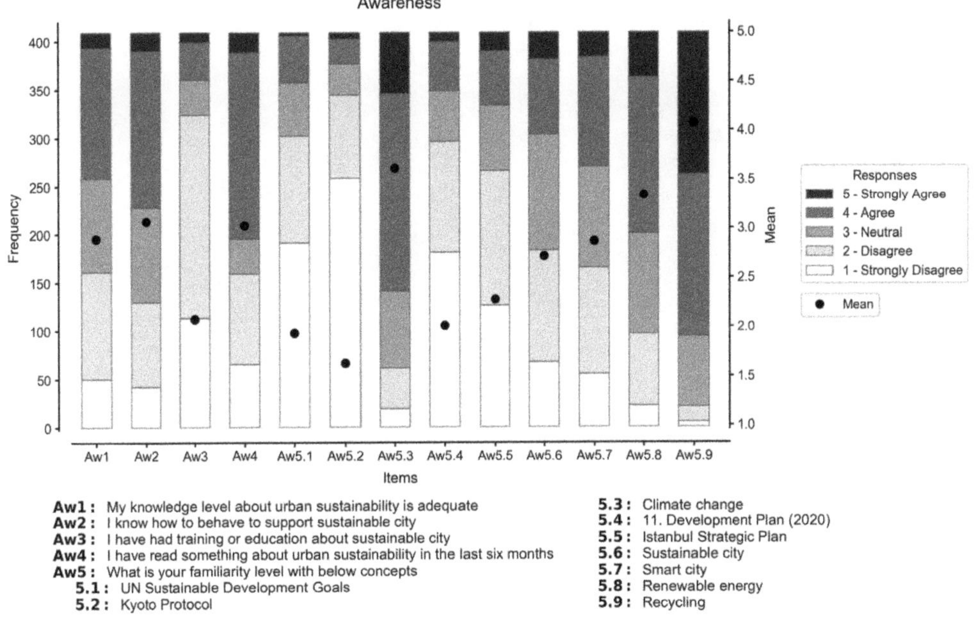

Aw1 : My knowledge level about urban sustainability is adequate
Aw2 : I know how to behave to support sustainable city
Aw3 : I have had training or education about sustainable city
Aw4 : I have read something about urban sustainability in the last six months
Aw5 : What is your familiarity level with below concepts
   5.1 : UN Sustainable Development Goals
   5.2 : Kyoto Protocol

5.3 : Climate change
5.4 : 11. Development Plan (2020)
5.5 : Istanbul Strategic Plan
5.6 : Sustainable city
5.7 : Smart city
5.8 : Renewable energy
5.9 : Recycling

Figure 3. Awareness questions.

By looking at the results, it became apparent that while respondents tended to report themselves knowledgeable about urban sustainability (Aw1) and they know how to behave sustainably (Aw2), they admitted to having a lack of education or training on these issues (Aw3). However, they reported also that they have read something about sustainable cities in various news channels (Aw4), which is slightly contradictory. The answers to Aw1, Aw2, and Aw4, therefore, appear to provide a coherent narrative, explaining that although there are some opportunities for individuals to gain ideas about urban sustainability behavior through media or via the internet, education or training opportunities are insufficient and need improvement. The items in the fifth question (Aw5) present an important trend. Regarding the first three items (Aw5.1 to 5.3), it is possible to deduce that people within Turkey are relatively unfamiliar with more technical and global topics such as The United Nations Sustainable Development Goals (Aw5.1) or the Kyoto Protocol (Aw5.2), but most of them have heard about popular topics such as climate change (Aw5.3). However, it was surprising to find that familiarity with the Kyoto Protocol was lacking, as it could also be viewed as a hot debate topic. When it comes to the last six items (Aw5.4 to 5.9), respondents were asked to grade their familiarity level regarding the 11th Development Plan of the Turkish Government (2019–2023) [11] (Aw5.4), Istanbul strategic plan [12] (Aw5.5), the sustainable city concept [8] (Aw5.6), the smart city concept (Aw5.7), renewable energy (Aw5.8), and recycling (Aw5.9). It can be seen that as the specificity and locality of the item increase, the familiarity level increases as well. It is also noteworthy that people are slightly more familiar with the term 'smart city' than the term 'sustainable city'. One of the reasons for this could be the increase in popularity of smart city as a term in recent years as a result of rapid technological developments. Moreover, it is slightly difficult to pronounce the Turkish translation of sustainability, which is written as "surdurulebilirlik". Therefore, it is possible to surmise that an alternative, catchy keyword with higher advertising value would preferentially be used alongside the sustainability term in the Turkish context. Moreover, the general public is more likely to be concerned by local sustainability issues.

Determinant 2: Perception

Regarding the perception determinant, less variance and higher scores were observed within the questions (Figure 4).

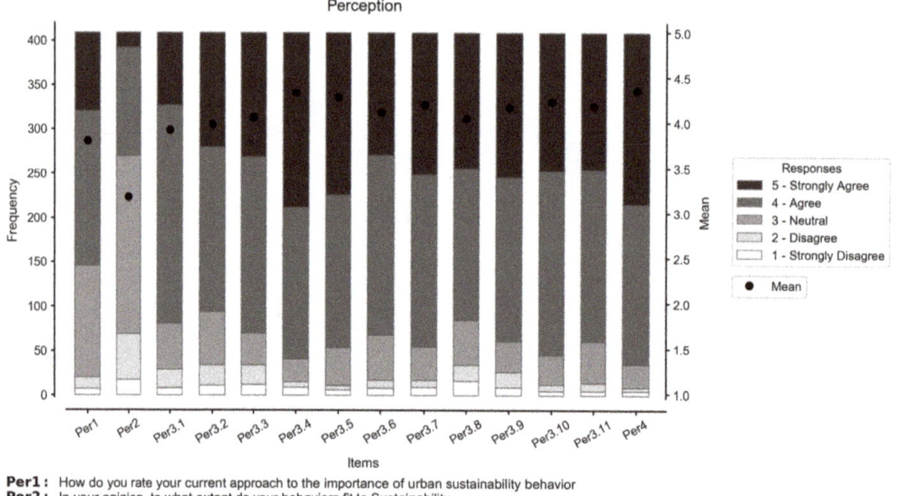

Per1: How do you rate your current approach to the importance of urban sustainability behavior
Per2: In your opinion, to what extent do your behaviors fit to Sustainability
Per3: To what extent do below expressions related with urban sustainability
   3.1: Enhancing public equity and engagement
   3.2: Providing cultural and public amenities
   3.3: Protection of the historical and local assets
   3.4: Protection of the environment and natural resources
   3.5: Reducing energy (electricity, water, fuel etc.) consumption
   3.6: Increasing the welfare
   3.7: Improving economic development
   3.8: Democracy and justice
   3.9: Governmental efficiency and transparency
   3.10: Providing infrastructural facilities
   3.11: Ensuring technological development and innovation
Per4: What do you think about the importance of urban sustainability for Istanbul

Figure 4. Perception questions.

It was found that 73% of the respondents believe in the importance of urban sustainability behavior (Per1). However, they admit that their behaviors do not conform to sustainability principles (Per2). It is important that people become good at distinguishing their perceptions about the importance of sustainability from their perceptions about their individual performances in behaving accordingly.

In question three (Per3.1 to Per3.11), a comprehensive list of urban sustainability related sub-areas were given and people were asked to grade the strength of their relationship with urban sustainability. Respondents appear to have a holistic perception about urban sustainability since they scored all items highly, although a higher score was observed for environmental aspects (Per3.4 and Per3.5) than social aspects (Per3.1 and Per3.2). It is therefore possible to state that people were more inclined to see sustainability more as an environmental issue than a social issue. The responses to the last question (Per4) showed there to be an overall consensus among the respondents on the importance of sustainability for Istanbul, which is expected to increase the chance of public participation in urban sustainability actions.

Determinant 3: Attitude

In terms of attitudes (Figure 5), although the overall attitude was found to be significantly high, it is possible to notice from the responses to questions Att1 and Att4 that a considerable proportion of the respondents still hold some remnants of an anthropocentric approach.

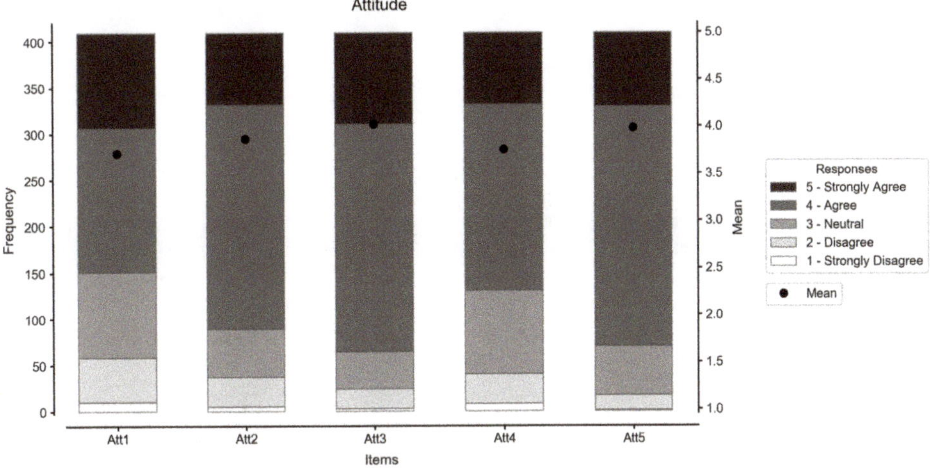

Att1 : We are approaching the limit of the number of people the earth can support
Att2 : I feel good when I contribute to Sustainable City
Att3 : I have the responsibility to take action in order to protect environment
Att4 : Economic problems are likely to occur if unsustainable urbanization continues
Att5 : Urban sustainability laws and regulations should be enforced more strongly

Figure 5. Attitude questions.

They cared somewhat less about the limits of the planet (Att1) and possible economic outcomes of unsustainability (Att4) when compared with other scores. On the other hand, they had a good sense of contributing to sustainability (Att2) and saw themselves as responsible for taking necessary actions (Att3). Moreover, they also thought that public authorities should enforce regulations more strongly (Att5). Their strong emphasis in this question was found to be a result of unsustainable urbanization policy and actions undertaken in Istanbul in recent years. It also showed a general discomfort with the actions that have been implemented. However, it is promising that they hold themselves responsible for urban sustainability actions and feel good about making contributions since it indicates that there are strong attitudes about pro-sustainability issues. Moreover, the consensus among the respondents on the importance of sustainability has the potential to be used in both supporting sustainable policy and actions and changing unsustainable practices, even if they act against the comfort of individuals or benefit specific power groups.

3.1.2. Data Preparation
Factor Analysis

All determinant questions were subjected to a factor analysis with varimax rotation. This helped to provide confirmation of the appropriateness of the questionnaire structure by deriving latent variables. Moreover, it was useful for providing empirical robustness for aggregate scales [39].

The results indicated that Bartlett's test coefficient was equal to 1943, which, along with a *p*-value of <0.05, showed that the data was suitable for factor analysis. Similarly, a KMO value of 0.867 (higher than the adequacy limit of 0.600) confirmed suitability of the data for factor analysis. In addition, a Scree plot method was used to choose the number of factors that could/should be considered. As a result, three factors with eigenvalues larger than 1.0 were retained [39].

In the final step, the three factors were rotated using varimax orthogonal rotation; the corresponding factor loadings are shown in Table 1.

Table 1. Factor loadings for all determinant items.

Variable	Factor 1	Factor 2	Factor 3	Communality	Cronbach's Alpha
Aw1	*0.804*	0.112	0.280	0.738	0.87
Aw2	*0.784*	0.068	0.314	0.718	
Aw3	*0.659*	0.240	−0.016	0.492	
Aw4	*0.694*	0.070	0.294	0.573	
Aw5	*0.671*	0.306	0.175	0.574	
Per1	0.186	*0.427*	0.098	0.227	0.62
Per2	0.302	*0.464*	−0.067	0.311	
Per3	0.032	*0.606*	0.231	0.421	
Per4	0.064	*0.558*	0.126	0.331	
Att1	0.092	−0.098	*0.451*	0.222	0.68
Att2	0.334	0.328	*0.520*	0.489	
Att3	0.243	0.269	*0.460*	0.343	
Att4	0.188	0.179	*0.585*	0.409	
Att5	0.057	0.288	*0.501*	0.337	
% Variance	0.213	0.118	0.111	0.442	

Note: Bold ones indicate the corresponding factor loadings.

The chosen three factors confirmed the exact grouping used in the questionnaire (i.e., awareness, perception, and attitude). As can be seen from Table 1, awareness items are grouped under factor 1, perception items are grouped under factor 2, and attitude items are grouped under factor 3. Cronbach's alpha statistics show that all scales were internally consistent, with the internal consistency score of the perception scale being the lowest. As a result, it is possible to state that the questionnaire used in this survey formed a reliable set of scales and provided a valid empirical base [40,41].

Scale Construction

Although the three separate groups of determinant questions (i.e., awareness, perception, and attitude) were created based on the insights provided from the existing literature, it was important to specify their distinct features empirically. This consisted of two steps.

In the first step, the questions that consisted of several expressions were converted into a scale. In other words, for Aw5 (which consisted of nine statements) and Per3 (which consisted of 11 statements), all statements were considered as equally weighted. Therefore, all answers were summed (from 1—strongly disagree to 5—strongly agree) and the sums were divided by the total number of statements (9 and 11, respectively) in order to obtain an overall score for these particular questions.

In the second step, all questions under each determinant groups—awareness (five questions), perception (four questions), and attitude (five questions)—were summed and the sums were divided by the total number of questions (i.e., 5, 4, and 5, respectively). By doing so, an overall average score for each of the awareness, perception, and attitude determinants were obtained.

3.2. Behavior

3.2.1. Items of Behavior

Behavior is the outcome of the socio-psychological processes that feature in the conceptual framework (Figure 1). While exploring urban sustainability understanding and behavior, the number of discrete questions under behavior were more numerous than the individual socio-psychological determinant scales since behavior was the central focus of this research. Similar to the process adopted for awareness, perception, and attitude, reported levels of behavior were measured and analyzed (Figure 6). Therein, 18 different aspects of behavior were considered within seven groups:

1. Personal (Q1.1 to 1.3)
2. Social (Q2.1 to 2.3)
3. Environmental (Q3.1 to 3.3)
4. Economic (Q4.1 to 4.3)
5. Governance (Q5.1 to 5.2)
6. Infrastructural (Q6.1 to 6.2)
7. Technological (Q7.1 to 7.2)

The results presented in Figure 6 show that respondents tended to report their efforts within their personal (Bh1.1) and social environments (Bh1.3) and admit that they were not able to attend informative activities frequently (Bh1.2). The reason behind this could be a lack of time and/or lack of informative activities related to urban sustainability.

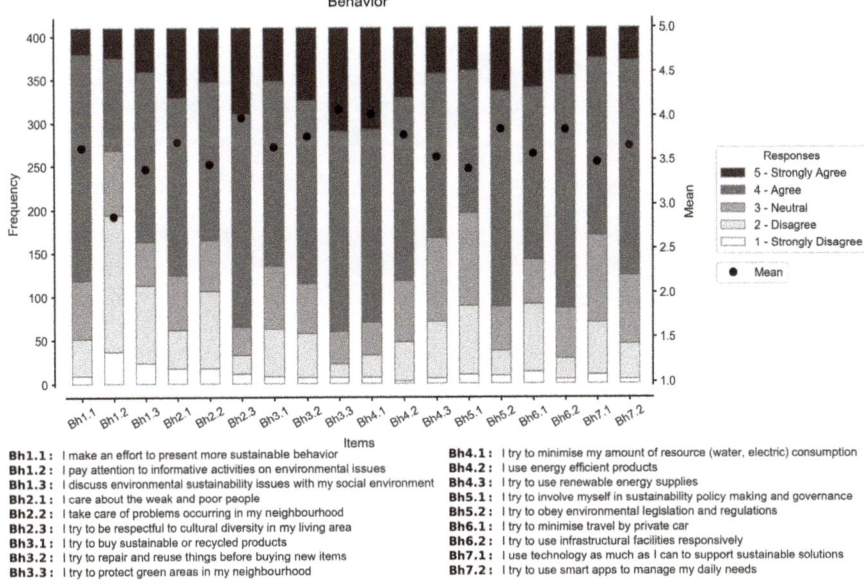

Bh1.1: I make an effort to present more sustainable behavior
Bh1.2: I pay attention to informative activities on environmental issues
Bh1.3: I discuss environmental sustainability issues with my social environment
Bh2.1: I care about the weak and poor people
Bh2.2: I take care of problems occurring in my neighbourhood
Bh2.3: I try to be respectful to cultural diversity in my living area
Bh3.1: I try to buy sustainable or recycled products
Bh3.2: I try to repair and reuse things before buying new items
Bh3.3: I try to protect green areas in my neighbourhood

Bh4.1: I try to minimise my amount of resource (water, electric) consumption
Bh4.2: I use energy efficient products
Bh4.3: I try to use renewable energy supplies
Bh5.1: I try to involve myself in sustainability policy making and governance
Bh5.2: I try to obey environmental legislation and regulations
Bh6.1: I try to minimise travel by private car
Bh6.2: I try to use infrastructural facilities responsively
Bh7.1: I use technology as much as I can to support sustainable solutions
Bh7.2: I try to use smart apps to manage my daily needs

Figure 6. Behavior questions.

Regarding the social behavior items (Bh2.1 to Bh2.3), higher scores were observed. The connection between social behavior items and cultural–moral values provided them with a strong basis. This indicates, therefore, that the advantage of these established cultural codes should be utilized in sustainability policies. A similar high trend in environmental behavior items (Bh3.1 to Bh3.3) was observed. Although the statement of "buying sustainable or recycled products" (Bh3.1) gained lower scores than other environmental behavior items, this aspect of sustainability behavior could be enhanced by providing better labelling and informative 'sustainability' signage on products. The responses about repairing and reusing things (Bh3.2) might also be related with cultural codes as well as economic benefits.

A higher score was observed in protecting the green areas (Bh3.3). This score is important since it denotes a need for value to be attached to green areas in urban living.

Economic behavior items (Bh4.1 to Bh4.3) achieved generally high scores, the lowest of the three using renewable energy supplies (Bh4.3). This might be due to either inadequate supply of, or respondents being insufficient informed about, renewable energy sources. This indicates that the sources of energy supplies used in household energy consumption could be presented more visibly. Under governance behavior items (Bh5.1 and Bh5.2), it was observed that participating in sustainability policy and governance (Bh5.1) achieved low scores. Therefore, it could be argued that the channels for participatory government should be enhanced, which would eventually be expected to result in a positive impact on overall sustainability behavior. On the other hand, people demonstrated a willingness to obey environmental regulations (Bh5.2). For infrastructural (Bh6.1 and Bh6.2) and technological (Bh7.1 and Bh7.2) items, a coherent response characteristic was observed with high mean values.

3.2.2. Data Preparation

Factor Analysis

In order to prepare the behavior data for inferential analysis, factor analysis with varimax rotation was once again conducted. By doing so, a reduction in the dimension was provided and empirically sound implicit factors were extracted. Bartlett's test coefficient of 3369 along with a p-value of <0.05 indicated that the data was suitable for factor analysis. Similarly, a KMO value of 0.938 (higher than adequacy limit of 0.600) confirmed the suitability of factor analysis. The Scree plot method was again used to choose the number of factors. Although three eigenvalues were above 1.0, initial analysis using three factors did not give a meaningful outcome. Factor loadings were closely distributed among three factors; therefore, it was not possible to place them under distinct groupings. Moreover, the cumulative variance score was lower than required. Hence, as the eigenvalue of the fourth factor (0.93) was very close to the threshold of 1.0 and the three-factor analysis did not give appropriate results, four factors were used for the second iteration. This proved successful: four factors rotated with varimax orthogonal rotation resulted in a meaningful outcome, the corresponding factor loadings being recorded in Table 2.

Table 2. Factor loadings for behavior items.

Variable	Factor 1	Factor 2	Factor 3	Factor 4	Communality	Cronbach's Alpha
Behavior 2.1	*0.550*	0.412	0.128	0.162	0.515	0.83
Behavior 2.3	*0.698*	0.314	0.187	0.064	0.625	
Behavior 3.3	*0.730*	0.097	0.152	0.289	0.649	
Behavior 4.1	*0.527*	0.022	0.428	0.278	0.538	
Behavior 5.2	*0.524*	0.177	0.318	0.321	0.510	
Behavior 1.1	0.304	*0.537*	0.245	0.289	0.525	0.80
Behavior 1.2	0.048	*0.671*	0.130	0.240	0.526	
Behavior 1.3	0.246	*0.718*	0.149	0.187	0.632	
Behavior 2.2	0.432	*0.491*	0.252	0.158	0.516	
Behavior 3.1	0.292	0.500	*0.518*	0.151	0.627	0.80
Behavior 3.2	0.289	0.054	*0.367*	0.191	0.258	
Behavior 4.2	0.258	0.287	*0.694*	0.195	0.669	
Behavior 4.3	0.129	0.333	*0.474*	0.394	0.507	
Behavior 5.1	0.124	0.461	*0.473*	0.236	0.507	
Behavior 6.1	0.171	0.248	0.192	*0.475*	0.353	0.75
Behavior 6.2	0.361	0.227	0.236	*0.467*	0.456	
Behavior 7.1	0.216	0.305	0.216	*0.482*	0.418	
Behavior 7.2	0.415	0.242	0.177	*0.445*	0.460	
% Variance	0.159	0.152	0.112	0.093	0.516	

Note: Bold ones indicate the corresponding factor loadings.

The factors that emerged based on the corresponding questions can be interpreted as:

- Factor 1: Socio-Environmental (Responsibility)—Behavior I
- Factor 2: Personal (Effort)—Behavior II
- Factor 3: Economic-Policy (Concerns)—Behavior III
- Factor 4: Infrastructural and Technological (Endeavors)—Behavior IV

Cronbach's alpha statistics showed that all scales were internally consistent [40,41]. Therefore, these factor groupings formed a reliable set of scales and provided good empirical bases. Moreover, it meant that multivariate analysis could be based on these scales.

Scale Construction

Based on the factor analysis findings, scales for four behavior groups and an overall behavior scale were required. All statements were treated as equally weighted; therefore, the scores for each question under the corresponding factor were summed (from 1—strongly disagree to 5—strongly agree) and divided by the total number of questions. By doing so, an average score for each of the factor groups was determined. In the next step, these four factor scores were summed and divided by four, assuming all factors had equal weights, to generate an overall behavior score for individuals.

3.3. Personality Traits

Based on the conceptual framework used in the study, personality traits were considered as one of the two main influencer groups on urban sustainability understanding and behavior. As explained in Topal et al. [35], the big five personality traits approach of Goldberg [42] was adopted in the conceptual framework, the traits being:

- Surgency (P1)
- Agreeableness (P2)
- Conscientiousness (P3)
- Emotional Stability (P4)
- Intellect (P5)

Five questions were used for the scale of each personality trait groups and each question was treated as equally weighted. Negatively framed questions were recoded for scaling purposes. The scores for each question under the corresponding trait group were summed (from 1—strongly disagree to 5—strongly agree) and divided by five (the total number of questions). Thus, the average score for each personality trait group was created.

3.4. Influencing Factors

As described in the conceptual framework, external influencing factors were the main variable group expected to have an impact on urban sustainability understanding and behavior. In accordance with the conceptual framework, influencing factors may vary according to different contexts and conditions, which provide the model with flexibility. For this research, 19 influencing factors that were considered to be critical to the current analysis were specified as a result of the systematic literature review performed by Topal et al. [31]. Figure 7 presents a general overview of the responses given to influencing factors, which will be discussed in detail in Section 6.3.2. It is important to note that six questions (i.e., IF2, IF13, and IF15 to IF18) are by design negatively framed with respect to favorable urban sustainability behavior.

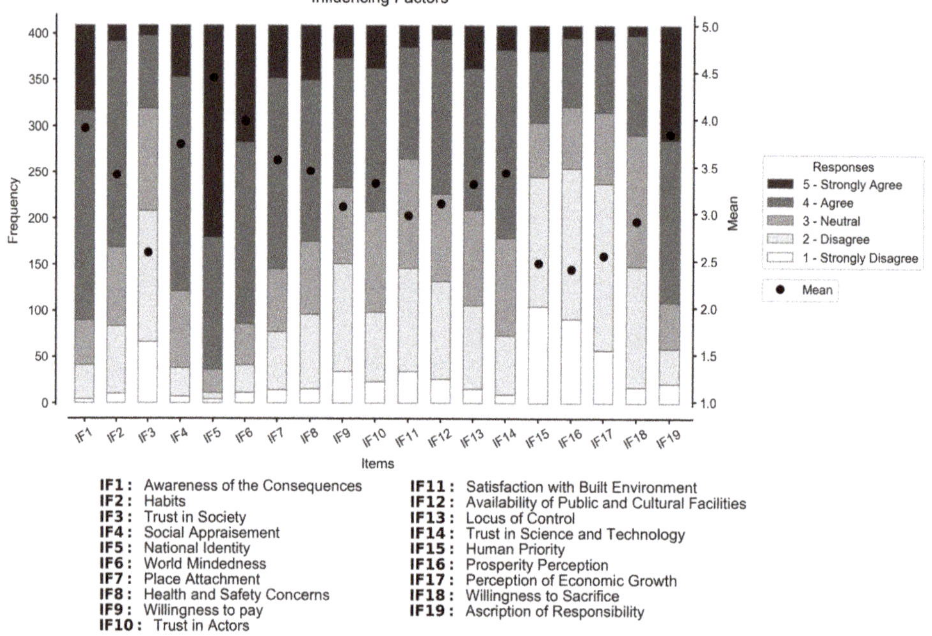

Figure 7. Influencing factor questions.

4. Bivariate Analyses and Results

4.1. Determinants of Understanding

Pearson's r correlation coefficients for each scale are given in Table 3. The correlation coefficients indicate that awareness, perception, and attitude were all positively correlated with each other. While the correlation between awareness–perception and perception–attitude were approximately equal (i.e., 0.37 and 0.35, respectively), the correlation between awareness–attitude was higher (0.49). Although the former finding was in line with what was expected, the latter finding was unexpected since it can be argued that as the distance between the socio-psychological determinants increase, the correlation value would be expected to decrease. However, that was not the case.

Table 3. Pearson correlation coefficients of determinants.

	Awareness	Perception
Awareness	1	
Perception	0.37 *	1
Attitude	0.49 *	0.35 *

* $p < 0.05$.

4.2. Determinants and Behavior

Pearson's r correlation test was used to analyze the relationship between behavior and determinants. The correlation coefficients are presented in Table 4, which shows that the correlations were statistically significant.

Table 4. Pearson correlation coefficients between determinants and behavior.

	Awareness	Perception	Attitude
Behavior	0.57 *	0.44 *	0.57 *

* $p < 0.05$.

The correlation coefficients of determinants with behavior were anticipated from the conceptual model to be high and the values from this research (0.57 for awareness, 0.44 for perception, 0.57 for attitude) confirmed this expectation to be valid. However, it can be argued that the effects of determinants are variable across behavior, behavior having a higher correlation coefficient value (0.57) with both awareness and attitude, whereas the correlation value with perception (0.44) is significantly lower.

4.3. Personality Traits

4.3.1. Personality Traits and Determinants

As shown in the conceptual framework, personality traits were expected to have an impact on urban sustainability understanding and behavior. Table 5 presents the Pearson correlation coefficients between personality types and determinants. As can be seen from the table, the bivariate relationships between all variables are significant ($p < 0.05$), except for that between attitude and personality IV.

Table 5. Pearson correlation coefficients between personality types and determinants.

	Personality I(P1)	Personality II(P2)	Personality III(P3)	Personality IV(P4)	Personality V(P5)
Awareness	0.33 *	0.33 *	0.31 *	0.16 *	0.43 *
Perception	0.26 *	0.20 *	0.22 *	0.12 *	0.27 *
Attitude	0.21 *	0.33 *	0.29 *	0.03	0.26 *

* $p < 0.05$.

For awareness, all personality types (except personality IV—emotional stability) have a strong correlation. The highest correlation was observed with personality V (intellect, 0.43), while personality I-II-III had broadly similar coefficients (0.31 to 0.33, respectively). In contrast, the coefficient for personality IV was notably lower than the other personality types. Therefore, it can be inferred that while intellect is an important predictor of awareness, emotional stability has limited impact.

For perception, personality I (surgency, which describes characteristics of quickness, cleverness, responsiveness, and spontaneity) and personality V had similar coefficients (0.26–0.27), followed by personality II and III (agreeableness, 0.20, and conscientiousness, 0.22, respectively). However, personality IV again differed from other personality types, with a markedly lower correlation value (0.12). Therefore, it is possible to conclude that people with characteristics of surgency and intellect hold better (i.e., more positive) perceptions about urban sustainability.

Regarding attitude, personality II had the highest significant coefficient (0.33), followed by personality III (0.29), personality V (0.26) and personality I (0.21). However, there was no significant relationship between personality IV and attitude. As a result, it can be said that people with a more agreeable and conscientious nature have better attitudes about urban sustainability.

4.3.2. Personality Traits and Behavior

Correlation values for behavior and personality types are presented in Table 6, in which it can be seen that the Pearson correlation coefficients were encouragingly high, all coefficients having significant positive values ($p < 0.05$).

Table 6. Pearson correlation coefficients between personality types and behavior.

	Personality I(P1)	Personality II(P2)	Personality III(P3)	Personality IV(P4)	Personality V(P5)
Behavior	0.30 *	0.43 *	0.46 *	0.20 *	0.43 *

* $p < 0.05$.

Although behavior had mostly strong bivariate relationships with all personality types, there were important differences. Personality III (conscientiousness) had the highest coefficient of 0.46, very closely followed by personality II and V (0.43 for each), while personality I (surgency) was more moderately correlated with behavior (0.30). On the other hand, personality IV proved again to be distinct from the other personality types with a markedly lower correlation coefficient (0.20).

To sum up, the correlational analyses showed that personality types of agreeableness, conscientiousness and intellect were strongly correlated with behavior. Surgency proved to be less of an influence on behavior, while emotional stability did not have a substantial influence on urban sustainability behavior. This is not surprising since sustainable behavior requires intellect to comprehend, agreeableness to obey the rules and conscientious to carry out the necessary actions and perform well.

4.4. Socio-Demographics and Influencing Factors

4.4.1. Socio-Demographics and Determinants and Behavior

In this section, the bivariate relationships between socio-demographic variables and urban sustainability understanding and behavior are explored. Due to the different nature of the data, *t*-test, Spearman correlation analysis and ANOVA tests were conducted.

Gender and residential status were the variables explored in the *t*-tests. For gender, the only significant difference (t = 2.51) found between male (m = 2.85, SD = 0.88) and female (m = 2.64, SD = 0.82) was with awareness. No other determinants or behavior had a statistically significant outcome. The residential status (i.e., landlord, tenant) did not result in a statistically significant difference in any of the determinants and behavior.

The socio-demographic factors presented in Table 7 were investigated for the Spearman analysis. The results showed that while age had a statistically significant negative relationship with awareness, education had statistically significant positive correlations with all of the variables, the highest correlation being with awareness (0.44). Income resulted in positive correlations with all determinants. The number of people living in a household was found to have a significant negative correlation with behavior. Regarding the years lived in Istanbul, positive relationships with attitude and behavior were found. Finally, it was found that the size of the house that people lived in was positively correlated with awareness and attitude.

Table 7. Spearman correlation coefficients between socio-demographics and determinants and behavior.

Social Demographics (SD)	Awareness	Perception	Attitude	Behavior
SD1—Age	−0.14 *	−0.06	−0.02	0.07
SD2—Education	0.44 *	0.21 *	0.16 *	0.22 *
SD3—Income	0.21 *	0.13 *	0.24 *	0.09
SD4—Household	0.04	0.04	−0.06	−0.15 *
SD5—Year living in Istanbul	0.00	0.03	0.18 *	0.17 *
SD6—Size of home	0.10 *	0.06	0.20 *	0.03

* $p < 0.05$.

For the other socio-demographic factors (occupation, the county that people live in, probable future residency, and political orientation) one-way ANOVA tests were performed to examine the relationships with determinants and behavior. Furthermore, a Tukey post-hoc test was conducted for each factor to specify the group differences.

In terms of occupation, statistically significant differences were found in awareness (f(9) = 5.88), perception (f(9) = 1.93), and attitude (f(9) = 2.97); these are shown in Figure 8.

For awareness, academics were found to have higher scores than students (1.58), workers (1.57), private sector employees (1.18), the self-employed (1.21), businessmen (1.28), the unemployed (1.40), housewives (1.72), and those who were retired (1.58). Moreover, housewives showed lower degrees of awareness than public servants (0.68) and private sector employees (0.54). For perception, academics obtained higher scores than workers (0.75), housewives (0.75), and unemployed people (0.84). Regarding attitude, private sector employees showed higher scores than housewives (0.29) and unemployed people (0.58).

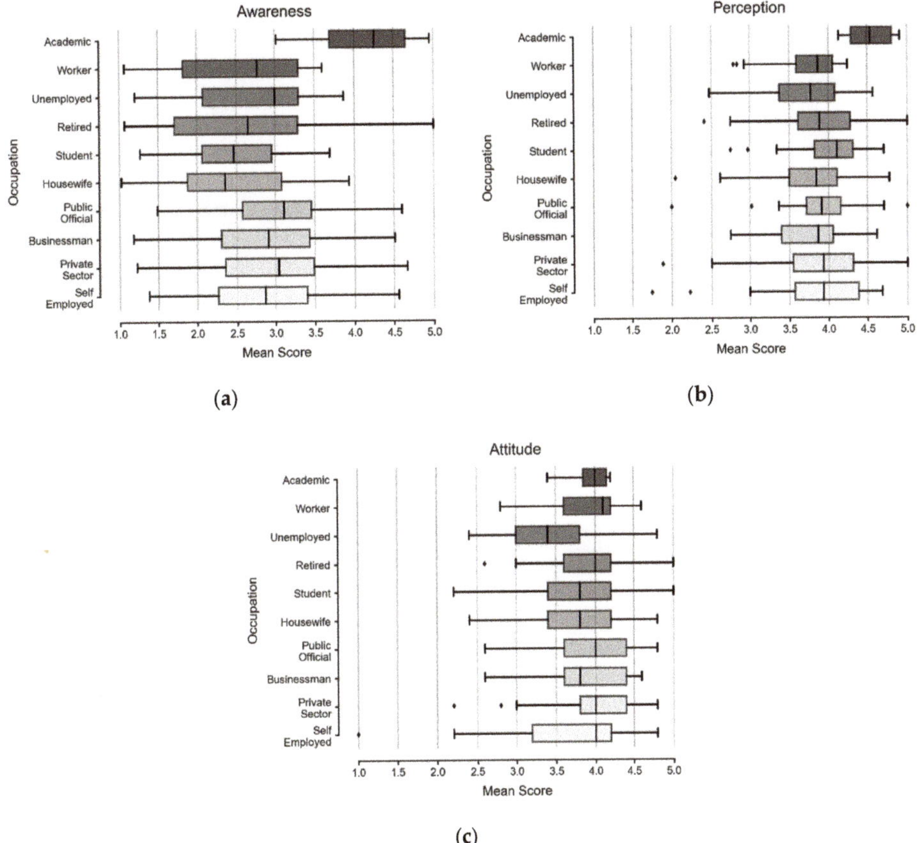

Figure 8. Box plots for occupation and determinants: (**a**) awareness; (**b**) perception; (**c**) attitude.

In terms of the county of residence, statistically significant differences were found in awareness (f(32) = 2.13), perception (f(32) = 2.40), attitude (f(32) = 5.53; see Table 8), and behavior (f(32) = 4.42; see Table 9). For awareness, Sariyer had a lower score than Beykoz (1.00), Gaziosmanpasa (0.99), and Umraniye (0.95). While Sisli district showed lower levels of perception than Bagcilar (0.53) and Kagithane (0.81), differences in attitude scores are noticeable in Table 8. For example, it is noticeable that Bahcelievler, Gaziosmanpasa, Kucukcekmece, Sultanbeyli, and Uskudar scored significantly higher in attitude than other counties.

Table 8. ANOVA mean differences between counties of residence for attitude.

	Bahcelievler	G.Pasa	K.Cekmece	Sultanbeyli	Umraniye	Uskudar
Avcilar	−0.95	−0.90	−0.94	−1.50		−0.71
Bakirkoy				−1.37		
Bagcilar	−0.80	−0.74	−0.78	−1.35	−0.51	−0.56
Beykoz	−0.82			−1.03		
Beyoglu		−0.76	−0.80	−1.37		−0.58
Besiktas	−0.91	−0.85	−0.89	−1.46		−0.67
Cekmekoy				−1.28		
Esenler	−0.88	−0.82	−0.86	−1.43		
Fatih	−0.95	−0.90	−0.94	−1.50	−0.67	−0.71
Gungoren				−1.36		
Sariyer	−0.79	0.74	−0.78	−1.34		−0.55
Sisli	−0.68		−0.66	−1.23		

Finally, for behavior, as seen in Table 9, Fatih and Sariyer had significantly lower scores than many other counties.

Table 9. ANOVA mean differences between counties of residence for behavior.

	Fatih	Sariyer	Sultanbeyli
Avcilar			−1.30
Bahcelievler	1.02	0.91	
Bakirkoy			
Bagcilar	0.67	0.56	
Beykoz		0.56	
Beyoglu			−1.21
Besiktas			−1.27
Esenler	0.91	0.80	
G.Pasa	0.96	0.85	
Gungoren	0.99	0.88	
K.cekmece	0.97	0.86	
Sultanbeyli	1.75	1.64	
Umraniye	1.06		
Uskudar	0.72	0.61	
Zeytinburnu	0.99	0.88	

In terms of future probable residency, no significant differences were found in any of the determinants and behavior. However, for political orientation, statistically significant differences were found in awareness ($f(9) = 9.77$), perception ($f(9) = 3.65$), attitude ($f(9) = 3.78$), and behavior ($f(9) = 5.93$), these being presented in Figure 9.

For awareness, the apolitical group resulted in lower scores than Ataturkist (0.96), left/socialist (0.86), and social democrat (0.77). What is more, the respondents who identified themselves as religious showed lower awareness than the groups of respondents who identified as liberal (1.51), nationalist (1.06), conservative democrat (1.31), left/socialist (1.65), and social democrat (1.55).

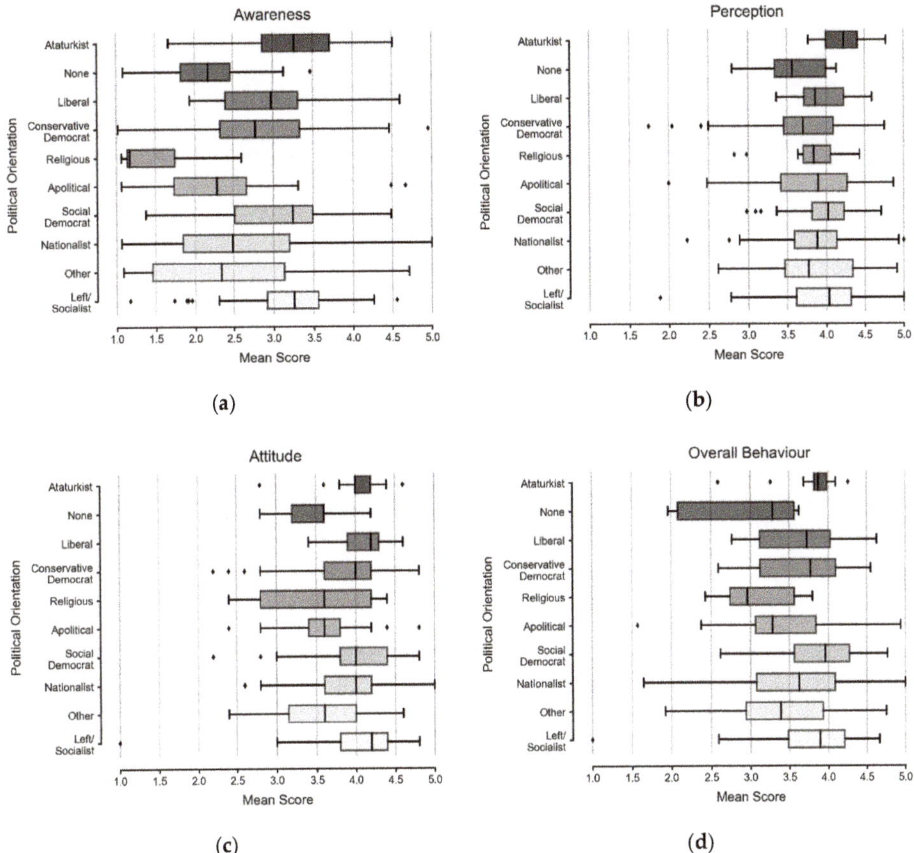

Figure 9. Box plots for political orientation and determinants and behavior: (**a**) awareness; (**b**) perception; (**c**) attitude; (**d**) behavior.

In terms of perception, the responses from the conservative democrat group resulted in lower scores than left/socialist (0.28), Ataturkists (0.54), and social democrats (0.33). On the other hand, the Ataturkist group had higher perception than the group without any ideology (0.70). Regarding attitude, the 'other ideology' group recorded lower scores than left/socialists (0.52) and social democrats (0.45). Similarly, the apolitical group had lower attitude scores than left/socialists (0.40).

Finally, for behavior, it was noticeable that people who do not prefer to be identified by a political ideology recorded lower scores than Ataturkists (0.89), left/socialists (0.90), social democrats (0.96), and conservative democrats (0.73). Left-socialist and social democrats were found to have higher behavior scores than most of the other ideologies, especially for the apolitical (0.45) and religious (0.68) groups.

4.4.2. Influencing Factors (IF) and Determinants

The Spearman correlation analysis (Table 10) shows the correlation between influencing factors (IF) and determinants. Therein, it can be seen that the awareness determinant was significantly correlated with the 12 influencing factors (IF 1-2-4-5-6-7-9-11-12-14-17-18) while perception had significant but comparatively lower correlation coefficients with the 11 influencing factors (IF 1-4-5-6-7-9-10-13-14-18-19). In contrast, attitude showed significant correlation with almost all of the influencing factors (only IF 13-19 were not).

It should be kept in mind that questions related to IF 2-13-15-16-17-18 were by design negatively framed.

Table 10. Spearman correlation coefficients between influencing factors and determinants.

Influencing Factors (IF)	Awareness	Perception	Attitude
IF 1—Awareness of Consequences	0.30 *	0.22 *	0.51 *
IF 2—Habit	0.10 *	0.09	0.13 *
IF 3—Trust in Society	0.08	−0.03	−0.12 *
IF 4—Social Appraisement	0.42 *	0.29 *	0.54 *
IF 5—National Identity	0.18 *	0.21 *	0.39 *
IF 6—World Mindedness	0.28 *	0.33 *	0.37 *
IF 7—Place Attachment	0.12 *	0.10 *	0.29 *
IF 8—Health and Safety Concerns	0.07	−0.04	0.15 *
IF 9—Willingness to Pay	0.33 *	0.20 *	0.31 *
IF 10—Trust in Actors	−0.04	−0.16 *	0.11 *
IF 11—Satisfaction with Built Environment	0.12 *	−0.04	0.11 *
IF 12—Availability of Public Facilities	0.11 *	−0.01	0.16 *
IF 13—Locus of Control	−0.03	−0.22 *	−0.01
IF 14—Trust in Science and Technology	0.25 *	0.27 *	0.36 *
IF 15—Human Priority	−0.09	−0.07	−0.40 *
IF 16—Prosperity Perception	−0.07	−0.05	−0.32 *
IF 17—Perception of Economic Growth	−0.11 *	−0.02	−0.38 *
IF 18—Willingness to Sacrifice	−0.13 *	−0.15 *	−0.16 *
IF 19—Ascription of Responsibility	0.00	0.15 *	0.07

* $p < 0.05$.

People who were more aware of the consequences of unsustainable actions often had more awareness and better perception; moreover, they held more positive attitude scores. As social appraisal of a behavior increased, so did the awareness, perception, and attitudes of people. If people had a stronger sense of national identity, they appeared to have reasonably strong awareness and perceptions, but also highly positive attitudes. People who considered themselves as more open-minded appeared to have good awareness, combined with positive perceptions and attitudes. Place attachment was another influencer that exhibited statistically significant positive correlation with awareness and perceptions of individuals, while its positive impact on attitudes was higher. Finally, health and safety concerns were found to have a significant correlation with attitudes of individuals; i.e., the stronger such a concern was felt by individuals, the stronger their attitude towards urban sustainability was likely to be.

Economic approaches of individuals deserve further attention. If, for example, individuals were more ready to pay for sustainability, they were more likely to have good awareness, combined with more developed perceptions and attitudes. On the other hand, if people held a more economically or personal prosperity-centered perspective over sustainability, they were found to have slightly lower awareness scores and considerably more negative attitudes. Similarly, people who were less willing to sacrifice their amenities (in the name of sustainability) resulted in them having negative scores for awareness, perception, and attitude. Finally, people who believed in humans having priority over nature were found to be negatively correlated with urban sustainability attitude.

As trust in a mediator (actor) increased, perceptions and attitudes about urban sustainability were found to become more positive. However, trust in society only resulted in better attitudes. In addition, people who had more trust in science and technology reported higher scores in awareness, perceptions, and attitudes. Counterintuitively, unsustainable habits were positively correlated with awareness and attitudes, albeit marginally. Even though this was an unexpected outcome, it might mean that, in general, the more conscientious people were about their wrong habits, the higher their awareness and attitudes.

The quality of the living environment was another factor to note. It was observed that people who have a greater satisfaction with the quality of the built environment also had greater awareness and better attitudes towards urban sustainability. Similarly, more cultural facilities in the vicinity of the area in which the respondents lived led to higher awareness and attitude scores. Interestingly, the locus of control factor was found to have a statistically significant correlation only with urban sustainability perceptions of individuals. People who felt they had less control over their sustainable built environment had lower sustainability perception scores. Similarly, ascription of responsibility resulted only in better perceptions.

4.4.3. Influencing Factors (IF) and Behavior

The Spearman correlation values between behavior and influencing factors (IF) are presented in Table 11. Statistically significant correlations therein are marked.

Table 11. Spearman correlation coefficients between influencing factors and behavior.

Influencing Factors (IF)	Behavior
IF 1—Awareness of Consequences	0.43 *
IF 2—Habit	0.23 *
IF 3—Trust in Society	0.21 *
IF 4—Social Appraisement	0.49 *
IF 5—National Identity	0.22 *
IF 6—World Mindedness	0.42 *
IF 7—Place Attachment	0.29 *
IF 8—Health and Safety Concerns	−0.01
IF 9—Willingness to Pay	0.47 *
IF 10—Trust in Actors	0.11 *
IF 11—Satisfaction with Built Environment	0.15 *
IF 12—Availability of Public Facilities	0.15 *
IF 13—Locus of Control	−0.03
IF 14—Trust in Science and Technology	0.29 *
IF 15—Human Priority	−0.16 *
IF 16—Prosperity Perception	−0.16 *
IF 17—Perception of Economic Growth	−0.15 *
IF 18—Willingness to Sacrifice	−0.17 *
IF 19—Ascription of Responsibility	0.17 *

* $p < 0.05$.

Sustainability behavior was significantly correlated with all factors, except two: health and safety concerns (IF8) and locus of control IF13. However, it can be seen that the factors of awareness of consequences (IF1), social appraisement (IF4), world mindedness (IF6), place attachment (IF7), willingness to pay (IF9), and trust in science and technology (IF14) had encouragingly high correlation coefficients with urban sustainability behavior.

5. Multivariate Analyses and Results

In this section, multivariate analyses based on the conceptual framework are reported. A series of multiple regression analyses were used to investigate the associations between the main variables of interest (i.e., socio-psychological determinants of awareness, perception, and attitude) and behavior.

5.1. Preparation for the Analysis

Variables that were either numerical or continuous were directly included in the analysis. Ordered categorical socio-demographic variables (age, education level, income level, number of people living in the household, length of time lived in Istanbul, and approximate size of houses) were also directly included in the analyses by converting them to numerical codes. Unordered categorical variables were converted into dummy variables for inclusion in the analysis. Probable future residency was recoded as 1 for 'continue to live in Istanbul' and 0 for 'continue to live elsewhere'. For gender, women were coded as 1 and men were coded as 0. Finally, for residential status, the landlord option was coded as 1 and the tenant option was coded as 0. Consequently, it was possible to include all predictor variables in the multiple linear regression tests, which are reported in the following sections.

5.2. Determinants and Behavior

Based on the conceptual framework, outcome variables for linear regression tests were specified as awareness, perception, attitude, and behavior. All influencing factors, personality types, and prepared socio-demographic questions were included as predictor variables. However, in accordance with the framework, each previous outcome variable was included in the following tests as a predictor variable. In other words:

i.   Awareness was included in the linear regression test of perception;
ii.   Awareness and perception were included in the linear regression test of attitude; and
iii.   Awareness, perception, and attitude were included in the linear regression test of behavior.

Table 12 shows the results of the series of multiple linear regression tests, including regression coefficients, R^2 values, and F statistics.

Table 12. Standardized linear regression coefficients for determinants and behavior.

	Awareness	Perception	Attitude	Behavior
Awareness		*0.176 ***	*0.153 ***	*0.218 ***
Perception			0.060	*0.128 ***
Attitude				*0.190 ***
Influencing Factors (IF)				
IF 1—Awareness of Consequences	*0.164 ***	0.002	*0.209 ***	0.079
IF 2—Habit	0.029	0.053	0.028	0.017
IF 3—Trust in Society	0.040	−0.061	*−0.102 ***	*0.111 ***
IF 4—Social Appraisement	*0.172 ***	0.082	*0.167 ***	*0.092 ***
IF 5—National Identity	0.031	0.089	*0.181 ***	0.063
IF 6—World Mindedness	0.005	*0.171 ***	*0.136 ***	*0.087 ***
IF 7—Place Attachment	−0.042	0.052	0.047	−0.023
IF 8—Health and Safety Concerns	0.035	*−0.091 ***	*0.083 ***	−0.031
IF 9—Willingness to Pay	*0.144 ***	0.085	*0.141 ***	*0.144 ***
IF 10—Trust in Actors	−0.075	−0.062	*0.094 ***	0.066
IF 11—Satisfaction with Built Environment	0.085	−0.057	−0.015	−0.003
IF 12—Availability of Public Facilities	0.051	−0.042	−0.016	−0.025
IF 13—Locus of Control	0.028	*−0.151 ***	*0.101 ***	−0.001
IF 14—Trust in Science and Technology	0.071	*0.173 ***	*0.081 ***	−0.073
IF 15—Human Priority	−0.020	0.009	−0.051	0.065
IF 16—Prosperity Perception	0.044	0.109	0.025	−0.027
IF 17—Perception of Economic Growth	−0.017	0.060	*−0.116 ***	0.014
IF 18—Willingness to Sacrifice	*−0.119 ***	*−0.151 ***	−0.040	−0.012
IF 19—Ascription of Responsibility	*−0.105 ***	*0.167 ***	*0.077 ***	0.059
Personality Traits (P)				
P1—Surgency	0.022	0.037	0.035	−0.020
P2—Agreeableness	0.078	−0.032	0.000	*0.085 ***
P3—Conscientiousness	0.022	0.067	−0.006	*0.131 ***
P4—Emotional Stability	0.021	0.014	−0.050	0.048
P5—Intellect	*0.149 ***	−0.042	−0.017	*0.084 ***
Social Demographics (SD)				
SD 1—Age	*−0.086 ***	0.014	−0.062	*0.095 ***
SD 2—Gender	*−0.074 ***	*0.101 ***	−0.034	0.001
SD 4—Education	*0.323 ***	0.038	−0.033	0.023
SD 6—Income	0.000	0.028	0.046	−0.045

Table 12. *Cont.*

	Awareness	Perception	Attitude	Behavior
SD 8—Household number	0.047	*0.110 **	−0.011	−0.069
SD 9—Year living in Istanbul	−0.088	−0.051	0.063	0.012
SD 10—Future residency	−0.006	0.020	0.024	−0.001
SD 12—Residential Status	*0.127 **	0.043	−0.031	0.006
SD 13—Size of House	0.006	−0.010	*0.090 **	−0.014
R^2—Model Fit	0.539	0.369	0.669	0.650
F	13.31	6.428	21.58	19.19
Error	0.67	0.79	0.57	0.59
Model Significance	$p < 0.01$	$p < 0.01$	$p < 0.01$	$p < 0.01$

Note: * $p < 0.05$. Bold ones indicate the statistically significant results.

Based on the results, it was observed that all models were significant, meaning that the outcome variables were successfully predicted (by predictor variables). Therein, perception had a weaker fit of 37%, awareness had a moderate fit (54%), whilst both attitude (67%) and behavior (65%) had good fits. Despite considerable error terms, which means that a substantial amount of unobserved variance existed the model, these findings can be considered valuable for theoretical purposes. Moreover, it can be observed that all determinants and behavior were influenced by several different factors while some factors were found to have an effect on more than one determinant. For example, awareness significantly predicted perception and attitude, yet no significant association was found between perception and attitude. Moreover, awareness, perception, and attitude were found to have significant positive associations with behavior in isolation. This relationship was in line with the theoretical premise proposed by the conceptual framework (see Topal et al. [35]).

5.2.1. Awareness

When exploring the awareness outcome variable, it can be seen that five influencing factors, one personality type, and four socio-demographic parameters, as predictor variables, were found to have significant predictive power. As expected, people who were more aware of the consequences of their behaviors (IF1) had higher urban sustainability awareness. Similarly, the effect of social appraisement (IF4) on awareness was considerable, which emphasized the importance of the social environment in sustainability policies. In addition, people who were found to be more willing to pay (IF9) or willing to sacrifice (IF18) were also found to have higher awareness. Therefore, people with altruistic characteristics might be more inclined to be aware of sustainability. Interestingly, ascription of responsibility (IF19) resulted in a negative effect on awareness. In terms of personality types, people with higher intellect (P5) were associated with higher levels of awareness. In this respect, it goes without saying that awareness requires a certain level of intellectual capacity, so this overall outcome seems reasonable. Regarding socio-demographics, age (SD1) was found to have a negative impact, which means the younger people are, the more aware they are. Since sustainability is a novel topic and the information channels to awareness require adaptation to technological developments, the finding that the younger generation tends to have more awareness is again reasonable. When it comes to gender (SD2), it was observed that males had greater awareness than females on average. In the Turkish context, where a more male-dominated culture is prevalent and males have better educational opportunities, this is understandable. As expected, there was a strongly positive influence of education (SD4) on awareness. Finally, it was found that people who live in their own house (SD12) had greater awareness than people who live in a rented property.

5.2.2. Perception

With regard to the perception outcome variable, the awareness determinant, six influencing factors, and two socio-demographic parameters as predictor variables were found to have significant predictive power. Among the other variables, awareness was found to have a strongly positive impact upon perception, which supports the theoretical conceptualization of the model. In terms of the influencing factors, world mindedness (IF6) was positively associated with perception, meaning that, in general, people who have a more global vision have a stronger urban sustainability perception. Health and safety concerns (IF8) also had a positive impact on perception, while a negative coefficient indicated that perception was negatively associated with those having less locus of control (IF13). Trust in science and technology (IF14) to solve environmental problems, on the other hand, was associated with a positive influence on perception. This highlights the critical role of science and technology within the field of urban sustainability. While willingness to sacrifice (IF18) and ascription of responsibility (IF19) were shown to positively affect perception, there was no significant predictive power of any personality type, meaning that urban sustainability perception of the individuals is independent from an individuals' personality. Regarding socio-demographics, it was found that the number of people living in the household (SD8) was positively associated with perception and females had higher perception than males (SD2) on average, which was contrary to the effect observed on awareness.

5.2.3. Attitude

Regarding the attitude outcome variable, a combination of the awareness determinant, 12 influencing factors and one socio-demographic parameter (as predictor variables) were found to have significant predictive power. However, neither perception nor the personality types were found to have any significant impact on attitude. In line with the conceptual model, awareness was found to have a positive (significant) association with attitude. In terms of influencing factors, awareness of the consequences (IF1), social appraisement (IF4), trust in actors (IF10), and trust in science and technology (IF14) were found to have a positive effect on the attitude of individuals. However, trust in society (IF3) resulted in a negative influence on attitude, which means that people who considered their society unsuccessful held better attitudes towards behavior. For national identity (IF5) and world mindedness (IF6), a positive association with attitude was observed. Therefore, it is possible to claim that having a global environmental approach and caring about the nation's future positively contributes to attitude. Health and safety concerns (IF8), willingness to pay (IF9), and ascription of responsibility (IF19) had a positive effect on attitude. However, it was observed that favoring economic development over urban sustainability (IF17) resulted in an individual holding a negative association with attitude. On the contrary, the thought of not having locus of control (IF13) resulted in a positive association with attitude. This result was counterintuitive and might be related to people's positive view of their attitudes, even though they think themselves unable to influence the sustainability of their built environment.

5.2.4. Behavior

Considering the behavior outcome variable, which is the final output of the socio-psychological processes presented in the conceptual model, awareness, perception and attitude determinants as predictor variables resulted in positive associations. Moreover, four influencing factors, three personality types, and one socio-demographic parameter were found to have significant predictive power as predictor variables. Of the influencing factors, trust in society (IF3) along with social appraisement (IF4) were found to have a positive association with urban sustainability behavior. This indicates the importance of society and the social environment. As a result, the success of any policy was deemed to be highly dependent on its acceptance by the society as a whole, which in turn has the potential of improving individual's actions. In addition, willingness to pay (IF9) and world mindedness (IF6) had positive significant impacts on behavior. This shows that having a world-minded approach contributes positively to sustainable behavior. In terms of personality types, agreeableness (P2), conscientiousness (P3), and intellect (P5) were positively associated with behavior. In other words, people who are more agreeable, more conscientious, or have a greater intellect were observed to have better (urban) sustainability behavior. Finally, age (SD1) was found to have a positive effect on behavior. Therefore, it can be concluded that older generations perform better when it comes to urban sustainability behavior, even though they tend to be less aware of the issues (see Section 5.2.1).

Taking into account this large number of influencing factors, it can be argued that urban sustainability behavior is a highly complex phenomenon. However, the explanatory power of 65% shows that the framework presented here has promising results.

6. Discussion

6.1. Urban Sustainability Understanding and Behavior

6.1.1. Determinants of Understanding

Urban sustainability understanding consisted of three individual determinant scales: awareness, perception, and attitude. While the perception and attitude scales resulted in very close mean values (3.86 and 3.87, respectively), the mean value of awareness was considerably lower (2.75). It is therefore possible to deduce that although people have a lower level of awareness about urban sustainability, their perceptions and attitudes are comparatively more positive. Since perception and attitude are assumed to be preceded by awareness, this points to the fact that there are other factors that have direct influence on each determinant. Correlation coefficients showed that awareness was better correlated with attitude (0.49) than perception (0.37). Finding a significant correlation between each determinant was an expected outcome from the previous studies of Guo et al. [43] and Tran [44]. When it comes to understanding, awareness had the highest correlation (0.86), followed by attitude (0.76) and perception (0.68). This finding highlights that awareness is the key factor for achieving a better urban sustainability understanding.

Looking at standardized regression coefficients based on the conceptual framework, awareness was found to have more predictive power on perception (0.173) than attitude (0.153), yet no significant relationship was found between perception and attitude. Whilst the former finding confirms the linear sequence of determinants given in the conceptual framework, the latter finding points out the highly intertwined relationship between perception and attitude determinants. Consequently, it is possible to deduce that increasing the awareness level of individuals should have a high potential impact on perception and attitudes (of urban sustainability).

6.1.2. Determinants and Behavior

The results for behavior indicated that both awareness and attitude have the same correlation coefficients (0.57), followed by perception (0.44). Regarding the standardized regression coefficients, awareness resulted in the highest predictive power (0.218), followed by attitude (0.190) and perception (0.128). These results showed that urban sustainability behavior is mainly predicted by awareness and attitude. This resonates with the findings of Barr [20], Buerke et al. [45], Cagáňová et al. [46], Guagnano et al. [47], and Peng et al. [48]. In order to enhance sustainability behavior, there should be a stronger focus on increasing the awareness of people by informative activities, training opportunities, public campaigns and advertisements, and social media [49]. Moreover, their attitude towards urban sustainability behavior needs to be enriched by encouraging sustainable practices, providing necessary laws and regulations, revealing the adverse environmental, economic, and social effects of unsustainability, stressing the criticality of the current situation for themselves and future generations, and reminding them of the responsibility of each and every individual to achieve a more sustainable future.

To sum up, behavior requires respective efforts in terms of socio-psychological determinants to achieve sustainable practices. Therefore, it is of the utmost importance to provide customized approaches based on the nature and type of the determinants and behavior. Lastly, it can be concluded that perception acts as a mediator between awareness and attitude and could be assumed as an intermediate determinant. Due to its intertwined meaning with awareness and attitude, it is more difficult to specify its individual impact (or contribution) on urban sustainability behavior.

6.2. Personality Traits

6.2.1. Personality Traits and Determinants

It was observed that the mean value of P3 and P2 (conscientiousness and agreeability, 3.88 and 3.85, respectively) are the highest, followed by P5 (intellect, 3.51) and P1 (surgency, 3.30), while P4 (emotional stability) holds the lowest mean value (3.09). Therefore, it can be concluded that the characteristic traits of the respondents are that they had an agreeable and conscientious nature with high intellect. This can be seen as a positive aspect since sustainable behavior is expected to be related with having (a) an agreeable nature, (b) a certain level of intellect, and (c) performed conscientious actions [50].

Considering the results of the regression test, it was found that only P5 (intellect) had a significant predictive power on the awareness determinant. This shows that although personality types have significant correlations with urban sustainability understanding, they have weaker predictive associations.

P1 (surgency) resulted in the highest correlation with awareness (0.33), followed by perception (0.26) and attitude (0.21). It is therefore evident that 'extroversion', quick-wittedness and confident sociability of the individuals have a strongly positive correlation with awareness. It would not be inappropriate to assume that this might be due to respondents' openness to new sources of information and innovation. Therefore, the informative channels for urban sustainability should be designed accordingly in order to attract those people who exhibit less surgency. P2 (agreeableness) resulted in exactly the same correlation with awareness (0.33) and attitude (0.33), followed by perception (0.20). Similarly, agreeable people were found to have more positive attitudes to, along with greater awareness of, urban sustainability. Similar results were found for P3 (conscientiousness) personality trait. However, it was found that P4 (emotional stability) had low degrees of correlation with awareness (0.16) and perception (0.12), while no significant correlation could be found with attitude. This striking finding suggests that the emotional stability trait of individuals has little or no relationship with their urban sustainability understanding. Finally, P5 (intellect) was found to have the highest correlation with Awareness (0.43), followed by perception (0.27) and attitude (0.26). From this, it can be further deduced that urban sustainability understanding of individuals increases with their intellectual capacity.

The results also indicated that all personality types, except P4 (emotional stability), have considerable positive correlations with socio-psychological determinants of urban sustainability understanding. Moreover, only P5 (intellect) had significant predictive power on the awareness determinant. This is logical since abstract thinking could facilitate the process of both gaining and interpreting the knowledge. However, this points to the additional need that informative policies should be designed in such a way that requires less intellectual effort in order to attract the attention and adherence of most citizens [51]. Keeping in mind the high mean values of the P2 (agreeableness) (3.85) and P3 (conscientiousness) (3.88) among respondents, the chance of successfully achieving urban sustainability behavior seems promising. To sum up, it is seen that although urban sustainability understanding is correlated with personality types of individuals [52], it is, on the whole, not predicted by them.

6.2.2. Personality Traits and Behavior

Contrary to determinants, personality traits of individuals were found to have higher correlations and more predictive power on urban sustainability behavior.

The Behavior results showed that the strongest correlations exist with P2 (agreeableness) (0.43), P3 (conscientiousness) (0.46), and P5 (intellect) (0.43). In line with the correlation coefficients, standardized regression coefficients revealed that P2 (agreeableness) (0.085), P3 (conscientiousness) (0.131) and P5 (intellect) (0.084) had significant predictive power on urban sustainability behavior. Based on these findings, it is possible to say that sustainability behavior is closely associated with an agreeable nature, conscientious practice, and intellectual effort. Moreover, it could be claimed that corresponding policies and regulations have more chance of success if they specifically target these key characteristics of individuals.

To conclude, it was observed in this research, but also elsewhere within the literature, that certain personality types have a close relationship with urban sustainability behavior [17,53–55]. This finding reveals the fact that political decisions as well as regulatory instruments need to take into consideration the differing personalities of the public. Moreover, it can be deduced that successful urban sustainability behavior could be achieved if these practices are adhered to. Finally, these political decisions and regulatory instruments should be shaped to be synergistic with conscientious traits in the public and provide an adequate intellectual background.

6.3. Socio-Demographics and Influencing Factors
6.3.1. Socio-Demographics and Determinants and Behavior
Age

Age has very little negative correlation with awareness and understanding, yet it is a significant predictor of awareness and behavior. While age has negative association with awareness, meaning that the younger generation has better awareness about urban sustainability, it has a positive relationship with behavior. It is therefore possible to conclude that although the younger generation is more aware of the sustainability as a result of having better access to digital informative channels [56,57], the formation of behavior and responsibility improves with the age of an individual [58,59]. In essence, this finding implies that the older generation needs more informative interventions while specific attention should be given to behavior formation of the younger generation in Turkey.

Gender

Gender had significant correlation with all determinants and behavior, the strongest correlation occurring with awareness. On the other hand, gender had predictive power only on awareness and perception. While male respondents had greater awareness than females [43], female respondents scored better in perception than males [60]. Contrary to the findings of several studies reporting the association between gender and sustainable behavior [58,61], the results of this study did not indicate any significant relationship. That said, the findings of the current study did reveal that females need to have more access to informative activities and specific actions are required to increase their knowledge and awareness level. Moreover, since the impact of women on the purchasing habits of the household is considerable, their positive perceptions could be an advantage for sustainability practices of the family unit in Turkey.

Education

Education had the strongest correlation with awareness (0.44) and resulted in significant association with awareness alone. The relationship between awareness and education echoes the findings of previous studies (for example, [43,62]) and is understandable considering the close relationship it holds also with gaining knowledge. On the other hand, the insignificant association of education with perception [60], attitude [63], and behavior [64,65] is contrary to the findings within the literature. Therefore, it could be inferred that current educational content in Turkey is inadequate to improve urban sustainability understanding and behavior. This includes better access to information tools, where the findings suggest that curricula of high schools and higher education should be improved and amended in order to enhance sustainability understanding and behavior.

Occupation

Considering the ANOVA test results along with the Tukey post-hoc test, it was found that statistically significant differences occurred among occupation types. The most noticeable differences were housewives having a weaker degree of awareness than public servants and private sector employees. Likewise, private sector employees had higher attitude scores than housewives, unemployed people, and academics, thereby showing higher levels of awareness and perception. These differences may simply be the result of people in certain occupations simply not having much time to pay attention to sustainability issues, as suggested by Barau [66].

Income

Income had moderate positive correlations with awareness and attitude, yet no apparent correlation was found with behavior (see Table 7). On the other hand, there was no significant association between income and any of the determinants and behavior presented in Table 12. Although income has been found to be both positively [43,63,64] or negatively [60,67] associated with urban sustainability understanding and behavior within the literature, it is evident from the findings of this research that there is no significant association within the Turkish context. This emphasizes the context-sensitive nature of sustainability behavior.

County of Residence

The ANOVA test results showed that the county of residence resulted in a statistically significant difference for determinants and behavior. It was observed that counties which contain residents form lower socio-economic groups had higher levels of awareness, perception, attitude, and behavior. For instance, Sariyer had lower awareness levels than Gaziosmanpasa (0.99) and Umraniye (0.95); Sisli had a lower perception score than Bagcilar (0.53); and finally, Kagithane (0.81), Sultanbeyli, and Gaziosmanpasa had better attitude scores than many other counties. Although income alone did not present any association, socio-economic condition seems to have an impact on sustainability understanding and behavior. While better perception, attitude, and behavior in these areas can be as a result of being exposed to the adverse effects of unsustainable urbanization, in line with the work of Maiello et al. [68], the better awareness results were unexpected. The widespread use of social media in all socio-economic spheres of Turkey may have been influential here.

Household Number

While the household number (i.e., occupancy rate) was not significantly correlated with any of the determinants, it did have a moderately negative correlation with behavior. On the other hand, the household number was positively associated with perception (0.110). It was found that large families have better perceptions, which means that they care more about urban sustainability. Therefore, it is important to reach the large family units to encourage them to engage in improved behavior. Moreover, these families could provide an opportunity to build upon the positive perceptions about sustainability—i.e., to engage individuals in the wider population. This finding resonates with the findings of Waitt et al. [61], which stresses the importance of household number in successfully advancing urban sustainability.

Length of Residency in Istanbul

Considering the length of residency, it was observed that positive correlations exist with attitude and behavior. However, no predictive power was found as a result of linear regression. As Rogers and Bragg [69] suggest, the length of residency is one of the major contributors to an individual forming a place attachment. Subsequently, this may also result in better urban sustainability understanding (and motivation), leading to responsible environmental behavior for people in Turkey.

Political Orientation

The ANOVA test results demonstrated that there were significant differences among different political orientations in relation to all determinants and behavior. However, perhaps remarkably, religious [43] and apolitical people were found to have lower scores (on determinants and behavior) than left/socialists, social democrats, and Ataturkist people. This is in line with the findings of Drews and van den Bergh [63], who reported that religiosity and political orientation have a direct relationship with sustainability understanding and behavior, not least in a Turkey context.

Residential Status

According to *t*-test results, no significant differences were found when considering residential status. However, when applying standardized regression coefficients, it could be seen that residential status had a significant impact on awareness. Additionally, landlords were found to have greater awareness (0.127) than tenants. This is contrary to the findings of Kang [51]. It is reasonable to suggest that investing in the purchase of a house increases the awareness of individuals about the sustainability of the property—not least in terms of the cost to run the property. Therefore, it might be a good idea to promote urban sustainability with real estate advertisements since it is likely to be the best time for people to engage with the subject, i.e., people are likely to be open to the receipt of information due to their financial investment.

Size of House

The size of house that people live in has a significant positive correlation with awareness and attitude. Similarly for linear regression results, the size of house has a predictive power on attitude (0.90). While larger houses are mentioned within the literature as the cause of unsustainable behavior, as a result of increased resource consumption [67], it was found within this study to have a positive impact on the attitude of an individual towards urban sustainability. This could be related to the socio-economic conditions of the individuals, since a larger income (allied with a larger surplus income) also had a positive correlation with attitude. However, it is possible to deduce that physical living conditions of the individuals have a positive impact on attitudes of individuals. In other words, improvement therein can enhance an individual's urban sustainability understanding and behavior. This should be carefully considered for urban planning policies in Turkey—not least when it comes to urban regeneration and sustainable retrofitting. Exemplar projects in terms of sustainability could make a step-change in this respect.

6.3.2. Influencing Factors and Determinants and Behavior

While the majority of the influencing factors have significant correlation with determinants and behavior, their predictive power varies. Details for each influencing factors will be given in the following sub-sections.

Awareness of Consequences

Awareness of consequences has been shown within this research to have strong positive correlations with all the determinants and behavior. It was found to be one of the most powerful influencing factors. Regarding the regression coefficients, it was found to have strong impacts on awareness and attitude. As stated in norm-activation theory by Schwartz and Howard [70], awareness of consequences leads to feeling of guilt. Moreover, it also appeals to the rational thinking of individuals. Therefore, making people aware of the consequences of unsustainable actions seems to have a huge impact on making them aware of urban sustainability, forming an appropriate attitude and behaving accordingly [14,44,47,57,71,72].

Habit

While unsustainable habits resulted in weak correlations with determinants, they have a moderate correlation with behavior. However, no significant associations were found. Therefore, habits may require consideration in terms of specific behavior types to determine whether there are further associations. While unsustainable habits can be accepted as a threat for sustainable practices, they could also be seen as opportunities if they are turned (nudged) into sustainable ones [17,57,58,73]. In so doing, it is important to specify what types of urban behavior are related with habits of the people and how they can they be channeled into sustainable forms through incentives and interventions.

Trust in Society

Trust in society had a moderate positive correlation with behavior, while in terms of the regression coefficients, it was found that it has a negative association with attitude and a positive relationship with behavior. It was observed that having confidence and trust in the sustainability performance of the society that people live in could have a considerable positive impact on their sustainability behavior. It is important to improve the overall performance of society in terms of an individual sustainable behavior, but this requires a significant level of trust being placed upon individuals. That said, there will always be people who do not take their civic duties and social responsibilities seriously—these individuals will behave badly, whatever the status quo [64,74].

Social Appraisement

Social appraisement (alongside trust in society) was found to be the other most influential factor due to its very high correlations with all determinants and behavior. Similarly, it was found to have predictive power on awareness, attitude, and behavior. It can be argued that if people believe that behaving in a sustainable manner is good for their social identity and their social environment approves of this behavior, they perform well in terms of sustainable behavior [66,72]. Moreover, social appraisement increases the awareness of individuals by their interaction within the society and in so doing this helps the formation of highly positive attitudes [66,75]. Consequently, society as a whole should be convinced of the importance of sustainability and likewise should be encouraged to behave sustainably, modifying the existing (and sometimes entrenched) social norms. In other words, sustainable behavior should become a new norm for Turkish society to ensure the success of sustainability policies. For example, only 15 years ago, the UK made it illegal for anyone to smoke in an enclosed public place and within the workplace; this is now widely accepted as the new norm and has brought with it substantial health benefits to individuals. Perhaps zero-emission cars and car-free neighborhoods/cities will be the new norm of the future for sustainable urban areas—time will tell.

National Identity

National identity had positive correlations with determinants, especially attitude and behavior. According to standardized regression coefficients, national identity was found to have predictive power on attitude. As presented in Section 3.1, most of the respondents identify themselves as nationalist, followed by conservative democrats, thus demonstrating the importance of national identity. Therefore, announcing (national) macro-sustainability strategies with emphasis on the national interest of Turkey could improve attitudes, which in turn could encourage people to participate [51,76].

World Mindedness

The high correlation coefficients with all determinants and behavior, along with the predictive power on perception, attitude, and behavior, indicates that having a world-minded approach is very important for achieving urban sustainability, understanding and behavior. It is therefore of the utmost importance to reference the international policy and practices of sustainability when making and announcing sustainability policies (and associated interventions). This echoes the work of Der-Karabetian et al. [77] who suggested that the simple act of letting people know about what international standards exist (for sustainability) and how they manifest into expected sustainability actions can lead to a positive impact—the same is true for improving people's urban sustainability understanding and behavior.

Place Attachment

While place attachment was found to have moderate positive correlations with determinants and behavior, no significant association was found as a result of the regression analysis. Although it is stated to have positive impact on sustainable lifestyles within the literature [66,69], no significant impact was found in the Turkey context. That said, it might also imply that the sense of belonging to the area people live in is low in many places internationally. Therefore, by providing a better urban landscape, i.e., high-quality built environment with more urban greenery, it would not be inappropriate to assume there would be a stronger place attachment resulting in more sustainable behavior.

Health and Safety Concerns

Health and Safety concerns had a significant correlation only with attitude, while the regression analysis demonstrated a significant negative association with perception and a positive association with attitude. In other words it can be seen that people who are worried about their health and safety mostly hold unfavorable perceptions about urban sustainability—this is a finding shared by Noonan et al. [78]. On the other hand, their attitudes also seem to be influenced positively from this concern [74]. Therefore, it might be deduced that health and safety concerns could be beneficial to urban sustainability (ultimately perhaps influencing behavior) as long as an attitude of self-preservation did not ensue.

Willingness to Pay

It was found that willingness to pay had strong positive correlations with urban sustainability understanding, i.e., all three determinants, and behavior. Similarly, regression results illustrate that it had a positive impact on awareness, attitude, and behavior. Thus, the more people improve their sustainability understanding and behavior, the more they are ready to pay, and vice-versa [49,67]. Sustainability policies should therefore be presented in a way that attracts people's attention and persuades them that it is necessary for the benefit of people and the planet to spend more in this arena. This requires the sustainability benefits (to Turkish society as a whole) to be identified. For example, this highlights the need for those who have the money to invest in more sustainable technologies, such as electric cars, in order to reduce carbon emissions and other greenhouse gas and meet international targets. This could be accompanied by a reduction in road tax for those who adhere to the change and an increase in car tax for those who keep their older (more polluting) vehicles. Likewise, sustainability grants, subsidies and loans would need to be brought in to help those who need it (i.e., those who are willing but unable to pay).

Trust in Actors

While trust in actors (such as public authorities, academics, NGOs) had a weak correlation with perception, attitude, and behavior, the regression analysis showed it had a significant positive impact on attitude. Therefore, in order to improve the attitudes of individuals and increase the success of policies, it is important that citizens should be confident with and trust in the actors [74,79]. Transparency and the public communication can be key factors for public authorities in Turkey to gain the trust of individuals.

Satisfaction with Built Environment

Satisfaction with the built environment had a weak to moderate correlation with awareness, attitude, and behavior, yet no significant relationship was found as a result of the linear regression tests. Although it was reported in the literature to have an impact on sustainability behavior [46,80,81], this was not the case for the Turkey context. The comments above made in relation to place attachment might therefore be echoed here.

Availability of Public Facilities

With very similar responses to satisfaction with the built environment, the availability of public facilities was also found to have a weak to moderate correlation with awareness, attitude, and behavior, and no significant association with the linear regression tests. However, this is contrary to what was previously reported in the literature, in which the availability of public facilities was highlighted as having an impact on the perceptions and behavior of individuals [46,74]. This could once again be related to the local context and conditions of the areas surveyed.

Locus of Control

The locus of control (defined as being determined by one's own behavior—internal control—as opposed to outside forces such as other people or fate—external control) was found to have a significant correlation only with urban sustainability perception, and importantly, not behavior, while the linear regression analysis showed it to have a significant association with perception and attitude. Within the literature, it is suggested that people who perceive themselves as having little or no control over the built environment (in which they live) tend to have a more negative sustainability perception [17,43,44,61,71]. In other words, it was expected that people living in Turkey would tend to behave more sustainability if they had an internal locus of control; however, this was not borne out strongly by the results.

Trust in Science and Technology

Trust in science and technology was found to have a strong correlation with all determinants and behavior. Moreover, it had significant positive association with perception and attitude [63]. Therefore, it is possible to deduce that when people have confidence in science and technology to solve environmental problems, they have more positive perceptions and attitudes. However, this confidence may result in reluctance to take actions and behave accordingly. Therefore, while technology should be utilized to improve the sustainability understanding of individuals, it is important to stress both its limits and the significance of individual practices towards the sustainability agenda. People should know also that technology only helps to improve sustainability as long as they collaborate; hence, its direct links with sustainable behavior cannot be ignored.

Human Priority

Having a human priority mindset holds negative correlations, especially with attitude and behavior. On the other hand, no significant associations were found with any of the determinants and behavior. Therefore, it can be assumed to be beneficial to convince people about the importance and priority of nature and its balance with humanity [60,75].

Prosperity Perception

Favoring personal prosperity over urban sustainability was found to be negatively correlated with attitude and behavior. On the other hand, it did not result in any predictive power on any of the determinants and behavior. The strongest correlation was found with attitude, which means that a person who prioritizes his/her own interests has strongly negative attitudes towards urban sustainability [63]. Keeping in mind the self-centered nature of some people within society, it is therefore important to stress both the short- and long-term benefits (and prosperity) of making sustainable choices rather than the short-term impacts alone.

Perception of Economic Growth

Similar to prosperity, the perception of economic growth (in relation to urban sustainability) was found to have strong negative correlations with attitude, alongside weaker negative correlations with awareness and behavior. Moreover, standardized regression showed it to have a negative association with attitude. Therefore, it can be deduced that people who favor economic development generally hold negative attitudes towards sustainability [63]. In order to overcome this, the economic benefits of urban sustainability need to be explicitly addressed in the relevant policies and interventions.

Willingness to Sacrifice

Willingness to sacrifice was found to have significant moderate correlations with all three determinants and behavior. Moreover, standardized regression coefficients showed that it had strong predictive power on awareness and perception. People who are not ready to forego certain actions associated with their patterns of living are therefore expected to have lower levels of awareness and perception, and accordingly, are not expected to perform well in terms of their behavior. Along with the other factors, it is important to encourage people to make sacrifices willingly, rather than reluctantly (by force) [61]. In order to persuade them, other positive perspectives and future benefits of creating sustainable urban areas should once again be emphasized.

Ascription of Responsibility

Although the ascription of responsibility was found to have a significant (weak) correlation only with perception and a weak correlation with behavior, it was found to have strong predictive power on awareness, perception, and attitude [16,47,82]. However, unlike other determinants, the impact of the ascription of responsibility was negative on awareness. This result means that although people who take responsibility may hold relatively low awareness, they are likely to have better sustainability perceptions and attitudes.

6.3.3. Summary

It can be concluded that sustainability understanding and behaviors have a very complex and context-sensitive relationships with a wide range of influencing factors. As stated in Topal et al. [35], this context-related nature of urban sustainability behavior requires distinctive efforts with a flexible method to understand, which is implemented in this study. According to the findings, while socio-demographic variables were found to be mostly influential on the awareness and perception of individuals, other influencing factors had a considerable impact on all determinants (i.e., awareness, perception, and attitude) and behavior. Amongst the many influences, the most influential factors on determinants of urban sustainability were identified to be education, age and gender, awareness of consequences, social appraisement, world mindedness, willingness to pay, locus of control, trust in science and technology, willingness to sacrifice, and ascription of responsibility. On the other hand, for behavior, it was found that factors of age, habit, trust in society, social appraisement, world mindedness, and willingness to pay and trust in science and technology had the most significant associations.

6.4. Recommendations

The findings of the study could have significant policy implications. By providing a systematic and organized approach, the framework has proved capable of evaluating the sustainability understanding and behavior in an urban context.

The fundamental outcome of the research is that each of the determinants of urban sustainability understanding must be treated separately, and customized approaches should be developed in order to improve sustainability behavior. Moreover, political decisions and corresponding regulative instruments must consider the impact of personality types within society and individual personality traits on their likely efficacy. Importantly, the wide range of different and distinct factors have intertwined effects (i.e., they are interdependent in different and complex ways), meaning that all should be specifically identified and assessed diligently, both individually and in combination, via a holistic approach based on the local context and conditions.

Regarding policy implications, those responsible for governance must treat each and every element of urban sustainability understanding and behavior as equally important. People's awareness about sustainability should be targeted as a precursor for improving their perceptions and attitudes in relation to general and specific sustainability issues. It is of crucial importance to target the awareness of individuals by improving their knowledge via: (i) educational content in schools and universities, and (ii) advertisements and training opportunities with the help of institutional tools. Based on the findings of the impact of personality types on sustainability understanding and behavior, it is possible to deduce that the content of these efforts should not be intellectually demanding if they are to resonate with the public. However, subjects should be prepared conscientiously, and provided with necessary justifications, in order to result in positive behavioral outcomes.

As noted within the literature, many variables are important in the prediction of urban sustainability awareness, perception, attitude, and behavior. However, particular situations need specific approaches. In the Turkish context studied herein, while females and older generations require more informative interventions, younger generations need encouragement to improve their behavior. Therefore, inclusive activities that target the youth have better chances of success if they result in practical outcomes such as: (i) public contests (or challenges), either in physical or digital world, and (ii) sustainability projects in high school and university curricula. Since people who are exposed to unsustainable outcomes and negative consequences in their living environment have better understanding and behaviors, municipalities should clearly indicate the connections between these outcomes and unsustainable practices in order to direct people towards desired actions.

In terms of the social environment at the micro- and macro-level, cultural codes, trust, social appraisement, and identity were found to be distinctive variables. In order to gain the trust of individuals, governmental actors have to provide transparent and inclusive policy-making procedures. With the help of district municipalities and digital tools, such as social media, participation of the public in decision-making processes about sustainability issues needs to be enabled. Similarly, up-to-date conditions and achievements can be shared to gain the trust about the ongoing practices. By doing so, individuals can be encouraged to participate in sustainable behaviors in their private spheres. Moreover, in order to improve trust in society, neighborhood-based sustainability activities organized by local authorities should be organized. With the help of these micro efforts, it would be possible to change the socio-cultural norms in favor of sustainability, which in turn could boost social appraisement mechanisms in relation to sustainability issues. It is of the utmost importance to provide an effort–reward system to support such micro social improvements.

7. Conclusions

7.1. Key Findings

The main aim of this research was to explore the urban sustainability understanding and behavior of the public. This study is unique since it has adapted a novel conceptual framework and applied it to the context and local conditions of Istanbul in Turkey. The influencing factors considered constitute a comprehensive list with an urban focus. Contrary to earlier research (which adopts either environmental or psychological approaches), this research has adopted a synthesized approach that puts urban sustainability at the very core. Overall, the conceptual framework proposed was found to be promising in terms of its ability to identify urban sustainability understanding and behavior, while also being able to cast light on the impact of a range of personality traits and influencing factors. The main findings of the research can be summarized as follows:

- Determinants: the statistical analysis on the 535 responses to sets of detailed questions on urban sustainability showed that the average awareness score of respondents was lower than perception and attitude scores. While respondents were found to be more aware of local sustainability issues than anticipated, they readily identified that educational opportunities needed improvement. Although sustainability was perceived as more of an environmental issue, respondents were able to distinguish between the importance of sustainability and the appropriateness of their behavior to advancing the cause of sustainability. Respondents were also found to hold positive attitudes in relation to taking responsibility for and contributing to urban sustainability. Bivariate analysis showed that although the correlations between awareness–perception and perception–attitude were broadly similar, the correlation between awareness–attitude was found to be stronger. Regarding the predictive power revealed by the multivariate linear regression analysis, awareness was found to impact upon both perception and attitude, yet there was no significant association between perception and attitude.

- Behavior: the overall behavior scores of the respondents produced a promising mean value (3.63). Furthermore, the bivariate analysis showed strong correlations with both awareness and attitude. The multivariate linear regression analysis demonstrated that all three determinants (awareness, perception, and attitude) had significant predictive power on behavior.
- Personality traits: the respondents as a whole were found to have both agreeable and conscientious natures. Surgency and intellect personality types were found to be highly correlated with awareness, while agreeableness and conscientiousness had strong correlations with awareness and perception. Similarly, behavior was found to have the strongest correlation with conscientiousness, closely followed by agreeableness and intellect. Additionally, people who had more intellectual personality traits resulted in them having better (sustainability) awareness and behavior. Likewise, an agreeable and conscientious personality was found to have positive impacts on sustainability behavior.
- Influencing factors: it was noticeable that while the factors of awareness of consequences, social appraisement, national identity, world mindedness, willingness to pay and trust in science and technology had the strongest correlations with the three determinants and behavior, having a human priority mindset and favoring personal prosperity resulted in strong negative correlations with attitude. Additionally, place attachment was strongly correlated with behavior. In terms of predictive associations, awareness of consequences, social appraisement, world mindedness, willingness to pay, locus of control, trust in science and technology, willingness to sacrifice, and ascription of responsibility were observed to be the most influential factors on the determinants of awareness, perception, and attitude. Furthermore, attitude was the most easily predicted determinant by influencing factors. In terms of behavior, it was found that habit, trust in society, social appraisement, world mindedness, willingness to pay, and trust in science and technology had the most predictive power.
- Socio-demographic factors: while males were more aware of urban sustainability, age was negatively correlated with awareness. Interestingly, education and income were positively correlated with all of the determinants, yet behavior had no correlation with income and a positive correlation with education. In terms of occupation, housewives and unemployed people had weaker awareness and attitudes than those in other occupations. Moreover, people who live in socio-economically more deprived areas produced higher scores for the determinants and behavior. Similarly, both apolitical and religious people were found to have weaker urban sustainability understanding and behavior than Ataturkist, left/socialist, and social democrats. In terms of predictive power, older ages and being female were found to have negative impact on awareness. However, while age had a positive impact on behavior, being female had a positive impact on perception. Moreover, landlords were found to be more aware while inhabitants of bigger houses had better attitudes.

7.2. Recommendations for Future Research

As a result of this study, some potential avenues for future research on urban sustainability understanding and behavior have been identified. These are:

- The understanding and behavior of urban sustainability have been explored according to the main determinants specified in the conceptual framework. However, there are several sub-determinants that could be considered for deeper investigation, such as knowledge, concern, value–belief, and personal norms [35]. Moreover, further investigation into the different urban sustainability behavior types (such as economic, environmental, and social urban sustainability behaviors) would provide valuable insights and a customized approach.

- Although quantitative approaches provide ease, accuracy, and generalizability of analysis, it can be valuable to explore urban sustainability understanding and behavior of individuals in qualitative ways. For instance, in-depth interviews, focus group discussions, or even observations, could provide alternative insights.
- The scope of the current study was the macro-scale, which concentrates on the whole city. However, the conceptual framework could be applied to local contexts and at micro-scales, such as the district, county, or even neighborhood level. This would have the potential to enable local authorities to better identify local needs and problems.
- There would be a benefit in conducting further intra- and international comparisons among urban areas in order to identify different urban characteristics and particular differences among various geographical regions and cultures. Therein, it would be interesting to see how developed and developing countries differ, and to interrogate those distinguishing characteristics that are most influential on urban sustainability behavior within each setting.
- The proposed conceptual framework provides a flexible and adaptable option for different case study areas due to the wide range of influencing factors that interact with determinants and behavior. Although an extensive list of factors has been tested within this research, it is possible to customize the influencing factors for different regions.

Author Contributions: Abstract and introduction, H.F.T., D.V.L.H. and C.D.F.R. Methodology, H.F.T. Main body of research with analyses and results, H.F.T. Discussion, H.F.T., D.V.L.H. and C.D.F.R. Conclusions, H.F.T., D.V.L.H. and C.D.F.R. Writing—original draft preparation, H.F.T. Supervision, revision, editing, and proofreading, D.V.L.H. and C.D.F.R. All authors have contributed to the work reported. All authors have read and agreed to the published version of the manuscript.

Funding: The first author would also like to thank his sponsor, The Ministry of National Education, Republic of Turkey, for funding his doctoral studies at the University of Birmingham under scholarship of Higher Education. C.D.F.R. and D.V.L.H. received the financial support of the UK EPSRC under grant EP/J017698/1 (Transforming the Engineering of Cities to Deliver Societal and Planetary Wellbeing, known as Liveable Cities) and EP/P002021 (From Citizen to Co-innovator, from City Council to Facilitator: Integrating Urban Systems to Provide Better Outcomes for People, known as Urban Living Birmingham).

Institutional Review Board Statement: The study was conducted with ethical approval from the University of Birmingham (reference number ERN_19-0513A).

Informed Consent Statement: Informed consent was obtained from all subjects involved in the study.

Data Availability Statement: Not applicable.

Conflicts of Interest: The authors declare no conflict of interest.

Appendix A.

Appendix A.1. General Information

The questionnaire starts with a wide range of socio-demographic questions. This helps to better understand the general characteristics, background information and representativeness of the respondents within the study area. Moreover, it was necessary to capture this information in order to subsequently compare and contrast the effects of the demographic characteristics on urban sustainability understanding and behavior. Hence, the questions first looked to identify the most frequently asked characteristics within any questionnaire: age, gender, education level, occupation, and monthly income. Furthermore, since the study was conducted at a city scale, respondents were asked which county they lived in—this facilitated making comparisons between different regions. Household characteristics were also investigated through use of specific questions, such as: (i) number of people living in the household, (ii) residential status, and (iii) approximate size of home. These were followed by questions that identified the length of time resident in Istanbul and probable (or likely) future residency. Finally, political orientations of individuals were solicited.

Appendix A.2. Sustainability Understanding

Socio-psychological determinants of urban sustainability understanding were investigated using three sub-sections containing individual questions. All questions herein employed the standardized Likert type scale (see [83]). This section started with questions that assess the general 'awareness' of respondents of/to sustainability. As such, respondents were asked to rate their knowledge about urban sustainability, sustainable behavior, and familiarity with different sustainability concept(s). In the second sub-section, the 'perceptions' of individuals about urban sustainability and sustainable behavior were evaluated. This sub-section consisted of four questions that ask about the respondent's approach to urban sustainability and sustainable behavior, how they rate the importance of urban sustainability, and the appropriateness of their behavior in terms of sustainability. In the last sub-section, the 'attitudes' of the respondents were explored. Respondents were asked to rate five statements relating to: (i) environmental limits; (ii) (contributing to) the sustainable city; (iii) environmental responsibility, (iv) economic problems (related to unsustainable urbanization), and (v) sustainability laws and regulations.

Appendix A.3. Sustainability Behavior

In this section respondents were asked to rate (using a Likert scale) 18 statements which belong to the different core elements of sustainability of personal, social, environmental, economic, governance, infrastructural, and technological aspects of sustainable behavior. This section started by assessing the individual efforts respondents took in respect to adopting sustainable behavior—in their personal area and also their social environment. Following this, their social behaviors were evaluated by asking about their practices pertaining to social issues within their living areas. Their environmental behaviors were subsequently evaluated by identifying relevant topics, such as 'repair, reuse, recycle', and green areas. For economic behavior, both resource efficiency and minimization behavior were assessed. Likewise, governance behaviors were determined by asking whether respondents participate in or are paying due obedience to related policies and regulations. Finally, the actions of individuals were evaluated in terms of both infrastructural and technological sustainability areas.

Appendix A.4. Influencing Factors

In this section, respondents were asked to rate (using the same Likert scale) 19 influencing factors of urban sustainability understanding and behavior. These questions were again specified based upon the literature review performed by Topal et al. [31]. The influencing factors were: awareness of consequences; habits; concerns for social appraisement, and for health and safety; trust in actors, society, and science and technology; national and global identities; place attachment; willingness to pay and sacrifice; prosperity, economy, and human priority perceptions; availability of public facilities; satisfaction with the built environment; locus of control; and ascription of responsibility.

Appendix A.5. Personality Traits

Five personality types, also specified according to findings from the literature search, were investigated here. The five major personality types adopted were [84]: surgency, agreeableness, conscientiousness, emotional stability, and intellect. Each group consisted of five statements to be rated in using a five-point Likert scale.

References

1. Brundtland, G.H. *Report of the World Commission on Environment and Development: Our Common Future*; United Nations: New York, NY, USA, 1987.
2. Mensah, J. Sustainable development: Meaning, history, principles, pillars, and implications for human action: Literature review. *Cogent Soc. Sci.* **2019**, *5*. [CrossRef]
3. Morelli, J. Environmental Sustainability: A Definition for Environmental Professionals. *JES* **2011**, *1*. [CrossRef]
4. United Nations. *World Population Prospects: The 2017 Revision*; United Nations Department of Economic and Social Affairs: New York, NY, USA, 2017; Volume 33, pp. 1–66.
5. United Nations Department of Economic and Social Affairs. *Population Division World Urbanization Prospects: The 2014 Revision*; United Nations Department of Economic and Social Affairs: New York, NY, USA, 2015.
6. United Nations Department of Economic and Social Affairs. *Population Division World Urbanization Prospects: The 2018 Revision*; United Nations Department of Economic and Social Affairs: New York, NY, USA, 2019; ISBN 978-92-1-148319-2.
7. Turkiye Istatistik Kurumu (TUIK) Adrese Dayali Nufus Kayit Sistemi. Available online: https://data.tuik.gov.tr/Bulten/Index?p=Adrese-Dayal%C4%B1-N%C3%BCfus-Kay%C4%B1t-Sistemi-Sonu%C3%A7lar%C4%B1-2020-37210&dil=1 (accessed on 13 April 2021).
8. World Bank. *Turkey—Sustainable Cities Project: Environmental and Social Management Framework (English)*; World Bank Group: Washington, DC, USA, 2016.
9. Sari, V.İ.; Kindap, A. *Türkiye'de Kentsel Yaşam Kalitesi Göstergelerinin Analizi*; Ankara University: Ankara, Turkey, 2018.
10. Candas, E.; Flacke, J.; Yomralioglu, T. Understanding Urban Regeneration in Turkey. *ISPRS Int. Arch. Photogramm. Remote Sens. Spat. Inf. Sci.* **2016**, *XLI-B4*, 669–675. [CrossRef]
11. Strateji ve Bütçe Başkanlığı, Türkiye Cumhuriyeti Cumhurbaşkanlığı, National Development Plans. Available online: https://www.sbb.gov.tr/kalkinma-planlari/ (accessed on 2 June 2021).
12. Istanbul Development Agency. *2014–2023 Istanbul Regional Plan*; Istanbul Development Agency: Istanbul, Turkey, 2014; p. 90.
13. Blake, J. Overcoming the 'value-action gap' in environmental policy: Tensions between national policy and local experience. *Local Environ.* **1999**, *4*, 257–278. [CrossRef]
14. Huijts, N.M.A.; Molin, E.J.E.; Steg, L. Psychological factors influencing sustainable energy technology acceptance: A review-based comprehensive framework. *Renew. Sustain. Energy Rev.* **2012**, *16*, 525–531. [CrossRef]
15. Steg, L.; Vlek, C. Encouraging pro-environmental behaviour: An integrative review and research agenda. *J. Environ. Psychol.* **2009**, *29*, 309–317. [CrossRef]
16. Stern, P.C. New environmental theories: Toward a coherent theory of environmentally significant behavior. *J. Soc. Issues* **2000**, *56*, 407–424. [CrossRef]
17. Kollmuss, A.; Agyeman, J. Mind the Gap: Why do people act environmentally and what are the barriers to pro-environmental behavior? *Environ. Educ. Res.* **2002**, *8*, 239–260. [CrossRef]
18. Turaga, R.M.R.; Howarth, R.B.; Borsuk, M.E. Pro-environmental behavior: Rational choice meets moral motivation. *Ann. N. Y. Acad. Sci.* **2010**, *1185*, 211–224. [CrossRef]
19. Abrahamse, W.; Steg, L.; Vlek, C.; Rothengatter, T. A review of intervention studies aimed at household energy conservation. *J. Environ. Psychol.* **2005**, *25*, 273–291. [CrossRef]
20. Barr, S. Factors Influencing Environmental Attitudes and Behaviors: A U.K. Case Study of Household Waste Management. *Environ. Behav.* **2007**, *39*, 435–473. [CrossRef]
21. Gifford, R. The Dragons of Inaction: Psychological Barriers That Limit Climate Change Mitigation and Adaptation. *Am. Psychol.* **2011**, *66*, 290–302. [CrossRef]
22. Bamberg, S. How does environmental concern influence specific environmentally related behaviors? A new answer to an old question. *J. Environ. Psychol.* **2003**, *23*, 21–32. [CrossRef]
23. Stern, P.C.; Dietz, T. The value basis of environmental concern. *J. Soc. Issues* **1994**, *50*, 65–84. [CrossRef]
24. Batur, İ.; Koç, M. Travel Demand Management (TDM) case study for social behavioral change towards sustainable urban transportation in Istanbul. *Cities* **2017**, *69*, 20–35. [CrossRef]
25. Balaban, O. *A Matter of Capacity: Climate Change and the Urban Challenges for Turkey*; Cambridge University Press: Cambridge, UK, 2017; Volume 56.
26. Kaygusuz, K.; Toklu, E. Energy issues and sustainable development in Turkey. *J. Eng. Res. Appl. Sci.* **2012**, *1*, 25.
27. Yalçıntaş, M.; Bulu, M.; Küçükvar, M.; Samadi, H. A Framework for Sustainable Urban Water Management through Demand and Supply Forecasting: The Case of Istanbul. *Sustainability* **2015**, *7*, 11050–11067. [CrossRef]
28. Egercioğlu, Y. Urban Transformation Processes in Illegal Housing Areas in Turkey. *Procedia Soc. Behav. Sci.* **2016**, *223*, 119–125. [CrossRef]
29. Okumuş, G. A Geographical Information System Based Urban Sustainability Evaluation Model Proposal in Neighbourhood Scale. *J. Plan.* **2017**, *27*, 193–204. [CrossRef]
30. Yurttaş, A.; Çağlar, A. The attitudes of governmental officialin terms of sustainable environment. *Int. Electron. J. Environ. Educ.* **2019**, *9*, 142–156.
31. Topal, H.F.; Hunt, D.V.L.; Rogers, C.D.F. Urban Sustainability and Smartness Understanding (USSU)—Identifying Influencing Factors: A Systematic Review. *Sustainability* **2020**, *12*, 4682. [CrossRef]

32. Sharifi, A.; Dawodu, A.; Cheshmehzangi, A. Limitations in assessment methodologies of neighborhood sustainability assessment tools: A literature review. *Sustain. Cities Soc.* **2021**, *67*, 102739. [CrossRef]

33. Bayulken, B.; Huisingh, D.; Fisher, P.M.J. How are nature based solutions helping in the greening of cities in the context of crises such as climate change and pandemics? A comprehensive review. *J. Clean. Prod.* **2021**, *288*, 125569. [CrossRef]

34. Morano, P.; Tajani, F.; Anelli, D. Urban planning decisions: An evaluation support model for natural soil surface saving policies and the enhancement of properties in disuse. *Prop. Manag.* **2020**, *38*, 699–723. [CrossRef]

35. Topal, H.F.; Hunt, D.V.L.; Rogers, C.D.F. Exploring Urban Sustainability Understanding and Behaviour: A Systematic Review towards a Conceptual Framework. *Sustainability* **2021**, *13*, 1139. [CrossRef]

36. Lynch, A.J.; Mosbah, S.M. Improving local measures of sustainability: A study of built-environment indicators in the United States. *Cities* **2017**, *60*, 301–313. [CrossRef]

37. Wakita, T.; Ueshima, N.; Noguchi, H. Psychological Distance Between Categories in the Likert Scale: Comparing Different Numbers of Options. *Educ. Psychol. Meas.* **2012**, *72*, 533–546. [CrossRef]

38. Johnson, R.B.; Christensen, L. *Educational Research: Quantitative, Qualitative, and Mixed Approaches*; SAGE Publications Inc.: New York, NY, USA, 2019; ISBN 1-5443-3785-X.

39. Field, A.P.; Miles, J.; Field, Z. *Discovering Statistics Using R*; Sage: London, UK; Thousand Oaks, CA, USA, 2012; ISBN 978-1-4462-0046-9.

40. Cordano, M.; Welcomer, S.A.; Scherer, R.F. An analysis of the predictive validity of the new ecological paradigm scale. *J. Environ. Educ.* **2003**, *34*, 22–28. [CrossRef]

41. Hair, J.F. (Ed.) *Multivariate Data Analysis*, 7th ed.; Pearson: Harlow, UK, 2014; ISBN 978-1-292-02190-4.

42. Goldberg, L.R. The development of markers for the Big-Five factor structure. *Psychol. Assess.* **1992**, *4*, 26. [CrossRef]

43. Guo, D.; Cao, Z.; DeFrancia, K.; Yeo, J.W.G.; Hardadi, G.; Chai, S. Awareness, perceptions and determinants of urban sustainable development concerns—Evidence from a central province in China. *Sustain. Dev.* **2018**, *26*, 652–662. [CrossRef]

44. Tran, K.C. Public perception of development issues: Public awareness can contribute to sustainable development of a small island. *Ocean Coast. Manag.* **2006**, *49*, 367–383. [CrossRef]

45. Buerke, A.; Straatmann, T.; Lin-Hi, N.; Müller, K. Consumer awareness and sustainability-focused value orientation as motivating factors of responsible consumer behavior. *Rev. Manag. Sci.* **2017**, *11*, 959–991. [CrossRef]

46. Cagáňová, D.; Stareček, A.; Horňáková, N.; Hlásniková, P. The Analysis of the Slovak Citizens' Awareness about the Smart City Concept. *Mob. Netw. Appl.* **2019**, 2050–2058. [CrossRef]

47. Guagnano, G.A.; Stern, P.C.; Dietz, T. Influences on attitude-behavior relationships: A natural experiment with curbside recycling. *Environ. Behav.* **1995**, *27*, 699–718. [CrossRef]

48. Peng, G.C.A.; Nunes, M.B.; Zheng, L. Impacts of low citizen awareness and usage in smart city services: The case of London's smart parking system. *Inf. Syst. e-Bus. Manag.* **2017**, *15*, 845–876. [CrossRef]

49. Himmel, S.; Zaunbrecher, B.S.; Wilkowska, W.; Ziefle, M. The Youth of Today Designing the Smart City of Tomorrow. In *Human-Computer Interaction. Applications and Services*; Kurosu, M., Ed.; Springer International Publishing: Cham, Switzherland, 2014; Volume 8512, pp. 389–400, ISBN 978-3-319-07226-5.

50. Ajzen, I. *Attitudes, Personality, and Behavior*; McGraw-Hill Education: London, UK, 2005; ISBN 0-335-22400-8.

51. Kang, S. Communicating sustainable development in the digital age: The relationship between citizens' storytelling and engagement intention. *Sustain. Dev.* **2019**, *27*, 337–348. [CrossRef]

52. Pavalache-Ilie, M.; Cazan, A.-M. Personality correlates of pro-environmental attitudes. *Int. J. Environ. Health Res.* **2018**, *28*, 71–78. [CrossRef]

53. Busic-Sontic, A.; Czap, N.V.; Fuerst, F. The role of personality traits in green decision-making. *J. Econ. Psychol.* **2017**, *62*, 313–328. [CrossRef]

54. Grob, A. A structural model of environmental attitudes and behaviour. *J. Environ. Psychol.* **1995**, *15*, 209–220. [CrossRef]

55. Klöckner, C.A. A comprehensive model of the psychology of environmental behaviour—A meta-analysis. *Glob. Environ. Chang.* **2013**, *23*, 1028–1038. [CrossRef]

56. He, G.; Boas, I.; Mol, A.P.J.; Lu, Y. E-participation for environmental sustainability in transitional urban China. *Sustain. Sci.* **2017**, *12*, 187–202. [CrossRef]

57. Tononi, M.; Pietta, A.; Bonati, S. Alternative spaces of urban sustainability: Results of a first integrative approach in the Italian city of Brescia. *Geogr. J.* **2017**, *183*, 187–200. [CrossRef]

58. Hsu, J.L.; Feng, C.-H. Evaluating environmental behaviour of the general public in Taiwan: Implications for environmental education. *IJCED* **2019**, *21*, 179–189. [CrossRef]

59. Rajapaksa, D.; Gifford, R.; Torgler, B.; Garcia-Valiñas, M.; Athukorala, W.; Managi, S.; Wilson, C. Do monetary and non-monetary incentives influence environmental attitudes and behavior? Evidence from an experimental analysis. *Resour. Conserv. Recycl.* **2019**, *149*, 168–176. [CrossRef]

60. Wong, T.K.; Wan, P. Perceptions and determinants of environmental concern: The case of Hong Kong and its implications for sustainable development. *Sustian. Dev.* **2011**, *19*, 235–249. [CrossRef]

61. Waitt, G.; Caputi, P.; Gibson, C.; Farbotko, C.; Head, L.; Gill, N.; Stanes, E. Sustainable Household Capability: Which households are doing the work of environmental sustainability? *Aust. Geogr.* **2012**, *43*, 51–74. [CrossRef]

62.   Branchini, S.; Meschini, M.; Covi, C.; Piccinetti, C.; Zaccanti, F.; Goffredo, S. Participating in a Citizen Science Monitoring Program: Implications for Environmental Education. *PLoS ONE* **2015**, *10*, e0131812. [CrossRef]

63.   Drews, S.; van den Bergh, J.C.J.M. Public views on economic growth, the environment and prosperity: Results of a questionnaire survey. *Glob. Environ. Chang.* **2016**, *39*. [CrossRef]

64.   Newton, P.; Meyer, D. Exploring the Attitudes-Action Gap in Household Resource Consumption: Does "Environmental Lifestyle" Segmentation Align with Consumer Behaviour? *Sustainability* **2013**, *5*, 1211–1233. [CrossRef]

65.   Rajapaksa, D.; Islam, M.; Managi, S. Pro-Environmental Behavior: The Role of Public Perception in Infrastructure and the Social Factors for Sustainable Development. *Sustainability* **2018**, *10*, 937. [CrossRef]

66.   Barau, A.S. Perceptions and contributions of households towards sustainable urban green infrastructure in Malaysia. *Habitat Int.* **2015**, *47*, 285–297. [CrossRef]

67.   Harlan, S.L.; Yabiku, S.T.; Larsen, L.; Brazel, A.J. Household Water Consumption in an Arid City: Affluence, Affordance, and Attitudes. *Soc. Nat. Resour.* **2009**, *22*, 691–709. [CrossRef]

68.   Maiello, A.; Battaglia, M.; Daddi, T.; Frey, M. Urban sustainability and knowledge: Theoretical heterogeneity and the need of a transdisciplinary framework. A tale of four towns. *Futures* **2011**, *43*, 1164–1174. [CrossRef]

69.   Rogers, Z.; Bragg, E. The Power of Connection: Sustainable Lifestyles and Sense of Place. *Ecopsychology* **2012**, *4*, 307–318. [CrossRef]

70.   Schwartz, S.H.; Howard, J.A. A normative decision-making model of altruism. In *Altruism Help. Behav*; Rushton, J.P., Sorrentino, R.M., Eds.; 1981; pp. 189–211, ISBN-10: 0898591554.

71.   Bamberg, S.; Möser, G. Twenty years after Hines, Hungerford, and Tomera: A new meta-analysis of psycho-social determinants of pro-environmental behaviour. *J. Environ. Psychol.* **2007**, *27*, 14–25. [CrossRef]

72.   Uren, H.V.; Dzidic, P.L.; Roberts, L.D.; Leviston, Z.; Bishop, B.J. Green-Tinted Glasses: How Do Pro-Environmental Citizens Conceptualize Environmental Sustainability? *Environ. Commun.* **2019**, *13*, 395–411. [CrossRef]

73.   Fransson, N.; Gärling, T. Environmental concern: Conceptual definitions, measurement methods, and research findings. *J. Environ. Psychol.* **1999**, *19*, 369–382. [CrossRef]

74.   Macke, J.; Casagrande, R.M.; Sarate, J.A.R.; Silva, K.A. Smart city and quality of life: Citizens' perception in a Brazilian case study. *J. Clean. Prod.* **2018**, *182*, 717–726. [CrossRef]

75.   Zhang, L.; Chen, L.; Wu, Z.; Xue, H.; Dong, W. Key Factors Affecting Informed Consumers' Willingness to Pay for Green Housing: A Case Study of Jinan, China. *Sustainability* **2018**, *10*, 1711. [CrossRef]

76.   Polese, F.; Barile, S.; Caputo, F.; Carrubbo, L.; Waletzky, L. Determinants for Value Cocreation and Collaborative Paths in Complex Service Systems: A Focus on (Smart) Cities. *Serv. Sci.* **2018**, *10*, 397–407. [CrossRef]

77.   Der-Karabetian, A.; Cao, Y.; Alfaro, M. Sustainable Behavior, Perceived Globalization Impact, World-Mindedness, Identity, and Perceived Risk in College Samples from the United States, China, and Taiwan. *Ecopsychology* **2014**, *6*, 218–233. [CrossRef]

78.   Noonan, D.; Zhou, S.; Kirkman, R. Making Smart and Sustainable Infrastructure Projects Viable: Private Choices, Public Support, and Systems Constraints. *Urban Plan.* **2017**, *2*, 18–32. [CrossRef]

79.   Granier, B.; Kudo, H. How are citizens involved in smart cities? Analysing citizen participation in Japanese "Smart Communities". *Inf. Polity* **2016**, *21*, 61–76. [CrossRef]

80.   Holdsworth, S.; Kenny, D.; Cooke, J.; Matfin, S. Are We Living with Our Heads in the Clouds? Perceptions of Liveability in the Melbourne High-Rise Apartment Market. In *Energy Performance in the Australian Built Environment*; Rajagopalan, P., Andamon, M.M., Moore, T., Eds.; Springer: Singapore, 2019; pp. 181–198, ISBN 978-981-10-7879-8.

81.   Norouzian-Maleki, S.; Bell, S.; Hosseini, S.-B.; Faizi, M.; Saleh-Sedghpour, B. A comparison of neighbourhood liveability as perceived by two groups of residents: Tehran, Iran and Tartu, Estonia. *Urban For. Urban Green.* **2018**, *35*, 8–20. [CrossRef]

82.   Karp, D.G. Values and their effect on pro-environmental behavior. *Environ. Behav.* **1996**, *28*, 111–133. [CrossRef]

83.   Norman, G. Likert scales, levels of measurement and the "laws" of statistics. *Adv. Health Sci. Educ.* **2010**, *15*, 625–632. [CrossRef]

84.   Norman, W.T. Toward an adequate taxonomy of personality attributes: Replicated factor structure in peer nomination personality ratings. *J. Abnorm. Soc. Psychol.* **1963**, *66*, 574. [CrossRef] [PubMed]

 *sustainability*

Article

Overview and Exploitation of Haptic Tele-Weight Device in Virtual Shopping Stores

Aqeel Farooq [1,*], Mehdi Seyedmahmoudian [1], Ben Horan [2], Saad Mekhilef [1] and Alex Stojcevski [1]

1   School of Science, Computing and Engineering Technologies, Swinburne University of Technology, Melbourne, VIC 3122, Australia; mseyedmahmoudian@swin.edu.au (M.S.); smekhilef@swin.edu.au (S.M.); astojcevski@swin.edu.au (A.S.)

2   School of Engineering, Deakin University, Geelong, VIC 3216, Australia; ben.horan@deakin.edu.au

*   Correspondence: afarooq@swin.edu.au or aqeelfarooq92@gmail.com

Abstract: In view of the problem of e-commerce scams and the absence of haptic interaction, this research aims to introduce and create a tele-weight device for e-commerce shopping in smart cities. The objective is to use the proposed prototype to provide a brief overview of the possible technological advancements. When the tele-weight device is affixed over the head-mounted display, it allows the user to feel the item's weight while shopping in the virtual store. Addressing the problem of having no physical interaction between the user (player) and a series game scene in virtual reality (VR) headsets, this research approach focuses on creating a prototype device that has two parts, a sending part and a receiving part. The sending part measures the weight of the object and transmits it over the cellular network to the receiver side. The virtual store user at the receiving side can thus realize the weight of the ordered object. The findings from this work include a visual display of the item's weight to the virtual store e-commerce user. By introducing sustainability, this haptic technology-assisted technique can help the customer realize the weight of an object and thus have a better immersive experience. In the device, the load cell measures the weight of the object and amplifies it using the HX711 amplifier. However, some delay in the demonstration of the weight was observed during experimentation, and this indirectly altered the performance of the system. One set of the device is sited at the virtual store user premises while the sending end of the device is positioned at the warehouse. The sending end hardware includes an Arduino Uno device, an HX711 amplifier chip to amplify the weight from the load cell, and a cellular module (Sim900A chip-based) to transmit the weight in the form of an encoded message. The receiving end hardware includes a cellular module and an actuator involving a motor gear arrangement to demonstrate the weight of the object. Combining the fields of e-commerce, embedded systems, VR, and haptic sensing, this research can help create a more secure marketplace to attain a higher level of customer satisfaction.

Keywords: PCB shield; HX711; amplifier chip; Sim900A; e-commerce; virtual store; firmware; embedded system; virtual reality and haptic sensing

Citation: Farooq, A.; Seyedmahmoudian, M.; Horan, B.; Mekhilef, S.; Stojcevski, A. Overview and Exploitation of Haptic Tele-Weight Device in Virtual Shopping Stores. *Sustainability* **2021**, *13*, 7253. https://doi.org/10.3390/su13137253

Academic Editor: Tan Yigitcanlar

Received: 29 May 2021
Accepted: 21 June 2021
Published: 29 June 2021

1. Introduction

Smart cities are a decisive upheaval, and the perception of a smart city is akin to a rummage sale for the urban and developed vision in which diverse electronically progressive approaches, practices, and sensor networks are cast-off to collect data and manage assets within a city including transportation systems, water supply networks, traffic system, information system, power grids, and other community services by the local government [1–3]. The goals of smart cities are to improve the quality of life by using technologically advanced tools, techniques, and methods. Data visualization, modeling, management, and e-commerce sectors are essential fragments of smart cities, and planning professionals are looking for state-of-the-art real-time simulations. In this current "what if?" situation, impact analysis requires a huge number of resources, and both virtual reality (VR) and virtual stores are the potential tools to address the upcoming challenges in smart cities [4,5].

193

Developers have been working for many years to create new VR-based systems. However, due to the inclusion of haptic devices, sensor actuators, and improved realistic graphics in the last decade, the developed VR systems have undergone a technical evolution with improvements in telepresence and immersion [6]. A VR platform generates a computer-simulated environment that is fostered and assisted by the head-mounted display (HMD), which utilizes a manipulator in an immersive environment. VR is broadly used for cognitive purposes in various industries in the fields of engineering, architecture, and medicine [7–9]. It is assimilated with other fields and innovations as a supporting technique. For instance, surgeons who treat patients undergoing VR have gained admiration [10,11].

Software makers are also in the perspective of introducing virtual stores where consumers can enter the virtual stores and visualize the dimensions and usability of an item prior to purchasing it [12]. A virtual shopping store is a computer-simulated store where a handler can envisage, feel, and interact with the products and items in the store. Implementation of the virtual store can be a breakthrough in the e-commerce market. Other famous terms for virtual stores are cybernetic stores and computer-simulated 3D spatial stores [13]. In current e-commerce stores, one of the recorded fraudulent activities is the delivery of an ordered item different from that itemized in the e-commerce store [14–17]. Needless to say, a virtual store can spark a revolution in the e-commerce industry by not only altering the way of shopping but also by making it secure. Haptic devices make physical interaction possible between a human and a computing device by means of sensors and actuators. Humans can interrelate with haptic devices to deliver programming instructions and, correspondingly, schedules can be employed to extract essential information via a haptic interface [18–20].

The tele-weight device is evaluated in this project. Overall, the tele-weight device project is split into two sections. One part of the device measures and records the object weight ordered on the virtual store and sends the encrypted value of the weight as a short message service (SMS) over the cellular network. The extra end receives the encrypted SMS, decrypts it, and illustrates the weight materially using motor gear (actuator) motion. The objective is to improve the immersive shopping experience of the e-commerce user.

VR headsets (e.g., Oculus Rift, HTC Vive, etc.) provide an interactive 3D graphical experience to the wearer, but at present there is no way to interact physically with the running simulation. This study addresses the problem of creating interaction and introduces a tele-weight device that allows the weight to be shown materially. This is done to improve the immersive experience of the user while shopping in a VR store. Load cells are common devices used to measure the weight of an object and have been used in this research [21]. Load cells measure the weight of an object (usually of order 10^{-6} V), and the measured weight is amplified using an HX711 amplifier. Upon amplification, the obtained weight is transmitted to the receiver end where the e-commerce customer can realize the weight of the object to have a more immersive VR experience.

The objective of this research is to present a haptic device named "tele-weight" whose resolution is to enable consumers to feel the weight of the product ordered in the virtual shopping store. Moreover, the customers can feel the weight physically upon selecting an item in the virtual store. By having a weight-based assessment along with the displayed and delivered item, the customers will be able to identify the authenticity of the merchandise. For the prototype highlighted in this research, the weight is measured from one end and then the attained value is sent to the receiver end hardware with the help of the cellular module. This exploration work focuses on fashioning a prototype that can demonstrate weight realization in combination with the immersive ability to visualize the object. Weight realization, together with immersive VR, can enhance the shopping experience of users. To create a simple example, one set of the hardware is assumed to be in a warehouse in China while the other part of the device is observed to be in a technical store in Jeddah.

2. Literature Review

2.1. Smart Cities, Sustainability, and Inclusion of Virtual Reality

Smart cities are described according to the classification system model and established on six pillars: smart social and mobility system, intelligent urban management, smart economic environment, dedicated lawful structure, smart e-governance, and sustainability of lifestyle. Gómez et al. define a smart city to be a city occasioned from urban development exploitation sensors and scientifically unconventional tools and systems [22]. Azraff Bin Rozmi, et al. [23] state that sustainability deals with the intricate relationship between the survival of human life in accordance with changes in the economy, culture, government policies, and infrastructure in an urban milieu [24].

Proposals for smart cities are frequently rehabilitated, bearing in mind the cyclic process in harmony with the decision-makers views to implement pioneering solutions anticipated by connoisseurs and relevant mavens. Alessandro et al. endorsed that the sequence of a system with the connectivity of haptic devices, machine learning, VR, and artificial intelligence needs to be adopted while creating forthcoming verdicts in smart cities [25,26]. Gaffary, et al. [27] propose that the advent of VR and haptic devices can effectively assist urban planners in smart cities in decision making and introducing innovations. The interaction between the components in a smart city framework is shown in Figure 1.

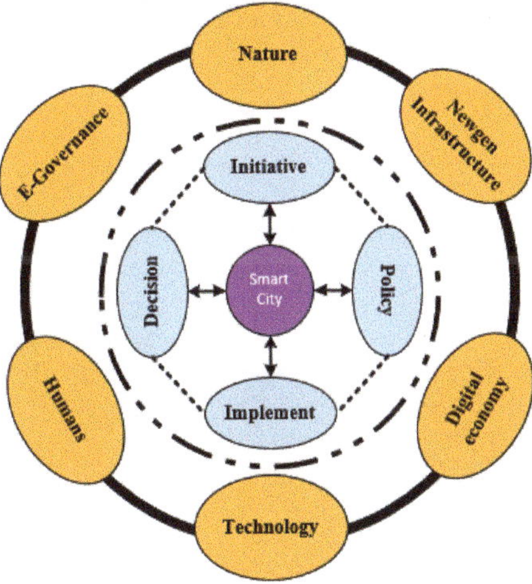

Figure 1. Smart city framework.

According to Akhtar, et al. [28], VR systems will be an integral part of the smart city concept and might assist in replacing mass tourism with digital tourism. Paszkiewicz, et al. [29] term the usage of VR systems as the best option to teach industry 4.0 standards in engineering education. Nasralla [30] uses the term virtual smart cities and states that a sustainable VR patient rehabilitation system should be created using IoT technology and machine learning on time series to identify features that support sustainability and rapidly detect problems without interrupting rehabilitation services. Ospina-Bohórquez, et al. [31] suggest that the synergy between VR and multi-agent systems is a pathway to accomplish sustainability in a smart city framework.

2.2. Proprioceptive Haptic Devices

Rossi, et al. [32] revealed that haptic devices operate based on the instructions provided to them and work on the proprioceptive feedback topology. Proprioception, from the Latin terms *"prospruis"* (one's own) and *"căpio"* (to attain), is the aptitude of the maneuver to obtain the activity information of the body. The authors further emphasize that the inclusion of proprioceptive feedback into the haptic interface can avoid the visual contact problem [33–35]. In their work, the researchers introduced a novel design termed *"HapPro design"* for the haptic device. The device is in the form of an arm that can be used on persons with disabilities to assist them in performing their tedious responsibilities. In designing the *"HapPro"*, both the linear and angular physical quantities are considered. The control strategy of the *"HapPro"* includes haptic feedback on the human EMG signal that helps control the soft hand placed on persons with disabilities. Experimental protocol and results present improvements in control with and without logarithmic mapping. With logarithmic mapping, efficiency is boosted from near 40% to around 80%, thereby doubling the improvement.

Sierra M, et al. [36] showed the application of the haptic interface in prosthetics. Shi, et al. [37] proposed the application of the haptic interface device and used both the purely mechanical system and the coupled hydraulics-based mechanism. The results indicate that the hydraulic and fluid flowing systems can be used in haptic feedback. Pistohl, et al. [38] show the control of a cursor on the desktop screen via EMG sensor signal with and without a proprioceptive feedback approach. The EMG signal is measured from the muscles of the left hand, and proprioceptive feedback is obtained from the right arm. The results showed that additional feedback is not required when the proprioceptive approach is used. However, in the present research, a haptic device is introduced that has the ability to assist customers in virtual stores since the weight of the object being ordered can be realized to have some influence on the perception of the product. This approach is useful to improve the satisfaction level of e-customers.

2.3. Concept of "Heaviness and Lightness": Load Cell Development

The notion of "heaviness and lightness" was first deliberated in the 6th century by the ancient Greek philosopher Plato. According to Georgakopoulos and Quigg [39], Plato termed weight to be the natural propensity of an object, and this physical property of heaviness and lightness is associated with all objects on earth. Later, Euclid of Alexandria, renowned as the "father of geometry," became the foremost Greek mathematician who described weight to be "the heaviness or lightness of substances relative to other substances and is measured by a balance." Brain Bowers specified that in the 17th century, a major advancement was recorded when Galileo proposed a scheme to measure the weight of a stirring object. In due course, Galileo figured out that weight was proportionate to the extent of substance in an entity. In modern science, the weight (vector quantity; the direction toward the earth's center) is thought to be the force resulting from gravitational action, whereby mass (scalar quantity, no direction) is equal to the matter quantity in an object. Sumit and Akhter [40] detailed that the primary maneuver to ration the weightiness of a body was christened as "Spring Balance" and was formed in 1770 by Richard Salter in Britain [41].

According to Ackerman and Seipel [42], in 1843 a British physicist named Sir Charles Wheatstone assembled a device dubbed as the "Wheatstone Bridge." The created device was capable of computing an unknown resistance magnitude. Far along in the 19th century, strain gauges were industrialized. By means of the latest expertise at that time, the perception of fabricating a load cell with a Wheatstone bridge arose into training. Replacing one leg resistor in the Wheatstone bridge with piezoelectric substantial resulted in varying voltages that corresponded to the applied weight. Bar, et al. [43] depicted that "variation in weight resulted in voltages across the Load cell terminals" is the approach that is eliminated nowadays in almost every weight measuring device. As the object is placed on one of the bridge's leg, the corresponding voltage fluctuates and the modification in voltage is amplified and processed in a microprocessor device across some typical scale to acquire the exact weight value of the object. This principle of load cell composition is used in the tele-weight device.

2.4. Virtual Reality Stores for E-Commerce

VR has recently become a trend in shifting technology. Over the last few years, it has been used for cognitive purposes. However, topical enhancements in half-life [44] and gaming engines have shifted the trend in fashioning VR stores to provide a better shopping experience. According to Hung, et al. [45], virtual stores are gaining in popularity and people are choosing to have a virtual shopping environment. Huang, et al. [46] suggested that considering all the impending trends and gaining customer satisfaction is imperative for the e-commerce market to meet errand requirements. Scarle, et al. [47] indicated that fraudulent activities are gaining a new level along with the trend.

One primary problem highlighted in this research is that the current virtual store experience is not immersive, and even with the utilization of haptic sensing e-commerce customers are unable to have a better overview of the object prior to purchasing. The key goal of this research is to introduce the idea of a tele-weight device that can enhance the user's immersive experience by letting the customer realize the weight of an item ordered in the virtual store. The idea is implemented in the form of a prototype, and the relevant details and results are discussed in this work.

With the help of the tele-weight device, an approach is indicated to allow e-commerce users to realize the weight of an object to have a better immersive experience while shopping in VR stores. The tele-weight device can be used in future HMD devices to achieve a better immersive shopping experience.

This research aims to create a tele-weight device that can be mounted over the HMD display and can display the weight of the item from the virtual world to the user, thereby allowing the user to have haptic interaction with the objects from the VR series game simulation. To demonstrate the accuracy of the tele-weight device, a prototype is created and an experiment is performed to collect data. The difference between the sent and collected weight demonstrates the overall accuracy of the device.

3. Methodology and Analysis

3.1. Working Principle of the Tele-Weight Device

The original aim for the tele-weight device is to be affixed over the HMD display and show the weight of the object. From the virtual store perspective, the cart item weight can be demonstrated to the user and, in this way, the customer can realize and interact with the item's equivalent weight in a haptic manner (Figure 2). The tele-weight device shows the weight of the item ordered by the customer in the virtual store. First, a user chooses the item, perceives it on the VR display, and then the sensor system takes command from the virtual store PC (sending end) and displays the equivalent weight (receiving end). This approach can also determine the shipping weight and cost prior to purchasing an item.

To determine the accuracy of this approach, a prototype comprised of two parts was created. At the receiving end side, a motor and gear arrangement was used to show the weight to the user. Therefore, the difference between the obtained weight and shown weight must be obtained to figure out the effectiveness.

3.2. Hardware Design for Performance Evaluation

The tele-weight haptic device prototype created for this research is divided into two portions. For the sake of simplicity, one side is labeled the sending end and the other side the receiving end. The circuitry at the *sending end* side (Figure 3) comprises a programmed set of Arduino family device, an HX711 amplifier chip, and a Global System for Mobile Communications (GSM) module [48]. The sending end circuit is programmed to send encrypted weight coded text messages over a cellular network, and these messages are received and decoded by the receiving (other) end. The circuit at the receiving end (Figure 4) involves a programmed (firmware installed) Arduino family processor, a custom-built motor driver, a Sim900A module, and a dc brushless motor with a gear chain mounted on the shaft. The projected archetype can demonstrate the mass equivalent of up to 5 kg.

The weight of the object under 5 kg is sent in an encrypted format to the receiving side by pressing a button.

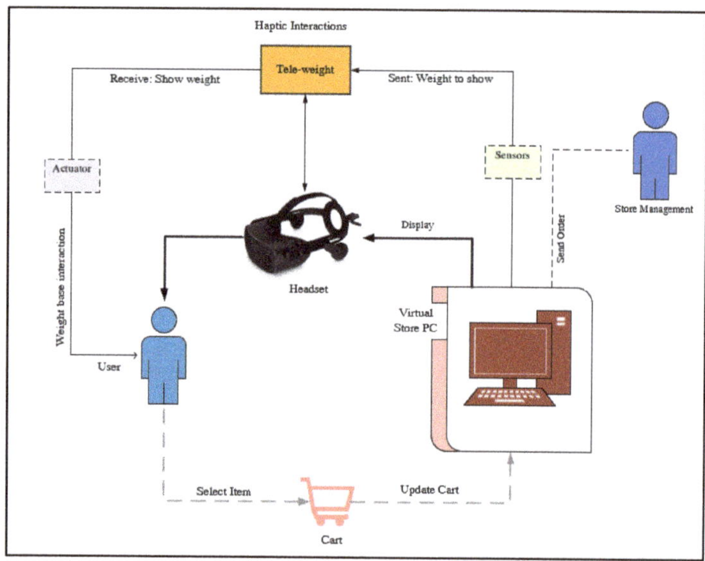

Figure 2. Tele-weight: virtual store concept diagram.

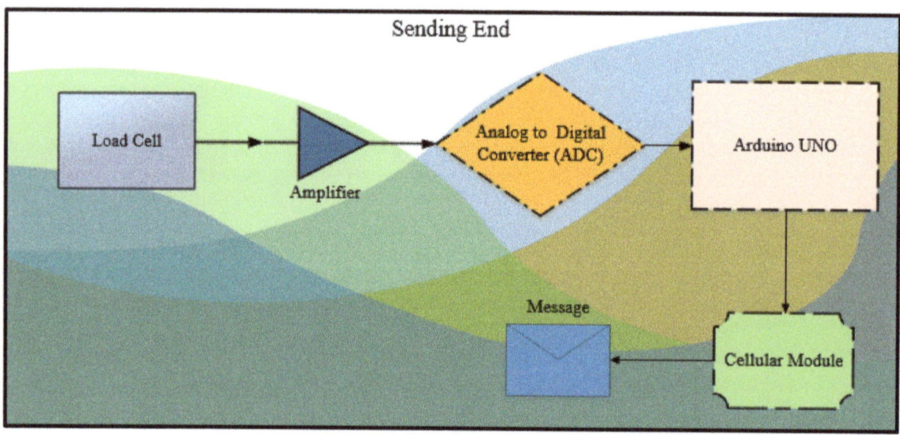

Figure 3. Diagram for proposed sending end (tele-weight hardware).

A limited weight displaying prototype is created because a weight over 5 kg can make the headset heavy, causing neck injury. Haptically feeling the weight of the item together with immersive visualization can allow the user to get a better idea about the in-hand (holding) experience of the item. This function not only enhances the customer's shopping experience by adding sustainability in the approach but also allows them to interact physically with the equivalent load in their hands. For this project, the approach to evaluate the created prototype is shown in Figure 5. At first, a prototype is created and then experimentation is performed on the hardware in the form of repeated trials to obtain estimation regarding the confusion matrix, accuracy, coverage range, and time factor.

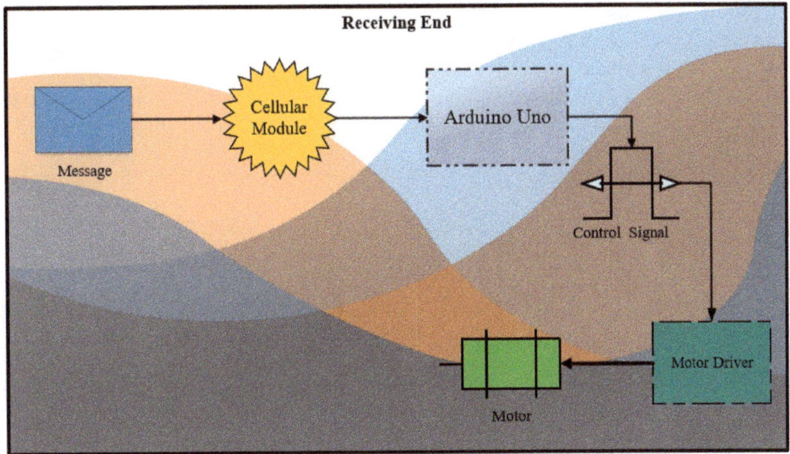

Figure 4. Diagram for proposed receiving end (tele-weight hardware).

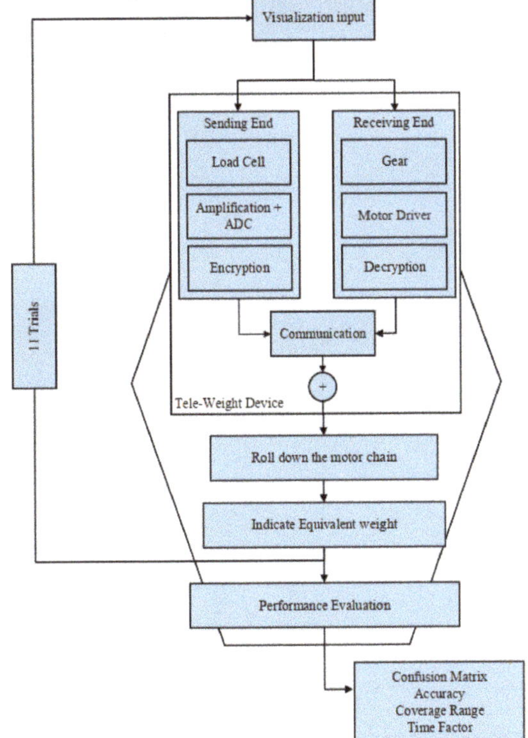

Figure 5. Overview (diagram) of the experimental approach.

4. Results

A tele-weight device is proposed in this research. The objective of this device is to assist the customers in feeling and realizing the weight of an object. The customer can order any item on a VR store by visualizing the item, and during the process of visualization, the gear chain on the motor can make the customer feel the weight of the object at their hand.

Tele-weight is a haptic device that can interact with the customer according to their actions on a computer. The sending end prototype (Figure 6) weighs the object, and the circuitry includes a weight sensor, a microprocessor, and a GSM message sending module. The message sending module sends the encrypted message, which is received by the receiving side (Figure 7) and decrypted. The corresponding weight is then shown accordingly.

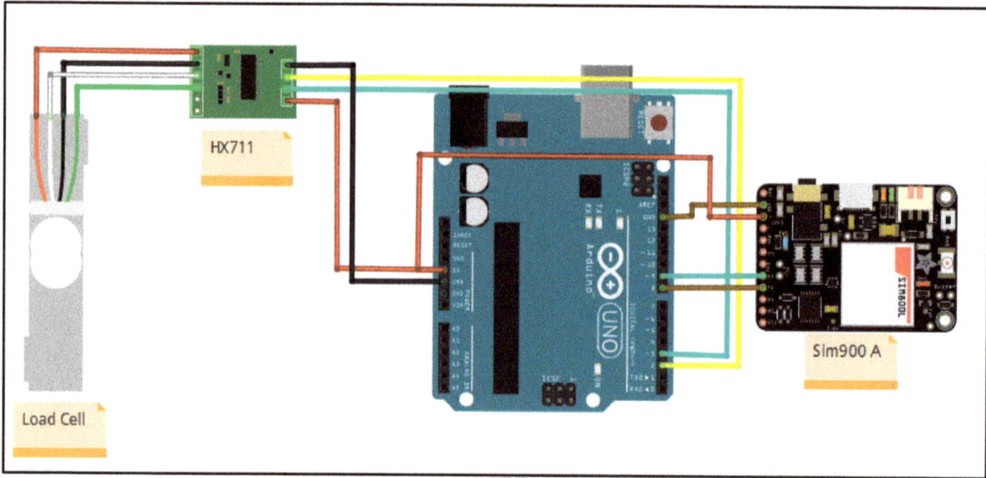

Figure 6. The schematic diagram for the sending end (tele-weight hardware).

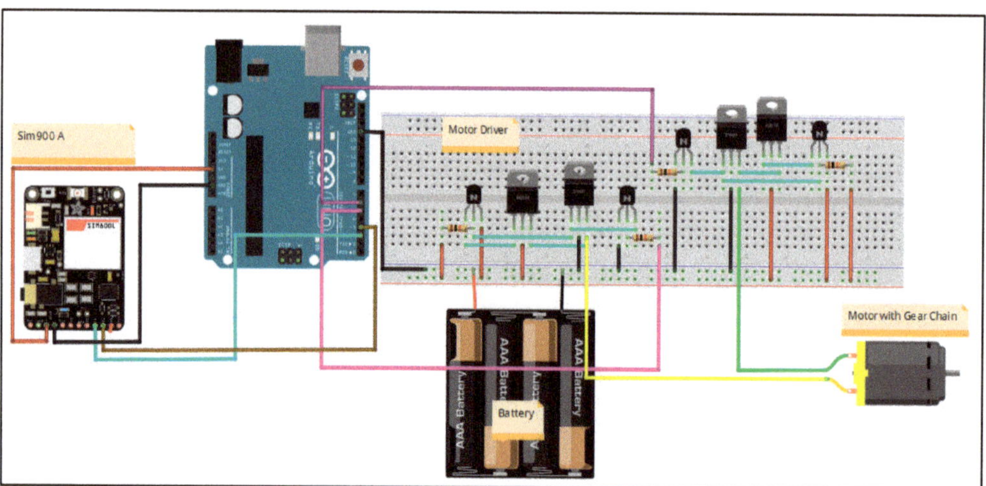

Figure 7. The schematic diagram for the proposed receiving end (tele-weight hardware).

Some degree of error was recorded in reading the measurement and in the demonstration end. A total of eleven trials were recorded, and the current prototype was observed to have some sort of error and thus requires tuning. Owing to the uniform distribution of the weight on the gear, inaccuracy in grams was recorded in the measurement. As the prototype system can show any value between 0 and 5 kg, the results of ten trials are shown in Figure 8.

As the sensors used in the prototype has a limit of 5 kg, therefore, trials were performed from 0 to 5 kg weight with a 0.5 kg decrease in each interval. In the first trial, the weight was set to be 5 kg at the sending end side and receiving end demonstrated a weight of 4.93 kg with 98.6% accuracy. The highest accuracy of 99.71% was achieved in the second trial where 4.5 kg weight was displayed with a positive error of 0.29% and a 4.513 kg weight value at the receiving end. The lowest accuracy of 96% was obtained on the tenth trial where a 0.5 kg weight was displayed on receiving end as 0.48 kg with a 4% error margin. The measured weight, displayed weight, and percentage error are shown for eleven trials and, from the Figure 8 results, it can be seen that the error never exceeded 4% for any of the individual trials. Another important thing to notice is that as the weight increases, the error value goes down comparatively.

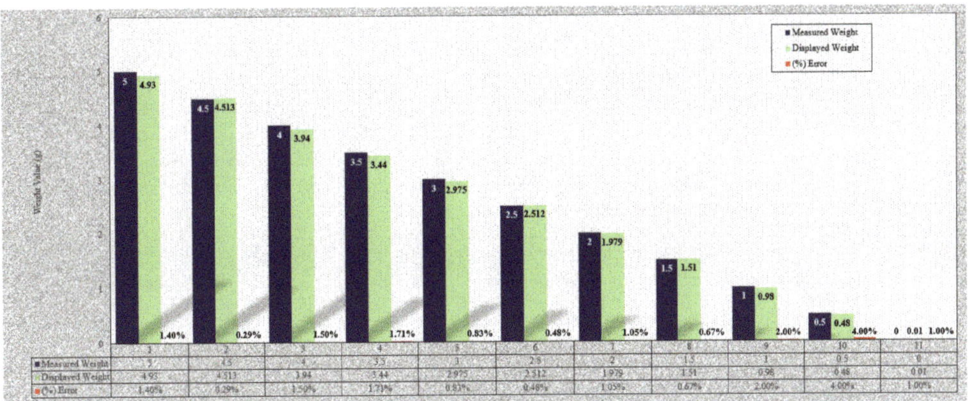

Figure 8. Results from ten trials having a value between 0 and 5 kg.

5. Performance Evaluation of Haptic Tele-Weight

5.1. Performance Evaluation I: Confusion Matrix and Accuracy

This evaluation is based on the eleven trials considering the values from 0 to 5 kg with an interval increment of 0.5 kg. A confusion matrix is represented for Actual (A) and predicted (P). When all the trials are combined, the overall measured weight is 27.5 kg, and the overall weight demonstrated by the motor gear arrangement is 27.26 kg. The confusion matrix is as follows:

$$Confusion\ matrix = \begin{matrix} & n = 11 & P:YES & P:NO \\ & A:YES & \begin{bmatrix} 27.269 & 0.231 \\ A:NO & 0.045 & 0.186 \end{bmatrix} \end{matrix}$$

The overall observed accuracy for the experiment performed in eleven trials is

$$Accuracy = \frac{Obtained\ demonstrated\ values\ by\ gear\ chain\ from\ 11\ trials}{Total\ measured\ weight\ from\ 11\ trials}$$

$$Accuracy = \frac{27.26}{27.50} \times 100\% = 99.89\%$$

The eleven trial values are ignored as they are at no load, whereas the relative index wise accuracy from the graph for the ten trials is

$$Accuracy\ [10] = \{\ 98.7\%,\ 99.71\%,\ 98.5\%,\ 98.28\%,\ 99.16\%,\ 99.52\%,\ 98.95\%, 99.33\%, 98\%, 96\%\}$$

The overall average accuracy from the relative method (index-based) of accuracy computation is 98.60%. Therefore, the accuracy is between 98.60% and 99.89% for loads under 5 kg.

5.2. Performance Evaluation II: Coverage Range

A generic architype was created to elucidate the enactment of the scheme. For the prototype, the warehouse was considered to be in China and the e-commerce VR store user was in Jeddah. The VR store user was interested in knowing the weight of an object before purchasing the item. Another possible reason for obtaining the weight is to estimate the shipment-related issues. However, the only target of the current research is improving the immersive experience. The item was in China, and the user can feel the object weight even with a distance of miles because the system uses cellular technology (GSM module), therefore making it independent of range.

5.3. Performance Evaluation III: Time Factor

The time factor is the time required to perform one maneuver (one complete process) and is calculated using Equation (1). Measurement time is the time required by the load cell circuit to measure the object weight and is denoted by T_m in this work (order of microseconds, µs). $T_{cellular}$ is the time required by the cellular grid to deliver an encrypted message to the receiving end. The mean value of $T_{cellular}$ is 1 s. Receiver display time (T_{rs}) is the time the receiver end circuit takes to decrypt the received message and show the weight in the motor gear arrangement.

$$T.F = T_m + T_{cellular} + T_{rs} \tag{1}$$

$$T_{rs} = 0.1 \times weight \; (in \; grams)$$

$$T_m = \frac{1}{No. \; of \; operations \times Clock \; Frequency \; (CPU)}$$

For the object having 1 kg of weight, the time factor (T.F) is

$$T.F = T_m + T_{cellular} + T_r$$

$$T.F = \frac{1}{8 \; \text{MHz} \times 50} + 1 + 0.1 \times 1000 = 101 \; s = 1.68 \; \text{min}$$

The time factor varies with the weight of the object With a larger object, the time factor is higher. For the object with 1 kg of weight, the weight takes 1.68–2 min.

6. Conclusions

The concept of smart cities is evolving, and different governments are trying to transform their cities into smart policy cities using available advancements. One technology being adapted worldwide is VR, and with its recent progression virtual e-commerce stores are becoming more prevalent. Virtual stores allow the user to interact with the products, but there is no way to feel the weight of the object before it is purchased. In this research, a device termed as a haptic tele-weight is proposed that will allow virtual store customers to feel the weight of an object prior to their purchase. In this way, the customer satisfaction level is increased and the delivery of an unintended item can be prevented. The objective of this research is to propose, design, and implement a prototype of the tele-weight device that can be used to display the weight of the object at the remote end. The tele-weight device has two parts, a sending end and a receiving end. At the sending end, the circuit computes, encrypts, and transmits the object weight and then the receiving end decrypts and demonstrates the weight of the object. Another intention of the approach is to enhance the e-commerce customer's level of user experience by allowing them to have immersive contact with the item in the virtual store.

A haptic tele-weight device is anticipated that utilizes the mobile cellular module to transmit encoded messages among its subparts. The accuracy and performance of the tele-weight device are marginally good, and an onboard microprocessor and circuit elements are used for communication. The crucial objective of this research is to improve the immersive experience of e-commerce shoppers. The computed time factor is not fixed in magnitude and depends largely on the weight of the object. The average accuracy of the created prototype is 98.60–99.89%.

The team of researchers working on this system is hopeful for the industrial version of the prototype. Many weight scales are available these days, but none complies with the VR store. In this research, a prototype of the tele-weight device is suggested to have a better user experience of immersive shopping. Many stockholders are interested in selling different products, but products in the shop often become inaccessible. This approach is handy and useful for the customer to have haptic immersive interaction with the item, allowing the weight to be placed on the hand of a person to have an improved pre-purchase experience. In this way, the trust of customers can be obtained as well. Moreover, every prospective customer can sense the weight and approximate the figure of the object (to have an estimation of the feeling of the item in the hand). A future recommendation is to create an industrial version of the prototype that works with the HMD.

Author Contributions: Conceptualization, A.F., M.S. and A.S. Prototype experiment, A.F. and M.S., writing—Introduction, Literature review, Methodology, A.F., M.S. and A.S., writing—Results, Conclusion, A.F. and M.S., editing B.H. and S.M., Project administration, M.S., A.S. and S.M. All authors have agreed to the manuscript published version. All authors have read and agreed to the published version of the manuscript.

Funding: This research work receives no external funding.

Institutional Review Board Statement: Not applicable.

Informed Consent Statement: Not applicable.

Data Availability Statement: No dataset is analyzed. Experiment is performed on prototype (tele-weight) device to obtain data points.

Conflicts of Interest: The authors declare no conflict of interest.

References

1. Carè, S.; Trotta, A.; Carè, R.; Rizzello, A. Crowdfunding for the development of smart cities. *Bus. Horiz.* **2018**, *61*, 501–509. [CrossRef]
2. Farooq, A.; Alhalabi, W.; Alahmadi, S.M. Traffic systems in smart cities using LabVIEW. *J. Sci. Technol. Policy Manag.* **2018**, *9*, 242–255. [CrossRef]
3. Farooq, A.; Seyedmahmoudian, M.; Stojcevski, A. A Wearable wireless sensor system using machine learning classification to detect arrhythmia. *IEEE Sens. J.* **2021**, *21*, 11109–11116. [CrossRef]
4. Jamei, E.; Mortimer, M.; Seyedmahmoudian, M.; Horan, B.; Stojcevski, A. Investigating the Role of Virtual Reality in Planning for Sustainable Smart Cities. *Sustainability* **2017**, *9*, 2006. [CrossRef]
5. Wu, J.; Song, S.; Whang, C.H. Personalizing 3D virtual fashion stores: Exploring modularity with a typology of atmospherics based on user input. *Inf. Manag.* **2021**, *58*, 103461. [CrossRef]
6. Loureiro, S.M.C.; Guerreiro, J.; Ali, F. 20 years of research on virtual reality and augmented reality in tourism context: A text-mining approach. *Tour. Manag.* **2020**, *77*, 104028. [CrossRef]
7. Man, D.W.K. Virtual reality-based cognitive training for drug abusers: A randomised controlled trial. *Neuropsychol. Rehabil.* **2020**, *30*, 315–332. [CrossRef] [PubMed]
8. Alhalabi, W.; Farooq, A.; Alhudali, A.; Khafaji, L. Smart Electrical Design of Medical Center to Vary Field Parameters: Sensor Network in Improving Health Care. *J. Eng. Appl. Sci.* **2019**, *14*, 879–886. [CrossRef]
9. Li, L.; Yu, F.; Shi, D.; Shi, J.; Tian, Z.; Yang, J.; Wang, X.; Jiang, Q. Application of Virtual Reality Technology in Clinical Medicine. *Am. J. Transl. Res.* **2017**, *9*, 3867–3880. Available online: https://pubmed.ncbi.nlm.nih.gov/28979666 (accessed on 16 May 2021).
10. Spiegel, B.; Fuller, G.; Lopez, M.; Dupuy, T.; Noah, B.; Howard, A.; Albert, M.; Tashjian, V.; Lam, R.; Ahn, J.; et al. Virtual reality for management of pain in hospitalized patients: A randomized comparative effectiveness trial. *PLoS ONE* **2019**, *14*, e0219115. [CrossRef]
11. Scapin, S.; Echevarría-Guanilo, M.E.; Junior, P.R.B.F.; Gonçalves, N.; Rocha, P.K.; Coimbra, R. Virtual Reality in the treatment of burn patients: A systematic review. *Burns* **2018**, *44*, 1403–1416. [CrossRef]
12. Pizzi, G.; Vannucci, V.; Aiello, G. Branding in the time of virtual reality: Are virtual store brand perceptions real? *J. Bus. Res.* **2020**, *119*, 502–510. [CrossRef]
13. Shamszaman, Z.U.; Ara, S.S.; Chong, I.; Jeong, Y.K. Web-of-Objects (WoO)-Based Context Aware Emergency Fire Management Systems for the Internet of Things. *Sensors* **2014**, *14*, 2944–2966. Available online: https://www.mdpi.com/1424-8220/14/2/2944 (accessed on 15 May 2021). [CrossRef] [PubMed]

14. Kolhar, M. E-commerce Review System to Detect False Reviews. *Sci. Eng. Ethics* **2018**, *24*, 1577–1588. [CrossRef]
15. Martínez-Navarro, J.; Bigné, E.; Guixeres, J.; Alcañiz, M.; Torrecilla, C. The influence of virtual reality in e-commerce. *J. Bus. Res.* **2019**, *100*, 475–482. [CrossRef]
16. Pizzi, G.; Scarpi, D.; Pichierri, M.; Vannucci, V. Virtual reality, real reactions?: Comparing consumers' perceptions and shopping orientation across physical and virtual-reality retail stores. *Comput. Hum. Behav.* **2019**, *96*, 1–12. [CrossRef]
17. Farshid, M.; Paschen, J.; Eriksson, T.; Kietzmann, J. Go boldly!: Explore augmented reality (AR), virtual reality (VR), and mixed reality (MR) for business. *Bus. Horiz.* **2018**, *61*, 657–663. [CrossRef]
18. Torabi, A.; Khadem, M.; Zareinia, K.; Sutherland, G.R.; Tavakoli, M. Application of a Redundant Haptic Interface in Enhancing Soft-Tissue Stiffness Discrimination. *IEEE Robot. Autom. Lett.* **2019**, *4*, 1037–1044. [CrossRef]
19. Yun, S.; Park, S.; Park, B.; Ryu, S.; Jeong, S.M.; Kyung, K.-U. A Soft and Transparent Visuo-Haptic Interface Pursuing Wearable Devices. *IEEE Trans. Ind. Electron.* **2019**, *67*, 717–724. [CrossRef]
20. Chen, D.; Song, A.; Tian, L.; Yu, Y.; Zhu, L. MH-Pen: A Pen-type Multi-mode Haptic Interface for Touch Screens Interaction. *IEEE Trans. Haptics* **2018**, *11*, 555–567. [CrossRef] [PubMed]
21. Khobaib, K.; Hornowski, T.; Rozynek, Z. Particle-covered droplet and a particle shell under compressive electric stress. *Phys. Rev. E* **2021**, *103*, 062605. [CrossRef]
22. Gómez-Carmona, O.; Casado-Mansilla, D.; López-De-Ipiña, D. Multifunctional Interactive Furniture for Smart Cities. *Proceedings* **2018**, *2*, 1212. [CrossRef]
23. Bin Rozmi, M.D.A.; Thirunavukkarasu, G.S.; Jamei, E.; Seyedmahmoudian, M.; Mekhilef, S.; Stojcevski, A.; Horan, B. Role of immersive visualization tools in renewable energy system development. *Renew. Sustain. Energy Rev.* **2019**, *115*, 109363. [CrossRef]
24. Silva, B.N.; Khan, M.; Han, K. Towards sustainable smart cities: A review of trends, architectures, components, and open challenges in smart cities. *Sustain. Cities Soc.* **2018**, *38*, 697–713. [CrossRef]
25. Masera, M.; Bompard, E.F.; Profumo, F.; Hadjsaid, N. Smart (Electricity) Grids for Smart Cities: Assessing Roles and Societal Impacts. *Proc. IEEE* **2018**, *106*, 613–625. [CrossRef]
26. Ismagilova, E.; Hughes, L.; Dwivedi, Y.K.; Raman, K.R. Smart cities: Advances in research—An information systems perspective. *Int. J. Inf. Manag.* **2019**, *47*, 88–100. [CrossRef]
27. Gaffary, Y.; Le Gouis, B.; Marchal, M.; Argelaguet, F.; Arnaldi, B.; Lecuyer, A. AR Feels "Softer" than VR: Haptic Perception of Stiffness in Augmented versus Virtual Reality. *IEEE Trans. Vis. Comput. Graph.* **2017**, *23*, 2372–2377. [CrossRef] [PubMed]
28. Akhtar, N.; Khan, N.; Khan, M.M.; Ashraf, S.; Hashmi, M.; Khan, M.; Hishan, S. Post-COVID 19 Tourism: Will Digital Tourism Replace Mass Tourism? *Sustainability* **2021**, *13*, 5352. Available online: https://www.mdpi.com/2071-1050/13/10/5352 (accessed on 15 May 2021). [CrossRef]
29. Paszkiewicz, A.; Salach, M.; Dymora, P.; Bolanowski, M.; Budzik, G.; Kubiak, P. Methodology of Implementing Virtual Reality in Education for Industry 4.0. *Sustainability* **2021**, *13*, 5049. Available online: https://www.mdpi.com/2071-1050/13/9/5049 (accessed on 16 May 2021). [CrossRef]
30. Nasralla, M.M. Sustainable Virtual Reality Patient Rehabilitation Systems with IoT Sensors Using Virtual Smart Cities. *Sustainability* **2021**, *13*, 4716. Available online: https://www.mdpi.com/2071-1050/13/9/4716 (accessed on 16 May 2021). [CrossRef]
31. Ospina-Bohórquez, A.; Rodríguez-González, S.; Vergara-Rodríguez, D. On the Synergy between Virtual Reality and Multi-Agent Systems. *Sustainability* **2021**, *13*, 4326. Available online: https://www.mdpi.com/2071-1050/13/8/4326 (accessed on 16 May 2021). [CrossRef]
32. Rossi, M.; Bianchi, M.; Battaglia, E.; Catalano, M.G.; Bicchi, A. HapPro: A Wearable Haptic Device for Proprioceptive Feedback. *IEEE Trans. Biomed. Eng.* **2018**, *66*, 138–149. [CrossRef] [PubMed]
33. Carpenter, C.W.; Tan, S.T.M.; Keef, C.; Skelil, K.; Malinao, M.; Rodriquez, D.; Alkhadra, M.A.; Ramírez, J.; Lipomi, D.J. Healable thermoplastic for kinesthetic feedback in wearable haptic devices. *Sens. Actuators A Phys.* **2019**, *288*, 79–85. [CrossRef] [PubMed]
34. Zhang, S.; Fu, Q.; Guo, S.; Fu, Y. Coordinative Motion-Based Bilateral Rehabilitation Training System with Exoskeleton and Haptic Devices for Biomedical Application. *Micromachines* **2019**, *10*, 8. Available online: https://www.mdpi.com/2072-666X/10/1/8 (accessed on 16 May 2021). [CrossRef] [PubMed]
35. Frohner, J.; Salvietti, G.; Beckerle, P.; Prattichizzo, D. Can Wearable Haptic Devices Foster the Embodiment of Virtual Limbs? *IEEE Trans. Haptics* **2019**, *12*, 339–349. [CrossRef]
36. Sierra, S.D.M.; Múnera, M.; Provot, T.; Bourgain, M.; Cifuentes, C.A. Evaluation of Physical Interaction during Walker-Assisted Gait with the AGoRA Walker: Strategies Based on Virtual Mechanical Stiffness. *Sensors* **2021**, *21*, 3242. Available online: https://www.mdpi.com/1424-8220/21/9/3242 (accessed on 15 May 2021). [CrossRef] [PubMed]
37. Shi, G.; Palombi, A.; Lim, Z.; Astolfi, A.; Burani, A.; Campagnini, S.; Loizzo, F.G.C.; Preti, M.L.; Vargas, A.M.; Peperoni, E.; et al. Fluidic Haptic Interface for Mechano-Tactile Feedback. *IEEE Trans. Haptics* **2020**, *13*, 204–210. [CrossRef] [PubMed]
38. Pistohl, T.; Joshi, D.; Ganesh, G.; Jackson, A.; Nazarpour, K. Artificial Proprioceptive Feedback for Myoelectric Control. *IEEE Trans. Neural Syst. Rehabil. Eng.* **2014**, *23*, 498–507. [CrossRef] [PubMed]
39. Georgakopoulos, D.; Quigg, S. Precision Measurement System for the Calibration of Phasor Measurement Units. *IEEE Trans. Instrum. Meas.* **2017**, *66*, 1441–1445. [CrossRef]
40. Sumit, S.H.; Akhter, S. C-means clustering and deep-neuro-fuzzy classification for road weight measurement in traffic management system. *Soft Comput.* **2018**, *23*, 4329–4340. [CrossRef]
41. Zhang, D.; Zheng, X.; Di Ventra, M. Local temperatures out of equilibrium. *Phys. Rep.* **2019**, *830*, 1–66. [CrossRef]

42. Ackerman, J.; Seipel, J. Effects of independently altering body weight and mass on the energetic cost of a human running model. *J. Biomech.* **2016**, *49*, 691–697. [CrossRef] [PubMed]
43. Bar, V.; Brosh, Y.; Sneider, C. Weight, Mass, and Gravity: Threshold Concepts in Learning Science. *Sci. Educ.* **2016**, *25*, 22.
44. Breuer, R.; Sewilam, H.; Nacken, H.; Pyka, C. Exploring the application of a flood risk management Serious Game platform. *Environ. Earth Sci.* **2017**, *76*, 93. [CrossRef]
45. Hung, S.-W.; Cheng, M.-J.; Chiu, P.-C. Do antecedents of trust and satisfaction promote consumer loyalty in physical and virtual stores? a multi-channel view. *Serv. Bus.* **2018**, *13*, 1–23. [CrossRef]
46. Huang, G.Q.; De Koster, R.; Yu, Y. Editorial: Online-to-offline ecommerce operations management (EOM). *Transp. Res. Part E Logist. Transp. Rev.* **2020**, *138*, 101920. [CrossRef]
47. Scarle, S.; Arnab, S.; Dunwell, I.; Petridis, P.; Protopsaltis, A.; De Freitas, S. E-commerce transactions in a virtual environment: Virtual transactions. *Electron. Commer. Res.* **2012**, *12*, 379–407. [CrossRef]
48. Rathore, R.; Singh, A.; Choudhary, H. Design, development, and calibration of bipedal force-plate for post prosthesis gait rehabilitation. *Mater. Today Proc.* **2021**, *44*, 4873–4877. [CrossRef]

 *sustainability*

Article

Framing Corporate Social Responsibility to Achieve Sustainability in Urban Industrialization: Case of Bangladesh Ready-Made Garments (RMG)

Polin Kumar Saha [1,*], Shahida Akhter [2] and Azizul Hassan [3]

[1] Environment and Sustainability Professional, Research Manager at Global Research and Marketing (GRM), Mirpur 1, Dhaka 1216, Bangladesh

[2] Department of Development Studies, BRAC Institute of Governance and Development, BRAC University, Dhaka 1212, Bangladesh; litun_tahsin@yahoo.com

[3] Tourism Consultants Network, The Tourism Society, 3 Port House, Square Rigger Row, London SW11 3TY, UK; azizulhassan00@gmail.com

* Correspondence: polin.msls2009@gmail.com or polinksaha@gmail.com

Abstract: According to both scholars and society, Corporate Social Responsibility (CSR) plays a controversial role in terms of corporate management towards sustainability. The business has presently become an integrated with the society and takes in a complex form of global demand for sustainability management. Due to the globalization of business, it is difficult to form a common sustainability model for CSR while its approach could be an opportunity for achieving sustainability. Evidently, a strong connection is found between the CSR approach and achieving urban sustainability since a smart urbanization requires a well-planned industrialization process that could be accelerated by the proper application of CSR. However, the CSR concept has already been initiated and developed in Western developed countries. Nowadays, it is being widely practiced by the developing countries as well. The developing country Bangladesh takes CSR issues seriously. Therefore, the study seeks the sustainability prospects of CSR by considering sustainability challenges in the rapid development of the "Ready-made Garments (RMG)" industry and corporate sector in Bangladesh. Finally, this paper explores some strategic paths to initiate sustainable development by framing the conditions and challenges of CSR.

Keywords: urbanization; sustainability; corporate social responsibility; ready-made garments; framework for strategic sustainable development; Bangladesh

Citation: Saha, P.K.; Akhter, S.; Hassan, A. Framing Corporate Social Responsibility to Achieve Sustainability in Urban Industrialization: Case of Bangladesh Ready-Made Garments (RMG). *Sustainability* **2021**, *13*, 6988. https://doi.org/10.3390/su13136988

Academic Editor: Tan Yigitcanlar

Received: 11 May 2021
Accepted: 17 June 2021
Published: 22 June 2021

Publisher's Note: MDPI stays neutral with regard to jurisdictional claims in published maps and institutional affiliations.

1. Introduction

Corporate Social Responsibility (CSR) is an umbrella term which recognizes several issues such as companies' responsibility for their impact on society and the natural environment, sometimes excluding legal compliance and the accountability of individuals; companies' responsibility for the behavior of others with whom they do business (e.g., within supply chains); and the obligation to control its bonding with the wider community. Whether for motives of economic viability or to add value to the society, CSR calculates to various social initiatives ranging from community development and environmental protection to various socially responsible business practices [1,2]. Theoretically, CSR is still controversial, unclear and complex in its traditional approaches [3]. In practice, CSR theories try to address four dimensions related to revenue, such as social aspects, political demand and ethical values, wherever it takes into consideration in a company. However, environmental demand is significantly missing in those dimensions, since a new theory needs to develop an integrated model for achieving sustainability.

There is a dearth of CSR studies, and the significance of dealing with CSR as a strategy has already been executed, but in the developing countries such as Bangladesh, the concept is still an emerging one, especially in the ready-made garments (RMG) industry [4].

Here, the RMG platform is used for scrutinizing CSR because the current urbanization of Bangladesh has been developed mostly by the RMG industries. By conducting this research, the study objective is to find out the current CSR conditions in Bangladesh's RMG industry as well as its prospects towards sustainability management through existing CSR planning for the future of smart urbanization.

The implementation of CSR in a new environment faces decent challenges in several industries due to carelessness and because of disrespect for CSR by the producers and a continuous lack and abuse of workers' rights, the sustainability of the RMG industry is now in question. The industry is being threatened by its unsustainable infrastructure leading to fire, building collapse, mass death, labor unrest and violence [5]. Consequently, practicing CSR activities is very important for companies in developing countries such as Bangladesh, especially in the RMG industry [4]. It is not only the industry who is responsible to assure the social responsibilities, but there must also be contributions from the societies where they perform their activities. The study also focuses on the integration process of sustainability into CSR in general considering the existing CSR challenges in Bangladesh. Sustainability indicators, or the process of integrating sustainability into CSR, help organizations to simplify, analyze, quantify and communicate the whole planning system of this complicated issue [6].

The aim of this study is to find out the challenges of CSR and seek for sustainability prospects of CSR by considering those challenges in the Bangladeshi RMG industry. Specifically, the concrete research questions are: first, what are the existing CSR challenges in the Bangladeshi RMG industry and should they work towards corporate sustainability management? Second, and finally, in what ways could CSR be more strategic in line with sustainability management?

1.1. Corporate Social Responsibility

Ref. [7] states that organizations need to incorporate environmental and social indicators into organizational management practices in order to improve sustainability performances, even in the processes of voluntary legislation and guidance. The lack of a CSR approach may be alternatively defined as the Corporate Social Irresponsibility (CSI). However, both the promotion of CSR and reducing the impact of CSI may occur through the process of conceptualization and institutionalization [8]. This process involves a continuous negotiation process at the national and international levels. Therefore, the theories of sustainability opportunity influence the ultimate extended performance of CSR. In other words, corporate sustainability is also the ultimate destination of the organization where CSR is the intermediate stage of action towards sustainability, standing on three bottom line pillars of profit, people and the planet (3P) [9].

1.2. Globalization and Corporate Social Responsibility Challenges

"Globalization along with a parallel shift to outsourcing and offshore manufacturing has resulted to the concept of globally as a supply chain accordingly" [10]. Globalization happens mostly because of corporate societies' influence in the global economy. Due to technological, political, economic and cultural changes in society, the global business sector has become integrated. At the present period of urbanization versus globalization, international business is becoming more popular beyond the country boundary, where the implications of the business demand corporate sustainability based on the CSR activities [11]. In this case, the consumption patterns of an organization are a strong indicator for corporate sustainability that assess to what extent CSR contributes to what level of sustainability is achieved [12]. Meanwhile, global trade wars can be threatening issue on CSR and the global economy in the twenty-first century, as demonstrated by Figure 1 below.

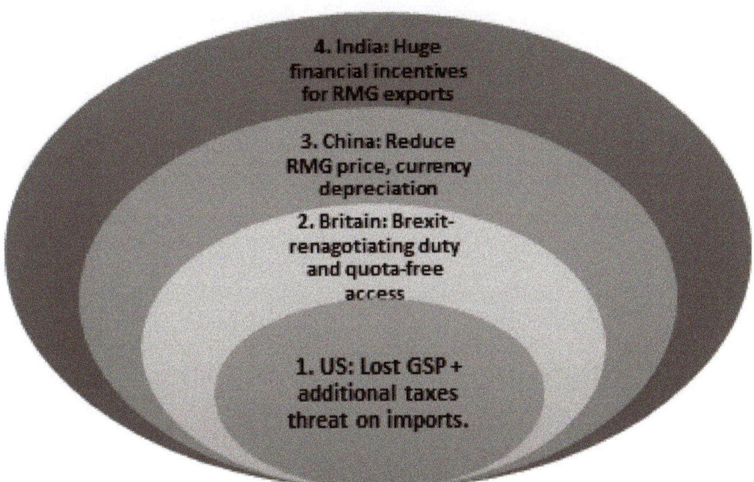

Figure 1. Global trade threat on CSR adapted from the Textile Today Bangladesh [13].

However, concerning the global phenomenon of corporate societies there is an ethical question raised by the author of [14]: will CSR reach full fruition as it becomes aligned, integrated and fully institutionalized within company strategies and operations? In the era of globalization, corporate societies have been operating in a market-based economy. Considering the major urban economy, various types of complexities imposed by different countries, for instance, India and Bangladesh, are the major impediments to the growth of bilateral trade and textile clothing value chain engagement between two countries such as India. However, both countries put a number of non-tariffs on textile and clothing products despite the agreement of cooperation between them due to their domestic CSR alongside social growth. Operations in market-based economies beyond borders needs proper regulation besides ethnicity. CSR in a market society has been discussed for decades, long before globalization became a buzz word [15]. While corporate societies are taking initiatives in their own interest, external regulation, e.g., government, institutional, is not a key issue to focus. In addition, there are some other factors which can challenge all the CSR initiatives taken by corporate societies. Day by day, corporate societies from different parts of the world are becoming integrated with each other. This integration brings a new challenge to the CSR initiatives.

Therefore, many social challenges arise, which are primarily transnational issues. These types of transnational social challenges cannot be regulated by the existing rules. For example, the greenhouse gas problem has been originated by developed countries, but the poorer countries such as Bangladesh are paying the price. In this case, even is Bangladeshi corporate houses reduce CFC (greenhouse gas) production significantly, would it be able to improve the situation in a significant manner towards urban sustainability? There is no one to give the answer.

To sum up, the global phenomenon of CSR is a great concern that there should have global regulation in addition to the corporate initiatives. In a global context, a corporate house runs through different cultural, political and geographical conditions. Therefore, beyond-border corporate activities demand regulation in addition to the responsible activities. On the other hand, the globalized form of the economy makes corporate culture complex and companies always think about their own sustainability rather than their social obligations and responsibilities.

1.3. Corporate Social Responsibility Practices in Bangladesh

In 1985, the UN adopted some basic CSR practice that were underway by the 1990s owing to the concern of international buyers from all nations that everyone should be considered as consumers regardless their incomes and social standing. At least stimulated by the potential of success, companies are increasingly adopting CSR practices, such as corporate philanthropy, Cause-Related Marketing (CRM), employee volunteerism, marginal support programs and other initiatives [16].

A forum of more than two hundred and fifty consumer organizations from all over the world stated that there should be four fundamental rights as a consumer, e.g., the human well-being and the right of proper environment. Ensuring the well-being of stakeholders and society as a whole, while maximizing the creation of shared value for the business owners, is what CSR aims to achieve [17]. However, researchers have stated that CSR offers various positive outcomes anticipated by the sellers and buyers such as customer trustworthiness, optimistic company image, cost decrease and achieving competitive advantage, all of which boost business performance [10]. According to the author of [18] has mentioned that CSR must be implemented: first, to eradicate extreme hunger and poverty; second, employment generating vocational skills; third, to ensure environmental sustainability; fourth, to promote of education, gender equality and empowering women; and, finally, to battle human immunodeficiency germ, developed immune deficiency disorder, malaria and other disease. These are the essential development pathways for us towards ensuring a smart city in the future.

1.4. Investment Situation in the Ready-Made Garment Industry in Bangladesh

In Bangladesh, the ready-made garments (RMG) sector has a greater potential than any other sector in terms of employment and foreign exchange earnings to reduce poverty and make a contribution to the national economy. This development of RMG has been initiated in most of the urban industrialization cases of the country. Along with its potential, the sector is also experiencing new challenges which can be the future determinants of its sustainability [19]. The journey of the RMG industry started in the late 1970s and since then has played a key role in the economy [20]. Within a very short period of time, it has become the largest export earner of the country through a major positive forward thrust in the early 1990s and socioeconomic status has increased in spite of lacking environmental practices. The RMG of Bangladesh globally exported over USD 24.49 billion in 2014–2015, compared to an amount of USD 12 million in 1984–85. The hold of RMG in the total export earnings reached more than 81% in the fiscal year of 2013–2014 and it shares almost 16% of the total GDP. The trend of the RMG industry's contribution is strongly predicted to be increased in the upcoming year [19]. This growth brings RMG industry under the microscope to find out its CSR activities and implications.

1.5. Framework for Strategic Sustainable Development

The Framework for Strategic Sustainable Development (FSSD) had been used as the theoretical approach to intervene the whole process of data analysis. The framework has five unique components that are interrelated in functioning and the whole function is the collective output of this five-level framework (Figure 2).

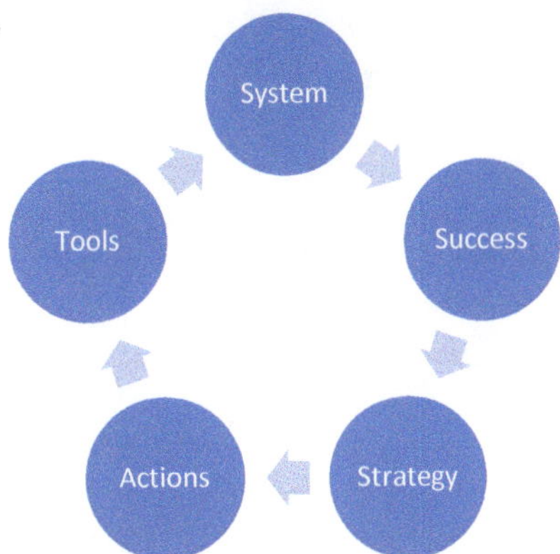

Figure 2. Five-level framework for strategic sustainable development [21].

A brief introduction of the five levels of the framework is as follows:

System: Principles to understand the services entire the organization and forming the whole organizational system towards more strategic as well as sustainable operations.

Success: Principles to define the success of the entire organizational system—basic principles "System Conditions" are a set of 'rule of thumb' practices to achieve success [22,23]. "System conditions" to achieve success are briefly illustrated as:

In a sustainable society, nature is not subject to systematically increasing . . .

. . . concentrations of substances extracted from the Earth's crust,

. . . concentrations of substances produced by society,

. . . degradation by physical means

and, in that society . . .

. . . people are not subject to conditions that systematically undermine their capacity to meet their needs.

Strategy: Strategy to reach at success which is defined through 'system conditions', which means sustainability principles.

Actions: Actions in corresponding to the assessment by tools and strategic guidelines executed by the principles of sustainability.

Tools: Tools that facilitate compliance with some principles and can be properly reported through the execution of sustainability principles (defined in 'success' level) in a system. Tools also can help to assess the performance of organizational policies to measure the success of any kind of tools that have been using in an organization to achieve success as well as facilitate capacity building of the people on human resource management, empowerment and knowledge management [22,23].

1.6. ABCD Analysis

The strategic planning tool "ABCD" had been first developed by an international NGO, the Natural Step International. Beginning with the above framework, the tool was used here to assess the performance of Corporate Social Responsibility in RMG industries in order to make it a more strategic way for movement towards sustainability in urban development. This planning had four sections of analysis and the findings out of this analysis are then incorporated into organizational policy and planning for achieving sustainability within different modes of actions (short, mid and long term). The meaning of four sections of this planning process had been guided in the following ways (Figure 3):

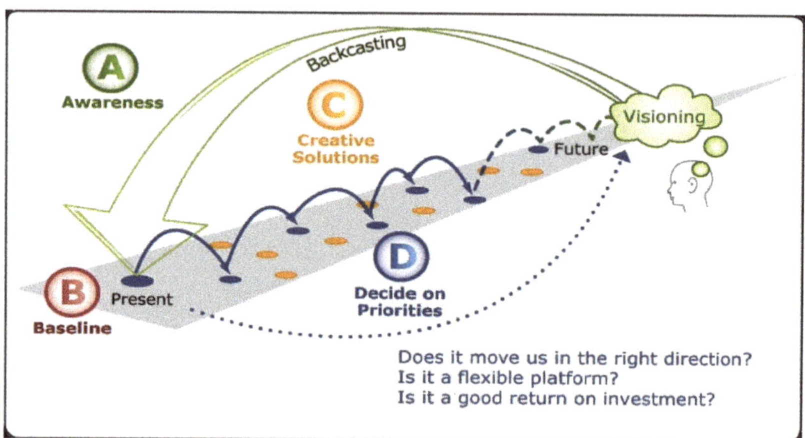

Figure 3. Diagram of the ABCD process [24].

"A" denotes Awareness and Visioning of an organization towards sustainability.

"B" indicates Baseline Mapping, which is conceptualized through the designated four sustainability principles to identify the 'sustainability gap' out of the major impacts of organizational activities.

"C" means the creative solutions that are developed in line with the findings out of "B", the baseline mapping.

"D" introduces the decision on priority basis depending on what type of problems are identified in the baseline findings.

1.7. Conceptual Framework of the Study

Based on overview of the corporate sustainability challenges, the following conceptual path can be shown for the study (Figure 4):

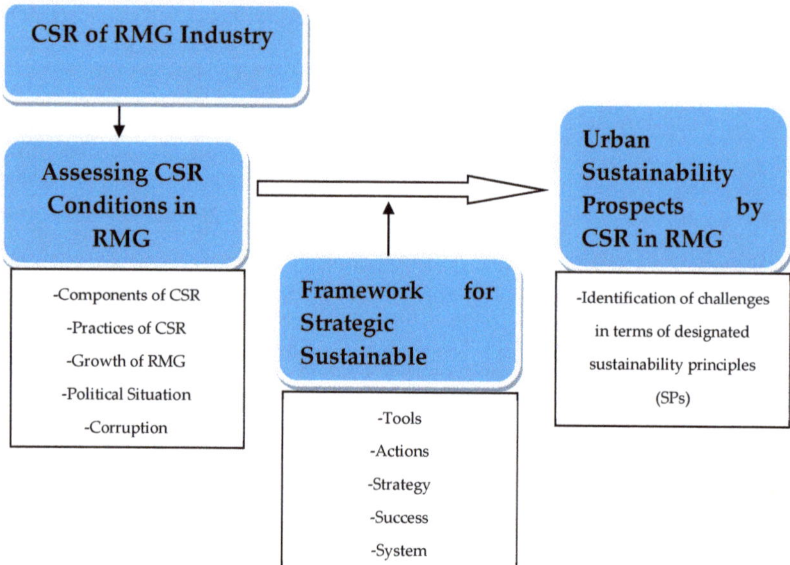

Figure 4. Conceptual framework of the study (Primary Source).

From the above framework, it can be shown that the study would analyze the CSR conditions based on five significant indicators, e.g., existing CSR components, CSR practices, growth of RMG, political situation and corruption. This assessment supports the study for final output, which is a scope of integrating sustainability into CSR considering its implementing challenges in Bangladesh. Here, the Framework for Strategic Sustainable Development (FSSD) is the intervening variable used to assess the system of CSR and its definition of success based on the designated four sustainability principles. Furthermore, the framework builds up some strategic guidelines for corporate sustainability practices through the existing CSR planning. The designated ABCD tool can be defined at the tool section of the FSSD. Therefore, among the five levels of the FSSD, two levels (system and success) are applied in assessing the CSR condition. On the other hand, the remaining three levels (strategy, actions and tools) are applied as the intervention in framing the corporate CSR towards sustainability. Details of the FSSD are discussed in the following Methods section.

2. Methods

The study followed a qualitative approach to find out the possible answer and in-depth knowledge about CSR challenges in Bangladesh. The paper attempted to fill in the void of CSR literature by investigating the underlying reasons of its socially lacking in responsible business practices [5]. The study was developed on literature reviews and a case approach of RMG industries in Bangladesh. The case was designed in single embedded with a type of instrumental analysis of the selected corporate sectors in Bangladesh. This instrumental case was specially focused on the theoretical explanation of the particular issues linked with several departments of an organization. All of the data used for analysis and discussions collected from the secondary resources. The discussion was developed on existing CSR theories to reach our conclusion.

2.1. Sample Size and Data Collection

The RMG industry and corporate sector is considered as the study experimental unit. A sample of 15 stakeholder groups had been selected in the study. The selected stakeholders were the internal CSR management representatives of the ready-made garments sectors. All the stakeholders were selected from three categories of garments industries such as large, medium and small types of industries in the Dhaka City area. Five stakeholders were selected randomly from each type of industry. Among all the samples of data sources, non-governmental groups of stakeholders had been selected. Identification of key stakeholders was a concern of the right approach to review CSR activities since they were on the way forward to sustainability through their existing CSR policies. A qualitative questionnaire tool is used to conduct interviews with the stakeholders. In the data collection, both on-site and off-site KII (Key Informant Interviews) were done to answer the research questions. For off-site data collection, telephone interviews were conducted or emailed responses were collected. Apart from the methods, the official websites and secondary materials of the stakeholder groups were used to gather information as well as the validation of the data. All the data had been analyzed thematically under the sustainability framework.

2.2. Framework for Strategic Sustainable Development (FSSD) and ABCD Analysis

Data analysis followed the FSSD and ABCD processes of analysis. By using the framework in any complex system, a good strategic analysis could be organized for moving towards organizational sustainability. To organize all kinds of information and thoughts about sustainability, the framework was applied in framing the whole system of the organizational function either in services or any kind of consumption matters. On the other hand, the ABCD planning method had been integrated in the analysis of the FSSD where the organizations were able to understand their clear vision for growth towards a sustainability-oriented approach [25].

3. Results and Analysis

3.1. ABCD: 'A'—Awareness

Components of CSR

In the corporate sector, CSR had been described as a process and a set of obligatory activities of the firm to utilize its resources to contribute to the society. CSR was a process where corporations take actions to combine an integrated social and environmental effort in their business principles. Socially responsible business did not only indicate the fulfillment of the social legal expectations, but it was also something more to invest in a corporation's human capital [26].

Many corporations in the world have taken CSR actions as an aid or a compliance-based agenda. This was a very congested view of CSR. The following figure cited from the policy formulation study on reporting to The Millennium Development Challenge through Private Sector's Involvement in Bangladesh, 2009 and after that SDGs have also showed the components of CSR and the greedy component's predicate on community development into CSR. The social value's elements are shown below (Figures 5 and 6).

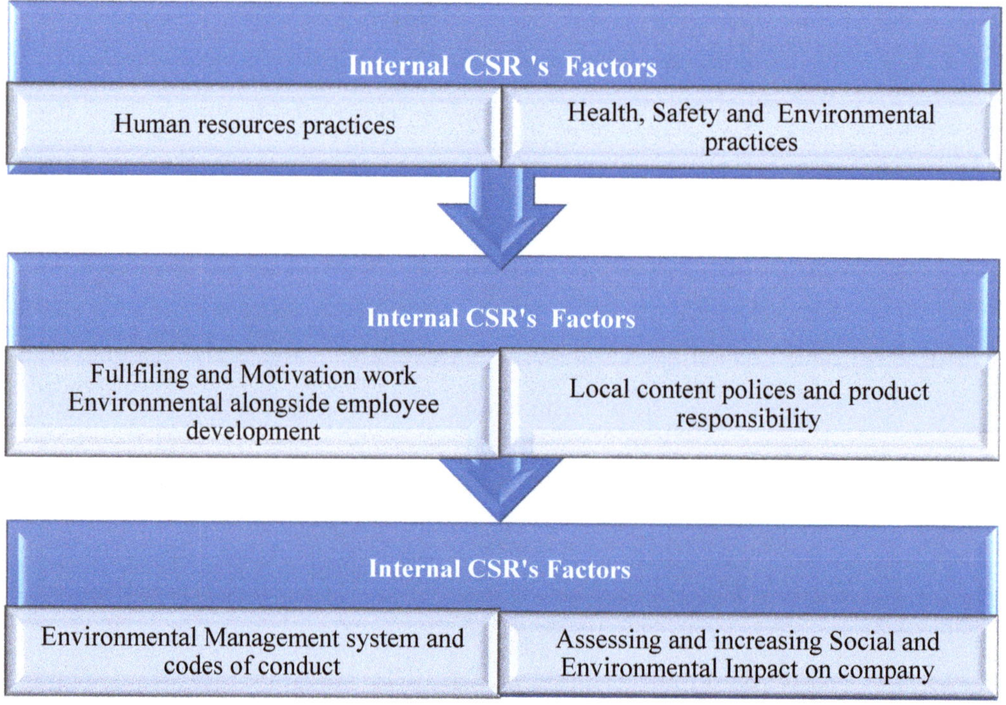

Figure 5. The social value's elements—internal (primary sources).

<table>
<tr><td>**External CSR Elements**</td><td>•Community Development Programmmes
•Shopsorship corporate SOCIAL responsibility</td></tr>
<tr><td>**External CSR Elements**</td><td>•Supply chain management
•Educational program at various levels</td></tr>
<tr><td>**External CSR Elements**</td><td>•Competitive markets need better prices to retain the market price
•Corporation's sustainability is a great concern in the bussiness world's resposible behaviour on environmental guarantee the sustainability</td></tr>
</table>

Figure 6. The social value's elements—external (primary sources).

3.2. 'B' and 'C'—Baseline Mapping and Compelling Measures of CSR

CSR was highly in demand for market-oriented factors where no single way had been implied for its successful operation. Toward the long-term success of CSR operations, legal restriction was very flexible and manipulation was needed depending on capitalistic factors [27].

CSR came about mainly through companies' codes of conduct, which were operated by the directions of a company's commitments the obtained certificates and labor standards [28]. In this case, CSR was designated as "social upgrading" from the side of social, economic and some ethical issue including workers' payment, occupational health and safety, work environment and human rights [28]. Because of the overlapping nature in CSR, the code of conduct was very confusing in practice; therefore, its establishment was needed for a correct path of framework to analyze every sector of organizational operations associated with green social banking upgrading. Under the guidance of B and C, the performance of CSR components used in the corporate sector are as follows.

3.2.1. Internal Positive Significance

The RMG industry is a capital earning source of foreign currency where most employees of the major cities (e.g., Dhaka, Chittagong, Naryanganj, Gazipur) in Bangladesh are involved in the RMG sector directly or indirectly. It finds that the present urbanization was being inspired to be conducted with CSR compliance for safe working processes, environmental management and providing healthy work environments. The garments industry of Bangladesh helped the country's people in many ways to reduce the systematic barriers to top management positions and barriers to workers meeting their needs. It even relied on inputs of companies that also did not create obstacles for its employees to meet their needs. The customers and employees respected the traditions of work and they actually liked it because they thought it to ultimately represent Bangladesh. The RMG industry has a good financial position that would allow some necessary investment if they could guarantee a Return on Investment (ROI). The RMG industry of Bangladesh has achieved a positive global reputation through its sensitive services with low product-cost in the long term. In the service dimension of RMG, it is a very good platform to conduct with CSR compliance in the whole process of smart urbanization.

3.2.2. Internal Negative Significance

The RMG industry had no strong compliance to CSR. Different certification processes and their maintenance were still without the enforcement labor laws in practice. The major considerable issues had to look at CSR compliance as: first, all work place hazards were not taken proper care of; second, all genuine stakeholders' input usually were not incorporated into work traditions; third, all stakeholder concerns had not been documented; finally, overly long working hours did not exist in its entire success.

The treatment of customers and employees was not done in a proper professional manner that kept everybody satisfied, content and happy. The RMG industry did not reduce dependence on fossil fuels nor did it use them efficiently. Thus, energy efficient transportation was not being used in the industry. No obvious measures were taken to substitute persistent and unnatural compounds with ones that were normally abundant or that break down more easily in nature (oil, remnants, chemicals detergents, dyes and other pesticides, pp bags, nylon, aluminum foils). The industry was still using recyclable carton packaging, bleached paper and other packing materials. The waste was not well-managed depending on the municipality that took care of it after separation. The RMG industry did not draw resources from well-managed eco-farms.

3.2.3. Political Barriers in Bangladesh

As a crosscutting issue, the political movement of Bangladesh influenced the desirable outcome of CSR approaches in the corporate sectors. "Strike (*Hartal*)" was a Bangladeshi political phenomenon that deserves to be discussed in a broader level to find out its necessity as a political element. *Hartal* observed a wide spread of violations among two political parties' activists, polices and general people [29]. On the other hand, Bangladesh saw its biggest terrorist attack in July 2016 (claimed by Islamic State of Iraq and al-Sham—ISIS) at the Holey Artisan restaurant in Dhaka, where twenty people—mostly foreigners—were killed. Prior to July 2016, several incidents of foreigners and bloggers being targeted by ISIS and Al-Qaeda took place. It stopped all kind of factory production, export shipment and daily business in all sectors [30]. Increasing security challenges have hampered at least some investment and trade opportunities. It caused immeasurable losses to the country's economy and workers continue to protest working conditions and low wages. In other unpredictable consequences of these violations, the major issues included unsafe public movement, increased working hours, lower or delayed salaries, loss of competitiveness, factory insecurity, negative image of industries and raising the overall cost of production. Therefore, these are all factors that have affected the success of CSR implementation direct or indirectly.

3.2.4. Corruption Barriers in Bangladesh

Bangladesh is a new country on the world map, but the presence of corruption is not new. In 1757, the first step of corruption was seeded by the East India Company. Therefore, it was a business organization who paved the world of corruption in Bengal history. "Since the independence, Bangladesh's leader is often condemned with the high incidence of corruption, but even since the return of democracy 1990s, neither has able to take effective action to addressee the systematic issue which allow corruption to flourish" [31]. The business community had put up with it and benefited from it. One survey of World Bank stated that about 2–3% of GDP is lost due to corruption each year. The per capita income of the nation's people might be doubled if the government was able to prevent corruption [32]. The ethics and issues of conflicts of interest had not been accorded a priority for legislative and administrative reform. However, when the private business joined together to make deals with public officials for procurement contracts, concessions and privatizations were then a considerable issue in front of the corrupted administration [32]. Though the government had established an Anti-Corruption Commission (ACC), it failed to come out from the hand of political party's influence. However, the trend of such unethical practices in the corporate sectors dominated CSR planning and implementation, considering political or self-interest rather than overall attention to the community.

3.2.5. Lack of Life Security System in Garments Industry and Threats to the International Market

Although Bangladesh is recognized as one of the major apparel exporters in international market, the country has also been criticized for its poor CSR practices due to the low-quality life security in the industry. Corporations could be said to be socially account-able if they would not significantly damage their stakeholders, employees, customers, their investors, suppliers or local community. In the operation of the industry, they must be rectified whenever any damage is discovered [33]. Human rights and safety issues in the RMG workplace have also been defined by the industry's reputation compared to progress in the global market. In several cases, an abrupt accident stood up against on CSR, and international buyers have become strongly behind the cases. The path of success has become threats to life in the RMG industry (Figure 7).

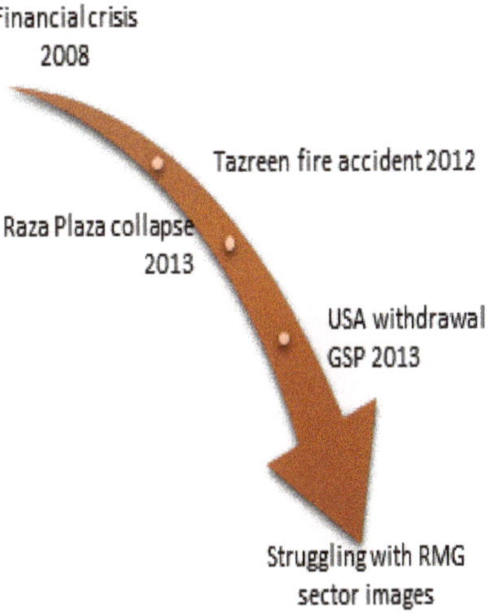

Figure 7. Indication of the path of the success with CSR in RMG industry adapted from Textile Today Bangladesh [13].

However, after the financial crisis of 2008, 112 garments workers had died and several hundred workers had been wounded by the Tazreen fire accident, and in the Rana plaza collapse, which was the biggest disaster, more than 1100 people were killed and over 2000 people were injured. The Trans-Pacific Partnership Agreement and the withdrawal of generalize system of preference by the USA were terrible for the RMG industry in Bangladesh. The crucial and unplanned damage to the image of the RMG sector may still be haunting the heart of readymade factory makers.

3.2.6. Environmental Management and Disability in CSR

A study [34] concluded that the conditions in which the RMG workers are set in causes health hazards such as fainting, eye strain, headache, jaundice, chest pain, asthma, fungal infections, helminthiasis, diarrhea, hepatitis, food poisoning, malnutrition, etc. The working conditions in several factories are subpar, and the absence of ventilation and light, lack of safe water for drinking and poor sanitation enables the development of diseases and viruses. Unfortunately, most of the factory owners are not aware of CSR in environmental sectors. Other difficult evidence about most of the garment factories stands at Ashulia and Tongi in Bangladesh. Unfortunately, there is normally no water management treatment plant in the dyeing industry. In Ashulia, liquid poison is mixed near the rivers and the water is black in the dry season. According to stakeholders and the local community, the factories are close to the river Bongshai and this river is a kind of dustbin for industrial waste. In this regard, all the factories inside or outside EPZ can be blamed in spite factory owners denying it. However, it is impossible to use the water of the Bongshai river. Moreover, the long-term ecological balance in the environment is reducing day by day as a result of weak environmental management.

The Bangladeshi RMG industry needs to adopt cleaner and improved technology and management strategies for a better environment in order to achieve its CSR goals. Considering the buyers' demand and requirements, industry owners have set priority to implement the guidelines from the world-renowned organizations for eco-friendly production. However, the crucial challenges that are most important to gain sustainable environment in RMG's sectors are related to CSR. The implementation of CSR practices can be of immense help for developing the city's economy, society and the environment of a country such as Bangladesh.

The urban challenges are: first, a lack of waste management and low technology; second, occupational health and safety measures with training and capacity development; third, resource-inefficient production processes; fourth, "GREEN" factory building and fire safety; fifth, chemical management best practices; sixth, chemical hazard assessment, prioritization and action; sixth, efficiency of captive power generator has to be increasesd; seventh, carbon emission reduction and implementation of an Environmental Management System; eighth, identifying potential to manage and mitigate environmental pollution; ninth, waste management and recycling; and finally, control and reduce environmental flow.

Moreover, these challenges can be supported to build up an expensive CSR for sustainable management in the RMG world. On another important note, the sustainability aspect is the greatest issue in SDGs whereas the Bangladesh RMG industry needs a comprehensive strategy to be sustainable for environmental issues. CSR-based policy makers need to understand the global impact of production and consumption patterns. In order to achieve sustainable environment, stakeholders must design SDGs in the RMG industry in Bangladesh according to needs of practices such as: first, rain water harvesting; second, condensate recovery boiler; third, reduce, reuse, recycle (3R) water; fourth, water-efficient dyes and chemicals; fifth, the use of renewable energy (solar panel); sixth, Prismatic skylight and T5 and LED light; and finally, sustainability reporting.

To the point, those practice in several areas and according to CPD [35], "The UN 2017 report on its Sustainable Development Goals shows that the rate of progress in many areas, including sustainable energy and infrastructure, is far slower than needed to meet 2030 targets, and urges governments and all stakeholders to make more efforts. Sustainably run companies also have to fight entrenched short-term shareholder thinking, particularly activist investors keen to gain fast returns". Since CSR focused on the environment can become a very important tool for RMG, particularly in response to the current globalization process, the compliance issue with the achievement of SDGs could be accomplished through framing CSR design. However, there is still a lack of concern about human and environmental standards that could result in trade barriers negatively affecting the industry in Bangladesh. Bangladesh's RMG sector is a globally competitive business sector, but the lack of CSR in this industrialization can be threatening not only for the country but also the achievement of sustainability globally. The lethargic implementation of CSR in the RMG industry may introduce trade barriers as well as the country's GDP (Gross Domestic Product). More efforts to strengthen CSR in Bangladesh are needed to overcome international challenges and compliance that is now fundamental for this booming sector.

3.3. 'D'—Priority Areas of Identified Problems and for Further Actions with CSR

In the previous section of compelling measures, the following working areas of CSR had been considered. These areas were both the internal and external approach of the current CSR operations.

3.3.1. Internal

(1) Compliance with CSR was measured for the safety environment and accidents and injuries at the workplace. (2) HR Management that had been measured in an efficient labor force for increased productivity and profit for the business corporations. (3) Waste Management where environment friendly initiatives of corporations were conducted saved

energy and earned more profit. These types of corporations got priority to access into the foreign market. The following components were important to consider into RMG industrial waste management. For garments: related solid and liquid waste through washing, dyeing and knitting; for food and beverages: through cooking and food service; for energy: fuel (personal and delivery cars); for packaging: product packaging and parcels; for hygiene: all about materials in practices relation to the occupational health and safety of workers; for lighting and other electronics: bulbs, projectors and fluorescent lights; and finally for stationary: papers, inks, printed materials and others office necessities.

3.3.2. External

(1) Supply chain management of CSR covered competitive markets which needed better pricing to retain the market share. A better supply chain could assure a better market position. (2) Social Investment of CSR considered corporate involvement with the community and its development activities gained media attentions and opened a window for earning more profit. (3) Environmental CSR indicated a corporation's sustainability that was a great concern in the business world's responsible behavior toward the environment and guaranteed the sustainability.

4. Discussion

4.1. CSR in the RMG Industry in Bangladesh in Line with the Existing Challenges

Most of the corporate businesses of Bangladesh are family-owned business as started from their first generation, and then converted by the RMG industry. On the other hand, these businesses are mostly operated from the major cities in Bangladesh, predominantly from Dhaka. However, in the 1960s and 1970s, when the marketing as well as social intellectuals were in progress of discussing CSR, they emphasized the social duties close to the marketing actions rather than on an overall social responsibility of the firm [36]. Different empirical studies also show that many small and medium enterprises fall under the informal sector without having the proper structures and resources to address social and environmental issues. This weakness drives the top management of local RMG companies to maximize their profits rather than considering the triple bottom line (people, planet and profit) issues [31,36].

The RMG industry of Bangladesh has been tremendously developed depending on the exports to developed countries. As a part of global market, it is difficult to ignore the CSR alongside environmental issues. It is evident from the RMG industry that the status of labor rights practices, environmental management, ex-post monitoring, risk management, low carbon technology and transparency in corporate governance are not yet satisfactory, largely due to the absence of poor enforcement of laws, weak participation of stakeholders to acquire their rights, lack of enforcement of backdated industrial laws and regulations, weak unions, lack of consumers' right groups and high levels of corruption. Within the regulatory authorities, these factors make CSR vulnerable in Bangladesh's RMG industry [31,37].

At present, there are still not sufficient uses of ethical business policies and CSR elements in the RMG sector. That is why not only child labor but also adult labor involved with RMG industry is facing discrimination. Though there has been a revision of the labor law in 2006, it is still based on the 1969 labor law. These forty-year-old labor laws are not sufficient to raise the CSR standard or to protect labor rights.

A good working environment is a key element to be assured by the socially responsible actions of corporations and organizations. A proper working environment and infrastructure is a fundamental right of workers. The RMG industry portrays a scenario where proper working environment is lagging far behind. The highest proportion of employment in the RMG industry is dominated by 80% women among 4.4 million workers. The provision for a woman-friendly work environment is rare in Bangladesh's RMG industry. An ILO paper [38] showed a concrete lack of CSR actions for women's work environment in the RMG industry such as inadequate childcare facilities on the factory premises; difficulties and irregularities getting maternity leave and benefits; scarcity of appropriate number of toilets; and congested work place in unsafe buildings.

The triple bottom line (people, profit and the planet) is one of the fundamental bases of CSR initiatives that are critically important for the urban sustainability of any country. Profit and maximum indicators in human approach are very important for any industry and that is why, an integrated path for sustainable CSR is necessary for ensuring a smart city. In the Resul section, issues about the integrated policies have been identified with the help of designated sustainability framework.

Employees are the life blood of any organization and industry. They are the most important element of stakeholder groups. Without strong stakeholder contribution, it is not possible to achieve the desirable standard of the CSR approach. By its name of CSR in garments industry, the outcome is expected from the relations between workers and their environment instead of making profit only. Therefore, a consideration of workers' demand into the CSR planning is a vital issue in order to improve the balance among social and economic return of the industry [39].

The RMG industry is a low-technology, labor-intensive industry that mostly depends on ready-to-export, turn-key factory units. Although the industry employs a great number of factory workers, lack of education and upgrading in jobs mean that workers have short shelf-life and are unceremoniously terminated after the end of productive years [40]. A labor union is a kind of institution that can facilitate the freedom platform to the worker. Unfortunately, there is no freedom of RMG associations in Bangladesh. Although union activities exist as a form of federations, their activities are prohibited at the factory level [41].

The study of CSR also suggests that without stakeholder participation, it is very difficult to integrate sustainability into social responsibilities. Stakeholder participation creates a platform where both parties can decide about their compliances. A consumer rights group can assure the mobility of stakeholder participation with corporation's CSR initiatives. On the other hand, labor unions in the industrial sector can play a significant role to protect their rights.

In more analytical findings, the CSR study suggests that without stakeholder participation, corporations can take advantage of CSR to retain their own interests, ignoring other parties' legal rights. To protect their own rights and ensuring more profit, RMG factories have been situated in Export Processing Zones (EPZs) in Bangladesh and have banned all kinds of union activities. Labors from EPZs agree with the government to create Workers Representative Welfare Committee. These types of committees are weak in nature and cannot make RMG factories implement CSR activities. On the other hand, getting the affiliation is very lengthy process.

CSR practice and profitability is embedded with the industrial principle. It is a natural phenomenon that industries firstly focus on profit, then their economic sustainability and their overall process of CSR activities to reach for their business goal. It is evident from the current financial market analysis, after the year 2005, that the Bangladeshi RMG industry [42] has been facing huge competition because of the elimination of the rigid quota system. The base of RMG product of Bangladesh is low-cost items. Low-cost items give lower profits compared to other types of products. On the other hand, lack of proper infrastructure, training and skilled labor productivity are becoming lower than other competitive countries. The scopes of training and skills development for Bangladeshi's RMG worker are very poor and a half a million skillful employees are foreigners. Investment in skilled labor development has not become a part of corporate culture and strategy in Bangladesh. Lower productivity leads to the cost of lower wages in order to gain better profit margins [38].

Bangladeshi political culture and corruption are a simultaneous challenge to each other. The involvement of corruption and political leaders in the corporate sector gives Bangladesh the position as one of most significantly corrupted countries in the world. According to the TIB (Transparency International Bangladesh) report [43], corruption is a form of indirect taxation. Because of the consequences of corruption, business establishment cost has become very high in Bangladesh. This widespread situation creates vulnerability for industries for achieving their sustainability. This corrupted culture also maintains a strong liaison between political leaders and labor unions. The RMG industry also has the same experiences. Indeed, the significance of corruption is not underestimated in Bangladesh, even it is considered as the regular way of life. Due to the presence of corruption in the country, all areas of business life are weakening the management ability of the business sector [44]. On the one difficult note that environmental management system is not strong and it has become a misgiving issue for employees and residential areas to increase in different diseases and threaten life, for instance, the Rana plaza collapsing due to corruption

To sum up, we would like to draw up recommendations to raise the CSR issues in a reasonable level. It is evident from the study findings that there is no baseline for the CSR standard all over the world. Corporations are a part of society, and as a part of society, they must obey the social duty, which is named by CSR. Regarding CSR implementation, the country must play a significant role. Modern society is talking about the state's roles in justifying the importance of CSR initiatives. In this regard, corporate governance also plays a significant role. In addition to corporate governance, the corporations should have to ensure the mass stakeholders' involvement with the corporate policy. The initiation CSR is a debatable concept itself. Corporations and the RMG industry in Bangladesh should ensure the implementation of CSR initiatives in more sensible way to send a clear statement that business is not beyond the society.

4.2. Strategic Pathways towards a Sustainable RMG Industry in Bangladesh

The RMG industry could make a tangible difference by contracting with eco-farms to provide customers with the best quality of eco-product. By planning this in company's CSR strategy, the RMG industry would champion the market through triggering this initiative, which might pave the road for others to join. Setting up new market values might also help the government make brave decisions in this regard. However, and since there were many operational strategies to establish and maintain CSR policy in the RMG industry, many industries could work together to make the shift easier, smoother and tangible. The strategic path of CSR could be a foundation under the basis of the ABCD analysis found in the "Results" section.

4.2.1. Awareness (A)

In this way, an organization will evaluate its present understanding on sustainability and identifies the common language to create a visionary platform for future sustainability actions. The approach of 'A' can be influenced by all sorts of present actions for social, ecological and environmental development within the business context. The understanding of sustainability may be back casted from the designated sustainability principles [45]. Here, four sustainability principles scrutinize the whole system of organizational actions to understand sustainability. Those are already mentioned in the previous section of the sustainability framework.

However, throughout the understanding of organizational sustainability, the visioning for sustainability is to be set up as an ambitious goal. People are to be encouraged about sustainability all over their understanding process of sustainability in existing work.

4.2.2. Baseline and Compelling Measures (B and C)

In other way, this analysis also runs the evaluation and monitoring of organizational services to see how the prospective positive changes can be introduced towards sustainability based on those sustainability principles. The analysis identifies the critical sustainability issues, values and implications through different monitoring tools within the organization.

It is the process of potential brainstorming which certainly creates innovative solutions to get rid of the existing unsustainable practices in organizations. To be more innovative organizations also go back to their sustainability vision produced from "A" part. This process can be termed as "Backcasting", which is very helpful in developing strategy to solve the identified problems (sustainability gaps of B section).

4.2.3. Priority Areas of Action (D)

The decisions are to be a good return on investment in regarding economic, social and environmental returns. This decision can take many actions flexibly, which are categorized in different mode of actions, e.g., short-term, mid-term and long-term measurements.

However, under the analysis of the ABCD model, the combined findings of CSR could be integrated into the Framework for Strategic Sustainable Development (FSSD). The CSR process ability to achieve sustainability can be analyzed using the four sustainability principles of FSSD, developed by The Natural Step. These principles are used as a protocol of the FSSD and subsequently develop the cyclic functions of the framework. The analytical discussion of the CSR in line with these principles is as follows.

The first sustainability principle works to reduce human contribution to the systematic increases in concentrations of substances from the Earth's crust. With this guidance, organizations take initiative to alternate some materials from their usage which are scarce in nature and also which can be replaced by any other abundant materials leading away from the dependency on fossil fuels. CSR can address fossil fuel use by advocating the use of solar energy, which includes wind, geothermal, biomass and photovoltaic energy. This option is verified through many certifications process including Green-e certified renewable energy certificates. The CSR process facilitates mining through the requirement of close loop cycling of some metals to reduce the amount of metals extracted. In reality, CSR process fails to address metal scarcity and does not think about its poor metal management practices that have the adverse impact to the community.

The second sustainability principle is to eliminate human contribution to the systematic increases in concentrations of substances produced by society. Today's practices of RMG certainly produce compounds whether solid, liquid or gaseous that breaks down the natural system. The CSR's ideology is based on the idea that an organization needs to become part of biological and technical social responsibility systems. The CSR has the scope to promote reducing and eventually eliminating the waste from production and the product itself through the 'waste equals food' concept. The organization promotes closed loop cycling that reduces the amount of resources used to produce waste. A CSR strategy may promote ppm (parts per million) limitations that address some metals' usage, but it does not specifically address many essential substances such as GHGs emissions in the implementation process. It is indirectly accounted for through the energy saving and material selection portion.

The third sustainability principle guides to reduce human contribution to the continuous physical degradation of nature through many forms of modification, introductions and over-harvesting. CSR process focus on sustainable wood harvesting by using the guidelines from a set of Forest Stewardship Council (FSC). The uses and discharge of water are also a part of CSR conduction adopted by water stewardship guidelines. However, the present practices of CSR fail to address marine resources such as overfishing, trolling and other forms of overharvesting and also different agricultural impacts; those are specifically responsible for the physical degradation of the environment. Though the extraction of the mined materials, transportation and distribution system are indirectly involved to the physical degradation of the nature, but there are no rules or limitations in CSR regarding these to protect the physical quality of the nature.

The fourth sustainability principle says about our contribution as much as while society can move towards the meeting of human needs in our local community to global needs based on the process of substitution and dematerialization dealings taken within the first three principles. The CSR process of an organization may involve the efficient use of resources and this activity of the CSR is termed as "eco-effectiveness of the product". The CSR is also very conscious about the human needs and rights in the society that mostly satisfy the sustainability principles. In the tool system it has been maintaining the ten

(10) principles of the UN global compact: A Pledge for Social Responsibility & Corporate Ethics. However, CSR does not deal with some social issues in this principle such as no clear conception of social benefits (e.g., no boundaries of small scale vs. big scale projects and, as a result, small scale projects and/or poor countries feel lack of interest to use CSR), company reputation, border transitions of supply chain, etc.

5. Conclusions

CSR plays an important role in promotion for attaining business success, and CSR has many benefits for human rights, labors standards, environmental impacts, corruption reduction, workplace relations and gender discrimination in the workplace. That means social wellbeing included in the human approach should work in diverse ways for the sustainable environment of a city. In reality, CSR is not addressing many issues such as the challenges within the sustainable principles if it is considered for work towards sustainability. So, an unclear picture of the CSR objective can be termed as "Blind Spots" which explore the first research questions on CSR challenges and the scopes for work towards sustainability management at the corporate level.

Answering this research question, the paper has explored the current CSR challenges which exist in both Bangladesh RMG industry, or in general for the next design phase of urban sustainability. The current CSR issues in Bangladesh are still in the young age. It has to go in a long way to serve the society in a sustainable way within the optimum capacity of an organization. This paper has focused on current CSR conditions based on not only the existing social conditions, but also other sustainability factors that need to be addressed in the existing CSR. The study [5] claims that the "majority of RMG suppliers focus on profit maximization without complying required social compliance issues, but the real scenario is not the same. The suppliers are now conscious about CSR implementation as the most significant mechanism to survive in the market; if they do not go for CSR implementation, they will lose business order and have to leave the market".

On the basis of the first research questions, the findings suggest some ways to integrate sustainability in corporate CSR, which is the exploration of the second research question. In this regard, the study finds out some significant and concrete reasons behind the drawbacks of CSR issues globally. For framing the corporate sustainability achievements, a 'ABCD' analysis can be used in the traditional CSR practices. Moreover, the FSSD guides whole corporate process to measure the violated sustainability principles over the CSR approach. The FSSD principles help CSR strongly to be more strategic including all other scattered challenges for movement towards urban sustainability such as political situation, lacking knowledge in environmental management, a huge presence of corruption throughout the administrative matters, which include absence of strong consumer/stakeholder groups, inadequate ethical business principles, insufficient laws to protect corporate violations and absence of proper monitoring are mentionable. Therefore, the future research directions could be a direction towards sustainable CSR in other business platforms where these specific challenges are lensed through the disciplinary sustainability framework. In addition, the realization degree of corporate social responsibility is directly proportional to the degree of investment is not explained here due to the study limitations. The future investment vs. CSR in the RMG industry can be another full-phase research question for the future.

Author Contributions: Conceptualization, Methodology, Formal Analysis, P.K.S.; Investigation, Writing, Review and Editing, P.K.S. and S.A.; Validation and Project Administration, A.H. All authors have read and agreed to the published version of the manuscript.

Funding: The Research received no external funding.

Institutional Review Board Statement: The study excludes ethical review and approval due to the lack of humans and animals involved in the study.

Informed Consent Statement: Informed consent was obtained from all subjects involved in the study.

Data Availability Statement: Collected primary data and review of secondary literature.

Conflicts of Interest: The authors declare no conflict of interests.

References

1. Du, S.; Bhattacharya, C.B.; Sen, S. Maximizing business returns to corporate social responsibility (CSR): The role of CSR communication. *Int. J. Manag. Rev.* **2010**, *12*, 8–19. [CrossRef]
2. Blowfield, M.; Frynas, J.G. Editorial Setting new agendas: Critical perspectives on Corporate Social Responsibility in the developing world. *Int. Aff.* **2005**, *81*, 499–513. [CrossRef]
3. Garriga, E.; Melé, D. Corporate social responsibility theories: Mapping the territory. *J. Bus. Ethics* **2004**, *53*, 51–71. [CrossRef]
4. Zabin, I. An investigation of practicing carroll's pyramid of corporate social responsibility in developing countries: An example of Bangladesh ready-made garments. *IOSR J. Bus. Manag.* **2013**, *12*, 75–81. [CrossRef]
5. Azmat, F. Understanding Responsible Entrepreneurship of Micro-Business in Bangladesh. 2008. Available online: http:// citeseerx.ist.psu.edu/viewdoc/download?doi=10.1.1.532.4881&rep=rep1&type=pdf (accessed on 20 May 2020).
6. Singh, R.K.; Murty, H.R.; Gupta, S.K.; Dikshit, A.K. An overview of sustainability assessment methodologies. *Ecol. Indic.* **2009**, *9*, 189–212. [CrossRef]
7. Adams, C.A.; Frost, G.R. Integrating sustainability reporting into management Practices. *Account. Forum* **2008**, *32*, 288–302. [CrossRef]
8. Windsor, D. Corporate social responsibility and irresponsibility: A positive theory Approach. *J. Bus. Res.* **2013**, *66*, 1937–1944. [CrossRef]
9. Mulkhan, U. Corporate Sustainability Reporting: A Content Analysis of CSR Reporting in Indonesia. *J. Perspekt. Bisnis* **2013**, *1*, 73–89.
10. Salma, U. Can Corporate Social Responsibility Be Used as a Marketing Tool by the Readymade Garment Suppliers in Bangladesh? 2016. Available online: http://hig.diva-portal.org/smash/get/diva2:942579/FULLTEXT01 (accessed on 20 August 2020).
11. Kolk, A.; Van Tulder, R. International business, corporate social responsibility and sustainable development. *Int. Bus. Rev.* **2010**, *19*, 119–125. [CrossRef]
12. Málovics, G.; Csigéné, N.N.; Kraus, S. The role of corporate social responsibility in strong sustainability. *J. Socio Eco.* **2008**, *37*, 907–918. [CrossRef]
13. Textile Today. Recent Challenges in Bangladesh RMG Industry and Possible way Outs. 2017. Available online: https://www. textiletoday.com.bd/recent-challenges-in-bangladesh-rmg-and-possible-way-outs/ (accessed on 10 June 2020).
14. White, A.L. New wine, new bottles: The rise of non-financial reporting. *Bus. Soc. Res.* **2005**, *6*, 1–6.
15. Ford, R.C.; Richardson, W.D. Ethical decision making: A review of the empirical Literature. *J. Bus. Ethics* **1994**, *13*, 205–221. [CrossRef]
16. Christofi, M.; Leonidou, E.; Vrontis, D.; Kitchen, P.; Papasolomou, I. Innovation and cause-related marketing success: A conceptual framework and propositions. *J. Serv. Mark.* **2015**, *29*, 354–366. [CrossRef]
17. Cisco. 2017 Corporate Social Responsibility Report. 2017. Available online: https://www.cisco.com/c/dam/assets/csr/pdf/ CSR-Report-2017.pdf (accessed on 12 June 2020).
18. Rani, P.; Khan, M.S. Corporate social responsibility (CSR): An analysis of Indian Banking sector. *Int. J. Appl. Res.* **2015**, *1*, 304–310.
19. Rakib, M.A.; Adnan, A. Challenges of ready-made garments sector in Bangladesh: Ways to overcome. *BUFT J.* **2015**, *3*, 77–90.
20. Haider, M.B. An overview of corporate social and environmental reporting (CSER) in developing countries. *Issues Soc. Environ. Acc.* **2010**, *4*, 3–17. [CrossRef]
21. Saha, P.K.; Seal, L. A strategic approach to Environmental Management Systems (EMS): An assessment of Sustainability in EMS to move toward Sustainability. *Int. J. Environ. Sci.* **2011**, *2*, 1093–1102.
22. Broman, G.I.; Robèrt, K.H. A framework for strategic sustainable development. *J. Clean. Prod.* **2017**, *140*, 17–31. [CrossRef]
23. Robèrt, K.H.; Broman, G.; Waldron, D.; Ny, H.; Byggeth, S.; Cook, D.; Johansson, L.; Oldmark, J.; Basile, G.; Haraldsson, H.V.; et al. *Strategic Leadership towards Sustainability*; Blekinge Institute of Technology: Karlskrona, Sweden, 2007; p. 221.
24. The Natural Step International. *Planning for Sustainability: A Starter Guide*; The Natural Step International: Stockholm, Sweden, 2012.
25. Saha, P.K. Strategic sustainability through a product development tool. In *Green Design, Materials and Manufacturing Processes, Proceedings of the 2nd International Conference on Sustainable Intelligent Manufacturing, Lisbon, Portugal, 26–29 June 2013*; SIM: Singapore, 2013; pp. 209–214.
26. European Commission. *Promoting a European Framework for Corporate Social Responsibility*; Office for Official Publications of the European Communities: Luxemburg, 2001.
27. Banerjee, S.B. Corporate Social Responsibility: The good, the bad and the ugly. *Cri. Soc.* **2008**, *34*, 51–79. [CrossRef]
28. De Neve, G. Power, Inequality and Corporate Social Responsibility: The Politics of Ethical Compliance in the South Indian Garment Industry. 2009. Available online: https://assets.publishing.service.gov.uk/media/57a08b6fe5274a31e0000b56/60620_ Power_inequality.pdf (accessed on 30 May 2020).
29. Odhikar. Bangladesh: Human Rights Defenders Trapped in a Polarised Political Environment. 2013. Available online: http: //odhikar.org/ (accessed on 1 June 2020).

30. Bangladesh-Market Challenges, 2018. Issue on Trade and Industry. Available online: https://www.export.gov/article?series=a0 pt0000000PAtGAAW&type=Country_Commercial__kav (accessed on 10 August 2020).
31. Mahmood, S.A.I. Public procurement and corruption in Bangladesh. Confronting the challenges and opportunities. *J. Pub. Adm. Pol. Res.* **2010**, *2*, 103–111.
32. Ahmad, M. Governance, Structural Adjustment & the State of Corruption in Bangladesh. In *Background Paper for the National Forum of the Structural Adjustment Participatory Review Initiative*; SAPRI Bangladesh: Dhaka, Bangladesh, 2001.
33. Campbell, J.L. Why would corporations behave in socially responsible ways? An institutional theory of corporate social responsibility. *Acad. Manag. Rev.* **2007**, *32*, 946–967. [CrossRef]
34. Nahar, N.; Ali, R.N.; Begum, F. Occupational Health Hazards in Garment Sector. *Int. J. Bio. Res.* **2010**, *1*, 1–6.
35. Centre for Policy Dialogue (CPD). Companies Are Shifting to Responsible and Sustainable Business Practices. 2017. Available online: http://rmg-study.cpd.org.bd/companies-shifting-responsible-sustainable-business-practices/ (accessed on 30 May 2020).
36. Maignan, I.; Ferrell, O.C. Corporate social responsibility and marketing: An integrative framework. *J. Acad. Marrk. Sci.* **2004**, *32*, 3–19. [CrossRef]
37. Rahim, M.M. Legal Regulation of Corporate Social Responsibility: Evidence from Bangladesh. *Common Law Word Rev.* **2012**, *41*, 97–133. [CrossRef]
38. Khondker, B.H.; Razzaque, A.; Ahmed, N. *Exports, Employment and Working Conditions: Emerging Issues in the Post-MFA RMG Industry*; International Labor Office: Dhaka, Bangladesh, 2005.
39. Bode, N. Global Actors, Local Governance: Corporate Social Responsibility in the Indian Garment Industry. Bachelor's Thesis, Leiden University, Leiden, The Netherlands, 2013.
40. Dhaka Tribune. Can RMG Navigate All the Risks? 2020. Available online: https://www.dhakatribune.com/opinion/op-ed/2020/05/12/can-rmg-navigate-all-the-risks (accessed on 30 May 2020).
41. The International Federation for Human Rights (FIDH). Bangladesh Labor Rights in the Supply Chain and Corporate Social Responsibility. 2008. Available online: http://www.fidh.org/en/ (accessed on 30 May 2020).
42. Bangladesh Garment Manufacturers and Exporters Association (BGMEA). About Garments Industry. 2014. Available online: http://www.bgmea.com.bd/home/about/AboutGarmentsIndustry (accessed on 20 May 2020).
43. Hassan, M.K. *The Shadow Economy of Bangladesh: Size Estimation and Policy Implications*; Transparency International Bangladesh Research Report: Dhaka, Bangladesh, 2011; pp. 1–21.
44. Zakiuddin, A.; Haque, W. *Corruption in Bangladesh: An Analytical and Sociological Study*; Transparency International Bangladesh: Dhaka, Bangladesh, 1998.
45. Holmberg, J.; Robèrt, K.H. Backcasting—A framework for strategic planning. *Int. J. Sustain. Dev. World Ecol.* **2000**, *7*, 291–308. [CrossRef]

sustainability

Article

Data-Driven Analysis on Inter-City Commuting Decisions in Germany

Hui Chen [1,*], Sven Voigt [2] and Xiaoming Fu [2]

[1] School of Chinese Language and Literature, Beijing Foreign Studies University, Beijing 100089, China
[2] Institute of Computer Science, University of Göttingen, 37077 Göttingen, Germany; sven.voigt1@aol.com (S.V.); fu@cs.uni-goettingen.de (X.F.)
* Correspondence: chenhui@bfsu.edu.cn

Abstract: Understanding commuters' behavior and influencing factors becomes more and more important every day. With the steady increase of the number of commuters, commuter traffic becomes a major bottleneck for many cities. Commuter behavior consequently plays an increasingly important role in city and transport planning and policy making. Although prior studies investigated a variety of potential factors influencing commuting decisions, most of them are constrained by the data scale in terms of limited time duration, space and number of commuters under investigation, largely owing to their dependence on questionnaires or survey panel data; as such only small sets of features can be explored and no predictions of commuter numbers have been made, to the best of our knowledge. To fill this gap, we collected inter-city commuting data in Germany between 1994 and 2018, and, along with other data sources, analyzed the influence of GDP, housing and the labor market on the decision to commute. Our analysis suggests that the access to employment opportunities, housing price, income and the distribution of the location's industry sectors are important factors in commuting decisions. In addition, different age, gender and income groups have different commuting patterns. We employed several machine learning algorithms to predict the commuter number using the identified related features with reasonably good accuracy.

Keywords: commuting; employment; housing price; GDP; income; big data; prediction

Citation: Chen, H.; Voigt, S.; Fu, X. Data-Driven Analysis on Inter-City Commuting Decisions in Germany. *Sustainability* **2021**, *13*, 6320. https://doi.org/10.3390/su13116320

Academic Editor: Tan Yigitcanlar

Received: 20 April 2021
Accepted: 26 May 2021
Published: 2 June 2021

Publisher's Note: MDPI stays neutral with regard to jurisdictional claims in published maps and institutional affiliations.

1. Introduction

With the urbanization development, commuting is becoming an increasingly important part of modern society. It is well-known that during morning and evening peak commuting periods on weekdays, roads become highly congested due to a large number of commuters, causing severe overheads to the transport infrastructure systems [1]. In the recent past, the number of inter-city commuters in Germany increased substantially (27.9%), from 2,442,630 in 2004 to 3,123,924 in 2014, while the country's whole population had a slight decrease (from 81,646,474 to 81,450,370) during the same period [2]. The growth of inter-city commuters can lead to personal, environmental and societal changes such as increased traffic loads and frequent congestion, more road/railway work, higher levels of pollution, lower life satisfaction and the need for subsidies [3]. It has been demonstrated that urban planning will be highly associated with commuting costs, and NO_x and CO_2 emissions from road traffic [4,5]. With the current discussion on environmental protection and sustainable societies, we believe that it is of high importance to understand inter-city commuting in more detail. It is especially vital to understand the volume and patterns of people's inter-city commuting (a commuting mode that typically connects the residents of the periphery with big cities) and to find the underlying infrastructural bottlenecks and suggest possible responses, as the majority means of inter-city/regional commuting are by car [6].

In the scope of our study, inter-city commuters are socially insured employees whose work municipality differs from their residential municipality [7]. As commuters typically

227

base their family and job location planning on several factors, we focus not only on the economic structure of the city but also on the living standard and commuting patterns, which have been largely ignored in previous studies. More specifically, we aim to conduct a data-driven analysis of the potential factors behind inter-city commuting decisions in Germany: the labor and real estate situation (without relying on questionnaires and surveys), commuting patterns, cities' economic structure such as gross domestic product (GDP) and industry sectors.

In this work, we use only publicly available data so that the data sources are easily available and our results can be replicated. By integrating multiple datasets from different sources from over two decades, we study features that have not been considered or not available but are very important for understanding the inter-city commuting behavior, such as GDP, various housing purchasing/rental prices information, the job market in different industry sectors and computing patterns. In addition, with our time-series data, we leverage machine learning approaches to perform commuter prediction—with reasonably good performance—which is not seen in the previous efforts.

Section 2 presents related works. After Section 3 describes our data sources and methods, Section 4 provides our in-depth analysis results on these data including commuter prediction results, and Section 5 discusses additional issues. Section 6 is the conclusion.

2. Literature Review

Over the decades, sociologists, economists, geographers and computer scientists have studied commuting from different angles. With the increasing importance of inter-city commuting, one focus of these studies has been the influencing factors of inter-city commuting decisions.

First, income has been found as a determinant factor for long-haul commuting [8–11]. For instance, Dauth and Haller [11] showed that the willingness to pay for a shortened commuting distance is no lower than the income increase for the people who seek a job change for the same commuting distance.

Second, location is another determinant factor for commuting decisions. Clark [12] observed that households prefer to move closer to the workplace if they lived far from the workplace before, and the commuting time is significant for relocation decisions. Kalter [9] noted that most long-haul commuters come from small municipalities. Eckey et al. [13] as well as Haas and Hamann [14] found that workers in west Germany are more willing to commute than those from east Germany. Andersson et al. [15] showed rural-to-urban long-distance commuting is rapidly increasing in Sweden, and rural residents working in large cities are better paid, better educated and younger than workers in rural municipalities.

Third, commuting distances play an important role in commuting decisions. Instead of focusing on residential or workplace location alone, Simpson [16] modeled both workplace and residential locations and found such a joint model considering commuting distances between two locations can well explain the commuting behavior. Levinson [17,18] also established that there is an interdependence between the workplace and residential locations. Kalter [9] showed long-haul commuters tend to remain in their current livingplace workplace combination.

Fourth, different types of work influence commuting decisions differently. Huinink and Feldhau [19] showed that women with a part-time job and long-distance commute will have much less fertility intention than women with full-time or self-employed jobs. Ding and Bagchi-Sen [10] found that workers in different industry categories have varying distances they are willing to commute. Eckey et al. [13] found that in general, blue-collar workers are more willing to commute than white-collar ones. However, Haas and Hamann [14] noted that the most highly qualified employees tend to commute.

Fifth, gender differences play a regulatory role in commuting decisions. It has been found that male workers (80.5%) are more willing to commute than female workers [9], and males commute longer than their female partners [20]. Reuschke [21] showed that the vast majority (87.6%) of female commuters are childless; 35% of female commuters have a second residence due to their partners. However, for female workers fertility intention does not play a significant role in the decision to commute, while getting pregnant has a high negative correlation with commuting [19].

Other factors related to commuting decisions that have been studied include age [9] educational background [9], nationality [13], housing costs [22], household com- position (with one or two workers) [23] and levels of well-being [24]. For example, Kalter [9] found that workers who are younger or with high school diplomas are more willing to commute. Eckey et al. [13] showed Germans are more willing to commute than foreigners in Germany. Mitra and Saphores [22] found that housing costs have a strong influence on long-distance commuting. Dickerson et al. [24] showed that longer commutes are not generally associated with lower levels of well-being.

An overview of different datasets, methods and factors studied in related literatures is given in Table 1.

Table 1. Datasets, Methods and Factors Considered in Previous Literatures.

Literature	Material (Data)	Method	Factors Considered
Clark [12]	556 residential relocations in Milwaukee metropolitan area, USA (1962–1963)	Probability model & tests	Short-haul commutes, workplace's attraction, relocation willingness
Simpson [16]	Household transportation survey data in Greater London, UK (1971–1972) and Metropolitan Toronto Travel survey data of 3508 households in Toronto, Canada (1979)	Regression	Commuting distance, job opportunity, skilled or not, family status, age, job changed or not
Kalter [9]	The "Socio-Economic Panel (SOEP)–West" data of Germany in 1985	Explanatory model	Costs of commuting and migration, real estate and labor markets
Levinson [18]	Travel survey data of 8000 households in Montgomery County, Washington DC, USA in 1991	Regression	Family status, housing type, age, gender, income, sector, employer's attitude on home office, location within the city, commuting time
Clark et al. [23]	Survey data of 2000 households in greater Seattle area in USA (1989–1990, 1992–1994 and 1996–1997)	Probability model	Commuting distance, residential location, work-place location, computing time
Eckey et al. [13]	Data of 142,129 commuters in Germany (2003–2005)	A traffic prognosis program VISUM	Commuting distance, commuting time, gender, professional types (white vs. blue-collar), housing supply, income
Haas & Hamann [14]	Two datasets about German commuters in 1995–2005	Basic comparison	Educational levels, commuting distance, region is east or west Germany, employment situation
Reuschke [21]	2007 questionnaires on 4 metropolises in Germany (Munich, Stuttgart, Düsseldorf, Berlin) in spring 2006, plus telephone interview on 20 commuters in spring 2009	Logistic regression	Family status and living situations, number of residential locations

Table 1. *Cont.*

Literature	Material (Data)	Method	Factors Considered
Dargay & Clark [8]	Survey data from National Travel Surveys (NTSs) of UK in 1995–2006	Econometric models	Gender, age, employment status, household composition, commuting distance
Huinink et al. [19]	Survey data from Family Panel of Germany in 2008–2009	Regression, panel model	Fertility behavior, gender, employment status, education, partnership status, intention of having and the number of children
Dickerson et al. [24]	Survey on 16,000 individuals in UK in 1996–2008	Linear fixed-effects (FE)model	Commuting time, transport mode, age, hours worked, household income, marital status, children number, university degree or not
Andersson et al. [15]	Micro data for all inhabitants in Sweden spanning two decades	Logit model	Commuting distance, workplace/residence changed or not, income, age, gender, highest degree, family status, sector, occupation type
Mitra & Saphores [22]	Survey data of 18,012 households in California, USA in 2012	A generalized structural equation model	Socio-economic variables, vehicle ownership, land use, and housing costs
Ding & Bagchi-Sen [10]	Longitudinal Employer–Household Dynamics (LEHD) data set of Buffalo, New York in 2014	Regression	Income, age, sector
Dauth &Haller [11]	Dataset on the employment biographies of German workers with geo-coordinates places of residence and work of Germany in 2000–2014	Statistics, correlation analysis	Income, place of residence, place of work, employment status of each worker
Chidambara & Scheiner [20]	Survey data of 4775 households in Germany in August 2012–July 2013	Regression analysis	Economic power, car access, labor and domestic work-sharing and preferences on work-sharing

To summarize, while sociologists mostly focus on the reasons behind commuting on a personal basis primarily based on surveys and questionnaires, economists focus on the trend of commuting at an aggregate level and emphasize more on the economic backgrounds and cost benefits for the commuters and regions using statistical data. The major data sources of both types of studies are panels and questionnaires, in addition to statistical data, and could be complemented by integrating multiple datasets available from heterogeneous sources, which form the starting point of this paper.

3. Materials and Methods

3.1. Data Sources

We scraped the commuting data, employment data including industry sector data, unemployment rate and income data from the Federal Employment Agency [7], the house and apartment price data from Immobilenscout24 [25] and the distance data from Google Maps API [26] for each city and county in Germany, plus GDP data from GovData [27] per county-level. In total, we collected and computed 16 categories of data, and an overview of these data is shown in Table 2. They represent four perspectives (*labor market, economic structure, real estate market* and *commuter pattern*) which are of potential relevance for commuting decisions. In addition, auxiliary information such as age range, gender, nationality and GPS coordinates are included where available.

Table 2. Potential Factors Influencing Commuting Decisions.

Labor Market	Economic Structure	Real Estate Market	Commuting Patterns
Jobs (primary sector) Jobs (secondary sector) Jobs (tertiary sector) Unemployed	GDP GDP per worker GDP per resident Median income (place of work) Median income (place of residence)	Apartment rent price Apartment buy price House rent price House buy price	Incoming commuters Outgoing commuters Commuting distance

For a better understanding of these data, besides their basic structure and some extreme cases, we chose four cities in State Lower Saxony (Göttingen, Braunschweig, Hannover and Wolfsburg) as examples. The sum of these represents roughly the industry distribution of Germany: Hannover is the capital of State Lower Saxony; both Wolfsburg and Braunschweig are known for their industry which has been expanded since the 1990s (leading to an increased need in workforce); Göttingen is a representative German university campus city and most known for its university.

3.1.1. Commuting Patterns

The commuting data on a municipality basis consist of about 14,000 municipalities from over two decades. Table 3 shows the basic statistics of commuters from the perspective of the total 11,385 German municipalities in 2017. It shows the commuter distribution is heavily unbalanced: a small number of cities have high numbers of commuters and heavily outweigh many small cities. With a mean of 2820 incoming and 3010 outgoing commuters, the median (50%) is only 232 incoming and 651 outgoing commuters. The 75% quartile of the incoming (outgoing) commuters is only 40.4% (63.2%) of the mean. There is also an extremely high standard deviation throughout the whole dataset.

Table 3. Commuters on a Municipality Level 2017 (Source: Federal Employment Agency).

	Incoming	Outgoing	Foreigners	Germans	Female	Male	<20	20–25	≥55	noCommuting	Business
Count	11,385	11,385	11,385	11,385	11,385	11,385	11,385	11,385	11,385	11,385	11,385
Mean	2820	3010	952	8352	4315	5015	218	713	11815	6315	630
Std	16,191	12,944	13,973	101,292	52,705	61,942	2729	9176	21,618	104,644	7719
Min	0	0	0	0	0	0	0	0	0	0	0
25%	43	234	5	212	110	138	7	15	57	15	13
50%	232	651	24	708	343	406	23	54	155	78	42
75%	1139	1901	140	2241	1111	1300	70	182	496	452	150
Max	411,672	423,964	694,052	5,993,872	5,997,872	3,614,232	175,175	538,684	1,268,705	6,283,373	434,147

Typically, a county consists of a central city and more affordable peripheries (e.g., towns and villages), which generally do not provide as many jobs as the central city. Thus, on average, there are more incoming than outgoing commuters in the central cities. On the contrary, there are fewer incoming commuters than outgoing commuters in the peripheries.

For commuting distance, we used the Google Maps API to scrape the coordinates of all cities and counties in Germany. We then classify some cities as metropolitan regions based on GDP, and calculate the nearest metropolitan area for each city. The distances from cities to their nearest 289 metropolis are calculated as follows (Table 3).

$$d_{lat} = lat_2 - lat_1$$
$$d_{long} = long_2 - long_1$$
$$a = sin\left(\tfrac{d_{lat}}{2}\right)^2 + cos(lat_1) \cdot cos(lat_2) \cdot sin\left(\tfrac{d_{long}}{2}\right)^2 \qquad (1)$$
$$c = 2 \cdot atan2(\sqrt{a}, \sqrt{(1-a)})$$
$$distance = R \cdot c$$

where R is the approximate radius of the earth in km (6373), and lat_1, $long_1$, lat_2 and $long_2$ are the lateral and longitudinal GPS coordinates of the two cities, respectively.

Using the coordinates, we are able to calculate the mean commuting distance for households living in each city. We use a weighted mean to take into account the number of commuters. For each city we calculate:

$$P_i = c_i \cdot d_i \ for \ every \ i \in (0, \ldots, count(workplaces)$$
$$mean_i = \frac{\sum P_i}{\sum d_i} \tag{2}$$

where c_i is the number of commuters between the current city and workplace i, and d_i is the distance between the two cities. Therefore, $mean_i$ is the mean distance between the city and the workplace in combination with the number of commuters.

The ratio of incoming and outgoing commuters to the resident population expressed as a percentage in the four example cities are shown in Table 4: Wolfsburg has the highest percentage of incoming commuters, at 64%; the second highest, though standing at only 33% is Hannover; Braunschweig and Göttingen are very close with 27% and 26%, respectively. The outgoing commuters do not vary significantly for the four cities, ranging between 8% and 14%.

Table 4. Incoming and Outgoing Commuters for the Four Example Cities. (Source: Federal Employment Agency. [†] Except For This Column, All Other Data Are Meant for 2017. * The Data for Göttingen After 2013 Were Based on the Whole County of Göttingen; the Municipality Göttingen Alone had 120,000 Residents in 2017).

City	Residences 1994 [†]	Residents	Incoming	Outgoing	Incoming %	Outgoing %
Braunschweig	256,000	250,000	65,000	35,000	26%	14%
Göttingen	128,000	330,000 *	90,000 *	250,000 *	27% *	8% *
Hannover	526,000	540,000	180,000	600,000	33%	11%
Wolfsburg	124,000	125,000	80,000	100,000	64%	8%

The county-level commuting data include the same type of municipality data, with additional information such as gender and nationality. Note that they do not distinguish places within the same county (e.g., the distance between Herzberg am Harz and Hann. Münden is 70 km, but both are in the same Göttingen county). As shown in the statistics in Table 5, like the municipality data, the data on the county level are also very unbalanced, with the mean deviating heavily from the median. This is again due to few (large) counties and many (small) counties.

Table 5 shows that there are more male commuters than female commuters (from the perspective of residence place), confirming the previous studies based on surveys and questionnaires [9,28,29]. It also shows that the number of commuters being native Germans is about 8–9 times of the number of commuters with foreign nationalities per German county on average in 2017, which is approximately the same ratio between the total number of native employees and that of foreign employees in Germany in the same year. Hence, we do not explore the nationality factor of commuters further here.

3.1.2. Labor Market

We scraped the employment (per sector) and unemployment data for each city and county from the Federal Employment Agency.

Figure 1 shows four distinct exemplary cities within geographical proximity with their six most important industry branches. We can see that among all employed workers, most (84%) of them work in the tertiary sector (e.g., corporate management, healthcare, education) including less than 1% in the higher education sector, and only 15% in the secondary sector (e.g., machine and vehicle technology, construction work).

Table 5. Incoming and Outgoing Commuters, County-Level, 2017 (Source: Federal Employment Agency).

(a) Incoming Commuters, 2017					
	Total	**Male**	**Female**	**Germans**	**Foreigners**
Count	79,803	79,803	79,803	79,803	79,803
Mean	884	552	359	787	90
Std	7182	4128	3088	6420	830
Min	0	0	0	0	0
25%	16	10	4	0	0
50%	33	23	9	21	4
75%	94	65	28	75	13
Max	384,943	215,965	166,978	328,890	55,623

(b) Outgoing Commuters, 2017					
	Total	**Male**	**Female**	**Germans**	**Foreigners**
Count	78,257	78,257	78,257	78,257	78,257
Mean	889	524	363	795	87
Std	5391	3140	2268	4790	716
Min	0	0	0	0	0
25%	16	10	4	0	0
50%	33	23	9	22	4
75%	97	66	31	79	13

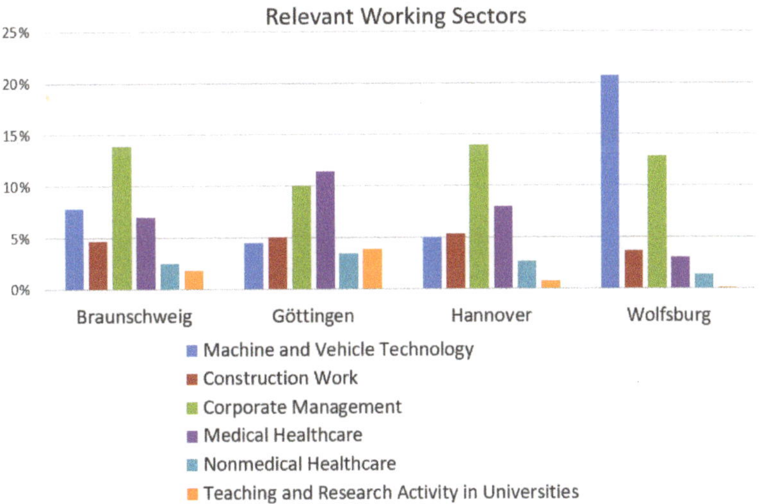

Figure 1. Industry Sectors for Example Cities (Source: Federal Employment Agency).

Table 6 shows some example cities with different unemployment situations in 2017, including several big cities and four cities in the state of Lower Saxony.

3.1.3. Economic Structure

We scraped GDP data from "GovData" [27] for German cities from 2000 to 2016, including GDP per city, per employee, per resident and per industrial sector. An example of GDP data is shown in Table 7, which leaves out the GDP per industrial sector for simplicity.

Table 6. Example Cities' Residents, Employment and Unemployment in 2017 (Source: Federal Employment Agency).

W/E	City	Residents	Employed	Unemployed	Unemployment Rate
	Germany	82,792,351	44,269,000	2,532,837	5.70%
W	West Germany (w/Berlin)	70,222,000	36,330,000	1,894,294	5.30%
E	East Germany (w/o Berlin)	12,571,000	7939	638,543	7.60%
	15 most populous cities:				
W	Berlin	3,613,495	1,426,462	168,991	9.00%
W	Hamburg	1,830,584	952,959	69,248	6.80%
W	Munich	1,456,039	850,395	35,718	3.90%
W	Cologne	1,080,394	553,442	48,227	8.40%
W	Frankfurt	746,878	564,826	23,307	5.90%
W	Stuttgart	632,743	405,383	15,581	4.70%
W	Düsseldorf	617,280	409,195	24,259	7.40%
W	Dortmund	586,600	231,529	34,100	11.10%
W	Essen	585,393	240,680	33,699	11.40%
E	Leipzig	581,980	262,537	22,946	7.70%
W	Bremen	568,006	273,068	28,027	9.70%
E	Dresden	551,072	258,758	19,074	6.60%
W	Hannover	535,061	329,083	25,163	6.80%
W	Nuremberg	515,201	305,674	17,096	6.00%
W	Duisburg	498,110	171.054	31,309	12.50%
W	4 cities in Lower Saxony:	123,914	118,922	3380	4.90%
W	Wolfsburg Braunschweig Göttingen (County)	248,023	127,827	8039	5.80%
W	Hannover	328,036	127,748	9953	5.90%
W		535,061	329,083	25,163	6.80%

Table 7. Gross Domestic Product per Region in Euros (Source: GovData).

Year & Key	Region	GDP	GDP per Employee	GDP per Resident
2016				
DG	Deutschland	3,144,050,000,000	72,048	38,180
01	Schleswig-Holstein	89,824,608,000	65,114	31,294
01001	Flensburg	3,712,513,000	62,017	42,827
-	-	-	-	-
2015				
DG	Deutschland	3,043,650,000,000	70,669	37,260
01	Schleswig-Holstein	86,689,473,000	63,975	30,473
01001	Flensburg	3,596,366,000	60,891	42,152

Table 8 shows exemplar median incomes for cities and counties with the highest and lowest median income. This shows an income disparity in Germany: after more than two decades of the German reunification [30], the median income of eastern Germany still is 19% lower than in the west; the top ten cities with the highest income are all in western Germany, while all of the five regions with the lowest income are in eastern Germany. Due to the continuous large amounts of workers moving from east Germany to west Germany [31] we conjecture that the median income difference between a large city and its adjacent regions will also influence the commuting behavior, which will be examined in the next section.

For each county/city we obtained data about the median income of employees from the Federal Employment Agency, including the median incomes of men, women and the residents in each region (city/Stadt or county/Landkreis). They are further split into three age groups, "15 to 25", "25 to 55" and "55 to 65" years old, and three educational levels, "no professional degree", "recognized professional degree" and "academic degree". A small example of the data can be seen in Table 9.

Table 8. Cities with the Highest and Lowest Median Income (Source: Federal Statistics Office [32]).

West/East Germany	County/City	Median Income
	Germany	2609 €
W	West Germany	2721 €
E	East Germany	2216 €
	Regions with highest median income:	
W	Ingolstadt	4635 €
W	Erlangen	4633 €
W	Wolfsburg	4622 €
W	Böblingen	4596 €
W	Ludwigshafen am Rhein	4534 €
W	Stuttgart	4351 €
W	Munich	4227 €
W	Darmstadt	4185 €
W	Frankfurt am Main	4182 €
W	Leverkusen	4170 €
	Regions with lowest median income:	
E	Altenburger	2218 €
E	Land Elbe-Elster	2215 €
E	Vorpommern-Rügen	2194 €
E	Erzgebirgskreis	2191 €
E	Görlitz	2183 €

Table 9. Median Gross Income (Euro/Month) in 2017 (Source: Federal Employment Agency).

Key	Region	Place of Work	Men	Women	Place of Residence
00000	Deutschland	3024	3207	2706	3027
01001	Flensburg, Stadt	2885	3077	2559	2647
01002	Kiel, Landeshauptstadt	3189	3382	2962	3030
01003	Lübeck, Hansestadt	2931	3033	2762	2895
01004	Neumünster, Stadt	2733	2800	2552	2700
01051	Dithmarschen	2768	2926	2297	2855
16077	Altenburger Land	2069	2100	1979	2182

The dataset contains the aggregated information of all employees working in each region for the "place of work" field, including incoming commuters but excluding outgoing commuters; whereas, "place of residence" includes everybody living in the city and excludes incoming commuters. Interestingly, even though it differs on a regional level, on average men are earning 500 € more than women per month. This may be a possible factor to explain the observation in [13], where men are found to be typically more willing to commute than women.

Overall, we can see that the median gross income for the "place of work" is higher than the income for the "place of residence". This further implies that commuting has a positive impact on income; therefore, it strengthens the conjecture that commuting contributes to the income discrepancy between men and women (https://statistik.arbeitsagentur.de/Statistikdaten/Detail/201712/iiia6/beschaeftigung-sozbe-qheft/qheft-d-0-201712-xls.xls?blob=publicationFile&v=1, accessed on 11 April 2021).

3.1.4. Real Estate Market

We scraped the house and rental prices via the ImmobilienScout24 API [25]. Table 10 shows an example of apartment rental prices.

Table 10. Prices to Rent Apartments (Source: ImmobilienScout24).

City & County	Square Meters	Price (€)	Price (€/sqm)
Aachen	70	668	10
Aachen (County)	79	560	7
Ahrweiler (County)	82	624	7.8
Munich	76	1658	23
Munich (County)	80	1315	17.5
Münster	76	880	11.8
Zwickau (County)	62	303	5

With the differentiation between cities and counties we have 419 data points. In this example, we include Munich because it indicates the vast difference between the house and apartment prices in Germany.

Since we have only the present housing price data, we add additional data pre-processing to incorporate additional knowledge from other sources and try to reflect the changes over the time as much as possible. For example, we superimpose an increase in housing prices by 21.7% from 2015 to 2018 (this information is from the Federal Statistics Office [32].

3.2. Methods

We use statistical methods to pre-process the data to get an overall view of different potential factors including their dynamic characteristics (where available).

We use linear regression to analyze the influence of factors like housing prices, GDP and median income on commuting decisions, taking housing prices as a specific example.

We use correlation to understand the potential factors related to commuting. To measure the correlation between variables x and y, the Pearson's correlation coefficient is given by:

$$r = \frac{\sum_{i=1}^{n}(x_i - \bar{x})(y_i - \bar{y})}{\sqrt{\sum_{i=1}^{n}(x_i - \bar{x})^2}\sqrt{\sum_{i=1}^{n}(y_i - \bar{y})^2}}$$

where x_i and y_i are the values of x and y for the ith individual

We use the following machine learning algorithms to predict the commuter number using the identified related features.

- *Linear regression*: an easy regression approach used to predict a continuous output (here, commuter number) where there is a linear relationship between the features of the dataset and the output variable. It assumes the input features to be mutually independent.
- *Decision trees*: this approach first splits the dataset into smaller subsets and then makes predictions based on what subset a new example would fall into; it re-cursively runs this process until a good match is found. Decision trees make no assumptions on distribution of data and work well with colinearity between input features.
- *Random forest*: a random forest aggregates a multitude of decision trees during the training time, each of which independently derives a prediction, then returns the mean prediction (regression) of the individual trees. It is one of the most accurate machine learning algorithms available and works well for many datasets.

4. Results

4.1. Commuter Dynamics in Four Exemplar Cities

In the four example cities in State Lower Saxony (Göttingen, Braunschweig, Hannover and Wolfsburg), the number of commuters gradually increases, as shown in Figure 2 for 1994–2013 (except for Hannover which has only data for 1994–2001 and no data for 2002–2013), and Figure 3 for 2014–2018.

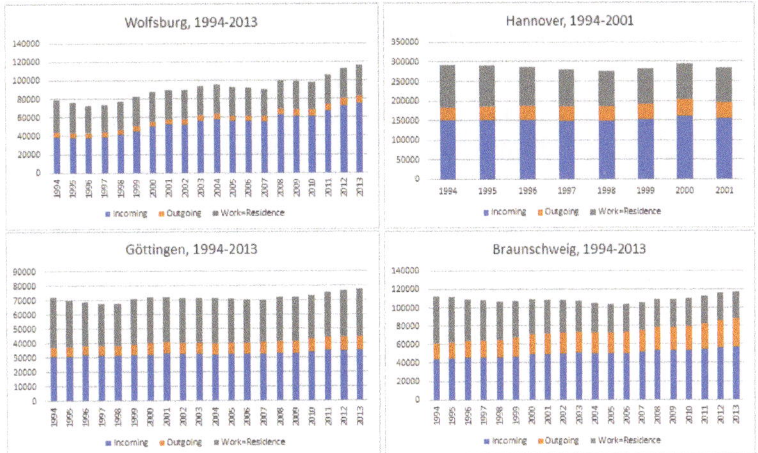

Figure 2. Number of Commuters from 1994 to 2013 (Source: Federal Employment Agency).

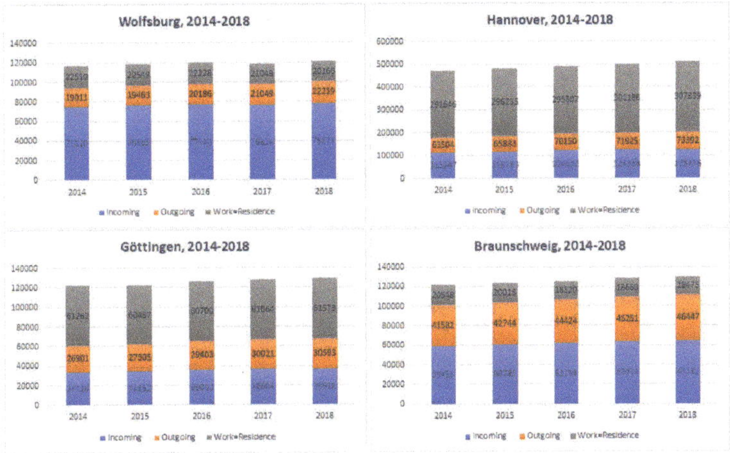

Figure 3. Number of Commuters from 2014 to 2018 (Based on Source: Federal Employment Agency).

The blue line shows the number of incoming commuters, the yellow line shows the number of outgoing commuters and the gray line shows the number of employees living in the same places where they work over the years. We see all commuter numbers increase but there are differences from each other. Wolfsburg denotes the most visible increase, almost doubling its incoming commuters during 1994–2013 due to its increased employment opportunities. Braunschweig and Göttingen's increases of incoming commuters are more subtle but still easily observable. Hannover, on the other hand, seems to stagnate. As a small university town, Göttingen's increase in incoming commuters is smaller.

The number of outgoing commuters stays almost the same for Wolfsburg, Hannover and Göttingen. Braunschweig, however, witnesses a big increase. Due to its closeness to Wolfsburg and the massive increase in incoming commuters to Wolfsburg, it is likely due to the strengthened industry in Wolfsburg and that many employees tend to commute there.

Starting from 2014, the Federal Employment Agency provides additional information on how many people live in the same region as they work. It now treats the county of Göttingen as Göttingen instead of the city of Göttingen, as in the 1994–2013 data. This leads to an increase in both incoming and outgoing commuters for Göttingen in 2014.

From the information on "work (place) = residence (place)" in Figure 3, we can see that both Göttingen and Braunschweig have the biggest proportion of their non-commuting employees. Incoming commuters in Braunschweig grew close in number to the non-commuting employees in 2016–2017. Both Wolfsburg (as an industry city) and Hannover (as the state capital) have more incoming commuters than non-commuting employees. Incoming commuters in Wolfsburg are nearly twice the number of non-commuting employees.

To sum up, we can see that the increase of commuters depends heavily on the city's industry and economic development and the relationship with the adjacent cities.

4.2. Housing Prices: Statistics

Housing prices are important for commuting decisions [33,34].

We collected data for over 400 cities in 2019. For most of them, we have house and apartment prices, as well as the rental prices for each type. Furthermore, we have the mean living space and with that can calculate the mean price per sqm. This is the most important part of the data since it allows us to compare the cities based on their living price per sqm.

Housing prices differ greatly for German cities. Many regions in eastern Germany are known for having cheap property, as there are not as many jobs as in western Germany. In the industry sector data from the Federal Employment Agency, we see that there are a total of 150,000 reported jobs, while there are 630,000 reported jobs in western Germany. The regions differ heavily in mean income as well. The median income for western Germany is 2700 e while the eastern Germany median is 2200 e. Therefore, it makes sense that the property prices for eastern Germany are lower than in western Germany. Due to the way Immobilienscout24 returns the data, we could not classify the advertisements to eastern or western Germany. However, if we look at the cheapest property prices, we can verify that most of them are regions in eastern Germany. This can be seen in Table 11.

Table 11. Cheapest Housing Prices per Square Meter in Germany (Source: ImmobilienScout24).

City	Apartment Prices		House Prices	
	Rent	Buy	Rent	Buy
Parchim	3.5 €	3565 €	5.8 €	1102 €
Grafschaft Bentheim	3.8 €	1750 €	6.8 €	2395 €
Jerichower Land	3.9 €	1397 €	5.4 €	935 €
Frankfurt (Oder)	3.9 €	1437 €	5.7 €	1540 €
Mansfeld-Südharz	4.1 €	1224 €	5.2 €	872 €

The five regions with the cheapest apartment rental prices, except for Grafschaft Bentheim which is on the western border to the Netherlands, are in eastern Germany. Apart from some small secluded regions, this trend continues throughout our data.

It is well known that Munich is the most expensive city in Germany [35], followed by Frankfurt and Stuttgart; these three cities are important metropolises for the German industry. In Figure 4, we see the most expensive mean prices per sqm for buying or renting a house or apartment.

The bars indicate the renting price, and the graphs denote the buying price. We see that Munich is the most expensive city to both rent or buy an apartment or a house. It reflects the property market well, having Munich, Stuttgart, Frankfurt, Hamburg, Berlin, Cologne and Mainz in the top 20 most expensive properties in all four categories.

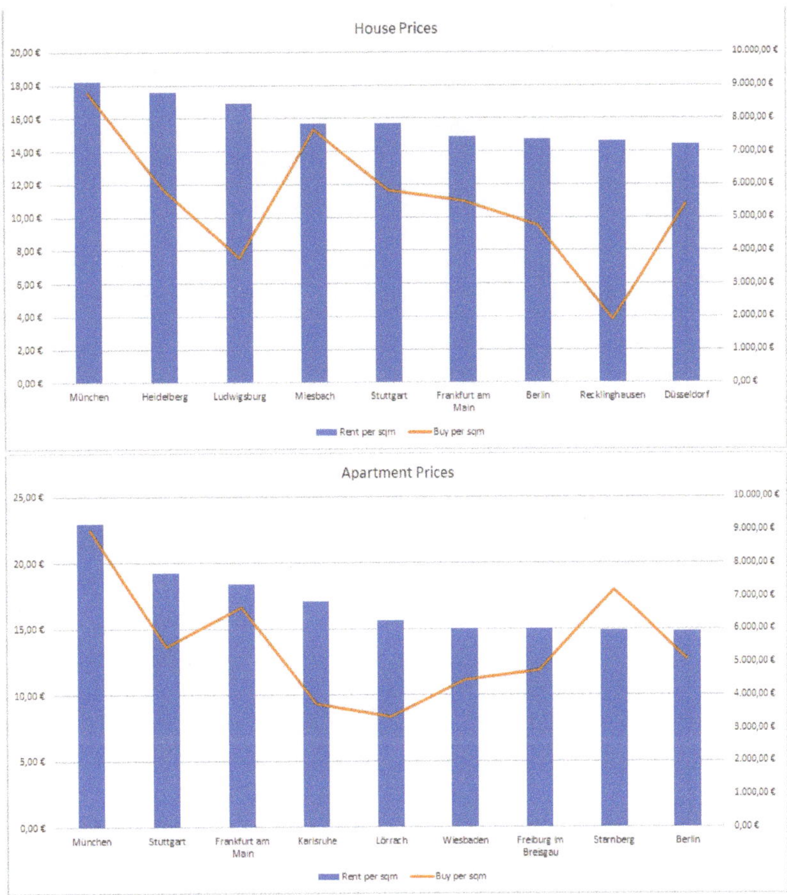

Figure 4. Highest House and Apartment Prices (EURO/sqm) (Source: ImmobilienScout24).

4.3. Commuting Distances: Statistics

Using the calculation method in Section 3.2, the statistical commuting distance data are computed in Table 12.

Table 12. Basic Statistics of the Average Commuting Distances (Calculated Based on Data Source: Federal Employment Agency).

	Average Distance (km)
Mean	77.7
Std	29.2
Min	23.5
25%	57.6
50%	71.8
75%	91.5
Max	181.3

The average commuting distance of 77 km is from the data on a regional level and therefore does not account for short-haul commuters. The minimal commuting distance is mostly for commuting that is between cities within the same county. Because the Federal Employment Agency lists them as different areas, they have a very short commuting distance with a very high amount of commuters. The maximal value of 183.2 km is for Birkenfeld where many employees are commuting to Bad Kreuznach, which is 140 km away. With both the mean and the median at about 70 km, we can see that these data are balanced and represent the long-distance commuters well. The exact distribution of commuting distances can be seen in Figure 5. The diagram shows the number of cities corresponding to the average commuting distance. The x-axis shows the intervals the cities belong to. These buckets have a size of 15 km each. The y-axis denotes the number of cities that are part of the respective bucket. For example, the column on the far left has 84 cities with an average commuting distance of 53 km to 68 km. The orange line represents the cumulative total, which is almost at 50% after the first two bars. It further shows that most of the commuters are commuting medium distances, between 38 km and 98 km, accounting for 75% of the total data. We can also see that only 10% of the cities have very long or very short distance averages below 38 km, or over 113 km. Our results seem to deviate a bit from Schulze [36] who found that most of the commuters commute up to 25 km. The reason is that Schulze used a different data source which can directly compute commuting distances, including for both intra-regional/city and inter-city commuters. With the Federal Employment Agency dataset, we have only aggregated information about inter-city commuters; due to the data provider's privacy restrictions we had to calculate the commuting distance ourselves.

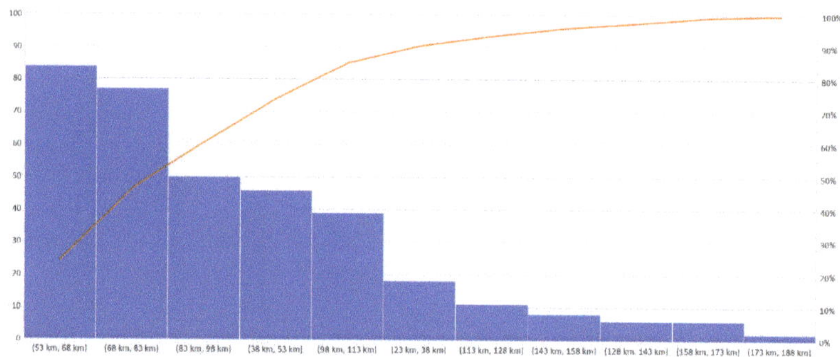

Figure 5. Pareto Chart of Commuter Distance.

Overall, the commuter data are not very balanced with many small regions with few commuters, and a smaller amount of big cities with very many commuters. Additionally, the type of city plays a key role in the observable commuting patterns. With our regional data, we are able to validate the findings of previous studies, e.g., by confirming that male commuters outnumber female commuters.

4.4. Housing Prices vs. Commuters: Linear Regression Results

We investigate the influence of the housing prices in regards to the number of commuters. To illustrate this, we conduct regression studies on apartment rental prices vs. the ratios of incoming and outgoing commuters to the number of local employees. The cases of other prices (apartment buying prices, house rental prices, house buying prices) are similar and skipped here due to space limit.

Two simple ordinary least squares (OLS) linear regression models are built for analyzing the relationship between apartment rental price (€ per sqm) and the ratio of commuters (against the local employees). The fit plots are shown in Figure 6. Both models suffer from heteroscedasticity which we can detect from both White's test results (Table 13, p-value <0.05) and residual plots as shown in Figure 7. To fix the heteroscedasticity, we apply the heteroscedasticity-consistent covariance matrix estimator [33].

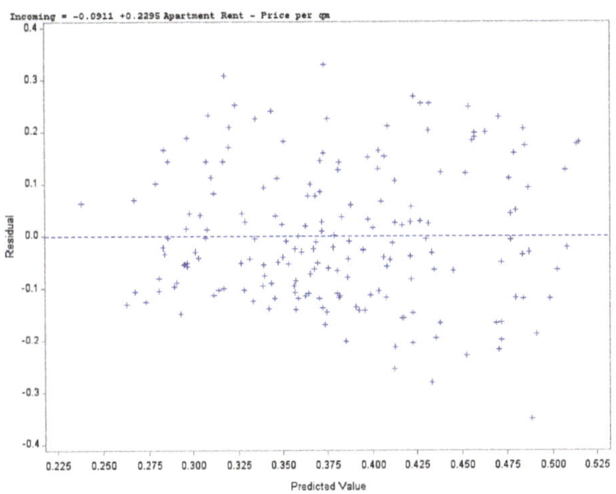

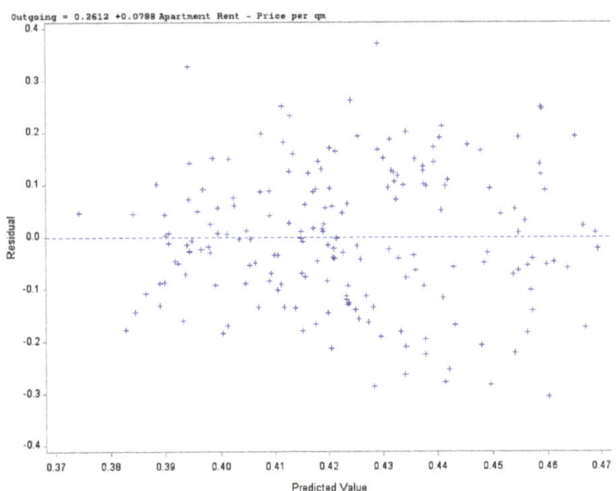

Figure 6. Apartment Rental Prices (€ per sqm) vs. Commuters (%).

Table 13. Linear Regression Result: Parameter Values Showing Heteroscedasticity.

Independent Value	Dependent Value	Intercept	Slope	ANOVA (Pr > F)	White's Test (Pr > ChiSq)
Apt rental price	Incoming Commuters	−0.09108	0.22949	<0.0001	**0.0069**
Apt rental price	Outgoing Commuters	0.26117	0.07885	0.0241	**0.012**

Using OLS linear regression for the log transformation of the apartment rental price (e per sqm), the result parameters are shown in Table 14; further model diagnosis reveals that the models' parameters are significant and there is no heteroscedasticity inside anymore (*p*-Value >0.05).

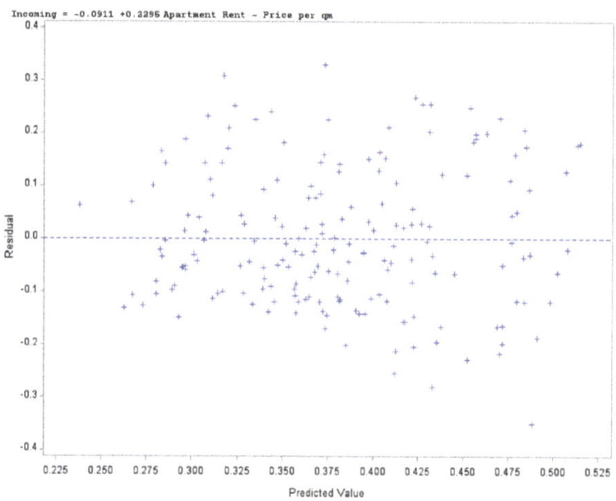

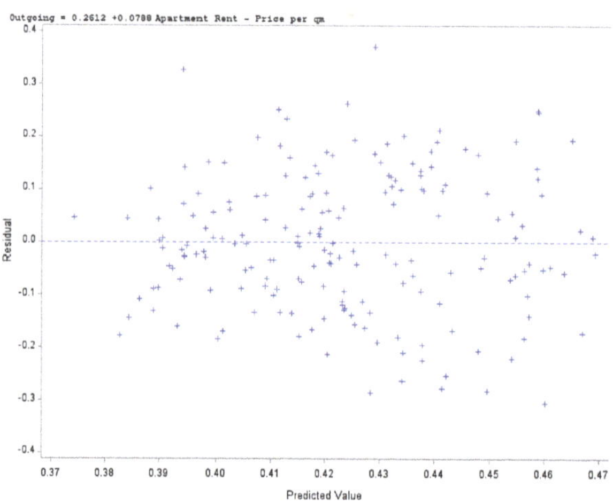

Figure 7. Apartment Rental Prices (€ per sqm) vs. Number of Commuters: Residuals.

Table 14. Linear Regression Result: Parameter Values after Fixing Heteroscedasticity.

Independent Value	Dependent Value	Intercept	Slope	ANOVA (Pr > F)	White's Test (Pr > ChiSq)
Log (Apt rental price)	Incoming Commuters	−0.13143	0.2506	<0.0001	**0.3465**
Log (Apt rental price)	Outgoing Commuters	0.21088	0.10406	0.0044	**0.1605**

We can see the relationship between the number of commuters and the logged unit price to rent an apartment in Figure 8. Both figures show an increasing trend, indicating a higher average number of corresponding commuters for a higher rental price. Furthermore, the number of incoming commuters increases faster with a higher rent cost than the number of outgoing commuters. The number of outgoing commuters also increases, likely due to being in bigger cities with more inhabitants.

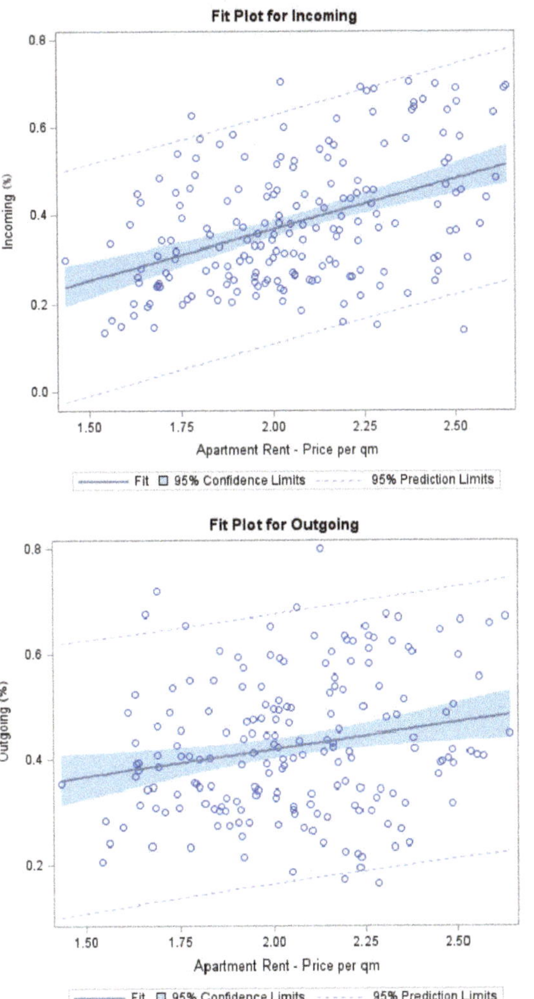

Figure 8. Logged Values of Apartment Rental Prices (e per sqm) vs. Commuters (%).

Again, we see that the incoming commuters increase quickly for higher apartment prices; the deviation is high for higher apartment prices due to the distribution of the data. Therefore, we rely on medium house prices and medium apartment prices.

Overall, the more expensive the real estate, the more employees will commute over long distances. This is in accordance with Boje et al. [34] who stated that according to location theory, rationally acting individuals compare the resulting benefit with the costs of commuting. If the costs outweigh the benefits, as they would have to pay a high percentage of his or her income for rent, they would give up renting in the workplace city and consider commuting instead. This behavior can be observed in our data, e.g., fewer employees in cities with low housing prices will decide to commute than in cities with higher housing prices.

4.5. Housing Prices and Income

We also analyze the relationship between housing prices and the median income. Similar to the previous subsection, our first results also show heteroscedasticity but can be fixed by the heteroscedasticity-consistent covariance matrix estimator; the results are omitted here again for the space limit, which explains that with an increasing median income, the apartment rent rises as well.

The result is expected, as it is logical that the real estate market and the median income are related to each other. Nonetheless, as the income has a strong link to the apartment and housing prices, it indicates a link to the commuter data as well.

4.6. GDP and Median Income

In this subsection we will take a closer look at our median income and GDP data.

While the individual city-level GDP data depict well the productivity of the city, the aggregated GDP information on the state level (Figure 9) shows a clear trend in the German economy distribution.

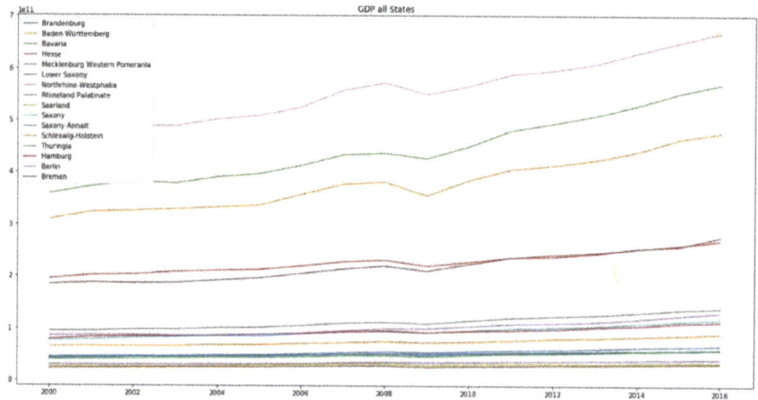

Figure 9. GDP of German States (2017).

As shown in Figure 9, North Rhine-Westphalia has the highest GDP, followed by Bavaria and Baden-Württemberg. North Rhine-Westphalia is well known for the Ruhrgebiet, which is a composite of industrial cities and thus a big metropolitan area. Bavaria also has important cities for the German industry like Nuremberg and Munich. Hamburg and Berlin are in the 4th and 5th place, respectively. This is no surprise, as these two cities are the biggest in Germany and hence have a great influence on the German economy. Overall, we see that the states in west Germany have higher GDP than their counterparts in east Germany.

4.7. Correlation Results

After analyzing all the data separately, we study their correlation with each other with a focus on the correlation with the commuting data.

To understand the most important reason behind commuting, we limit the correlation matrix to the 16 most important factors (see Table 2). The result is shown in Figure 10.

Beyond the highest correlations between the jobs in any two of the three industrial sectors, primary sector, secondary sector and tertiary sector, another high correlation is found between incoming commuters and outgoing commuters (in percentage of local employees). Except for commuting-related factors, the highest negative correlation is found between median income and metropolitan distance.

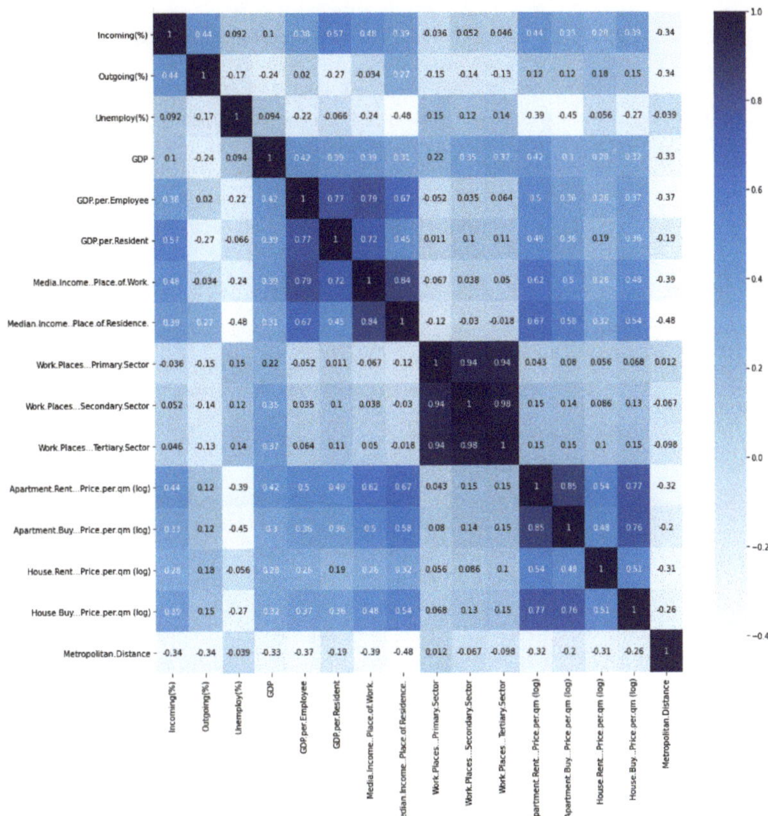

Figure 10. Correlation Matrix.

Now we examine the factors behind commuting based on this correlation matrix:

1. The matrix shows that the most important factor behind commuting is the GDP per resident of the city, as among all factors it has the highest Pearson's correlation coefficient with incoming commuters in percentage of the local employers (0.57) and the lowest (and negative) coefficient with outgoing commuters in percentage of the local employers (−0.22). This is somewhat surprising, as we expected that the median income and housing prices may have a more important influence on commuting decisions.

2. The median incomes of work and living places are also important. The median income in the place of work is highly influential on incoming commuters, as more employees may commute if they receive a higher income. How much they earn in their residence is influential to both commuting groups. The income in the place of residence is a main factor of commuting, either leaving the city or coming there, because if it is high, many people will decide to commute there; if it is low, more people will leave the region to work somewhere else.

3. The third most important factor for incoming commuters is the apartment price; more expensive apartments seem to be a factor related to employees commuting. A plausible reason behind this relation is that if the cost–benefit ratio of buying an apartment is bad, the employees may consider commuting over longer distances. For outgoing commuters, the distance to the next metropolitan area is very important. This means

that if the distance to the next metropolitan area increases, employees are less likely to leave their region to commute, given the cost–benefit ratio of long-haul commuting.

4. An interesting anti-correlation can be found between the outgoing commuters and the metropolitan distance. If the metropolitan distance increases, the outgoing commuters decrease, as their commuting distance would get longer and become most likely unprofitable.

5. A surprising high correlation can be found between commuters and the unemployment data. This has a big influence on both incoming and outgoing commuters. This may be related to the fact that most bigger cities tend to have a higher unemployment rate.

6. In regard to jobs (workplaces), the secondary and tertiary sectors are more influential on commuters than primary sectors, likely due to their high number of employees. For example, there were 82.3% jobs in the tertiary sector, and 17.2% in the secondary sector, in contrast to 0.5% in the primary sector as of 2017 [32]. Workplaces in the primary sector even show an anti-correlation with commuters, indicating that most farmers tend to not commute.

4.8. Commuter Prediction Results

As commuting is an important part of social life, for city and infrastructure planners, it is helpful to predict the commuting trend for the next years. Since the data collection of the Federal Employment Agency changed in 2013, we mainly focus on predictions using data from 1994 to 2012 to predict the number of commuters in 2013 for each city.

First, we generate our time series data using the *TimeSeriesSplit* function of scikit-learn, which splits the commuter data into different time frames. We then train a linear regression model (with heteroscedasticity detection and correction procedures), a decision tree model and a random forest model (with 100 decision trees as baseline), respectively, to predict the incoming and outgoing commuters for each city in 2013.

Three metrics of measuring the prediction accuracy are used here: (1) The mean absolute error (MAE) means that we are on average off by a certain number of commuters. (2) The mean squared error (MSE) measures the average of the squares of the errors; the closer to zero the MSE is, the better. (3) The root mean squared error (RMSE) is the root of the MSE and measures the accuracy of a forecast; again, the closer to zero the better, where a value of zero would mean that the prediction is perfect. The results for accuracies for incoming commuters and outgoing commuters in 2013 using the 1994–2012 data are shown in Table 15. The following observations can be made:

Table 15. Accuracy of Predicting the Number of Commuters in 2013.

Algorithm	Incoming Commuters			Outgoing Commuters		
	MAE	MSE	RMSE	MAE	MSE	RMSE
Linear regression	61.65	91,334.48	302.22	133.16	**316,451.25**	**562.54**
Decision tree	18.38	22,041.83	148.46	44.58	575,035.90	758.31
Random forest (with 100 decision trees)	**14.36**	**12,273.61**	**110.79**	**41.97**	504,505.48	710.29

In general, linear regression yields the worst performance as the input features do not hold collinearity; meanwhile, decision trees achieve much reduced MAE, MSE and RMSE. Random forest provides further improvements on prediction accuracy. An outlier is the MSE and RMSE are better for predicting outgoing commuters using linear regression compared to using decision tree or random forest algorithms, which may be attributed to the limited features available for the better balanced outgoing commuter data; more concrete reasons have to be found out.

Overall accuracy is reasonably good, considering the mean and median (50%-percentile) commuters numbers (see Table 3) of incoming commuters (2820 and 232) against its MAE (14.36 in the case of random forest, 18.38 for decision tree), and outgoing commuters (3010 and 651) against its MAE (41.97 for random forest, 44.58 for decision tree). This reflects only roughly 0.5–6.8% of absolute errors on average in the prediction.

The prediction accuracy for incoming commuters is generally better than that of outgoing commuters. This is affected by the highly unbalanced commuter data that the small numbers of incoming commuters in the cities are much more heavily distributed than large numbers, compared to outgoing commuters (see Table 3). When the overall incoming commuting number for a city is small, it is easier to predict with lower MAE than to predict the larger number.

We then examine how the number of decision trees affects the prediction accuracy. We try it at low as 10 and as high as 300 decision trees. The corresponding MAE can be seen in Figure 11.

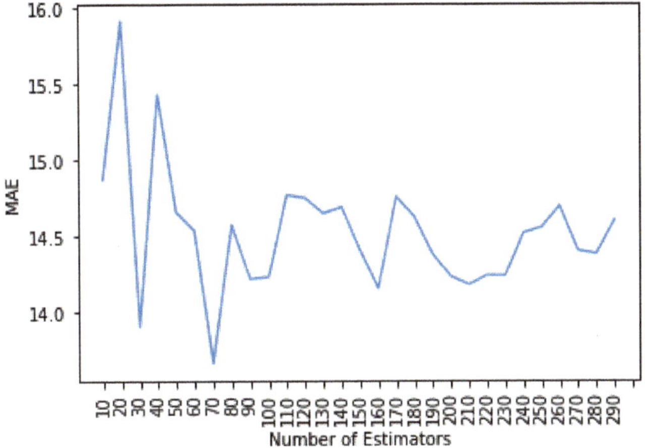

Figure 11. MAE in Relation to the Number of Estimators.

We see that the MAE fluctuates between 14.2 and 14.7 after 90 estimators. Hence, it does not make much sense to increase the number of trees over 100. The low MAE at 70 estimators is most likely due to the randomness of the trees.

The important feature of the decision trees (see Table 16) shows that the last two (i.e., year 2011 and year 2012) and the four to last year (i.e., year 2009) are the most important ones.

Table 16. Feature Importances.

([0.00225194,	0.00204321,	0.00835944,	0.00111139,	0.00299128,
0.00415498,	0.00459426,	0.00433323,	0.01164195,	0.00594237,
0.00293781,	0.0748592,	0.09929031,	0.04211062,	0.09771488,
0.13891466,	0.08747522,	0.12738761,	0.28188564])	

This result is expected, as we are working with a time series and the number of commuters of the next year is mostly influenced by the most recent data.

5. Discussion

Although this work focuses on the Germany case, we believe the methodology proposed in this paper can be extended for studying commuting behaviors in other countries, as most countries have published their per-city level employment, income, GDP and commuter information online, and there are abundant other sources like LinkedIn and Facebook as well as real estate market websites to gain access to further information.

Furthermore, it may be useful to include the total number of residents (rather than just socially insured employees) in the analysis, which covers the whole commuter population such as students, who may contribute to the peak hour congestion. Furthermore, more studies on commuting distances may be also useful to understand the commuting behavior from cost–benefit tradeoffs.

Additionally, the social, educational and medical facilities could be considered as potential additional factors. Including data like the number of hospitals, doctors or kindergartens, or even green areas and points of interest may be helpful for better under- standing commuter decisions and for the prediction of commuters. Furthermore, with the increase in housing prices over the last years, we think that it could be interesting to perform an in-depth analysis of the connection between the real estate market and commuters. We only had the house price data for one year, so looking at other historic data sources may reveal new information.

Our commuter prediction is currently only based on our time-series commuter data during 1994–2013, which can be extended for later (2014–2018) data which contain richer information such as housing prices, GDP and jobs in different sectors in each county or city. The results are still yet to be improved by future fine-tuning of the models and feature engineering, and subject to further analysis on how individual factors affect the performance of commuter predictability. Nonetheless, our initial results show that even with simple methods a reasonably good prediction can be achieved. This will bring value as it helps the city and infrastructure planners to better understand the commuting trend and deploy better countermeasures, e.g., for clogged roads or traffic jams in a short term, or developing alternative mobility options other than cars in a longer term.

Lastly, the current COVID-19 pandemic may significantly change commuting behavior. This may open a large body of new insights for future exploitation.

6. Conclusions

The question of what leads to commuting is a critical issue for modern society's development. Most prior studies focused on a small set of factors constrained by limited scale in terms of timespans, space and commuter numbers. To fill this gap, in this paper, we explored a big data approach, by collecting data from multiple publicly accessible sources and performing a systematic analysis on the potential influencing factors from four perspectives (the cities' economic structure, labor and real estate markets as well as commuting patterns). We found that the GDP, the median income and the price of buying or renting an apartment or a house in potential places for work and residence, as well as their distance to the next metropolitan area, are key factors in the decision to commute. We showed these main driving factors behind commuting in our data, confirming some findings in previous work and offering some new insights such as GDP, detailed categories of housing prices and job market in different sectors with the aid of much richer data sources. We hope that such a data-driven approach will open this field of study to more coverage in the future, as commuting is an important part of daily life in Germany (and worldwide).

Additionally, we leveraged several machine learning models to predict the number of commuters. Our results show it is possible to forecast the commuters quite precisely.

Author Contributions: Conceptualization, H.C. and X.F.; methodology, X.F. and H.C.; data collection and processing, S.V.; writing—original draft preparation, S.V. and H.C.; writing—review and editing, H.C. and X.F.; data curation, X.F.; supervision, H.C. and X.F. All authors have read and agreed to the published version of the manuscript.

Funding: This research was funded by the Fundamental Research Funds for the Central Universities (Grant No. 2020JJ014, YY19SSK05), and the European Union's Horizon 2020 research and innovation program under the Marie Sklodowska-Curie grant agreement No. 824019.

Institutional Review Board Statement: Not applicable.

Informed Consent Statement: Not applicable.

Data Availability Statement: All datasets related to this paper can be obtained in the URLs specified in references [7,25–27,32].

Conflicts of Interest: The authors declare no conflict of interest.

References

1. Hamedmoghadam, H.; Jalili, M.; Vu, H.L.; Stone, L. Percolation of heterogeneous flows uncovers the bottlenecks of infrastructure networks. *Nat. Commun.* **2021**, *12*, 1254. [CrossRef]
2. DGB. Mobilität in der Arbeitswelt: Immer Mehr Pendler, Immer Größere Distanzen. *Arb. Aktuell.* 2016. Available online: https://www.dgb.de/themen/++co++2dee53d6-d19c-11e5-9018-52540023ef1a (accessed on 11 April 2021).
3. Borck, R.; Wrede, M. Subsidies for intracity and intercity commuting. *J. Urban. Econ.* **2009**, *66*, 25–32. [CrossRef]
4. Lee, C. Metropolitan sprawl measurement and its impacts on commuting trips and road emissions. *Transp. Res. Part. D Transp. Environ.* **2020**, *82*, 102329. [CrossRef]
5. Takayama, Y.; Ikeda, K.; Thisse, J.-F. Stability and sustainability of urban systems under commuting and transportation costs. *Reg. Sci. Urban. Econ.* **2020**, *84*, 103553. [CrossRef]
6. Pendeln in Deutschland: 68% nutzen Auto für Arbeitsweg. Available online: https://www.destatis.de/DE/Themen/Arbeit/Arbeitsmarkt/Erwerbstaetigkeit/Tabellen/pendler1.html?nn=206552/ (accessed on 11 April 2021).
7. Bundesagentur für Arbeit. Available online: https://statistik.arbeitsagentur.de/ (accessed on 11 April 2021).
8. Dargay, J.M.; Clark, J. The determinants of long-distance travel in Great Britain. *Transp. Res. Part. A* **2012**, *46*, 576–587. [CrossRef]
9. Kalter, F. Pendeln statt Migration? *Z. Soziologie* **1994**, *23*, 460–476. [CrossRef]
10. Ding, N.; Bagchi-Sen, S. An Analysis of Commuting Distance and Job Accessibility for Residents in a U.S. Legacy City. *Ann. Am. Assoc. Geogr.* **2019**, *109*, 1560–1582. [CrossRef]
11. Dauth, W.; Haller, P. Is there loss aversion in the trade-off between wages and commuting distances? *Reg. Sci. Urban. Econ.* **2020**, *83*, 103527. [CrossRef]
12. Clark, W.A.V.; Burt, J.E. The impact of workplace on residential relocation. *Ann. Assoc. Am. Geogr.* **1980**, *70*, 59–66. [CrossRef]
13. Eckey, H.-F.; Kosfeld, R.; Türck, M. Pendelbereitschaft von Arbeitnehmern in Deutschland. *Raumforsch. Raumordn.* **2007**, *65*, 5–14. [CrossRef]
14. Haas, A.; Hamann, S.; Pendeln—Ein zunehmender Trend, vor allem bei Hochqualifizierten: Ost-West-vergleich. IAB-Kurzbericht. 2008. Available online: http://hdl.handle.net/10419/158268 (accessed on 31 March 2021).
15. Andersson, M.; Lavesson, N.; Niedomysl, T. Rural to urban long-distance commuting in Sweden: Trends, characteristics and pathways. *J. Rural. Stud.* **2018**, *59*, 67–77. [CrossRef]
16. Simpson, W. Workplace Location, Residential Location, and Urban Commuting. *Urban. Stud.* **1987**, *24*, 119–128. [CrossRef]
17. Levinson, D.M. Job and housing tenure and the journey to work. *Ann. Reg. Sci.* **1997**, *31*, 451–471. [CrossRef]
18. Levinson, D.M. Accessibility and the journey to work. *J. Transp. Geogr.* **1998**, *6*, 11–21. [CrossRef]
19. Huinink, J.; Feldhaus, M. Fertilität und Pendelmobilität in Deutschland. *Z. Bevölkerungswissenschaft* **2012**, *37*, 463–490.
20. Chidambaram, B.; Scheiner, J. Understanding relative commuting within dual-earner couples in Germany. *Transp. Res. Part. A Policy Pr.* **2020**, *134*, 113–129. [CrossRef]
21. Reuschke, R. Job-induced commuting between two residences—characteristics of a multilocational living arrangement in the late modernity. *Comp. Popul. Stud.* **2010**, *35*, 107–134.
22. Mitra, S.K.; Saphores, J.-D.M. Why do they live so far from work? Determinants of long-distance commuting in California. *J. Transp. Geogr.* **2019**, *80*, 102489. [CrossRef]
23. Clark, W.A.; Huang, Y.; Withers, S. Does commuting distance matter? Commuting tolerance and residential change. *Reg. Sci. Urban. Econ.* **2003**, *33*, 199–221. [CrossRef]
24. Dickerson, A.; Hole, A.R.; Munford, L.A. The relationship between well-being and commuting revisited: Does the choice of methodology matter? *Reg. Sci. Urban. Econ.* **2014**, *49*, 321–329. [CrossRef]
25. Immobilienscout24 API. Available online: https://api.immobilienscout24.de/ (accessed on 11 April 2021).
26. Google Maps API. Available online: https://developers.google.com/maps/documentation/?hl=de (accessed on 11 April 2021).
27. Govdata. Available online: https://www.govdata.de (accessed on 11 April 2021).
28. Sermons, M.; Koppelman, F.S. Representing the differences between female and male commute behavior in residential location choice models. *J. Transp. Geogr.* **2001**, *9*, 101–110. [CrossRef]
29. White, M.J. Sex differences in urban commuting patterns. *Am. Econ. Rev.* **1986**, *76*, 368–372.
30. Geib, T.; Lechner, M.; Pfeiffer, F.; Salomon, S.; Die Struktur der Einkommensunterschiede in Ost-und Westdeutschland ein Jahr nach der Vereinigung. ZEW Discuss. *Pap.* 1992. Available online: http://hdl.handle.net/10419/29430 (accessed on 31 March 2021).
31. Fendel, T. Migration and Regional Wage Disparities in Germany. *Jahrbücher Natl. Stat.* **2016**, *236*, 3–35. [CrossRef]
32. Federal Statistics Office (Statistisches Bundesamt). Available online: https://www.genesis.destatis.de/ (accessed on 11 April 2021).
33. MacKinnon, J.G.; White, H. Some Heteroskedasticity Consistent Covariance Matrix Estimators with Improved Finite Sample Properties. *J. Econom.* **1985**, *29*, 305–325. [CrossRef]
34. Boje, A.; Ott, I.; Stiller, S.; Entwicklungsperspektiven für die Stadt Hamburg: Migration, Pendeln und Spezialisierung. HWWI Policy Pap. 2010. Available online: https://econpapers.repec.org/RePEc:zbw:hwwipp:124 (accessed on 31 March 2021).

35. Kholodilin, K.A.; Mense, A. Wohnungspreise und Mieten steigen 2013 in vielen deutschen Großstädten weiter. *DIW Wochenber.* **2012**, *79*, 3–13.
36. Schulze, S.; Einige Beobachtungen zum Pendlerverhalten in Deutschland. HWWI Policy Pap. 2009. Available online: https://econpapers.repec.org/RePEc:zbw:hwwipp:119 (accessed on 31 March 2021).

 *sustainability*

Review

Exploring the Role of Digital Infrastructure Asset Management Tools for Resilient Linear Infrastructure Outcomes in Cities and Towns: A Systematic Literature Review

Savindi Caldera *, Sherif Mostafa, Cheryl Desha and Sherif Mohamed

Cities Research Institute, Griffith University, 170 Kessels Rd, Nathan, QLD 4111, Australia;
sherif.mostafa@griffith.edu.au (S.M.); c.desha@griffith.edu.au (C.D.); s.mohamed@griffith.edu.au (S.M.)
* Correspondence: s.caldera@griffith.edu.au

Citation: Caldera, S.; Mostafa, S.; Desha, C.; Mohamed, S. Exploring the Role of Digital Infrastructure Asset Management Tools for Resilient Linear Infrastructure Outcomes in Cities and Towns: A Systematic Literature Review. *Sustainability* 2021, 13, 11965. https://doi.org/10.3390/su132111965

Academic Editor: Tan Yigitcanlar

Received: 27 September 2021
Accepted: 26 October 2021
Published: 29 October 2021

Publisher's Note: MDPI stays neutral with regard to jurisdictional claims in published maps and institutional affiliations.

Abstract: Linear infrastructure such as roads, railways, bridges and tunnels enable critical functionality within and between metropolitan and regional cities and towns, facilitating the movement of goods and services, as part of vibrant, thriving economies. However, these asset types are typically challenged by costly asset management schedules and continually eroding maintenance and refurbishment budgets. These challenges are compounded by the increasing frequency and intensity of disruptive events such as fire, floods, and storm-surge that can damage or destroy property. The United Nations Sustainable Development Goal 9 (SDG-9) highlights the urgent need for enabling evidence-based decision making for infrastructure asset management (IAM). Around the world, digital engineering (DE) efforts are underway to streamline the capture, processing, and visualization of data for IAM information requirements, towards timely and evidence-based decision support that enables resilient infrastructure outcomes. However, there is still limited understanding about which IAM information can be digitized and the types of tools that can be used. This study sought to address this knowledge gap, through reviewing the extent of available and emerging linear infrastructure related DE technologies and their IAM information requirements. A systematic literature review elicited 101 relevant conceptual and empirical papers, which were subsequently evaluated with regard to the extent and characteristics of digital infrastructure asset management tools. Findings are discussed using three themes that emerged from the analysis: (1) DE tools and their IAM asset information requirements; (2) Interoperability and integration of DE tools across IAM platforms; and (3) Application of DE tools to enable resilient linear infrastructure outcomes. A 'Digital Technology Integration Matrix' is presented as an immediately useful summary for government and industry decision-makers, particularly in the field of disaster management preparedness and recovery. The Matrix communicates the synthesis of tools and likely end-users, to support effective data gathering and processing towards more timely and cost-effective infrastructure asset management. The authors conclude with a research roadmap for academics, including recommendations for future investigation.

Keywords: digital engineering; information requirements; infrastructure asset management; technology integration matrix

1. Introduction

There are urgent calls for improving asset management processes within the infrastructure asset management (IAM) sector, towards an improved resilience of infrastructure that services our metropolitan and regional cities and towns [1,2]. Such evidence-based decision-making is crucial to ensuring adequate long-term funding in often budget-constrained operating environments, particularly for continuous or 'linear' assets (for example, part of a road or rail network) that form the core connecting structures of our urban environments [3]. The length and complex nature of these assets are challenged by costly asset management schedules and continually eroding maintenance and refurbishment budgets, compounded by the increasing frequency and intensity of climate-related disruptive events that damage or destroy property [4].

IAM is described as *"the integrated, multidisciplinary set of strategies for sustaining physical assets, such as roads, dams, bridges, railways, manufacturing plants and pipelines"* [5]. Looking beyond typical planning design construction asset lifecycle interfaces, there are urgent calls to develop asset management models that facilitate continuous flows of information from design and construction through to asset operation, maintenance, and end-of-life repurposing or disassembly [6,7]. These traditional asset management processes are largely driven by user-led documentation through site visits, maintenance checks, and asset audits, which can be time consuming, labor intensive, and prone to human errors. The data schemas used by current asset management tools and platforms are also highly variable in terms of asset location referencing and asset hierarchy systems. Furthermore, traditional asset management systems tend to focus on discrete or individual asset management phases and tools, lacking information integration across all assets and life cycle phases [8].

The IAM sector is also experiencing a rapid emergence of technology-enabled design and practice, which is increasingly referred to as Digital Engineering (DE) [9]. The Australian Government defines DE as *"the convergence of emerging technologies such as Building Information Modelling (BIM), Geographic Information Systems (GIS) and related systems to derive better business, project and asset management outcomes"* [10]. At a sub-national level, the New South Wales Government [11] has published a DE framework to distinguish key elements, including technologies, digital twin, ways of thinking, procurement, skills, and resourcing, as shown in Figure 1.

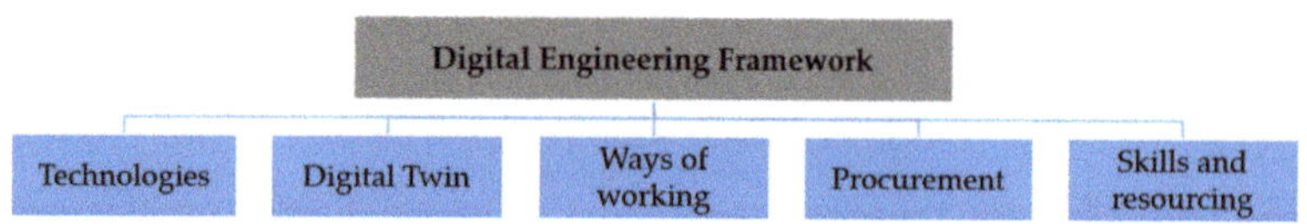

Figure 1. Digital engineering framework. Source: Adapted from [11].

These elements span the fields of Geographic Information Systems (GIS), Building Information Modelling (BIM), Civil Information Modelling (CIM), Bridge Information Modelling (BRIM) and others [12,13]. Within each element, there are a number of Asset Information Requirements (AIR) i.e., *"the information required to operate and maintain a built asset in line with an organisation's asset management strategy"* [14]. According to the standard Publicly Available Specifications (PAS) 1192-3 asset management standard, these AIRs logically exist within four categories as summarized in Table 1 [14].

Table 1. Information requirement categories and types.

Category	Information Requirement Types
Managerial	Type of asset, condition, location, warranties, maintenance plans, end of life processes, location
Technical	Engineering data, design parameters, operational data, interdependencies
Financial	original cost, operating costs, maintenance cost
Legal	ownership, maintenance demarcation, work instructions, risk assessments and control measures)

Within this IAM context, DE is considered an unprecedented opportunity to support decision-making, addressing project delivery challenges; information traceability and accountability through the lifecycle of assets, and allowing for faster delivery with resilient outcomes [10,15,16]. However, there is still widespread uncertainty and limited guidance about the rationale for digitizing information, and the types of tools that can be used to digitalize asset information requirements [17].

Responding to this knowledge gap, the authors asked: Amongst the spectrum of DE tools available, what AIRs could be digitalized to improve asset management decision making for resilient linear infrastructure outcomes? This question comprised two sub-questions: (1) what DE technologies exist to digitalize asset information requirements? and (2) what can we learn about priority AIRs from DE technology applications in linear infrastructure asset management? The authors subsequently undertook a systematic literature review (SLR) to elicit the spectrum of existing and emergent DE technologies and useful AIRs.

In the following sections, we summarize the methodology and key findings from the 101 conceptual and empirical papers in the SLR. We discuss the findings using the emergent key themes, and we present a matrix of the DE tools and corresponding AIRs that can be digitised to support practitioners and authorities in enabling resilient infrastructure outcomes. The paper concludes with the next steps in engaging with DE towards improved infrastructure asset management outcomes.

2. Research Methodology

This paper adopted a systematic literature review (SLR) which is a comprehensive and reproducible approach to synthesise the existing literature and contribute to the advancement of knowledge [18,19]. The SLR can combine existing evidence and create new knowledge. It has been recognised as a critical scientific approach to bridge the research-practice gap [19–21]. The research comprised a review of papers discussing DE that were published in reputable academic journals from 1998 to 2020. This period was selected to deeply reflect on relevant information related to DE research conducted over the last two decades. Tasks were undertaken in three key stages: (1) planning the review approach and identifying the relevant literature, (2) screening the literature resources using inclusion and exclusion criteria, and (3) descriptive and thematic analysis involving the extraction, synthesis, and documentation of the review. This approach to creating an evidence-based literature review has been established in similar research areas, including asset management [22] and digital engineering [7,23]. Table 2 presents an overview of the review protocol.

The authors acknowledge that the findings are conditioned to the chosen literature sampling criteria (e.g., search keywords, with specific inclusion and exclusion criteria, and excluding non-English research papers). Subsequently, we consider the 101 articles a thorough exploration, but not necessarily exhaustive due to these listed limitations.

Firstly, the purpose of the literature review was clearly defined, and the aims and objectives were developed to align with the overall purpose. The review protocol was created with all necessary review steps and details including time frame, databases, key search terms, and inclusion and exclusion criteria. Databases including ScienceDirect, Web-of-Science, Scopus, EBSCohost, and Google Scholar were searched within the timeframe of 1998–2020. An extensive range of search terms including, "asset management" with "infrastructure", "digital engineering", "information", "life cycle", "road", "rail", "tunnel", and "bridge" were used to develop the search strings, to search related full text, peer-reviewed journal articles [24]. In the process of reviewing articles, other cited articles were added (i.e., snowball sampling). Boolean connectors (AND, OR and NOT) were used in conjunction with the keywords to create additional search strings. To make this process more efficient, online resources extracted through online databases were used for the review [25].

Table 2. Overview of the review protocol (PRISMA checklist, reprinted with permission [19]).

Section/Topic	Checklist Item
Title	Exploring the role of digital infrastructure asset management tools for resilient linear infrastructure outcomes in cities and towns: A systematic literature review
Research questions	What AIRs could be digitalized to improve asset management decision making for resilient linear infrastructure outcomes?
Key word search	"Asset management" with "infrastructure", "digital engineering", "information", "life cycle", "road", "rail", "tunnel", and "bridge
Search protocol	An extensive range of search terms including, "asset management" with "infrastructure", "digital engineering", "information", "life cycle", "road", "rail", "tunnel", and "bridge" were used to develop the search strings, to search related full text, peer-reviewed journal articles were used to develop the search strings
Search strategy and selection	Title, year, keywords, abstract
Electronic database	ScienceDirect, Web-of-Science, Scopus, EBSCohost, and Google Scholar
Inclusion and exclusion criteria	Inclusion criteria—Full-text, peer-reviewed academic journal articles, from year 1998–2020 Exclusion criteria: Conference papers, dissertations, Book reviews, non-English publications and grey literature, peer-reviewed journal papers where a full text version was not available

Inclusion and exclusion criteria were established using the C-I-M-O (context-intervention-mechanism-outcome) framework [26]. These criteria guided the research team to deliberately select the relevant articles [27]. In selecting relevant articles, backward and forward reviews were carried out to capture an extensive range of relevant literature. The title and then the abstract were reviewed to ensure the articles were relevant to the study scope. After the initial metasearch, 910 articles were identified. Then, all duplicated articles were removed, and papers only aligned with linear infrastructure were stored. Of the total 175 articles discovered, 168 were assessed for eligibility and 101 articles met the inclusion and exclusion criteria of this study, as illustrated in Figure 2. The authors were guided by the PRISMA statement [19] and have incorporated the steps of inclusion, eligibility, screening, and identification to complement the systematic literature review process [21].

Some of the initially collected publications were excluded whether they were not directly connected to the research topic (e.g., digital asset management papers that referred to digital sources instead of infrastructure assets), and where they were irrelevant (e.g., where papers referred to a name of a digital asset instead of physical infrastructure assets). Full papers were then reviewed using an excel database that the first author had previously developed for other SLR studies [28,29] to code the key information.

Descriptive and thematic analysis was used as to categorise and synthesise the distribution and patterns of the reviewed literature. The descriptive analysis describes the research context, research distribution, types of data, methods, journal outlets, and geographic distribution. The thematic analysis highlights the emergent themes in the digital infrastructure asset management landscape and the knowledge gaps [30]. The articles were coded and categorised into several themes using the NVivo software. Two coders were involved to ensure internal validity through inter-coder reliability. After coding emergent themes, the coded outcomes were synthesised into summary tables on infrastructure asset information.

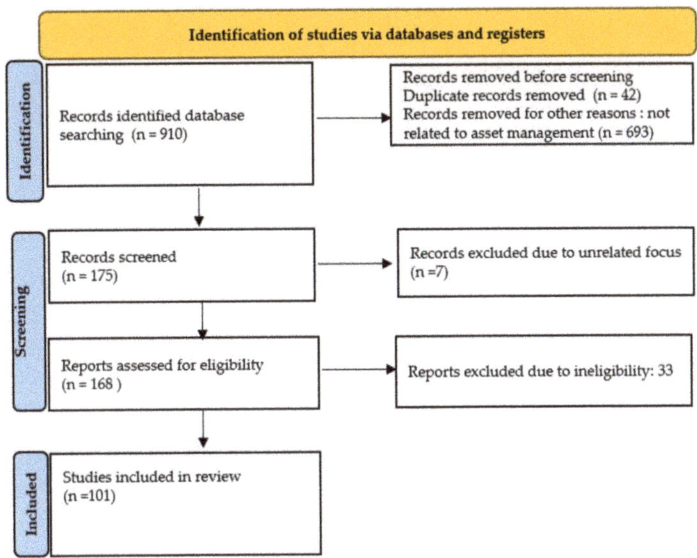

Figure 2. Flow of information through the different phases of a systematic review using PRISMA statement (Reprinted with permission from [19]).

3. Descriptive Analysis and Discussion

The descriptive analysis describes the research context, research distribution, types of data, methods, journal outlets and geographic distribution. The authors consider the researcher journey in exploring DE, through publication and authorship data. We then evaluate the typology of the publications according to focus area.

3.1. Publication and Authorship Data

Figure 3 presents the number of publications chronologically over the last two decades, highlighting this research field's comparatively novel nature, with an emerging narrative about this concept. From 2010, there was a marked increase in the number of papers annually, which could be attributed to increasing attention on digital technologies and its role in infrastructure asset management. There are then two time periods where publications suddenly declined (2015, 2019–2020). Such patterns could be due to a number of reasons, including initial focus on structure geometry and semantics and the shift in focus after 2015 to BIM and big data which needed more time for in-depth investigations, and subsequently COVID-19-related research and publication challenges in 2019–2020. It is also possible that near-ready papers from 2014–2015 were subsequently published on top of the existing publication rate.

Figure 4 indicates the geographic distribution and number of articles organised by the first authors. By Nation, researchers have been active in the United States of America. Regionally, most research was recorded in Europe, followed by Australasia.

There were 34 research papers about infrastructure assets in general (including two or more liner assets). A summary of research papers by asset types is provided in Figure 5, noting that these infrastructure types are not mutually exclusive (some publications focused on more than one type of infrastructure).

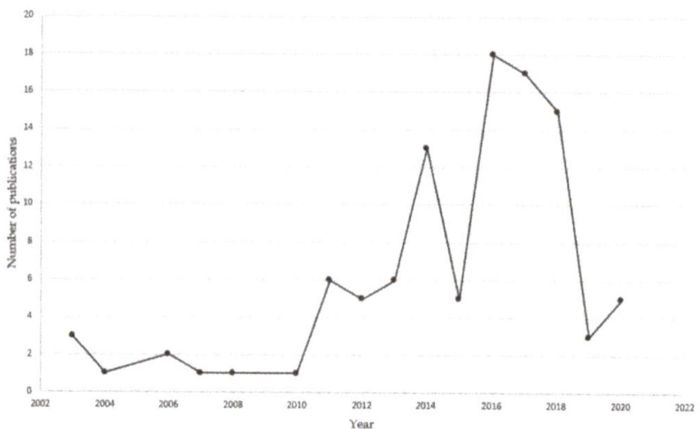

Figure 3. Publication distribution over the period of 1998 to 2020.

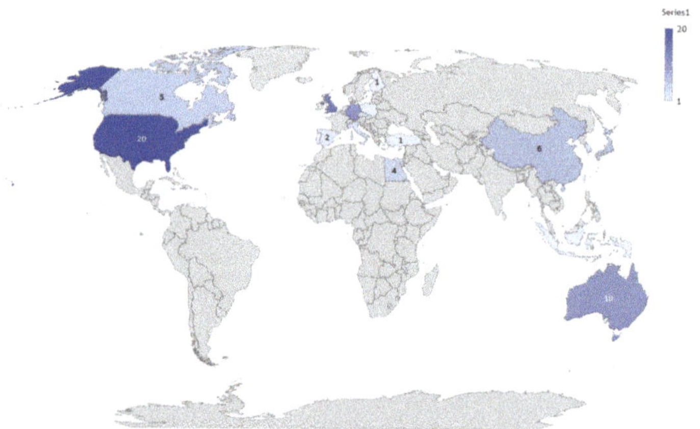

Figure 4. Geographical distribution of digital infrastructure asset management research (1998–2020).

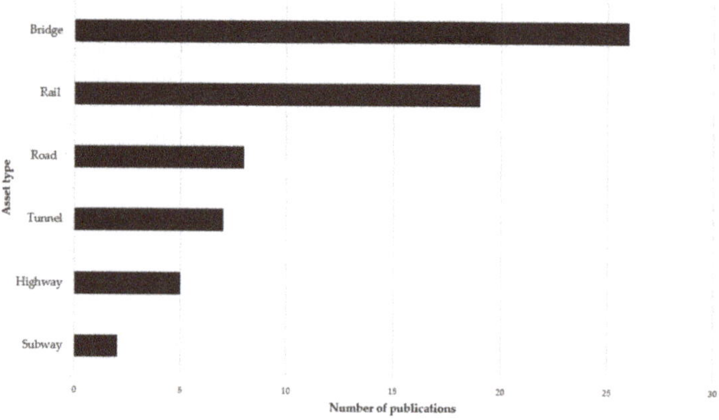

Figure 5. Types of assets used as focus/examples in the reviewed publications (1998–2020).

With regard to the sector of focus, most articles were from the building and engineering (39 per cent) construction management (28 per cent), and then transport, technology, and computer science disciplines. More than half of the publications were journal articles (61 per cent).

Of the total, 61 per cent of papers were from journal publications, including a significant proportion from the field of Construction. This indicates suitable outlets to publish digital infrastructure asset management-related articles and shows that the *Automation in Construction* journal has significantly contributed to this field of digital infrastructure asset management research. Papers were also published in fields such as economics ($n = 1$), information management/information systems ($n = 1$), manufacturing engineering ($n = 1$), project management ($n = 1$), and social sciences ($n = 1$); 39 per cent were conference papers representing conferences such as the International Conference on Computing in Civil and Building Engineering and the International Conference on Computing in Civil and Building Engineering.

3.2. Publication Typology (Approaches, Applications, and Models)

Figure 6 illustrates the key literature reviewed as a mind map, categorised according to approaches, applications, and models. The codes and the details of the key literature are provided in the Supplementary File.

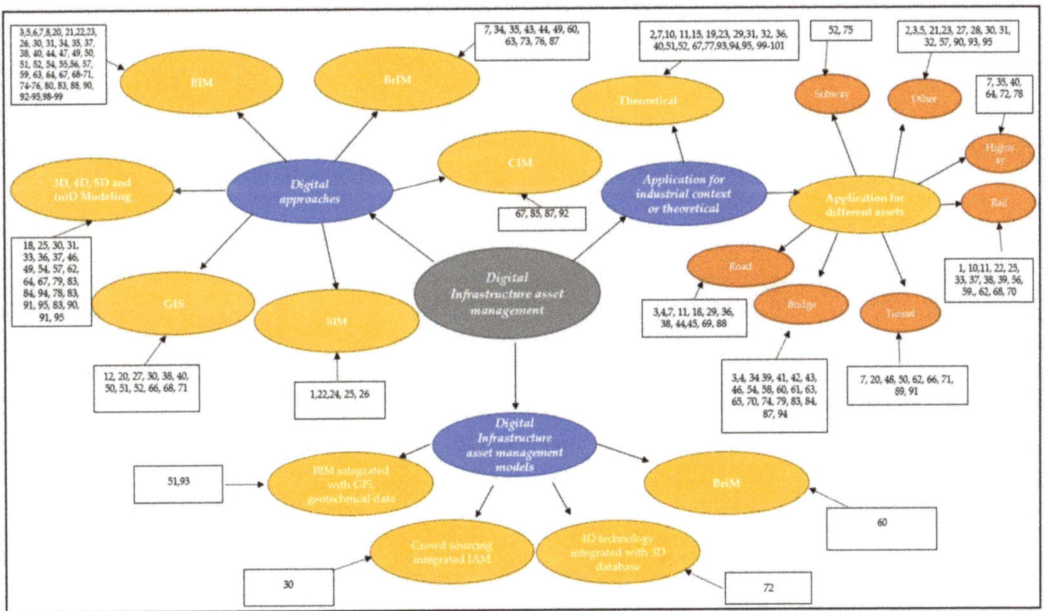

Figure 6. Themes associations within the digital infrastructure asset management domain (serial numbers are presented in Supplementary Materials).

4. Thematic Analysis and Discussion

In the following paragraphs, the thematic findings of the structured literature review are discussed under three key themes: (1) DE tools and their AIRs; (2) Interoperability and integration of DE tools across IAM platforms; and (3) Application of DE tools to enable resilient linear infrastructure outcomes.

4.1. DE Tools and Their AIRs

A variety of DE technologies have evolved over the past 20 years, determining potential integrative digital approaches in the transport infrastructure management domain [23].

These include Building information modelling (BIM), Geographic Information System (GIS), Computer Aid Designs (CAD), Civil/Construction Information Modelling (CIM), Bridge Information Modelling, 3D, 4D, and 5D modelling design, and other techniques [7]. The authors then grouped this range of tools into four themes, namely: data management, data sensing, data modelling/visualisation, and data monitoring inspired by categorisation of [7,31]. The resultant detailed list is attached in Appendix A.

Considering these tools/processes and the variety of information requirements, we synthesized the findings into a *'Digital Technology Information Matrix'* (Table 3) to support practitioners and decision-makers in choosing appropriate DE tools and processes for digitizing asset information requirements. Within the matrix, the four themes are the same as those used in the detailed mapping described above. The AIR categories are those used in the publicly available standard *PAS 1192-3 Asset Management Standard*.

The Matrix highlights that there are substantially more digital tools for data management, modelling, and visualisation, than tools for monitoring and sensing. The themes are discussed in more detail in the following paragraphs, with reference to the literature from the review.

4.1.1. Data Management, Modelling/Visualization

Data management is critical for capturing essential information requirements and enables better facility management practices at the operation phase of a project. The Internet of Things (IoT) and Artificial Intelligence (AI) were often referred to in the literature as helpful interventions to manage complex asset data. Eight tools were elicited from the literature that focused on managing data and modelling/visualising through digital platforms. These included as-built drawings and models, Asset Information Models (AIM), Building Information Modelling (BIM), Bridge Information Modeling (BrIM), Bridge Management System (BMS), Civil Integrated Management (CIM), Computer-aided design (CAD), and eRIM (electronic Requirements Information Management). Industry Foundation Classes (IFC) data formats were also often highlighted due to their features related to platform-neutral, open file format specifications that are not controlled by a single vendor or group of vendors [32].

For example, Asset Information Models (AIM) capture the data and information necessary to support asset management while offering graphical and non-graphical data and information. An AIM can be formed from existing asset information systems, from new information, or from information in a Project Information Models [33]. Previous research provides evidence in using Building information models in transport infrastructure to improve constructor business processes and the effective governance and value of information [6]. Zak and Macadam [34] add to this by providing evidence for using BIM in several virtual design and construction practices using the experiences with new technologies gained from the application of BIM related workflows. Successful implementation of BIM on bridge projects requires four key steps: (1) selecting high value BIM uses; (2) creating the BIM execution process; (3) evaluating the BIM deliverables; and (4) forming the infrastructure [35]. Blanco and Chen [36] provide evidence of using Building Information Modelling in the United Kingdom by the Transport Industry to show how this working approach achieves cost savings and environmental benefits.

Table 3. Digital Technology integration matrix (DTIM) for improved linear IAM.

Digital Engineering Tools/Processes (In Alphabetical Order)	*Asset Information Requirement Type*	Managerial (i.e.,: Type of Asset, Condition, Location, Warranties, Maintenance Plans, End of Life Processes, Location)	Technical (i.e.,: Engineering Data, Design Parameters, Operational Data, Interdependencies)	Financial (i.e.,: Original Cost, Operating Costs, Maintenance Cost)	Legal (i.e.,: Ownership, Maintenance Demarcation, Work Instructions, Risk Assessments and Control Measures)
	As-built drawings and models	●	●		
	AIM (Asset Integrity Management)	●	●	●	
	Building Information Modeling (BIM)	●	●	●	●
	Bridge Information Modeling (BrIM)	●	●	●	●
	BMS (Bridge Management System)	●	●	●	
Data management, modelling/visualisation	CiM (Construction/Civil Information Modelling)	●	●	●	●
	Computer-aided design (CAD)	●	●		
	Digital twin	●	●		
	eRIM (electronic Requirements Information Management)	●			
	Parametric Modelling		●		
	Multiscale Modelling		●		
	Multidimensional (nD) Modelling	●	●	●	●
	Geometrical Modelling		●		

Table 3. *Cont.*

Digital Engineering Tools/Processes (In Alphabetical Order)	*Asset Information Requirement Type*	Managerial (i.e.,: Type of Asset, Condition, Location, Warranties, Maintenance Plans, End of Life Processes, Location)	Technical (i.e.,: Engineering Data, Design Parameters, Operational Data, Interdependencies)	Financial (i.e.,: Original Cost, Operating Costs, Maintenance Cost)	Legal (i.e.,: Ownership, Maintenance Demarcation, Work Instructions, Risk Assessments and Control Measures)
	3D Modelling	•	•	•	
	4D Modelling	•	•	•	
	Virtual Prototyping Simulation (VPS)		•		
Sensing	Global Positioning System (GPS)		•		
	Geographic Information Systems (GIS)		•		
Monitoring	SHM (Structural Health Monitoring)	•	•		
	Petri-net model		•		
	Artificial neural networks	•	•		

Civil Integrated Management (CIM) (inspired by BIM for construction) was defined as a system which facilitates the collection, organisation, and managed accessibility to accurate data and information related to an infrastructure asset. This model quantified parameters such as errors, omissions, and information redundancy [7] to facilitate data integration throughout the asset life cycle. It is a commonly used approach for integration for data and construction activities and advances visualisation. The quality of as-built documents for electrical systems was quantified using a system information model.

While modelling and visualisation are considered as one of the most comprehensible forms of digital engineering, it is critical to understand the role of the computable data behind the model as well. Sankaran et al. [37] provide evidence for using CIM for 3D design and for terrain modelling and advance visualisation of structures. The implementation of CIM is influenced by factors including contract specifications, project delivery, and budget. Sankaran et al. [38] added to this conversation by assessing the modelling practices in large infrastructure projects, presenting key challenges and opportunities for integrating CIM for design and construction initiatives. Challenges to such outcomes include insufficient data collection techniques, limited expertise and competencies of the designers, and data incompatibility. The benefits of CIM was established through research by Yabuki et al. [39] including visualisation, automated clash detection and quantity take-off.

The concept of a 'digital twin' has been rapidly gaining popularity as a full digital representation of a physical asset, process, or system, as well as the engineering information that allows us to understand and model its performance [23]. It is a vital component of the DE framework, as shown in Figure 1. Typically, a digital twin can be continuously updated from multiple sources demonstrating high potential for asset lifecycle management [40]. The main advantages of digital approaches and its application for asset management and logistics were highlighted as reducing cost, delivery time, increased reliability, and flexibility [23]. Some research has pointed out the limitations of digital technologies such as computer-aided designs in terms of efficacy, cost, and resultant errors [41].

4.1.2. Data Sensing

Data sensing includes digital technologies such as GIS and GPS that leverages spatial capabilities. Fraga-Lamas et al. [42] presented a holistic approach to leverage the Industrial Internet of Things (IoT) for railway management building on previous research on predictive maintenance, smart infrastructure, advanced monitoring, and freight information systems the proposed approach. The web-based GIS was used for land and building management for data management while maintaining data integrity [43].

These approaches are critical in moving forward towards future proofing assets and networks, enabling the Architecture, Engineering, Construction, and Operations (AECO) sector to provide a platform for researchers and policymakers and practitioners to advance their knowledge and capabilities [2]. The use of sensors, mobiles, open data, and laser technology has received much attention from researchers worldwide. Sánchez-Rodríguez et al. [44] showed that laser scanning technologies along with tailored processing tolls offer data for structural functions yielding significant results. By using sensor technology, it was claimed that some observed data from the design and construction phases could inform asset register primary stages. Big data analytics have proven effective decision-making in highway infrastructure [45].

4.1.3. Data Monitoring

Through monitoring assets throughout their life cycle, asset managers can systematically follow-up tasks and capture data during changes to successfully deliver the project. For example, structural health monitoring provides a diagnosis of the structure state during the life of a structure [46,47]. Petri-net models enable different data calibration techniques and different data sources and can interact with one another to model element deterioration, inspection, and maintenance [48]. An artificial neural network (ANN) can be used to rapidly determine the fatigue life remaining at the site. Although identifying crack patterns that may arise in future, the ANN is used to assure a robust and reliable artificial model [49].

4.2. Interoperability and Integration of DE Tools across Platforms

While there is a range of DE tools available for digitalising asset information requirements, it is critical to ensure easy transfer of data files across different platforms [33]. This process needs to be efficient and to reduce the additional attempts to convert and re-convert files from across platforms. This phenomenon is described as interpretability which is "a characteristic of a product or system, whose interfaces are completely understood, to work with other products or systems, at present or in the future, in either implementation or access, without any restrictions" [33]. However, this has become a challenging situation for many commonly used electronic/digital processes.

4.2.1. Overcoming Task Duplication

In order to overcome the duplication of tasks and converting data from one format to another, improving interoperability has emerged as a targeted approach to enable the realisation of long-term strategic objectives, amelioration of data integration, augmented knowledge management, and enhanced performance measurement. This could be done through enriched training and competence development for facility managers and asset managers to better manage with the ad hoc, variable range of services [50]. More than 15 years ago, Halfawy et al. [51] defined requirements for standard data models. Through this research, the criticality of interoperability from an asset management perspective was re-iterated. In addition, the contribution of geographic information systems for efficient management of life cycle data was highlighted [52].

Around the same time, Shirole et al. [53] emphasised the complex nature of bridge data at the project level, and this situation has been aggravated by the increased number of "stove-piped" software applications augmented with a variety of accompanying file formats. This has proliferated over the years without proper consideration of functional interoperability (Shirole, Chen, and Puckett) [53]. It was suggested potential opportunities for means of leveraging bridge data from the design stage in contrast to the traditional approach to enhance the project viability of integrated, project delivery, and effective life-cycle management via a prototype integrated system. This system aims to exchange data and applications throughout the bridge life cycle [53]. These examples re-iterate the importance of the interoperability function of DE tools and data for easy access of users.

4.2.2. Taking Advantage of Multiple Systems

The reviewed literature contained a number of papers discussing hybrid systems that can help to achieve more effective outcomes when used as integrated approaches to asset management. Table 4 summarises five key integrated digital asset management models which demonstrate a variety of applications to improve communications, collaboration, visualisation, and data integration. These integrated tools support decision-making through addressing project delivery challenges, information traceability and accountability through the lifecycle of assets, and allowing for faster delivery with better outcomes.

With the increasing attention on social networking, crowdsourcing system for integrated IAM has the capability to cater to the requirements of multiple stakeholders through enhanced communication [1]. This system will enable the phases of management, operations and repair a series of interdependent infrastructure facilities. This show evidence of a system to enhance communication transparency and effectiveness of asset management, while reducing the risks in repair and maintenance phases.

Kurwi et al. [54] highlight the opportunity of using BIM with GIS as a potential coupled approach to enhance collaboration for better decision among key stakeholders. Furthermore, collaborative forms of procurement are recommended, along with the use of Building Information Modelling and Systems Information Modelling. GIS has gained much popularity among asset owners and managers and highlights the prospect of the development of iSRE for geological disposal projects as a meaningful tool for repository design [55]. BIM can also be augmented with VR and AR technologies for better functionalities, including virtual walkthrough, schedule visualisation, clash detection, and as-built modeling [56].

Liapi [57] suggested 4D technology that makes use of a comprehensive 3D project database for the visualisation of construction scheduling. Integration of geotechnical data with BIM processes is expected to provide significant cost and time savings in major infrastructure projects. Tawelian and Mickovski [58] establish this with case study evidence of design and construction road embankment in Scotland. Applying a real-time, dynamic, three-dimensional building information model for alignment of data was emphasised by Huang et al. [59] using a case study of west main tracks of the Qidu Switchyard of Taiwan Railways Administration.

Table 4. Integrated tools for digital infrastructure asset management (1998–2020).

Examples of Integrated Tools	Key Feature/Application	Key Reference/s
Crowdsourcing integrated IAM	• Enhance communication transparency • Improving the effectiveness of asset management	[1]
BIM integrated with GIS and geotechnical data	• Virtual design and construction practices using the experiences with new technologies gained from the application of BIM related workflows • Combining with geotechnical data • Achieves cost savings and environmental benefits • Collaborative forms of procurement • Financial and time management of infrastructure bridges.	[54,55]
BIM integrated with Virtual reality and augmented reality	• Enhance its functionalities, including virtual walkthrough, schedule visualization, clash detection, and as-built modeling. • Asset management functions during the maintenance and operation phase of facilities	[56]
4D technology integrated with 3D database	• Improving visualization of construction scheduling • Creating photo-realistic animations that can facilitate the dissemination of traffic measures	[57]
Mobile model-Based Bridge Lifecycle Management System	• Integrates 4D bridge models with Building Management Systems • Links all the information about the lifecycle stages of a bridge • Supports distributed databases and mobile location-based computing	[60]

The Model-Based Bridge Lifecycle Management System houses all relevant information about the lifecycle stages of a bridge to a 4D model of the bridge incorporating different scales of space and time to record events throughout the lifecycle with suitable levels of details (LoDs). Furthermore, this system supports distributed databases and mobile location-based computing [60].

Implementing digital engineering models, especially in multiple DE system is costly. Therefore, public–private partnerships (PPP) have emerged as a targeted mechanism for addressing infrastructure capital investment backlogs [61]. Due to the dynamic nature due to change of scope in construction projects, the original contractual value might vary during its phases which highlight the significant need for public and private sector asset owners to implement a cost contingency approach. Another author added to this dialogue by stating the importance of adopting cost contingency approaches [62]. It was emphasised that using a probabilistic instead of a deterministic approach will increase the capacity to accommodate the scope changes and enable to achieve best outcomes through using multiple DE systems.

4.3. Application of DE Tools to Enable Resilient Linear Infrastructure Outcomes

Drawing on the literature review, the authors have identified several infrastructure types to embed digital infrastructure into transport infrastructure, comprising bridges, roads, and railways. The following sections provide details on specific systems being applied for each of these infrastructure types.

4.3.1. Bridge Infrastructure

Bridge management and monitoring using digital technologies have been researched vastly over the last two decades. Among the key research findings, Jeong, Hou, Lynch, Sohn, and Law [46] present an information modelling framework for supporting bridge monitoring applications building on previous work on the OpenBrIM standards. This framework comprises the information relevant to engineering analysis and sensor network aiming for scalability, flexibility, and performance.

While data related to bridge management is complex as it contains detailed data related to the project life cycle, there is a number of accompanying file formats that have emerged, aiming for functional interoperability, and enabling such data to be optimised throughout the life cycle [53]. Shim et al. [63] used digital mock-up, parametric model combined with 4D and 5D simulation to model the construction stage of a bridge to improve bridge construction project. Considering the whole visualisation of bridge elements and associated information, Marzouk and Hisham [64] developed a BrIM framework to store data and inspect spreadsheets based on Structured Query Language statements. This framework also could integrate BrIM with advanced analytical calculations of bridge structural conditions. To ensure the data integrity throughout the asset life cycle, Karaman et al. [65] proposed a 3D control curve which captures bridge geometrical data, exchanges associated with analysis, design, detailing for fabrication, erection, and construction. The critical need for standard data models for the life cycle of rail was further emphasised by [32]. Mawlana et al. [66] have proposed a novel approach to develop reconstruction phasing plans while calculating the relevant stochastic spatiotemporal interactions. This proposed approach consists of a constraint-based system and 4D modelling to achieve a feasible sequence to support sections to be constructed or demolished.

4.3.2. Road and Highway Infrastructure

Advance technologies such as BIM have proven capabilities to support road design, planning, and maintenance during their asset life cycle. BIM can support in managing the flow of information aligned with the asset life cycle. Brous et al. [67] highlight the criticality of organising data structures and dealing with complex data to achieve a multi-faceted appreciation for data governance. The authors evaluated the conditions and factors for effective sustainable development to model efforts of data governance on data infrastructure. To address the dynamic nature of transport networks, a comprehensive asset integrity management approach was presented by Fuggini et al. [68] to replace existing the time-based strategy with the performance-based strategy to improve service availability and reduce cost. Trojanová [69] emphasised the importance of strategic asset management for road networks in Slovakia.

4.3.3. Railway Infrastructure

Durazo-Cardenas et al. [70] proposed an integrated approach that fuses asset monitoring, planning, and scheduling to be applied for a range of scenarios including complex systems with abundant sensors with monitoring systems. This system design also has the capacity for the automatic maintenance and resource sequence factoring the accounted costs. The importance of digitalisation and the use of smart technologies for future-proofing of infrastructure assets was emphasised by Love, Zhou, Edwards, Irani, and Sing [62]. Furthermore, Whyte et al. [71] added that asset information can be managed through digital platforms using relatively hierarchical, asynchronous, and sequential processes to manage complex projects.

To address the variations of the scoping in construction projects, the improved capability of cost contingency using probabilistic rather than a deterministic approach was highlighted [12]. To achieve this goal, a collaborative approach using BIM and System information modelling is important. In addition, laser scanning technology together with targeted processing tools can provide digital data for further structural operations [44]. Yang et al. [72] presented an integrated framework addressing diverse aspects such as core process integration, contingency management, climate change response and adaptation, resilience, and sustainability.

5. Conclusions and Future Outlook

This SLR has examined the role of DE technologies IAM for resilient linear infrastructure, resulting in a Digital Technology Integration Matrix to guide practitioners and authorities to choose appropriate tools to effectively manage infrastructure assets. This includes a newfound appreciation of the suite of existing and emerging DE tools, and the corresponding types of information requirements that can be digitised. While previous studies mostly focused on using one specific DE technology [6,17,33,73], this paper provides a holistic account of the range of available DE technologies and their suitability for different phases of the project life cycle. It also supports improved asset management decisions, through a novel 'Digital Technology Integration Matrix' that can facilitate continuous flows of information from design and construction through to asset operation, maintenance, and end-of-life repurposing or disassembly towards better business, project, and asset management outcomes. In the face of natural and man-made disruption, the SLR findings also demonstrate the importance of leveraging data and digital technologies for improved future-oriented disaster response and recovery decisions about capital works and maintenance spending.

The authors conclude the importance of a common understanding of available and emerging technologies to digitise asset information requirements, so that authorities and practitioners can better evaluate capture of infrastructure data for resilient outcomes. With this knowledge, industrial practitioners can identify new prospects in digital asset management not only towards increasing efficiency in asset management, but also for communication, collaboration, and data integration.

This synthesis of information to date will be beneficial to both academics and industry practitioners in obtaining valuable information on the influence of digital processes on infrastructure asset management, and thereby provide new pathways to mainstream digital infrastructure asset management. Furthermore, the synthesised mind map of the DE literature for digital IAM provides a guide for other researchers to further explore how DE can be adopted as a targeted approach to construct asset management models that are more data-driven, and accessible with improved level of interoperability. It was evident that technologies such as Building Information Modelling (BIM) have emerged as the prevailing digital approach to asset management due to their fully integrated systems for collecting, managing, and utilising building data across all phases of the asset life cycle [73]. Specifically, Building Information Modelling (BIM), as an intelligent 3D model-based process to inform and communicate project decisions, has demonstrated capacity in the design and construction stages of linear infrastructure [8]. Furthermore, the 'digital twin' has also been rapidly gaining popularity as a full digital representation of a physical asset, process, or system, as well as the engineering information that allows us to understand and model its performance. It is also a critical component of the DE framework and has the ability to be continuously updated from multiple sources demonstrating high potential for asset lifecycle management [40].

The SLR has implications for academics and industrial practitioners working in the DE domain. For the IAM sector, this SLR shows a clear opportunity to conduct further research studies that can be focussed on the research gaps related to:

- Evaluating barriers and enablers for integrating digital engineering for infrastructure asset management
- Applying hybrid digital engineering technologies and simulations within road, rail, bridge, and tunnel management
- Investigating the temporal, spatial and logical relationships between information categories and pathways to leverage DE tools to map these relationships

Herein, researchers could identify barriers for digital IAM and then collect data to investigate the potential of DE applications and simulations within the road, rail, bridge, and tunnel management. Additionally, similar studies will continue to test and validate the applicability of the suggested Technology Integrated Matrix (TIM).

Supplementary Materials: The following are available online at https://www.mdpi.com/article/10.3390/su132111965/s1, Table S1: Serial numbers explanation from the Figure 6.

Author Contributions: S.C. was involved in conceptualization, formal analysis, writing, reviewing and editing. S.M. (Sherif Mostafa) C.D. and S.M. (Sherif Mohamed) were involved in funding acquisition, project administration, reviewing and editing. All authors have read and agreed to the submitted version of the manuscript.

Funding: This research was funded by the 2018 CRI Priority Project Development Scheme, Cities Research Institute, Griffith University, Australia.

Conflicts of Interest: The authors declare no conflict of interest.

Appendix A

Table A1. The resultant detailed list.

Digital Process/Model Type	Key Features	Key Information Requirements/Data	Key Reference/s
Data Management, Modelling/Visualisation			
As-built drawings and models	• Facilitates data capturing and documentation • Enables better facility management practices at the operation phase of a project	• Dimensions, geometry, and location of all components of the project	[7,74]

Table A1. *Cont.*

Digital Process/Model Type	Key Features	Key Information Requirements/Data	Key Reference/s
AIM (Asset Integrity Management)	• Considers the dynamic nature of the transport network • Manages design, repair and maintenance planning of infrastructures	• Value of assets • Asset performance	[68]
Building Information Modeling (BIM)	• Generates, builds and manages rail infrastructure data throughout their lifecycle • Creates a field verifiable 'as-built' models • Manages assets and the network during its operations and maintenance • Enables information about component to be attached to a corresponding object • Offers financial and technical benefits to stakeholders	• BIM standard, BIM protocol, BIM guideline, BIM project • Non-geometric execution guide	[1,6,7,34,36,37,45,54,58,59,62,65,71,72,75–92]
Bridge Information Modeling (BrIM)	• Focuses on bridges as an extension of BIM • Captures unique features like roadway alignment and girder camber • Facilitates design, construction, and fabrication • Offers financial and technical benefits to stakeholders	• Warranty information, • cost (to replace, maintain etc.), • System visualisation • System performance information, • Locations of panels and valves that control equipment (e.g., electrical • Panel location, shut off valve location), • Sequence of operation (start-up/shut down information), • Maintenance history • BMS Operation • Monitoring/tracking • Location • Commissioning information • Design criteria	[46,64,78,84,93–95]

Table A1. *Cont.*

Digital Process/Model Type	Key Features	Key Information Requirements/Data	Key Reference/s
BMS (Bridge Management System)	• Enables better operability • Acts as a platform to connect key stakeholders • Facilitates management and maintenance of bridges network that organises all the management and maintenance activities	• General information (location, name, type, load capacity, etc.), • design information and physical properties of the elements • Inventory data, • Regular inspection records • Condition and strength assessment reports • Repair and maintenance records, • Cost records	[64,69,74,84,96,97]
CiM (Construction/Civil Information Modelling)	• Shares information for the life cycle of a building, structure, or asset • Models transport infrastructure in the design phase • Enables collection, organisation, and managed accessibility to accurate data and information related to a facility • Captures key digital technologies that provide managers with opportunities to use accurate data and information • Enablers advanced visualisation	• Operational data • Cost	[37,78]
Computer-aided design (CAD)	• Enables design, create, document and manage information • Replaces manual drafting with an automated process	• Design information • Space related characteristics	[7,12,41,50,71]

Table A1. *Cont.*

Digital Process/Model Type	Key Features	Key Information Requirements/Data	Key Reference/s
eRIM (electronic Requirements Information Management)	• Defines an information-centric, andprocess and service-oriented enterprise architecture approach to requirements management • Improves and more efficient and effective management of client requirements across all stages of a project	• Client requirements	[98]
Parametric Modelling	• Describes dependencies between the geometric entities of the different levels of detail	• Parametric geometry descriptions	[65,86]
Multiscale Modelling	• Provides the possibility for a stringent definition of dependencies between individual geometric elements on different • levels of detail	• Scale • Multiple levels of detail	[99]
Multidimensional (nD) Modelling	• Models parameters such as schedule, cost, and quality • Describes the whole process of the construction. • Allows direct extraction of any technical information such as object specifications and attributes from the nD model • Enables using and exchanging information based on nD model among different stakeholders • Improves the interoperability during the construction process.	• Cost Breakdown Structures (CBS) • Risk Breakdown Structures (RBS) • Schedule • Cost, and quality • Geometric and management information	[100]
Geometrical Modelling	• Defines dependencies between geometric entities on different LoDs	• Geometric information • Semantic information	[101,102]

Table A1. *Cont.*

Digital Process/Model Type	Key Features	Key Information Requirements/Data	Key Reference/s
3D Modelling	• Explores different design alternatives and cost/schedule • Compares different alternatives early in the design/construction process when they make the biggest impact on building life-cycle costs.	• Cost • Schedule	[44,50,53,63,80,83,102,103]
4D Modelling	• Displays animated stages in the geometry of a building that reflects consecutive activities in its construction schedule • Allows automatic generation of a computer animation file for the visualisation of project planning	• Geometric information • Photorealistic representations • Construction documents.	[57]
Virtual Prototyping Simulation (VPS)	• Simulates different construction scenarios in order to help planners identify optimal construction plans. • Assesses different scenarios and alternatives in the planning phase • Enables planners attain optimal plans for bridge construction projects	• Construction information • Geometric information • Equipment details	[104]
Digital Twin modelling	• Has the capability to be continuously updated from multiple sources • Including sensors and continuous surveying, to represent its near real-time status, working condition, or position	• Sensory data • Quality inspection information	[23]
Sensing			
Global Positioning System (GPS)	• Assists navigation • Assists the integration and delivery of information about disaster prevention and mitigation	• Spatial data/location data	[105]

Table A1. *Cont.*

Digital Process/Model Type	Key Features	Key Information Requirements/Data	Key Reference/s
Geographic Information Systems (GIS)	• Allows spatial utility information to besystematically visualised, analysed, and updated • Integrates building information modelling (BIM) models	• Spatial data/location data	[43,45,51,54,80,106–108]
Monitoring			
SHM (Structural Health Monitoring)	• Provides diagnosis of the state of the structure during the life of a structure	• Performance data	[46,47]
Petr-net model	• Enables different data calibration technique and different sources of data. • The modules interact with one another to model element deterioration, inspection and maintenance.	• Technical data	[48]
Artificial neural networks	• Enables nonlinear statistical data modelling • Models relationships between inputs and outputs	• Crack locations of bridges • Crack pattens and their widths	[49]

References

1. Ng, S.T.; Xu, F.J.; Yang, Y.; Li, H.; Li, J. A social networking enabled crowdsourcing system for integrated infrastructure asset management. In Proceedings of the Construction Research Congress, American Society of Civil Engineers (ASCE), New Orleans, LA, USA, 2–4 April 2018; pp. 322–331.
2. Roberts, C.; Pärn, E.; Edwards, D.; Aigbavboa, C. Digitalising asset management: Concomitant benefits and persistent challenges. *Int. J. Build. Pathol. Adapt.* **2018**, *36*, 152–173. [CrossRef]
3. Sun, Y.; Ma, L.; Robinson, W.; Purser, M.; Mathew, A.; Fidge, C. Renewal decision support for linear assets. In *Engineering Asset Management and Infrastructure Sustainability*; Springer: Amsterdam, The Netherlands, 2012; pp. 885–899.
4. Hayes, S.; Desha, C.; Burke, M.; Gibbs, M.; Chester, M. Leveraging socio-ecological resilience theory to build climate resilience in transport infrastructure. *Transp. Rev.* **2019**, *39*, 677–699. [CrossRef]
5. Aktan, A.E.; Bartoli, I.; Karaman, S.G. Technology Leveraging for Infrastructure Asset Management: Challenges and Opportunities. *Front. Built Environ.* **2019**, *5*, 61. [CrossRef]
6. Bradley, A.; Li, H.; Lark, R.; Dunn, S. BIM for infrastructure: An overall review and constructor perspective. *Autom. Constr.* **2016**, *71*, 139–152. [CrossRef]
7. Love, P.E.; Zhou, J.; Matthews, J.; Lavender, M.; Morse, T. Managing rail infrastructure for a digital future: Future-proofing of asset information. *Transp. Res. Part A Policy Pract.* **2018**, *110*, 161–176. [CrossRef]
8. Hampson, K.; Wang, J.; Drogemuller, R.; Wu, P.; Omrani, S.; Shemery, A. *Digital Asset Information Management (DAIM): Case Studies*; Final Case Studies Report, Project 2.51; Sustainable Build Environment National Research Centre: Bentley, WA, Australia, 2018.
9. Caldera, S.; Mohamed, S.; Mostafa, S.; Desha, C. Digital engineering for resilient road infrastructure outcomes: Evaluating critical asset information requirements. *J. Sustain. Dev. Energy Water Environ. Syst.* **2020**. [CrossRef]

10. Australian Government The Department of Transport Infrastructure. Cities and Regional Development. National Digital Engineering Policy Principles. Available online: https://www.infrastructure.gov.au/infrastructure/ngpd/files/Principles-for-DE_Template_2.pdf (accessed on 3 March 2019).

11. New South Wales Government, Transport for New South Wales (T.f.W.S.W). Digital engineering Framework. Available online: https://www.transport.nsw.gov.au/digital-engineering/digital-engineering-framework (accessed on 20 May 2020).

12. Love, P.E.; Zhou, J.; Matthews, J.; Luo, H. Systems information modelling: Enabling digital asset management. *Adv. Eng. Softw.* **2016**, *102*, 155–165. [CrossRef]

13. Huang, J.; Gheorghe, A.; Handley, H.; Pazos, P.; Pinto, A.; Kovacic, S.; Collins, A.; Keating, C.; Sousa-Poza, A.; Rabadi, G. Towards Digital Engineering—The Advent of Digital Systems Engineering. *arXiv Prepr.* **2020**, arXiv:2002.11672. [CrossRef]

14. PSCG. Asset Information Requirements. Available online: https://www.designingbuildings.co.uk/wiki/Asset_information_requirements_AIR (accessed on 20 November 2019).

15. Desha, C.; Caldera, S. Foundations and horizons: The critical role of international coursework to engage students in engineering for the 21st century. In Proceedings of the WEC2019: World Engineers Convention, Melbourne, VIC, Australia, 20–22 November 2019; p. 1261.

16. Caldera, S.; Mostafa, S.; Desha, C.; Mohamed, S. Integrating Disaster Management Planning into Road Infrastructure Asset Management. *Infrastruct. Asset Manag.* **2021**, *1*, 1–14. [CrossRef]

17. Jang, R.; Collinge, W. Improving BIM asset and facilities management processes: A Mechanical and Electrical (M&E) contractor perspective. *J. Build. Eng.* **2020**, *32*, 101540.

18. Okoli, C.; Schabram, K. A guide to conducting a systematic literature review of information systems research. *Sprouts Work. Pap. Inf. Syst.* **2010**, *10*, 1–39. [CrossRef]

19. Moher, D.; Liberati, A.; Tetzlaff, J.; Altman, D.G.; The PRISMA Group. Preferred reporting items for systematic reviews and meta-analyses: The PRISMA statement. *PLoS Med.* **2009**, *6*, e1000097. [CrossRef] [PubMed]

20. Rousseau, D.M.; Manning, J.; Denyer, D. 11 Evidence in management and organizational science: Assembling the field's full weight of scientific knowledge through syntheses. *Acad. Manag. Ann.* **2008**, *2*, 475–515. [CrossRef]

21. Denyer, D.; Tranfield, D. Producing a systematic review. In *The Sage Handbook of Organizational Research Methods*; Bryman, D.A.B.A., Ed.; Sage Publications Ltd.: Thousand Oaks, CA, USA, 2009; pp. 671–689.

22. Mostafa, S.; Chileshe, N.; Abdelhamid, T. Lean and agile integration within offsite construction using discrete event simulation: A systematic literature review. *Constr. Innov.* **2016**, *16*, 483–525. [CrossRef]

23. Macchi, M.; Roda, I.; Negri, E.; Fumagalli, L. Exploring the role of digital twin for asset lifecycle management. *IFAC PapersOnLine* **2018**, *51*, 790–795. [CrossRef]

24. Saunders, M.; Lewis, P.; Thornhill, A. *Research Methods for Business Students*; Pearson: Harlow, UK, 2012; Volume 6.

25. Levy, Y.; Ellis, T.J. A systems approach to conduct an effective literature review in support of information systems research. *Inf. Sci. Int. J. Emerg. Transdiscipl.* **2006**, *9*, 181–212. [CrossRef]

26. Briner, R.B.; Denyer, D. Systematic review and evidence synthesis as a practice and scholarship tool. In *Handbook of Evidence-Based Management: Companies, Classrooms and Research*; Oxford University Press: New York, NY, USA, 2012; pp. 112–129.

27. Booth, A.; Sutton, A.; Papaioannou, D. *Systematic Approaches to a Successful Literature Review*; Sage: London, UK, 2016.

28. Caldera, S.; Ryley, T.; Zatyko, N. Developing a marketplace for construction and demolition waste based on a systematic quantitative literature review. In Proceedings of the 1st Asia Pacific Sustainable Development of Energy Water and Environment Systems, Gold Coast, QLD, Australia, 6–9 April 2020.

29. Caldera, H.T.S.; Desha, C.; Dawes, L. Exploring the role of lean thinking in sustainable business practice: A systematic literature review. *J. Clean. Prod.* **2017**, *167*, 1546–1565. [CrossRef]

30. Ward, V.; House, A.; Hamer, S. Developing a framework for transferring knowledge into action: A thematic analysis of the literature. *J. Health Serv. Res. Policy* **2009**, *14*, 156–164. [CrossRef] [PubMed]

31. Belmont, D.; Hade, K.; Sefah, K. Digital Asset Performance Management—Taking Utility Asset Management to the Next Level. *Clim. Energy* **2020**, *37*, 9–14. [CrossRef]

32. Lee, S.-H.; Kim, B.-G. IFC extension for road structures and digital modeling. *Procedia Eng.* **2011**, *14*, 1037–1042. [CrossRef]

33. Xia, J. What Interoperability Really Means in a BIM Context? Available online: https://www.bimmodel.co/single-post/2016/09/05/What-Interoperability-really-means-in-a-BIM-context (accessed on 20 September 2019).

34. Zak, J.; Macadam, H. Utilization of building information modeling in infrastructure's design and construction. *IOP Conf. Ser. Mater. Sci. Eng.* **2017**, *236*, 012108. [CrossRef]

35. Marzouk, M.; Hisham, M.; Ismail, S.; Youssef, M.; Seif, O. On the use of building information modeling in infrastructure bridges. In Proceedings of the 27th International Conference on Applications of IT in the AEC Industry, Cairo, Egypt, 16–18 November 2010; pp. 1–10.

36. Blanco, F.G.B.; Chen, H. The implementation of building information modelling in the United Kingdom by the transport industry. *Procedia Soc. Behav. Sci.* **2014**, *138*, 510–520. [CrossRef]

37. Sankaran, B.; Nevett, G.; O'Brien, W.J.; Goodrum, P.M.; Johnson, J. Civil Integrated Management: Empirical study of digital practices in highway project delivery and asset management. *Autom. Constr.* **2018**, *87*, 84–95. [CrossRef]

38. Sankaran, B.; France-Mensah, J.; O'Brien, W.J. Data Integration Challenges for CIM in Large Infrastructure: A Case Study of Dallas Horseshoe Project. In Proceedings of the 16th International Conference on Computing in Civil and Building Engineering, Osaka, Japan, 6–8 July 2016.

39. Yabuki, N.; Aruga, T.; Furuya, H. Development and application of a product model for shield tunnels. In Proceedings of the ISARC. Proceedings of the International Symposium on Automation and Robotics in Construction, Montreal, QC, Canada, 11–15 August 2013.

40. Lu, Q.; Xie, X.; Parlikad, A.K.; Schooling, J.M.; Konstantinou, E. Moving from building information models to digital twins for operation and maintenance. *Proc. Inst. Civ. Eng. Smart Infrastruct. Constr.* **2020**, *172*, 1–11. [CrossRef]

41. Love, P.E.; Zhou, J.; Sing, C.-P.; Kim, J.T. Documentation errors in instrumentation and electrical systems: Toward productivity improvement using System Information Modeling. *Autom. Constr.* **2013**, *35*, 448–459. [CrossRef]

42. Fraga-Lamas, P.; Fernández-Caramés, T.M.; Castedo, L. Towards the Internet of Smart Trains: A Review on Industrial IoT-Connected Railways. *Sensors* **2017**, *17*, 1457. [CrossRef] [PubMed]

43. Ginardi, R.H.; Gunawan, W.; Wardana, S.R. WebGIS for Asset Management of Land and Building of Madiun City Government. *Procedia Comput. Sci.* **2017**, *124*, 437–443. [CrossRef]

44. Sánchez-Rodríguez, A.; Riveiro, B.; Soilán, M.; González-deSantos, L. Automated detection and decomposition of railway tunnels from Mobile Laser Scanning Datasets. *Autom. Constr.* **2018**, *96*, 171–179. [CrossRef]

45. Aziz, Z.; Riaz, Z.; Arslan, M. Leveraging BIM and big data to deliver well maintained highways. *Facilities* **2017**, *35*, 818–832. [CrossRef]

46. Jeong, S.; Hou, R.; Lynch, J.P.; Sohn, H.; Law, K.H. An information modeling framework for bridge monitoring. *Adv. Eng. Softw.* **2017**, *114*, 11–31. [CrossRef]

47. Okasha, N.M.; Frangopol, D.M. Computational platform for the integrated life-cycle management of highway bridges. *Eng. Struct.* **2011**, *33*, 2145–2153. [CrossRef]

48. Yianni, P.C.; Neves, L.C.; Rama, D.; Andrews, J.D.; Dean, R. Incorporating local environmental factors into railway bridge asset management. *Eng. Struct.* **2016**, *128*, 362–373. [CrossRef]

49. Fathalla, E.; Tanaka, Y.; Maekawa, K. Remaining fatigue life assessment of in-service road bridge decks based upon artificial neural networks. *Eng. Struct.* **2018**, *171*, 602–616. [CrossRef]

50. Pärn, E.; Edwards, D.; Sing, M. The building information modelling trajectory in facilities management: A review. *Autom. Constr.* **2017**, *75*, 45–55. [CrossRef]

51. Halfawy, M.R.; Vanier, D.J.; Froese, T.M. Standard data models for interoperability of municipal infrastructure asset management systems. *Can. J. Civ. Eng.* **2006**, *33*, 1459–1469. [CrossRef]

52. Halfawy, M.R. Integration of municipal infrastructure asset management processes: Challenges and solutions. *J. Comput. Civ. Eng.* **2008**, *22*, 216–229. [CrossRef]

53. Shirole, A.M.; Chen, S.; Puckett, J. Bridge Information Modeling for the Life Cycle. In Proceedings of the Tenth International Conference on Bridge and Structure Management, Buffalo, NY, USA, 20–22 October 2008.

54. Kurwi, S.; Demian, P.; Hassan, T.M. *Integrating BIM and GIS in Railway Projects: A Critical Review*; Loughborough University: Loughborough, UK, 2017.

55. Sugita, Y.; Kawaguchi, T.; Hatanaka, K.; Shimbo, H.; Yamamura, M.; Kobayashi, Y.; Fujisawa, Y.; Kobayashi, I.; Yabuki, N. Development of a design support system for geological disposal using a CIM concept. In Proceedings of the International Conference on Computing in Civil and Building Engineering, Osaka, Japan, 6–8 July 2016.

56. Salem, O.; Samuel, I.J.; He, S. Bim And Vr/Ar Technologies: From Project Development To Lifecycle Asset Management. In Proceedings of the International Structural Engineering and Construction, Angamaly, India, 14–15 May 2020.

57. Liapi, K.A. 4D visualization of highway construction projects. In Proceedings of the Seventh International Conference on Information Visualization, London, UK, 16–18 July 2003; pp. 639–644.

58. Tawelian, L.R.; Mickovski, S.B. The implementation of geotechnical data into the BIM process. *Procedia Eng.* **2016**, *143*, 734–741. [CrossRef]

59. Huang, S.-F.; Chen, C.-S.; Dzeng, R.-J. Design of track alignment using building information modeling. *J. Transp. Eng.* **2011**, *137*, 823–830. [CrossRef]

60. Hammad, A.; Zhang, C.; Hu, Y.; Mozaffari, E. Mobile model-based bridge lifecycle management system. *Comput. Aided Civ. Infrastruct. Eng.* **2006**, *21*, 530–547. [CrossRef]

61. Giglio, J.M.; Friar, J.H.; Crittenden, W.F. Integrating lifecycle asset management in the public sector. *Bus. Horiz.* **2018**, *61*, 511–519. [CrossRef]

62. Love, P.E.; Zhou, J.; Edwards, D.J.; Irani, Z.; Sing, C.-P. Off the rails: The cost performance of infrastructure rail projects. *Transp. Res. Part. A Policy Pract.* **2017**, *99*, 14–29. [CrossRef]

63. Shim, C.-S.; Yun, N.; Song, H. Application of 3D bridge information modeling to design and construction of bridges. *Procedia Eng.* **2011**, *14*, 95–99. [CrossRef]

64. Marzouk, M.; Hisham, M. Bridge information modeling in sustainable bridge management. In Proceedings of the ICSDC 2011: Integrating Sustainability Practices in the Construction Industry, Kansas City, MS, USA, 23–25 March 2012; pp. 457–466.

65. Karaman, S.G.; Chen, S.S.; Ratnagaran, B.J. Three-dimensional parametric data exchange for curved steel bridges. *Transp. Res. Rec.* **2013**, *2331*, 27–34. [CrossRef]

66. Mawlana, M.; Vahdatikhaki, F.; Doriani, A.; Hammad, A. Integrating 4D modeling and discrete event simulation for phasing evaluation of elevated urban highway reconstruction projects. *Autom. Constr.* **2015**, *60*, 25–38. [CrossRef]

67. Brous, P.; Herder, P.; Janssen, M. Governing Asset Management Data Infrastructures. *Procedia Comput. Sci.* **2016**, *95*, 303–310. [CrossRef]

68. Fuggini, C.; Manfreda, A.; Andrés, J.J.Á.; Pardi, L.; Holst, R.; Bournas, D.A.; Revel, G.M.; Chiariotti, P.; Llamas, J.; Gatti, G. Towards a comprehensive asset integrity management (AIM) approach for European infrastructures. *Transp. Res. Procedia* **2016**, *14*, 4060–4069. [CrossRef]

69. Trojanová, M. Asset management as integral part of road economy. *Procedia Eng.* **2014**, *91*, 481–486. [CrossRef]

70. Durazo-Cardenas, I.; Starr, A.; Turner, C.J.; Tiwari, A.; Kirkwood, L.; Bevilacqua, M.; Tsourdos, A.; Shehab, E.; Baguley, P.; Xu, Y. An autonomous system for maintenance scheduling data-rich complex infrastructure: Fusing the railways' condition, planning and cost. *Transp. Res. Part C Emerg. Technol.* **2018**, *89*, 234–253. [CrossRef]

71. Whyte, J.; Stasis, A.; Lindkvist, C. Managing change in the delivery of complex projects: Configuration management, asset information and 'big data'. *Int. J. Proj. Manag.* **2016**, *34*, 339–351. [CrossRef]

72. Yang, Y.; Ng, S.T.; Xu, F.J.; Skitmore, M. Towards sustainable and resilient high density cities through better integration of infrastructure networks. *Sustain. Cities Soc.* **2018**, *42*, 407–422. [CrossRef]

73. Ashton, H.; Hou, L. Bridge asset management: A digital approach to modelling asset information. In Proceedings of the 17th International Conference on Computing in Civil and Building Engineering, Tampere, Finland, 5–7 June 2018; pp. 1–8.

74. Abudayyeh, O.; Al-Battaineh, H.T. As-built information model for bridge maintenance. *J. Comput. Civ. Eng.* **2003**, *17*, 105–112. [CrossRef]

75. Brous, P.; Herder, P.; Janssen, M. Towards modelling data infrastructures in the asset management domain. *Procedia Comput. Sci.* **2015**, *61*, 274–280. [CrossRef]

76. Bruno, S.; De Fino, M.; Fatiguso, F. Historic Building Information Modelling: Performance assessment for diagnosis-aided information modelling and management. *Autom. Constr.* **2018**, *86*, 256–276. [CrossRef]

77. Cavka, H.B.; Staub-French, S.; Poirier, E.A. Developing owner information requirements for BIM-enabled project delivery and asset management. *Autom. Constr.* **2017**, *83*, 169–183. [CrossRef]

78. Costin, A.; Adibfar, A.; Hu, H.; Chen, S.S. Building Information Modeling (BIM) for transportation infrastructure—Literature review, applications, challenges, and recommendations. *Autom. Constr.* **2018**, *94*, 257–281. [CrossRef]

79. Fanning, B.; Clevenger, C.M.; Ozbek, M.E.; Mahmoud, H. Implementing BIM on infrastructure: Comparison of two bridge construction projects. *Pract. Period. Struct. Des. Constr.* **2014**, *20*, 04014044. [CrossRef]

80. Lee, P.-C.; Wang, Y.; Lo, T.-P.; Long, D. An integrated system framework of building information modelling and geographical information system for utility tunnel maintenance management. *Tunn. Undergr. Space Technol.* **2018**, *79*, 263–273. [CrossRef]

81. Leviäkangas, P.; Paik, S.M.; Moon, S. Keeping up with the pace of digitization: The case of the Australian construction industry. *Technol. Soc.* **2017**, *50*, 33–43. [CrossRef]

82. Love, P.E.; Ahiaga-Dagbui, D.; Welde, M.; Odeck, J. Light rail transit cost performance: Opportunities for future-proofing. *Transp. Res. Part A Policy Pract.* **2017**, *100*, 27–39. [CrossRef]

83. Love, P.E.; Simpson, I.; Hill, A.; Standing, C. From justification to evaluation: Building information modeling for asset owners. *Autom. Constr.* **2013**, *35*, 208–216. [CrossRef]

84. Sacks, R.; Kedar, A.; Borrmann, A.; Ma, L.; Brilakis, I.; Hüthwohl, P.; Daum, S.; Kattel, U.; Yosef, R.; Liebich, T. SeeBridge as next generation bridge inspection: Overview, information delivery manual and model view definition. *Autom. Constr.* **2018**, *90*, 134–145. [CrossRef]

85. Amann, J.; Singer, D.; Borrmann, A. Extension of the upcoming IFC alignment standard with cross sections for road design. In Proceedings of the International Conference on Civil and Building Engineering Informatics, Tokyo, Japan, 22–24 April 2015.

86. Ji, Y.; Borrmann, A.; Beetz, J.; Obergrießer, M. Exchange of parametric bridge models using a neutral data format. *J. Comput. Civ. Eng.* **2012**, *27*, 593–606. [CrossRef]

87. Borrmann, A.; Kolbe, T.H.; Donaubauer, A.; Steuer, H.; Jubierre, J.R.; Flurl, M. Multi-scale geometric-semantic modeling of shield tunnels for GIS and BIM applications. *Comput. Aided Civ. Infrastruct. Eng.* **2015**, *30*, 263–281. [CrossRef]

88. Breunig, M.; Borrmann, A.; Rank, E.; Hinz, S.; Kolbe, T.H.; Schilcher, M.; Mundani, R.-P.; Jubierre, J.R.; Flurl, M.; Thomsen, A. Collaborative Multi-Scale 3D City and Infrastructure Modeling and Simulation. In Proceedings of the Tehran's Joint ISPRS Conferences of GI Research, SMPR and EOEC 2017, Tehran, Iran, 7–10 October 2017; pp. 341–352.

89. DiBernardo, S. Integrated modeling systems for bridge asset management—Case study. In Proceedings of the Structures Congress 2012, Chicago, IL, USA, 29–31 March 2012; pp. 483–493.

90. Becerik-Gerber, B.; Jazizadeh, F.; Li, N.; Calis, G. Application areas and data requirements for BIM-enabled facilities management. *J. Constr. Eng. Manag.* **2011**, *138*, 431–442. [CrossRef]

91. Lee, S.-H.; Park, S.; Park, J.; Seo, K.-W. Open BIM-based information modeling of railway bridges and its application concept. In *Computing in Civil and Building Engineering 2014*; American Society of Civil Engineers: Reston, VA, USA, 2014; pp. 504–511. [CrossRef]

92. Sanchez, A.; Kraatz, J.A.; Hampson, K.D.; Loganathan, S. BIM for sustainable whole-of-life transport infrastructure asset management. In Proceedings of the 2014 IPWEA Sustainability in Public Works Conference, Tweed HeaCds, NSW, Australia, 27–29 July 2014.

93. Ali, N.; Chen, S.; Srikonda, R.; Hu, H. Development of concrete bridge data schema for interoperability. *Transp. Res. Rec. J. Transp. Res. Board* **2014**, *2406*(1), 87–97. [CrossRef]
94. Markiz, N.; Jrade, A. Integrating a fuzzy-logic decision support system with bridge information modelling and cost estimation at conceptual design stage of concrete box-girder bridges. *Int. J. Sustain. Built Environ.* **2014**, *3*, 135–152. [CrossRef]
95. Tanaka, F.; Hori, M.; Onosato, M.; Date, H.; Kanai, S. Bridge information model based on IFC standards and web content providing system for supporting an inspection process. In Proceedings of the 16th International Conference on Computing in Civil and Building Engineering, Osaka, Japan, 6–8 July 2016; pp. 1140–1147.
96. Bień, J. Modelling of structure geometry in Bridge Management Systems. *Arch. Civ. Mech. Eng.* **2011**, *11*, 519–532. [CrossRef]
97. Sung, Y.-C.; Chen, C.-C.; Chang, K.-C. Extended Bridge Inspection Module of a Life-cycle Based Bridge Management System. In Proceedings of the International Conference on Computing in Civil and Building Engineering, Osaka, Japan, 6–8 July 2016.
98. Jallow, A.K.; Demian, P.; Anumba, C.J.; Baldwin, A.N. An enterprise architecture framework for electronic requirements information management. *Int. J. Inf. Manag.* **2017**, *37*, 455–472. [CrossRef]
99. Vilgertshofer, S.; Borrmann, A. Using graph rewriting methods for the semi-automatic generation of parametric infrastructure models. *Adv. Eng. Inform.* **2017**, *33*, 502–515. [CrossRef]
100. Ding, L.; Zhou, Y.; Luo, H.; Wu, X. Using nD technology to develop an integrated construction management system for city rail transit construction. *Autom. Constr.* **2012**, *21*, 64–73. [CrossRef]
101. Gao, G.; Liu, Y.-S.; Wu, J.-X.; Gu, M.; Yang, X.-K.; Li, H.-L. IFC railway: A semantic and geometric modeling approach for railways based on IFC. In Proceedings of the 16th International Conference on Computing in Civil and Building Engineering, Osaka, Japan, 6–8 July 2016.
102. Sampaio, A.N.Z. Geometric modeling of box girder deck for integrated bridge graphical system. *Autom. Constr.* **2003**, *12*, 55–66. [CrossRef]
103. Le, T.; Jeong, H.D. Interlinking life-cycle data spaces to support decision making in highway asset management. *Autom. Constr.* **2016**, *64*, 54–64. [CrossRef]
104. Li, H.; Chan, N.K.; Huang, T.; Skitmore, M.; Yang, J. Virtual prototyping for planning bridge construction. *Autom. Constr.* **2012**, *27*, 1–10. [CrossRef]
105. Kellner, F.; Otto, A.; Brabänder, C. Bringing infrastructure into pricing in road freight transportation—A measuring concept based on navigation service data. *Transp. Res. Procedia* **2017**, *25*, 794–805. [CrossRef]
106. Borrmann, A.; Jubierre, J. A.; Jubierre, J. A multi-scale tunnel product model providing coherent geometry and semantics. In *Computing in Civil Engineering (2013)*; American Society of Civil Engineers: Reston, VA, USA, 2013; pp. 291–298. [CrossRef]
107. Koch, C.; Vonthron, A.; König, M. A tunnel information modelling framework to support management, simulations and visualisations in mechanised tunnelling projects. *Autom. Constr.* **2017**, *83*, 78–90. [CrossRef]
108. Li, X.; Lin, X.; Zhu, H.; Wang, X.; Liu, Z. A BIM/GIS-based management and analysis system for shield tunnel in operation. Available online: https://www.semanticscholar.org/paper/A-BIM-%2F-GIS-based-management-and-analysis-system-in-Li-Lin/b6030accd8eb2cb14e112b82260bf7b90b6a42c1 (accessed on 20 November 2019).

Review

Blockchain and Building Information Management (BIM) for Sustainable Building Development within the Context of Smart Cities

Zhen Liu [1,*], **Ziyuan Chi** [1,*], **Mohamed Osmani** [2] and **Peter Demian** [2]

[1]   School of Design, South China University of Technology, Guangzhou 510006, China
[2]   School of Architecture, Building and Civil Engineering, Loughborough University,
      Loughborough LE11 3TU, UK; m.osmani@lboro.ac.uk (M.O.); p.demian@lboro.ac.uk (P.D.)
*   Correspondence: liuzjames@scut.edu.cn (Z.L.); 201820143426@mail.scut.edu.cn (Z.C.)

Citation: Liu, Z.; Chi, Z.; Osmani, M.; Demian, P. Blockchain and Building Information Management (BIM) for Sustainable Building Development within the Context of Smart Cities. *Sustainability* 2021, *13*, 2090. https://doi.org/10.3390/su13042090

Academic Editor: Tan Yigitcanlar

Received: 3 January 2021
Accepted: 8 February 2021
Published: 16 February 2021

Publisher's Note: MDPI stays neutral with regard to jurisdictional claims in published maps and institutional affiliations.

Abstract: 'Smart cities' are a new type of city where stakeholders are jointly responsible for urban management. City Information Management (CIM) is an output tool for smart city planning and management, which assists in achieving the sustainable development of urban infrastructure, and promotes smart cities to achieve the goals of stable global economic development, sustainable environmental development, and improvement of people's quality of life. Existing research has so far established that blockchain and BIM have great potential to enhance construction project performance. However, there is little research on how blockchain and BIM can support sustainable building design and construction. Therefore, the aim of this paper is to explore the potential impact of the integration of blockchain and BIM in a smart city environment on making buildings more sustainable within the context of CIM/Smart Cities. The paper explores the relationships between blockchain, BIM and sustainable building across the life cycle stage of a construction project. This paper queries the Web of Science (WoS) database with keywords to obtain relevant publication, and then uses the VOSviewer to visually analyze the relationships between blockchain, BIM, and sustainable building within the context of smart cities and CIM, which is conducted in bibliometric analysis followed by micro scheme analysis. The results demonstrate the value of this method in gauging the importance of these three topics, highlighting their interrelationships and identifying trends, giving researchers an objective research direction. Those aspects reported in the paper constitute an original contribution.

Keywords: smart cities; blockchain; building information management (BIM); city information management (CIM); sustainable building; life cycle; VOSviewer

1. Introduction

The term 'smart cities' first appeared in the 1990s [1]. Since then, there have been several definitions of smart cities. In different development periods, various stakeholders will give different definitions. Among the more widely definition relates to the six dimensions—namely, smart people, smart economy, smart governance, smart mobility, smart life, and smart environment—to measure the development of smart cities [2]. Cities can make further adjustments and improvements according to their own actual development conditions, and form their own characteristic development ways. The British Standards Institution (BSI) [3] defines the smart city as 'the effective integration of physical, digital and human systems in the built environment to deliver sustainable, prosperous and inclusive future for its citizens'.

The development of smart cities takes 'citizen-focused and improving the quality of life of residents' [4] as the fundamental goal, and achieves the triad of economic, social and environmental development in the long-term development. Emerging technologies, as a tool for urban development, acquire data and knowledge all the time around the world, produce qualitative changes in substantial amounts of data, and promote innovation and

277

change [5]. Desouza et al. [6] identify three paths to the development of smart cities: greenfield, neighborhood and platform. However, policymakers make it clear that technology is not the core of policy making and urban development, but care for citizens is. The smart city strategy brings innovative changes to the region that helps the region improve its image and status, promote industrial development, attract high-quality talents, and contributes to the formation of a smart city model that is urgently needed for resilient settlements in a climate emergency [7]. Smart sustainable cities are currently the main development mode adopted by cities around the world, because this mode can best respond to environmental changes, ensure that the government is clean and honest, innovate the global network economy, and improve the quality of life [8].

Urban development has undergone multi-stage evolution, facing the challenges of retention and elimination, rebirth and decline, and the evolution and regeneration of multiple industries such as agriculture and industry [9]. These changes have formed today's urban strategy. Every urban renewal is a process of urban iterative optimization. Western cities have undergone five transformations, from urban reconstruction in the 1950s, urban revitalization in the 1960s, urban renewal in the 1970s, urban development in the 1980s, to urban regeneration in the 1990s. The reform of the city improves the policies according to the actual development challenges and opportunities of the city, so as to carry out the systematic implementation and management of the city and improve the quality of life of the city at the same time [5].

Taking the transformation of Barcelona's Poblenou smart cities as a case, its development focuses on social inclusion [10]. From a long-term perspective, the development of the general direction is forward-looking and sustainable. Being a knowledge-intensive city can bring the following advantages to the city:

- The influx of knowledge-based talents brings strong innovation momentum to the society and the ability to face the challenges of urban development [11];
- Promote the construction of infrastructure and life security, such as business environment, cultural and recreational facilities, medical education, housing and transportation [12];
- Promote investment in real estate, such as talent centers, business districts, and school districts; and in public sectors such as transportation, universities, cultural facilities, and tourism [13];
- Promote the establishment of mutual help relationship between university education and companies, so that the university can get practical work experience and the company can get knowledge and skills [14];
- Promote citizens to participate in urban development discussions through dialogue [15];
- As a soft asset, knowledge can effectively create value for cities and provide more appropriate solutions for urban management [16].

The case of Barcelona's Poblenou smart cities shows that the construction of a smart knowledge city requires the support of all social stakeholders, analyzes the current situation, and develops strategic plans. The formulation of the plan requires the establishment of a knowledge management framework based on different city characteristics, capabilities, and limitations. Cities can learn knowledge and expertise from each other; develop innovative solutions and sustainable strategies; and develop green, open, inclusive and sustainable cities [17].

Sustainability is an important pillar for the development of smart cities [18]. Smart devices and smart services have become parts of the smart cities [19], and the application of City Information Management (CIM) can provide ideas for effectively dealing with problems in urban construction [20]. CIM is not only an innovative and comprehensive new infrastructure construction, but also a carrier for realizing the implementation of big data new infrastructure, which plays a leading role in the realization of accelerating the digital and intelligent transformation of urban industries, helping the formation of new economic and technological forms, and in the realization of urban industrialization and industrial urbanization [21]. Building Information Management (BIM) is a key direction of information management in the future construction industry [22]. Its combined platform with GIS, data analysis tools, visualization tools, and parametric design tools constitutes CIM [23], which provides strong support for the comprehensive management of cities. As a part of smart city, smart buildings must also integrate sustainable development throughout the entire life cycle. In order to make the society more sustainable, various concepts of 'green', 'environmental protection', and 'sustainable development' have emerged, but there is a gap between the concept and the actual operation [24]. The realization of the sustainable development requires technical support, however, the construction industry has always lagged behind other sectors in the use of digital information technology [25,26].

An increasing body of work was published on BIM. Moreover, BIM has different functions for different stakeholders. However, due to the limited understanding of technology among practitioners and the current obstacles, the key functions of BIM cannot be brought into play, and the full potential of BIM has not yet been realized [27,28]. Interoperability is also challenged by decentralized collaboration in the construction industry. This opacity and lack of communication cannot give positive feedback to all stakeholders [19,29]. This will lead to disjointed teams with divergent priorities, which is not conducive to the overall progress of the project [25].

Blockchain is an encrypted distributed accounting technology and also is a decentralized database, which has the potential to address interoperability problems of barriers/challenges facing smart city [26,30]. Blockchain can store data securely and easily for query on the chain, providing support for the long construction cycle and reducing unnecessary work. Limited yet increasing studies explored the integration of blockchain and BIM in construction projects [31–33]. These studies, however, focus on applications in the financial aspects of construction (payment security, comprehensive project delivery) and security. Penzes [27] further emphasizes the importance of blockchain in the construction industry and describes its potential applications in Payment and Project Management, Procurement and Supply Chain Management, and BIM and Smart Asset Management. There are few related explorations on how blockchain and BIM affect the development of the entire cycle of sustainable building for smart cities and there is insufficient research on investigating the relationship between BIM, blockchain, and sustainable building. In addition, the potential benefits of integrating blockchain and BIM have not been fully realized in the construction industry. Hence, this paper explores the integration of blockchain and BIM to support sustainable building practices in smart cities.

2. Methodology

The adopted methodology queries the Web of Science (WoS) database with keywords to obtain relevant publication, and then uses the VOSviewer to visually analyze the relationships between blockchain, BIM and sustainable building within the context of smart cities, which is conducted in bibliometric analysis followed by micro scheme analysis. The bibliometric analysis is used via VOSviewer for conducting a macro analysis of the research field, which generates big pictures of the research focus and trend in the field. VOSviewer has the functions of visualizing keywords, co-authors, and citations. It can visually present research hotspots and trends year by year; and present their relevance or frequency through color changes and distance. The micro scheme analysis summarizes the development of sustainable building life cycle in smart cities and the application of blockchain and BIM, which presents three phases of the building life cycle—namely, design, construction, and operation—for verifying and refining the content obtained in bibliometric analysis.

WoS is selected as the database source, which is an internationally recognized database that reflects the level of scientific research. It covers a wide range of journals recognized as authoritative. WoS core collection is selected to search keywords for comprehensive publication retrieval based on 'topics'. In the field of smart cities, when searching for keywords related to sustainable building and blockchain at the same time, there are few papers that match all keywords. Keywords include 'Building Information Modeling', 'Sustainable Building', 'Sustainable Building Design', 'Sustainable Building Construction',

'Sustainable Building Operation' and 'Blockchain'. Therefore, the conjunction 'or' will be used flexibly to find the connection between design, construction, operation, blockchain, sustainable building and BIM. Blockchain is a state-of-the-art approach to BIM and for sustainable building development, hence this research surveys articles published between 2011 and 2020.

VOSviewer was used for bibliometric analysis. Few studies use keyword visualization methods amongst the studies reported in existing publications, most of which use logical diagrams [34–36], scene diagrams [37–40], roadmaps [41–43], and strategy diagrams [44–46]. The VOSviewer has been used to review the latest technology of BIM supported building performance [47], and at the same time, vosviewer is used as a Scientometric Analysis in scientific landscape of sustainable urban and rural areas research [48] and mapping two decades of autonomous vehicle research [49]. However, there is otherwise little research using keyword co-occurrence analysis in BIM. Such techniques have been used in patent analysis [50], databases [51], social networking sites [52], and the recording and prediction of subject changes in biological and medical computers [53]. The keywords research results demonstrate the effectiveness and reliability of keyword search, selection and processing [51,54]. Therefore, this paper attempts to use VOSviewer to visualize keywords to uncover relationships between blockchain, BIM, sustainable building development, and smart cities.

In the VOSviewer, the frequency gap between blockchain and other keywords is too large to appear in one view at the same time. Hence, other relevant keywords are visualized in addition to blockchain. There are 1305 research papers retrieved from the WoS keyword search, which are imported into VOSviewer, and 43,531 keywords are calculated by the software. It was found that 632 of these keywords are used in at least 13 papers. Then, unreated keywords are manually filtered out and 87 keywords selected, as shown in Figure 1. These keywords are automatically divided into three colors (red, green, and blue) by the VOSviewer, as three categories, and show the research hotspots in recent years.

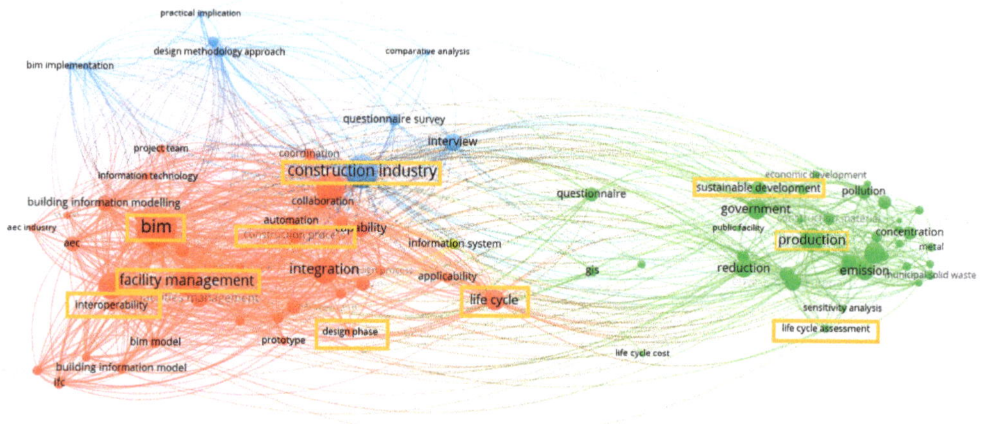

Figure 1. Visualization of research hotspots of major cluster sets and keywords in the construction industry via VOSviewer.

3. Results

3.1. Bibliometric Analysis

VOSviewer includes three forms of analysis: network visualization, overlay visualization and density visualization. The size of each circle represents the weight of the keyword. The distance between two circles represents the affinity between the two circles. If the affinity is stronger, the distance is shorter, and the weaker the affinity is, the distance is farther. The color of the circle represents the cluster to which it belongs, and different clusters are represented by different colors.

Figure 1 is a network visualization. BIM is the core in the red area, and the degree of closeness of keywords of 'Facility Management (FM)', 'Construction Process', and 'Design Phase' decreases. Most of the publication are FM related, and the construction and design part are weak. 'Life Cycle' connects the red and green sides. According to the data visualization analysis, the keywords in the green area taken 'Production' as the core, where the 'Sustainable Development' and 'Life Cycle Assessment' are associated with it. However, the keywords 'Construction Industry' and 'BIM' are far away from the keyword 'Sustainable Development', which means that it is less relevant. At the same time, the correlation with the whole life cycle is insignificant. This situation or a lack of construction industry development, reflected in the sustainable building development.

Figure 2 shows the visualization of research trend with time as the criterion, highlighting the keywords that connect with BIM as the core is the hotspot of research in the past five years. Thus, based on the bibliometric analysis, which are summarized in Figures 1 and 2, the following current picture of the field can be revealed:

- The integration of BIM and network systems can support the improvement of the construction and operation phases of the Architecture, Engineering and Construction (AEC) [55–58], which makes information retrieval and management easier, but is currently in the preliminary and intermediate stage.
- The use of BIM in the entire life cycle of a building has both potential and obstacles [30,58–61].
- The interoperability of data information throughout the building life cycle is still a challenge [62,63].
- In the research into construction industry and BIM, the FM stage received more attention than the building design, construction and maintenance phases. Life cycle issues should be addressed early on, but it has only been mentioned many times in recent years.

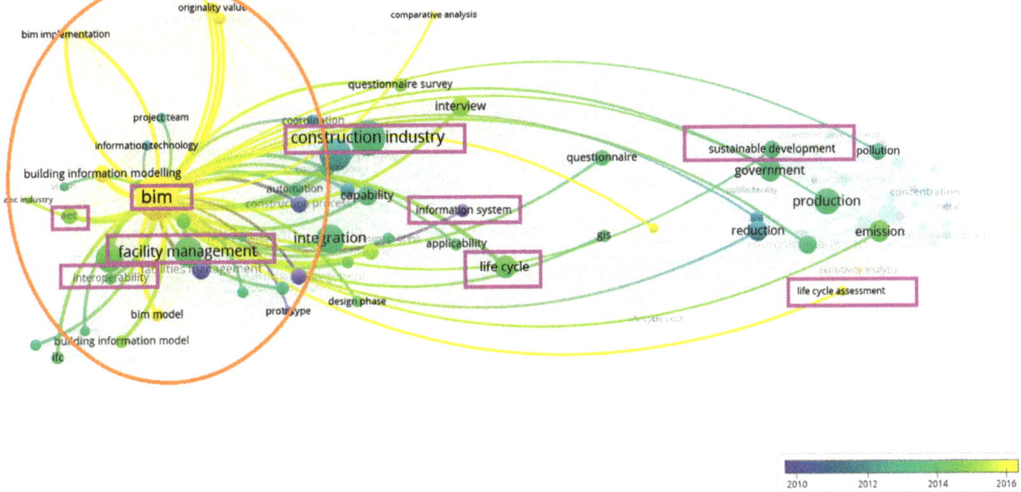

Figure 2. Visualization of research hotspots in the construction industry.

As an emerging technology, blockchain has been applied in many fields such as smart cities, agriculture, medical care, finance, transportation, the Internet of Things and other fields. As a platform, blockchain provides stakeholders in smart cities a mechanism to participate in urban construction. In recent years, blockchain has become popular, and explored widely [25,26]. The construction industry has also joined the wave of this digital technology. However, there are few studies on blockchain in the construction industry at present. Hence, it is necessary to search the keywords of blockchain and construction industry separately.

As such, 11 research papers with 463 keywords were selected from ScienceDirect. 59 of these keywords were used in at least two papers. Keywords unrelated to blockchain and construction have been manually filtered out, leaving 27 keywords, as shown in Figure 3. From the number of papers and Figure 3, it can be seen that there is less research on the integration of blockchain and construction industry, and the affinity between them is not clearly established. There have been few studies linking BIM, blockchain, and construction industry in the past three years. The integration of construction industry and blockchain is still in its infancy, and the speed of digital transformation is very slow, but a number of researchers have highlighted the potential of this integration [30,33].

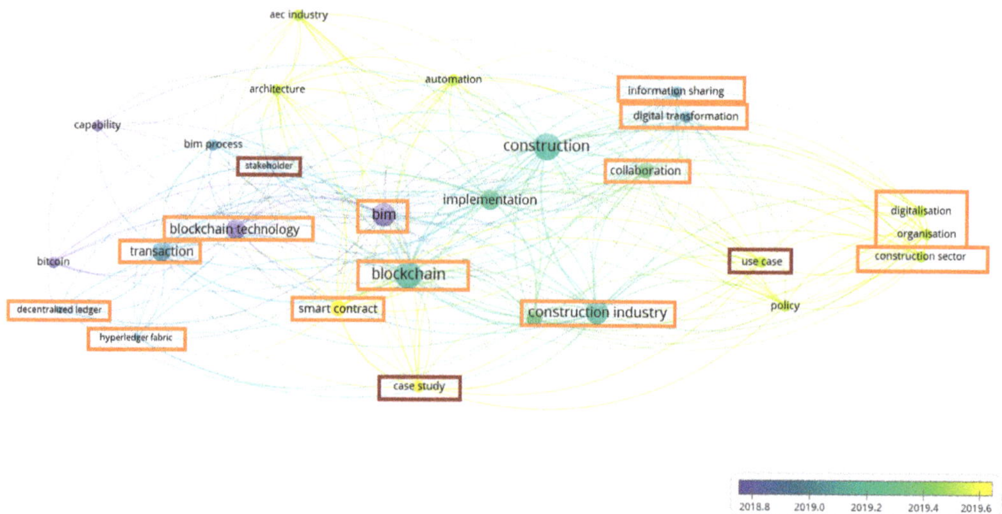

Figure 3. Visualization of research hotspots in the application of blockchain in the construction industry.

It can be seen from Figure 3 that the digital transformation of the construction industry has emerged in recent years. This will become the development trend of the construction industry in the future, and the use of blockchain in the construction industry is worth exploring.

3.2. Micro Scheme Analysis

After the macro analysis of VOSviewer keyword visualization, 61 articles associated with blockchain, BIM, and sustainable building development were specifically selected and included in the micro scheme analysis. The articles were mapped to particular building life cycle phases, including design, construction and operation. Organized according to three stages and comprehensive stages, and pointed out the research methods used in each publication and the research situation of blockchain, BIM and sustainable building.

As shown in Tables 1–4, judging from the number of publications, in the past three years, studies associated with blockchain, BIM and sustainable building are focused on the operation phase of the building life cycle than in the design and construction phase. In addition, the application of blockchain in the design and construction stage is insufficient, while the application in the operation aspect is gradually increasing in recent years, mainly studying the potential and challenges of blockchain in the construction industry [26,30,64]. The data storage of blockchain adopts a decentralized distribution way, which can reduce the risk of centralized storage. Even if one node of blockchain is lost, the overall data will not be affected [65]. The characteristics of blockchain can meet the needs of real-time communication of data information between different stakeholders on the basis [19].

The integration of blockchain and BIM in the construction industry can provide a more positive impetus for the future development of smart buildings [25,29]. As an important technology in construction projects, BIM is widely used in all stages of the construction life cycle. BIM has been deemed as an effective tool for the integration of natural systems and technical systems in building design [66]. BIM includes databases in multiple fields [22], providing a basis for editing and managing data throughout the life cycle of smart building [20]. It will help significantly reduce construction duplication and waste, reduce errors caused by traditional methods, and accelerate project delivery time [42]. Currently, various BIM performance models have been developed to assist in building design, construction, operation, and maintenance [45]. The applicability and focus of these vary, so that the needs of different BIM users are met [67].

Table 1. The application of blockchain, Building Information Management (BIM), and sustainable building in the design stage of building life cycle.

Source	Year	Research Method	Sustainable Building	Blockchain	BIM
Kuster et al.	2020	Literature and the NeOn methodology	*		
Rajaee et al.	2019	Questionnaire	*		*
Xue et al.	2019	Experimental tests			*
Ghosh	2018	Review			*
Olawumi et al.	2018	A Delphi survey	*		*
Fathi et al.	2016	Case studies	*		*
Kylili et al.	2015	Literature review	*		
Farias Stipo	2015	Literature reviews, case study and interviews	*		
Shoubi et al.	2014	Case study	*		*
Jrade et al.	2013	Modelling	*		*

* indicates that the literature contains the content.

Table 2. The application of blockchain, BIM, and sustainable building in the construction stage of building life cycle.

Source	Year	New Build	Renovation	Research Method	Sustainable Building	Blockchain	BIM
Manganelli et al.	2020		*	Model and a case study	*		
Passer et al.	2020	*		Review	*		*
Shrubsole et al.	2019	*		Literature review	*		
Olawumi et al.	2018	*		A Delphi survey	*		*
Chen et al.	2018		*	Literature review			*
Ghaffarianhoseini et al.	2017	*		Literature review	*		
Bretherton	2017		*	Case studies	*		
Capeluto et al.	2016		*	Comparative analysis	*		
Khaddaj et al.	2016		*	Literature review			*
Jones et al.	2015	*		Literature review	*		*

* indicates that the literature contains the content.

As shown in Table 1, research in the design phase is mostly reflected in the construction of environmentally friendly buildings [62,68–70] and integrated design processes [71]. It is proposed that designers should strengthen the ability to use BIM [62], promote the combination of BIM and theory, and find decisions that support sustainable building development through design [56], and pursue a balance between the environment, economy, and society [24,62]. The design phase is based on compliance with building codes and engineering codes, and several requirements should be considered when establishing BIM models: building energy requirements, ecological building parameters, quality control requirements, user comfort, user requirements, local climate conditions, and life culture [37,72]. The combination of intelligent sensor system and BIM can help develop a smart, humane and sustainable built environment, promote the realization of sustainable strategic initiatives in the built environment, and realize green buildings, safe and healthy lives surroundings [73]. The main research methods used in this phase are literature review, questionnaire survey, experiment, case analysis, and interview.

Table 3. The application of blockchain, BIM, and sustainable building in the operation stage of building life cycle.

Source	Year	FM	Maintenance	Research Method	Sustainable Building	Blockchain	BIM
Eicker et al.	2020	*		Case study	*		
Marmo et al.	2020	*	*	Case studies			*
Chen et al.	2020	*		Modeling			*
Olawumi et al.	2020	*		Literature review and questionnaire	*		*
Li et al.	2020	*		Modeling			*
Quinn et al.	2020	*		Case studies			*
Perera et al.	2020	*		Literature review and a use case analysis		*	
Sheng et al.	2020	*		A case study		*	
Elghaish et al.	2020	*		Experiment		*	*
Chong et al.	2020	*		A questionnaire and a case study		*	*
Marzouk et al.	2020	*		Case study	*		*
Kumar et al.	2020	*		Case studies and test	*		
Matarneh et al.	2019	*	*	Literature review			*
Cachat et al.	2019	*		Literature review			*
Ammari et al.	2019	*		Usability testing and questionnaire			*
Chen et al.	2019	*	*	Case study			*
Gao et al.	2019	*		Literature review			*
Gong et al.	2019	*		Case study			*
Bonci et al.	2019	*		Case study			*
Lokshina et al.	2019	*		Evaluation	*	*	*
Yilmaza et al.	2019	*		Expert reviews and case study			*
Bortoluzzi et al.	2019	*		Case studies			*
Chen et al.	2018	*		Literature review			*
Lin et al.	2018	*		Case study			*
Wong et al.	2018	*		Literature review and focus group			*
Pardis et al.	2018	*		Case studies and semi-structure interview			*
Araszkiewicz	2017	*		Case study			*
Chien et al.	2017	*		Experiment			*
Zadeh et al.	2017	*	*	Case studies			*
Nicał et al.	2016	*		Literature review			*
Aziz et al.	2016	*		Literature review			*
Oti et al.	2016	*		Example description and test case			*
Singh et al.	2011	*		Literature reviews, focus group and case study			*
Arayici et al.	2011	*		Case study			*

* indicates that the literature contains the content.

Table 4. The application of blockchain, BIM, and sustainable building in the whole stage of building life cycle.

| Source | Year | Building Life Stage | | | | | Research Method | Sustainable Building | Blockchain | BIM |
| | | Design | Construction | | Operation | | | | | |
			New Build	Renovation	FM	Maintenance				
Liu et al.	2020	*			*		Literature review	*		*
Liu et al.	2019	*	*		*		Questionnaire & semi-structured interviews	*	*	*
Tang et al.	2019	*	*	*	*	*	Literature review	*		*
Ustinovichius et al.	2018	*		*			Case study			*
Kim et al.	2018			*	*		Evaluation			*
Turk et al.	2017		*			*	Scenarios		*	*
Wong et al.	2015	*	*	*	*	*	Literature review	*		*

* indicates that the literature contains the content.

As shown in Table 2, an in-depth study of the obstacles and challenges faced by stakeholders in the integration of BIM and sustainability practices during the construction phase [56,60]. Based on BIM to restore heritage buildings [74], post-earthquake buildings [75,76], and refurbished buildings [77]. Promote the implementation of the BIM platform to optimize the construction process, improve the quality of project transformation, and effectively track the project with visual data [63]. This stage reflects urban renewal, which is an opportunity for sustainable development and intelligent growth of towns. The establishment of urban reconstruction plan indicators can carry out quality inspections on existing buildings, reduce the vulnerability of buildings, and contribute to the evaluation and decision-making of the seismic capacity of buildings [76,77]. The protection of the building does not mean sticking to the old appearance, but it can be rebuilt into a resilient, sustainable, and green building with innovative materials and technology [74]. The main research methods used in this phase include case studies, literature review, interviews, and comparative analysis.

Table 3 building operation phase discusses how urban renewal is more sustainable [34,61,78], by identifying and alleviating barriers to smart and sustainable practices in the built environment [61], and prioritizing them, which provides a basis for decision makers to make feasible decisions. Operating information is managed through the integration of BIM and FM systems. This traditional maintenance management is combined with digital technology to improve management performance and maintenance efficiency in the future operation and maintenance phase [34,35]. Chen et al. designed the framework into five modules, including project documents, personnel and contacts, FM planning and execution, technical performance evaluation, and safety and emergency management [44]. In addition, the potential and application of BIM in FM and maintenance are also studied [36,45,55]. From the perspective of equipment managers, evaluate and summarize the current BIM-O&M research and application progress, analyze research trends, and find research gaps and promising research directions in the future [45]. Using BIM to obtain real-time information [39,79] and conduct final inspection [40] during the completion of the project to strengthen communication and information management capabilities [80]. Araszkiewicz proposes the possible directions for further research on the digitalization of property management, and the influence of digitalization on the implementation process of the concept of intelligent sustainable building engineering [80]. Reference [26] proves the potential of blockchain in the construction industry and uses it for quality information management [81] and integrated project delivery (IPD) [32] in construction. Solve the problem of security of payment (SOP) through the integration of blockchain, BIM, and smart sensors [31]. The main research methods used in this stage include case analysis, literature review, focus group, interview, and questionnaire.

Table 4 is about the literature involving multiple phases of life cycle. It can be seen that there are few systematic comprehensive studies on the whole life cycle of a building according to the design–construction–operation phase in the literature. From a sustainability perspective, integrating sustainability into all stages of building is something that all stakeholders in the construction industry must consider and implement [62]. Integrating physical buildings with digital modeling provides a collaborative platform in a visual manner to help designers, architects, engineers, developers, and even end users build green and sustainable buildings throughout the project life cycle [66]. In order to achieve more effective low-carbon management, BIM tools need to incorporate the concept of 'reduce, reuse, and recycle'. There are many kinds of BIM models, but it is still necessary to improve the practicability, ease of use, and compatibility of BIM tools, so as to maximize their effectiveness and be used by more stakeholders without obstacles [28]. Blockchain provides reliable technology for building information management in the overall life cycle stage [33], which can reduce counterfeiting risks and costs without third-party management. This ensures the security, reliability and privacy of the data and establishes trust between the device and the user [82].

4. Discussion

4.1. Smart Cities and City Information Management (CIM)

The combination of information and infrastructure lays the foundation for smart city management, which greatly improves construction efficiency and reducing construction risks [20]. CIM in cities is similar to the role of BIM in architecture, which is the integration of all spatial data models. CIM helps optimize the management process [83], improve public services, and improve the quality of life of citizens [84]. In its unique database, all stakeholders jointly participate in decision-making in terms of project design, planning, operation, and maintenance, and comprehensively integrate information on the built environment, the legal environment, and the natural environment [85] to obtain an optimal solution. However, in the process of model information management, information integration is a complicated process, and there are many problems and challenges. As a distributed ledger, blockchain can safely store data and information in the chain; and the messages, it transmits have integrity, consistency, and reliability. Only authorized users have the right to make changes, and the rest are not allowed to make changes, ensuring the security of information.

In addition to information resources, infrastructure is also an important part of smart cities. As an important field in smart cities, the construction industry contains complex management, and construction projects involve the joint cooperation of multiple companies. From a sustainable perspective, buildings consume massive energy around the world. Under the major premise of ensuring the health of users, the life cycle of building should be maximized to achieve energy saving, water saving, material saving, and environmental protection [86]. For energy management, it can reduce energy waste by developing smart grids and regulating power generation on demand. For carbon management and water management, companies should give full play to their sense of corporate social responsibility, visually disclose carbon emissions and water transactions. For waste management, a unified waste management system based on blockchain will be developed to reuse or recycle construction waste to achieve the sustainability of construction projects. The openness, fairness, and transparency of blockchain can satisfy the decentralized management of construction quality information [81]. All stakeholders are encouraged to cooperate together, and collective interests are put before personal interests [30] to ensure the orderly implementation of decentralized management of the construction industry. Blockchain can efficiently process transactions, secure data, reduce labor costs, and improve transparency and security [87].

4.2. Blockchain and BIM

Although the use of blockchain in the construction industry has not yet been widely used, the integration of BIM and blockchain has shown a potential in the construction industry for sustainable development. BIM is a shared knowledge resource that can store important information collected in blockchain to ensure information security and obtain the required information from blockchain at any time. However, BIM information will be constantly updated by stakeholders. This will produce multiple new BIM files with a lot of duplicate information in the files. The Semantic Differential Transaction (SDT) can capture these changes and protect these changes in blockchain to reduce this duplication [29].

Moreover, BIM and blockchain are complementary platforms, and blockchain can make up for the shortcomings of BIM applications. For example, when different stakeholders cannot seamlessly collaborate and communicate easily when using the model, and cannot exchange effective information, blockchain will be a trusted means of collaboration. There are a large number of temporary companies participating in the construction asset management system, which is fragmented and complicated and troublesome in management. Blockchain can store all necessary data related to assets, providing a better asset life cycle [31].

Furthermore, BIM and blockchain data are transmitted electronically, but the complex collaboration framework of BIM cannot achieve complete security protection [25]. However, blockchain can safely store privacy-sensitive sensor data. Hence, the integration of blockchain and BIM can solve the obstacles of BIM development [33].

4.3. Blockchain and Sustainable Building

The characteristics of blockchain can avoid the centralized control of traditional management. The security of blockchain provides a trustworthy platform for stakeholders in the construction industry. There are practical applications or theoretical research on the three stages of the sustainable building life cycle: design–construction–operation.

In the design phase, it is very important to protect the design results from being violated. The management of digital copyright involves the scope of intellectual property rights. In order to prevent plagiarism and theft, blockchain establishes a complete chain of trust from the copyright holder to the service provider [88]. The watermark-based multimedia framework prevents tampering with the original content of the media [87]. During the design process, the design drawings and construction drawings will be continuously updated, and the unified opinions of the construction parties are also required. The blockchain system based on smart contracts [87] can update the latest drawing information to all parties without causing omissions. Distributed consensus protocol is adopted to ensure the consistency of data on the chain of participants [89], and real-time negotiation, modification, and information sharing can be carried out.

In the construction phase, in the face of the global pandemic or the future environment, electronic bidding is a desirable way. This method can protect the privacy of user identity, conduct transactions and submit bids anonymously, free from distance and weather, save time, and reduce fraud in black box operations [64]. It is important to prevent fraudulent digital signatures when signing construction contracts and approval procedures. It is conducive to mutual authentication and the establishment of trust [90].

In the operation phase, the building consumes huge energy during the use and maintenance stage, accounting for 90% of the building life cycle [91]. Operation is a long-term process, and maintenance is one of the most costly and longest stages in the process [34]. Therefore, it is very important to establish a circular and sustainable design–construction–operation system. Through the transparency of blockchain, the status of maintenance requests and the maintenance process can be known, so that the maintainer will know the time, place and maintenance components. By sharing information through blockchain, the maintenance department can monitor the update of information in real time and maintain in time. It can improve management performance and maintenance efficiency, reduce the cost of FM [41], and reduce economic losses and safety hazards caused by construction hazards.

In addition, the operational phase also includes various management systems. For building supply chain management, blockchain helps audit and track supply chain information. Under the premise of ensuring quality, it reduces operating costs, logistics costs and the risk of tampering with products [88]. There are also building management systems (BMS), which manage system components such as access control, building control and mobile applications [19].

Funds should not be sloppy during the entire operation phase. For the procurement system, the traceability of blockchain makes the supervision more transparent and easier, it can also formulate a detailed timetable to avoid default [88]. On delivery payment (transfer of large amounts of funds)/security of payment (SOP), security can be guaranteed and payment risks eliminated. The use of virtual currency eliminates the need for bank notes or cash, reducing labor costs. It can also reduce late payment and default settlement [31]. In integrated project delivery (IPD), blockchain assists in the establishment of decentralized, automated, and secure financial platforms [32], where data providers are paid and users receive data [92].

4.4. Sustainable Building, Blockchain, and BIM

The integration of digital technologies, such as BIM and blockchain, has contributed to the digital transformation of sustainable buildings [19]. It can be seen from the table in the results that there are few studies that discuss the three together.

In the design phase, information, construction knowledge, environmental, social and economic impacts need to be taken into overall consideration [93]. It is necessary to unify decision-makers, supervisors, investors, designers, construction parties, owners, and other stakeholders to participate in the design of BIM [8]. BIM integrates with sustainable building design process model [71] to collect a lot of unstable data information. Meanwhile, blockchain can balance privacy and accessibility to ensure the safe transmission of data to organizations or devices [94]. Liu et al. [95] established the relationship and role of BIM and blockchain in sustainable building design information management from the perspective of architects, which emphasizes how users (stakeholders of sustainable building design projects/BIM customers) manage.

With the assistance of BIM, the project was designed to comply with architectural and engineering specifications during the preliminary design [73]. In the process of simulation, the model is continuously optimized, and problems are solved in time. This can foreseeably avoid many problems encountered by traditional architectural design and construction [96]. Such as the order of construction, size, safety, placement, selection of materials, shapes, and environmental issues caused by spaces. Early planning for these issues will make the initial design more standardized, and each stakeholder can understand the design process, clarify the work content of all parties, and make the work more organized. At the same time, blockchain can provide interoperable protocols for all partners to ensure effective authorization interactions among decision makers, designers, and construction parties; and handle data interoperability issues well [97].

During the construction phase, renovation is more energy-efficient, environmentally friendly and cheaper than new ones [68]. However, BIM is not mature enough to maintain and update existing BIM in renovation projects, and data collection is a challenge [63]. Therefore, it is very important to provide a rich semantic database and integrate different information sources [36]. Using architectural design knowledge can improve the effectiveness of BIM reconstruction [98]. For worn-out buildings that do not meet legal requirements, using BIM for repairs has huge energy-saving potential [72].

In the operation phase, effective operation activities can improve the built environment and enable it to have more integrated functions [99]. Blockchain can realize transactions from sales and operations to finance and management, of which the smart contract can balance the profits of each partner and the interests of the company and users in an open and transparent manner [100]. The circular economy runs through the construction industry, retaining used building materials as a material library [101]. City managers and architects are familiar with the reuse of these building materials. These experiences can increase the market for reused materials, reduce the use of raw materials, focus on environmental benefits, and promote the sustainable construction of smart cities. BIM and blockchain can help analyze the most valuable materials for recycling [102].

The integration of FM, BIM, and blockchain can greatly improve the execution of construction projects and provide assistance for maintenance management and construction performance evaluation activities [35]. The implementation of BIM needs to adopt the bottom-up approach and ensure that all parties participate in planning and the participants have sufficient understanding and technical capacity [103]. Blockchain can assist BIM to efficiently use data to objectively evaluate projects, simulate building performance and full-cycle usage to improve the overall construction project quality, and achieve the goal of environmental protection and efficient resource conservation [47,104].

5. Conclusions

The aim of this paper is to explore the potential impact of the integration of blockchain and BIM in a smart city environment on making buildings more sustainable within the context of CIM/Smart Cities. This paper queries the Web of Science (WoS) database with keywords to obtain relevant publication, and then uses the VOSviewer to visually analyze the relationships between blockchain, BIM, and sustainable building within the context of smart cities and CIM, which is conducted in bibliometric analysis followed by micro scheme analysis. The main contributions and novelties of this paper are as follows: (1) This is the first attempt to explore the relationship between blockchain, BIM, and sustainable building under smart cities and CIM by using bibliometric analysis and scheme analysis. (2) A tool to visualize scholarly publications, VOSviewer, is used to conduct a macro analysis. The method of visualizing keywords can bring out insights into blockchain, BIM and sustainable building. This includes the relationship among the three, the development trend, the research hotspots, and the application, providing a reliable research method for the future research. (3) Compared with the existing similar publications, this paper discusses more comprehensively, including the relationship and application of blockchain and BIM, blockchain and sustainable construction, blockchain and BIM, and sustainable construction. (4) Compared with the existing similar publications, this paper incorporates the design–construction–operation of the building life cycle into the research. A complete life cycle study can help designers, constructors, decision makers, and supervisors who use information technology in the construction industry to make informed project decisions. Regarding CIM, although there are few research publications on CIM retrieved so far, the application of CIM in smart cities will be the trend in the future. The CIM basic platform is a new type of infrastructure for smart cities and an important tool to promote more information and intelligent city management. CIM integrates all urban spatial models, including BIM, as well as visualization, data analysis and parameterization functions. It can directly obtain information from the virtual model to improve urban development and provide services for citizens, which is of great significance to the more sustainable development of cities. The publications retrieved are represented by keywords, which makes the data reported in this paper limited to the content contained in the WoS database. Future research will explore the expansion of BIM to a wider range of CIM, and study the practical application of CIM in supporting more sustainable urban development.

Author Contributions: Conceptualization, Z.L. and Z.C.; Methodology, Z.L., Z.C., M.O. and P.D.; Software, Z.C.; Validation, Z.L., Z.C., M.O. and P.D.; Formal analysis, Z.C.; Investigation, Z.L. and Z.C.; Resources, Z.L.; Data curation, Z.L. and Z.C.; Writing—original draft preparation, Z.L. and Z.C.; Writing—review and editing, Z.L., Z.C., M.O. and P.D.; Visualization, Z.C.; Supervision, Z.L.; Project administration, Z.L.; Funding acquisition, Z.L. All authors have read and agreed to the published version of the manuscript.

Funding: This research was funded by Guangzhou City Philosophy and Social Science Planning 2020 Annual Project: grant number 2020GZYB12, and South China University of Technology Central University Basic Scientific Research Operation Funds (Social Science): (x2sjC2191370) grant number XYZD201928.

Institutional Review Board Statement: Not applicable.

Data Availability Statement: Publicly available datasets were analyzed in this study. This data can be found here: [https://login.webofknowledge.com].

Acknowledgments: The authors wish to thank all the people who support this research. Z.L. would like to thank singer Jay Chou who inspired him many years ago and combined his unique music style with Chinese and western elements to create a new music ecosystem that continues to affect the present and the future, which is the same with current blockchain and building information management are developing in smart cities towards sustainable development.

Conflicts of Interest: The authors declare no conflict of interest. The funders had no role in the design of the study; in the collection, analyses, or interpretation of data; in the writing of the manuscript, or in the decision to publish the results.

References

1. Yin, C.; Xiong, Z.; Chen, H.; Wang, J.; Cooper, D.; David, B. A literature survey on smart cities. *Sci. China Inf. Sci.* **2015**, *58*, 1–18. [CrossRef]
2. Ibrahim, M.; El-Zaart, A.; Adams, C. Smart sustainable cities roadmap: Readiness for transformation towards urban sustainability. *Sustain. Cities Soc.* **2018**, *37*, 530–540. [CrossRef]
3. BSI (British Standards Institution). Smart Cities Overview—Guide. Available online: http://shop.bsigroup.com/upload/Shop/Download/PAS/30313208-PD8100-2015.pdf (accessed on 8 December 2020).
4. Cardullo, P.; Kitchin, R. Smart urbanism and smart citizenship: The neoliberal logic of 'citizen-focused' smart cities in Europe. *Environ. Plan. C Polit. Space* **2019**, *37*, 813–830. [CrossRef]
5. Greene, S.; Pettit, K.L.S. *What If Cities Used Data to Drive Inclusive Neighborhood Change*; Urban Institute: Washington, DC, USA, 2016.
6. Desouza, K.C.; Hunter, M.; Jacob, B.; Yigitcanlar, T. Pathways to the Making of Prosperous Smart Cities: An Exploratory Study on the Best Practice. *J. Urban Technol.* **2020**, *27*, 3–32. [CrossRef]
7. Yigitcanlar, T.; Han, H.; Kamruzzaman, M.; Ioppolo, G.; Sabatini-Marques, J. The making of smart cities: Are Songdo, Masdar, Amsterdam, San Francisco and Brisbane the best we could build? *Land Use Policy.* **2019**, *88*, 104187. [CrossRef]
8. Angelo, H.; Vormann, B. Long waves of urban reform. *City* **2018**, *22*, 782–800. [CrossRef]
9. Camerin, F. From "Ribera Plan" to "Diagonal Mar", passing through 1992 "Vila Olímpica". How urban renewal took place as urban regeneration in Poblenou district (Barcelona). *Land Use Policy* **2019**, *89*, 104226. [CrossRef]
10. Noori, N.; Hoppe, T.; De Jong, M. Classifying Pathways for Smart City Development: Comparing Design, Governance and Implementation in Amsterdam, Barcelona, Dubai, and Abu Dhabi. *Sustainability* **2020**, *12*, 4030. [CrossRef]
11. Laitinen, I.; Piazza, R.; Stenvall, J. Adaptive learning in smart cities—The cases of Catania and Helsinki. *J. Adult Contin. Educ.* **2017**, *23*, 119–137. [CrossRef]
12. Yigitcanlar, T.; Baum, S.; Horton, S. Attracting and retaining knowledge workers in knowledge cities. *J. Knowl. Manag.* **2007**, *11*, 6–17. [CrossRef]
13. Huston, S.; Warren, C. Knowledge city and urban economic resilience. *J. Prop. Invest. Financ.* **2013**, *31*, 78–88. [CrossRef]
14. Hope, A. Creating sustainable cities through knowledge exchange. *Int. J. Sustain. High. Educ.* **2016**, *17*, 796–811. [CrossRef]
15. March, H.; Ribera-Fumaz, R. Smart contradictions: The politics of making Barcelona a Self-sufficient city. *Eur. Urban. Reg. Stud.* **2016**, *23*, 816–830. [CrossRef]
16. Gascó, M.; Trivellato, B.; Cavenago, D. How Do Southern European Cities Foster Innovation? Lessons from the Experience of the Smart City Approaches of Barcelona and Milan. In *Co-Creating Digital Public Services for an Ageing Society*; Springer Nature: Berlin, Germany, 2015; Volume 11, pp. 191–206.
17. Bibri, S.E.; Krogstie, J. Environmentally data-driven smart sustainable cities: Applied innovative solutions for energy efficiency, pollution reduction, and urban metabolism. *Energy Inform.* **2020**, *3*, 1–59. [CrossRef]
18. Haarstad, H. Constructing the sustainable city: Examining the role of sustainability in the 'smart city' discourse. *J. Environ. Policy Plan.* **2016**, *19*, 423–437. [CrossRef]
19. Lokshina, I.V.; Greguš, M.; Thomas, W.L. Application of Integrated Building Information Modeling, IoT and Blockchain Technologies in System Design of a Smart Building. *Procedia Comput. Sci.* **2019**, *160*, 497–502. [CrossRef]
20. Li, Y.-W.; Cao, K. Establishment and application of intelligent city building information model based on BP neural network model. *Comput. Commun.* **2020**, *153*, 382–389. [CrossRef]
21. CIM Brain and CIM Digital Twin New Infrastructure White Paper. Available online: https://mp.weixin.qq.com/s/R6maxZ2XtVRMgHUvTo7_EA (accessed on 15 December 2020).
22. Li, X.; Wu, P.; Shen, G.Q.; Wang, X.; Teng, Y. Mapping the knowledge domains of Building Information Modeling (BIM): A bibliometric approach. *Autom. Constr.* **2017**, *84*, 195–206. [CrossRef]
23. Falcão, G.; Beirão, J. Design Narrative and City Information Modeling. In *Advances in Intelligent Systems and Computing*; Springer Science and Business Media LLC: Berlin, Germany, 2020; pp. 142–147.
24. Passer, A.; Wall, J.; Kreiner, H.; Maydl, P.; Höfler, K. Sustainable buildings, construction products and technologies: Linking research and construction practice. *Int. J. Life Cycle Assess.* **2014**, *20*, 1–8. [CrossRef]
25. Nawari, N.O.; Ravindran, S. Blockchain and the built environment: Potentials and limitations. *J. Build. Eng.* **2019**, *25*, 100832. [CrossRef]

26. Perera, S.; Nanayakkara, S.; Rodrigo, M.; Senaratne, S.; Weinand, R. Blockchain technology: Is it hype or real in the construction industry? *J. Ind. Inf. Integr.* **2020**, *17*, 100125. [CrossRef]
27. Penzes, B. *Blockchain Technology in the Construction Industry*; Institution of Civil Engineers: Westminster, London, UK, 2018; pp. 1–53. [CrossRef]
28. Wong, J.K.W.; Zhou, J. Enhancing environmental sustainability over building life cycles through green BIM: A review. *Autom. Constr.* **2015**, *57*, 156–165. [CrossRef]
29. Xue, F.; Lu, W. A semantic differential transaction approach to minimizing information redundancy for BIM and blockchain integration. *Autom. Constr.* **2020**, *118*, 103270. [CrossRef]
30. Hunhevicz, J.J.; Hall, D.M. Do you need a blockchain in construction? Use case categories and decision framework for DLT design options. *Adv. Eng. Inform.* **2020**, *45*, 101094. [CrossRef]
31. Chong, H.-Y.; Diamantopoulos, A. Integrating advanced technologies to uphold security of payment: Data flow diagram. *Autom. Constr.* **2020**, *114*, 103158. [CrossRef]
32. Elghaish, F.; Abrishami, S.; Hosseini, M.R. Integrated project delivery with blockchain: An automated financial system. *Autom. Constr.* **2020**, *114*, 103182. [CrossRef]
33. Turk, Ž.; Klinc, R. Potentials of Blockchain Technology for Construction Management. *Procedia Eng.* **2017**, *196*, 638–645. [CrossRef]
34. Chen, C.; Tang, L. BIM-based integrated management workflow design for schedule and cost planning of building fabric maintenance. *Autom. Constr.* **2019**, *107*, 102944. [CrossRef]
35. Marmo, R.; Polverino, F.; Nicolella, M.; Tibaut, A. Building performance and maintenance information model based on IFC schema. *Autom. Constr.* **2020**, *118*, 103275. [CrossRef]
36. Matarneh, S.T.; Danso-Amoako, M.; Al-Bizri, S.; Gaterell, M.; Matarneh, R. Building information modeling for facilities management: A literature review and future research directions. *J. Build. Eng.* **2019**, *24*, 100755. [CrossRef]
37. Habibi, S. Micro-climatization and real-time digitalization effects on energy efficiency based on user behavior. *Build. Environ.* **2017**, *114*, 410–428. [CrossRef]
38. Kang, T.-W.; Choi, H.-S. BIM perspective definition metadata for interworking facility management data. *Adv. Eng. Inform.* **2015**, *29*, 958–970. [CrossRef]
39. Kim, K.; Kim, H.; Kim, W.; Kim, C.; Kim, J.; Yu, J. Integration of ifc objects and facility management work information using Semantic Web. *Autom. Constr.* **2018**, *87*, 173–187. [CrossRef]
40. Lin, Y.-C.; Lin, C.-P.; Hu, H.-T.; Su, Y.-C. Developing final as-built BIM model management system for owners during project closeout: A case study. *Adv. Eng. Inform.* **2018**, *36*, 178–193. [CrossRef]
41. Chen, W.; Chen, K.; Cheng, J.C.; Wang, Q.; Gan, V.J. BIM-based framework for automatic scheduling of facility maintenance work orders. *Autom. Constr.* **2018**, *91*, 15–30. [CrossRef]
42. Oti, A.; Kurul, E.; Cheung, F.; Tah, J. A framework for the utilization of Building Management System data in building information models for building design and operation. *Autom. Constr.* **2016**, *72*, 195–210. [CrossRef]
43. Wong, J.K.W.; Ge, J.; He, S.X. Digitisation in facilities management: A literature review and future research directions. *Autom. Constr.* **2018**, *92*, 312–326. [CrossRef]
44. Chen, L.; Shi, P.; Tang, Q.; Liu, W.; Wu, Q. Development and application of a specification-compliant highway tunnel facility management system based on BIM. *Tunn. Undergr. Space Technol.* **2020**, *97*, 103262. [CrossRef]
45. Gao, X.; Pishdad-Bozorgi, P. BIM-enabled facilities operation and maintenance: A review. *Adv. Eng. Inform.* **2019**, *39*, 227–247. [CrossRef]
46. Shen, W.; Hao, Q.; Xue, Y. A loosely coupled system integration approach for decision support in facility management and maintenance. *Autom. Constr.* **2012**, *25*, 41–48. [CrossRef]
47. Jin, R.; Zhong, B.; Ma, L.; Hashemi, A.; Ding, L. Integrating BIM with building performance analysis in project life-cycle. *Autom. Constr.* **2019**, *106*, 102861. [CrossRef]
48. Sheikhnejad, Y.; Yigitcanlar, T. Scientific Landscape of Sustainable Urban and Rural Areas Research: A Systematic Scientometric Analysis. *Sustainability* **2020**, *12*, 1293. [CrossRef]
49. Faisal, A.; Yigitcanlar, T.; Kamruzzaman, M.; Paz, A. Mapping Two Decades of Autonomous Vehicle Research: A Systematic Scientometric Analysis. *J. Urban Technol.* **2020**, 1–30. [CrossRef]
50. Noh, H.; Jo, Y.; Lee, S. Keyword selection and processing strategy for applying text mining to patent analysis. *Expert Syst. Appl.* **2015**, *42*, 4348–4360. [CrossRef]
51. Yu, X.; Yu, Z.; Liu, Y.; Shi, H. CI-Rank: Collective importance ranking for keyword search in databases. *Inf. Sci.* **2017**, *384*, 1–20. [CrossRef]
52. Biswas, S.K.; Bordoloi, M.; Shreya, J. A graph based keyword extraction model using collective node weight. *Expert Syst. Appl.* **2018**, *97*, 51–59. [CrossRef]
53. Faust, O. Documenting and predicting topic changes in Computers in Biology and Medicine: A bibliometric keyword analysis from 1990 to 2017. *Inform. Med. Unlocked* **2018**, *11*, 15–27. [CrossRef]
54. Meharwade, A.; Patil, G. Efficient Keyword Search over Encrypted Cloud Data. *Procedia Comput. Sci.* **2016**, *78*, 139–145. [CrossRef]
55. Bonci, A.; Carbonari, A.; Cucchiarelli, A.; Messi, L.; Pirani, M.; Vaccarini, M. A cyber-physical system approach for building efficiency monitoring. *Autom. Constr.* **2019**, *102*, 68–85. [CrossRef]

56. Jones, B. Integrated Project Delivery (IPD) for Maximizing Design and Construction Considerations Regarding Sustainability. *Procedia Eng.* **2014**, *95*, 528–538. [CrossRef]
57. Tang, S.; Shelden, D.R.; Eastman, C.M.; Pishdad-Bozorgi, P.; Gao, X. A review of building information modeling (BIM) and the internet of things (IoT) devices integration: Present status and future trends. *Autom. Constr.* **2019**, *101*, 127–139. [CrossRef]
58. Zadeh, P.A.; Wang, G.; Cavka, H.B.; Staub-French, S.; Pottinger, R. Information Quality Assessment for Facility Management. *Adv. Eng. Inform.* **2017**, *33*, 181–205. [CrossRef]
59. Nicał, A.; Wodyński, W. Enhancing Facility Management through BIM 6D. *Procedia Eng.* **2016**, *164*, 299–306. [CrossRef]
60. Olawumi, T.O.; Chan, D.W.; Wong, J.K.; Chan, A.P. Barriers to the integration of BIM and sustainability practices in construction projects: A Delphi survey of international experts. *J. Build. Eng.* **2018**, *20*, 60–71. [CrossRef]
61. Olawumi, T.O.; Chan, D.W. Concomitant impediments to the implementation of smart sustainable practices in the built environment. *Sustain. Prod. Consum.* **2020**, *21*, 239–251. [CrossRef]
62. Jrade, A.; Jalaei, F. Integrating building information modelling with sustainability to design building projects at the conceptual stage. *Build. Simul.* **2013**, *6*, 429–444. [CrossRef]
63. Khaddaj, M.; Srour, I. Using BIM to Retrofit Existing Buildings. *Procedia Eng.* **2016**, *145*, 1526–1533. [CrossRef]
64. Li, J.; Greenwood, D.; Kassem, M. Blockchain in the built environment and construction industry: A systematic review, conceptual models and practical use cases. *Autom. Constr.* **2019**, *102*, 288–307. [CrossRef]
65. Ge, C.; Liu, Z.; Fang, L. A blockchain based decentralized data security mechanism for the Internet of Things. *J. Parallel Distrib. Comput.* **2020**, *141*, 1–9. [CrossRef]
66. Bonenberg, W.; Wei, X. Green BIM in Sustainable Infrastructure. *Procedia Manuf.* **2015**, *3*, 1654–1659. [CrossRef]
67. Yilmaz, G.; Akcamete, A.; Demirors, O. A reference model for BIM capability assessments. *Autom. Constr.* **2019**, *101*, 245–263. [CrossRef]
68. Capeluto, I.G.; Ben-Avraham, O. Assessing the green potential of existing buildings towards smart cities and districts. *Indoor Built Environ.* **2016**, *25*, 1124–1135. [CrossRef]
69. Cheshmehzangi, A. Feasibility Study of Songao's Low Carbon Town Planning, China. *Energy Procedia* **2016**, *88*, 313–320. [CrossRef]
70. Kylili, A.; Fokaides, P.A. European smart cities: The role of zero energy buildings. *Sustain. Cities Soc.* **2015**, *15*, 86–95. [CrossRef]
71. Stipo, F.J.F. A Standard Design Process for Sustainable Design. *Procedia Comput. Sci.* **2015**, *52*, 746–753. [CrossRef]
72. Ustinovichius, L.; Popov, V.; Cepurnaite, J.; Vilutienė, T.; Samofalov, M.; Miedziałowski, C. BIM-based process management model for building design and refurbishment. *Arch. Civ. Mech. Eng.* **2018**, *18*, 1136–1149. [CrossRef]
73. Ghosh, S. Smart homes: Architectural and engineering design imperatives for smart city building codes. In Proceedings of the 2018 Technologies for Smart-City Energy Security and Power (ICSESP), Bhubaneswar, India, 28–30 March 2018; pp. 1–4. [CrossRef]
74. Dornelles, L.D.L.; Gandolfi, F.; Mercader-Moyano, P.; Mosquera-Adell, E. Place and memory indicator: Methodology for the formulation of a qualitative indicator, named place and memory, with the intent of contributing to previous works of intervention and restoration of heritage spaces and buildings, in the aspect of sustainability. *Sustain. Cities Soc.* **2020**, *54*, 101985. [CrossRef]
75. Bretherton, J. Christchurch's High Performance Rebuild. *Procedia Eng.* **2017**, *180*, 1044–1055. [CrossRef]
76. Zhen, X.; Furong, Z.; Wei, J.; Yingying, W.; Mingzhu, Q.; Yajun, Y. A 5D simulation method on post-earthquake repair process of buildings based on BIM. *Earthq. Eng. Eng. Vib.* **2020**, *19*, 541–560. [CrossRef]
77. Manganelli, B.; Tataranna, S.; Pontrandolfi, P. A model to support the decision-making in urban regeneration. *Land Use Policy* **2020**, *99*, 104865. [CrossRef]
78. Olawumi, T.O.; Chan, D.W. Critical success factors for implementing building information modeling and sustainability practices in construction projects: A Delphi survey. *Sustain. Dev.* **2019**, *27*, 587–602. [CrossRef]
79. Pishdad-Bozorgi, P.; Gao, X.; Eastman, C.; Self, A.P. Planning and developing facility management-enabled building information model (FM-enabled BIM). *Autom. Constr.* **2018**, *87*, 22–38. [CrossRef]
80. Araszkiewicz, K. Digital Technologies in Facility Management—The state of Practice and Research Challenges. *Procedia Eng.* **2017**, *196*, 1034–1042. [CrossRef]
81. Sheng, D.; Ding, L.; Zhong, B.; Love, P.E.; Luo, H.; Chen, J. Construction quality information management with blockchains. *Autom. Constr.* **2020**, *120*, 103373. [CrossRef]
82. Sharma, P.K.; Park, J.H. Blockchain based hybrid network architecture for the smart city. *Futur. Gener. Comput. Syst.* **2018**, *86*, 650–655. [CrossRef]
83. Melo, H.C.; Tomé, S.M.G.; Silva, M.H.; Gonzales, M.M.; Gomes, D.B.O. Implementation of City Information Modeling (CIM) concepts in the process of management of the sewage system in Piumhi, Brazil. In Proceedings of the IOP Conference Series: Earth and Environmental Science, SBE19 Brussels—BAMB-CIRCPATH "Buildings as Material Banks—A Pathway For A Circular Future", Brussels, Belgium, 5–7 February 2019; Volume 225, p. 012076. [CrossRef]
84. Dantas, H.S.; Sousa, J.M.M.S.; Melo, H.C. The Importance of City Information Modeling (CIM) for Cities' Sustainability. In Proceedings of the IOP Conference Series: Earth and Environmental Science, SBE19 Brussels—BAMB-CIRCPATH "Buildings as Material Banks—A Pathway For A Circular Future", Brussels, Belgium, 5–7 February 2019; Volume 225, p. 012074. [CrossRef]
85. Reitz, T.; Schubiger-Banz, S. The Esri 3D city information model. In Proceedings of the IOP Conference Series: Earth and Environmental Science, 8th International Symposium of the Digital Earth (ISDE8), Kuching, Malaysia, 26–29 August 2013; Volume 18, p. 12172. [CrossRef]

86. Vatalis, K.I.; Manoliadis, O.G.; Charalampides, G.; Platias, S.; Savvidis, S. Sustainability Components Affecting Decisions for Green Building Projects. *Procedia Econ. Financ.* **2013**, *5*, 747–756. [CrossRef]
87. Aggarwal, S.; Chaudhary, R.; Aujla, G.S.; Kumar, N.; Choo, K.-K.R.; Zomaya, A.Y. Blockchain for smart communities: Applications, challenges and opportunities. *J. Netw. Comput. Appl.* **2019**, *144*, 13–48. [CrossRef]
88. Mistry, I.; Tanwar, S.; Tyagi, S.; Kumar, N. Blockchain for 5G-enabled IoT for industrial automation: A systematic review, solutions, and challenges. *Mech. Syst. Signal Process.* **2020**, *135*, 106382. [CrossRef]
89. Huang, B.; Zhang, R.; Lu, Z.; Zhang, Y.; Wu, J.; Zhan, L.; Hung, P.C. BPS: A reliable and efficient pub/sub communication model with blockchain-enhanced paradigm in multi-tenant edge cloud. *J. Parallel Distrib. Comput.* **2020**, *143*, 167–178. [CrossRef]
90. Viriyasitavat, W.; Anuphaptrirong, T.; Hoonsopon, D. When blockchain meets Internet of Things: Characteristics, challenges, and business opportunities. *J. Ind. Inf. Integr.* **2019**, *15*, 21–28. [CrossRef]
91. Martínez-Rocamora, A.; Solís-Guzmán, J.; Marrero, M. Ecological footprint of the use and maintenance phase of buildings: Maintenance tasks and final results. *Energy Build.* **2017**, *155*, 339–351. [CrossRef]
92. Zhao, Y.; Yu, Y.; Li, Y.; Han, G.; Du, X. Machine learning based privacy-preserving fair data trading in big data market. *Inf. Sci.* **2019**, *478*, 449–460. [CrossRef]
93. Kuster, C.; Hippolyte, J.-L.; Rezgui, Y. The UDSA ontology: An ontology to support real time urban sustainability assessment. *Adv. Eng. Softw.* **2020**, *140*, 102731. [CrossRef]
94. Dagher, G.G.; Mohler, J.; Milojkovic, M.; Marella, P.B. Ancile: Privacy-preserving framework for access control and interoperability of electronic health records using blockchain technology. *Sustain. Cities Soc.* **2018**, *39*, 283–297. [CrossRef]
95. Liu, Z.; Jiang, L.; Osmani, M.; Demian, P. Building Information Management (BIM) and Blockchain (BC) for Sustainable Building Design Information Management Framework. *Electronics* **2019**, *8*, 724. [CrossRef]
96. Shoubi, M.V.; Shoubi, M.V.; Bagchi, A.; Barough, A.S. Reducing the operational energy demand in buildings using building information modeling tools and sustainability approaches. *Ain Shams Eng. J.* **2015**, *6*, 41–55. [CrossRef]
97. Khezr, S.; Yassine, A.; Benlamri, R. Blockchain for smart homes: Review of current trends and research challenges. *Comput. Electr. Eng.* **2020**, *83*, 106585. [CrossRef]
98. Xue, F.; Lu, W.; Chen, K.; Webster, C.J. BIM reconstruction from 3D point clouds: A semantic registration approach based on multimodal optimization and architectural design knowledge. *Adv. Eng. Inform.* **2019**, *42*, 100965. [CrossRef]
99. Nam, K.; Dutt, C.S.; Chathoth, P.; Khan, M.S. Blockchain technology for smart city and smart tourism: Latest trends and challenges. *Asia Pac. J. Tour. Res.* **2019**, 1–15. [CrossRef]
100. Chaudhary, R.; Jindal, A.; Aujla, G.S.; Aggarwal, S.; Kumar, N.; Choo, K.-K.R. BEST: Blockchain-based secure energy trading in SDN-enabled intelligent transportation system. *Comput. Secur.* **2019**, *85*, 288–299. [CrossRef]
101. Gepts, B.; Meex, E.; Nuyts, E.; Knapen, E.; Verbeeck, G. Existing databases as means to explore the potential of the building stock as material bank. *IOP Conf. Series: Earth Environ. Sci.* **2019**, *225*, 012002. [CrossRef]
102. Copeland, S.; Bilec, M.M. Buildings as material banks using RFID and building information modeling in a circular economy. *Procedia CIRP* **2020**, *90*, 143–147. [CrossRef]
103. Arayici, Y.; Coates, P.; Koskela, L.; Kagioglou, M.; Usher, C.; O'Reilly, K. Technology adoption in the BIM implementation for lean architectural practice. *Autom. Constr.* **2011**, *20*, 189–195. [CrossRef]
104. Hong, T.; Langevin, J.; Sun, K. Building simulation: Ten challenges. *Build. Simul.* **2018**, *11*, 871–898. [CrossRef]

Perspective

Green Artificial Intelligence: Towards an Efficient, Sustainable and Equitable Technology for Smart Cities and Futures

Tan Yigitcanlar [1,2,*], Rashid Mehmood [3] and Juan M. Corchado [4,5,6]

[1] School of Architecture and Built Environment, Queensland University of Technology, 2 George Street, Brisbane, QLD 4000, Australia
[2] School of Technology, Federal University of Santa Catarina, Campus Universitario, Florianopolis, SC 88040-900, Brazil
[3] High Performance Computing Center, King Abdulaziz University, Al Ehtifalat St, Jeddah 21589, Saudi Arabia; rmehmood@kau.edu.sa
[4] BISITE Research Group, University of Salamanca, 37007 Salamanca, Spain; corchado@usal.es
[5] Air Institute, IoT Digital Innovation Hub, 37188 Salamanca, Spain
[6] Department of Electronics, Information and Communication, Faculty of Engineering, Osaka Institute of Technology, Osaka 535-8585, Japan
[*] Correspondence: tan.yigitcanlar@qut.edu.au or tan.yigitcanlar@ufsc.br; Tel.: +61-7-3138-2418

Citation: Yigitcanlar, T.; Mehmood, R.; Corchado, J.M. Green Artificial Intelligence: Towards an Efficient, Sustainable and Equitable Technology for Smart Cities and Futures. *Sustainability* **2021**, *13*, 8952. https://doi.org/10.3390/su13168952

Academic Editor: Marc A. Rosen

Received: 24 May 2021
Accepted: 9 August 2021
Published: 10 August 2021

Publisher's Note: MDPI stays neutral with regard to jurisdictional claims in published maps and institutional affiliations.

Abstract: Smart cities and artificial intelligence (AI) are among the most popular discourses in urban policy circles. Most attempts at using AI to improve efficiencies in cities have nevertheless either struggled or failed to accomplish the smart city transformation. This is mainly due to short-sighted, technologically determined and reductionist AI approaches being applied to complex urbanization problems. Besides this, as smart cities are underpinned by our ability to engage with our environments, analyze them, and make efficient, sustainable and equitable decisions, the need for a green AI approach is intensified. This perspective paper, reflecting authors' opinions and interpretations, concentrates on the "green AI" concept as an enabler of the smart city transformation, as it offers the opportunity to move away from purely technocentric efficiency solutions towards efficient, sustainable and equitable solutions capable of realizing the desired urban futures. The aim of this perspective paper is two-fold: first, to highlight the fundamental shortfalls in mainstream AI system conceptualization and practice, and second, to advocate the need for a consolidated AI approach—i.e., green AI—to further support smart city transformation. The methodological approach includes a thorough appraisal of the current AI and smart city literatures, practices, developments, trends and applications. The paper informs authorities and planners on the importance of the adoption and deployment of AI systems that address efficiency, sustainability and equity issues in cities.

Keywords: artificial intelligence (AI); green AI; sustainable AI; responsible AI; ethical AI; explainable AI; AI regulation; green sensing; sustainable development goals; smart cities

1. Introduction: AI in the Smart City Context

The second digital and fourth industrial revolutions cultivated an innovation culture for the flourishing of many technological developments and breakthroughs [1,2]. For instance, the field of artificial intelligence (AI)—defined as algorithms that mimic the cognitive functions of the human mind to make decisions without being supervised [3]—has undergone remarkable exponential growth over the last couple of decades [4]. Today, AI is undoubtedly an in-trend, disruptive technology with countless applications, and even more prospects, for all industry sectors and areas of life—ranging from health to agriculture, engineering to finance, gaming to transportation, and so on [5]. Besides this, AI is one of the fundamental drivers of the global smart city movement [6].

Smart cities are widely seen as locations whereat digital technology and data are widely applied to generate efficiencies for economic growth, quality of life, and sustainability [7]. Today, in many urban policy circles and debates, concerning smart city

295

transformation, AI has become a subject of debate, particularly among urban policymakers and planners who search for technocentric solutions to alarming urbanization problems [8]. This popularity is due to the increasing recognition of technocentric solutions as a potential panacea to the complex and complicated urbanization challenges—ranging from quality of life to climate change, and safety and security to mobility and accessibility [9]. The effective use of big data, AI-powered smart urban technologies and platforms is predicted to benefit urban infrastructure and service efficiency, and to address or at least significantly ease these challenges [10,11].

As stated by Wang and Cao [12], technological advancements have generated an era wherein large volumes of data—i.e., big data—are collected via smart sensors deployed in cities, and are made available via various commercial and public channels. Due to the recent advances in AI techniques and ubiquitous computing, these data now feed into the services that improve quality of life, city operation systems, and the environment. In the context of cities, AI has various applications in areas that aim to create efficiencies in urban infrastructure and services [13]. The following are among the most prevalent AI-powered examples:

- Automated algorithmic urban decision-making (e.g., identification and penalization of traffic offences and tax evasions through smart sensors and machine learning-based data analytics) [14];
- Automated urban infrastructure assessment (e.g., monitoring urban infrastructure health through automated aerial mapping and deep-learning characterizations) [15–17];
- Autonomous urban post-disaster reconnaissance (e.g., detecting disaster damage and impact through synergistic use of deep learning and 3D point cloud features) [18];
- Autonomous and connected urban mobility (e.g., offering increased urban mobility through shared autonomous vehicles and autonomous shuttle bus fleets) [19,20];
- Urban descriptive, diagnostic, predictive and prescriptive analytics (e.g., gathering and interpreting urban air pollution data to describe what is the pollution level, why it happened, when it may occur again, and actions to influence future desired outcomes) [21];
- Urban security, safety, rescue and maintenance robots (e.g., emergency services operating rescue robots in risky and dangerous environments, such as natural disaster events, mining accidents and building collapses and fires) [22];
- Urban service agent chatbots (e.g., offering improved customer experiences with reduced waiting times to access services in different languages related to taxation, health services, public transport, family services, job opportunities and so on) [23].

In an attempt to generate the required efficiencies and proficiencies, many governments across the globe have started to deploy various AI system initiatives at the national, state and local levels [24,25]. The following are among the most common AI-driven applications [26]:

- AI process automation systems;
- AI-based knowledge management software;
- Chatbots/virtual agents;
- Cognitive robotics and autonomous systems;
- Cognitive security analytics and threat intelligence;
- Identity analytics;
- Intelligent digital assistants;
- Predictive analytics and data visualization;
- Recommendation systems;
- Speech analytics.

Despite the efforts made to adopt and deploy AI in the public sector, almost all of these initiatives have either struggled, failed, or lacked the adequate potential to generate ethical, responsible and sustainable solutions [27,28]. This is also the case for the smart city domain—the existing purely technocentric or algorithmic AI perspectives could not play a prominent role in smart city transformation [26]. Toronto Sidewalk, Masdar and Songdo are among the major smart city initiatives that have resulted in project cancellations or failures in living up to their smart urban future promises [29].

The main reason behind this failure is that AI system adoption practices are heavily technologically determined and reductionist in nature, and do not envisage and develop long-term, ethical, responsible and sustainable solutions [30]. Most government AI approaches have also overlooked urban, human and social complexities; subsequently, this has created conditions for new forms of societal control, and boosted inequality and marginalization among the layers of our societies [31]. Thus, the current practice of AI has generated as many constraints as prospects, where, at times, the constraints outweigh the prospects [32].

Against this backdrop, this perspective paper highlights the fundamental shortfalls in current AI system conceptualization and practice, and points to a novel approach— i.e., green AI that also accommodates green sensing—that moves away from short-term efficiency solutions to focus on a long-term ethical, responsible and sustainable AI practice that will help build sustainable urban futures for all through smart city transformation. Here, we note that, as this is a perspective piece reflecting the authors' opinions and interpretations of the topic, the methodological approach of the paper is not systematic; rather, the paper uses the literature to support the claims and views related to AI and smart city discourses, practices, developments, trends and applications.

2. Prospects and Constraints of AI for Smart City Transformation

The utilization of smart and innovative digital technologies has become a common approach to tackling urban crises—whether they are related to climate, pandemics, natural disasters or socioeconomic factors [33,34]. In recent years, advancements in AI—as one of the most prominent technologies of our time, with significant implications for our economy, society, environment and governance—have resulted in invaluable opportunities for cities to increase their infrastructural efficiencies and predictive analytic capabilities, and hence, to a degree, to improve the quality of life and sustainability in cities under the smart city brand [35]. According to Ullah et al. [36], today, AI is rapidly becoming a critical smart city element that helps in achieving necessary efficiencies and automation in order to deliver urban infrastructures, services and amenities.

Especially when coupled with other smart urban technologies, AI applications—e.g., chatbots and virtual agents, cognitive robotics and autonomous systems, cognitive security analytics, expert systems, identity analytics, intelligent digital assistants, knowledge management systems, predictive analytics and data visualization, process automation systems, recommendation systems, speech analytics, threat intelligence—provide new capabilities and directions for our cities, such as building the next generation smart cities, i.e., the "Artificially Intelligent Cities" of the future [37]. There is a vast array of literature on the prospects of AI for smart cities [38].

Nonetheless, as much as creating benefits—for instance, generating operational infrastructure or service efficiencies—AI technology also involves substantial risks and disruptions for cities and citizens, through the opaque decision-making processes and the privacy violations that are related to it; e.g., automating inequality, generating algorithmic bias due to bad or limited data and training, removing or limiting human responsibility, and lacking an adequate level of transparency and accountability [39]. Additionally, in comparison to the other technologies, AI involves some unique data-related challenges that include data acquisition, the large volume and the streaming of data, heterogeneous data, complex dependencies among the data, noisy and incomplete data, distributed data storage and processing, training data, and data privacy [12].

Some examples of AI mishaps that impact society, and that also diminish public trust in the AI solutions implemented as part of smart city projects, include, but are not limited to, the following:

- AI misdiagnosis of child maltreatment and the prescription of inappropriate solutions in Pittsburgh, PA, USA [40];
- Amazon's AI recruiting tool, which took biased decisions towards women [41];
- Bias towards people of color in the decisions made by AI algorithms used in US hospitals [42];
- Clearview AI's scandalous facial recognition image database developed with images from social media, which got hacked in 2020, leaving citizens of democratic countries with privacy threats, and citizens of autocratic regimes under a situation akin to an Orwellian nightmare [43];
- The malfunctioning of the Australian government's automated debt recovery program, called Robodebt, resulting in a scandal, as it had unlawfully taken AUD 721M from over 400,000 Australians [44].

One of the main reasons behind the failure of AI systems relates to the development and integration stages of AI in urban and public services. Pasquinelli [45] linked the underlying issues of AI—or in broader terms, how machines learn—to the development of AI systems when operators engage in training data, learning algorithms and model application stages. In these stages, operators could encounter three types of bias, namely:

- Just-world bias (e.g., a cognitive bias that assumes people get what they deserve, leading to failures in helping or feeling compassion for others or disadvantaged groups, such as poor or homeless people);
- Data bias (e.g., an error caused by certain elements of data being more heavily represented or weighted than other elements, leading to wrong decisions or inequity issues—such as for women, people of color or minorities);
- Algorithmic bias (e.g., a lack of fairness, originating from the output of an algorithmic system, with consequential unfavorable decisions, actions or externalities—such as a credit score algorithm denying a loan).

When an AI system containing such bias is integrated with an urban or public service, the failure of the service, or dissatisfaction with the service, is inevitable [46].

The growing concern over negative AI externalities and service failures, particularly in smart cities, proves the need for more ethical and regulated AI systems [47]. Subsequently, in recent years, attempts to provide a more holistic perspective on AI have resulted in a number of new AI conceptualizations [48]. These include "responsible AI", "ethical AI", "explainable AI", "sustainable AI", "green AI" and the like, the aim of which is to ensure the ethical, transparent and accountable use of AI applications in a manner that is consistent with user expectations, organizational values, environmental conservation and societal laws and norms [49]. It is also argued that such renewed approaches to AI will help maximize the desired smart city outcomes and positive impacts for all citizens, while minimizing the negative consequences [50].

3. The Green AI Approach for the Flourishing of Humans and the Planet

The effects of human activity—e.g., unsustainable and rapidly growing populations, urbanization, industrialization and consumerism—since the industrial revolution of 1850s, and particularly during the last five decades, have taken their toll on the environment [51–53]. As presented by Hunter and Hewson [54], the most catastrophic threats humanity is facing today include the following:

- Chemical pollution of the earth system, including the atmosphere and oceans;
- Collapse of ecosystems and loss of biodiversity;
- Decline of natural resources, particularly water;
- Global warming and human-induced climate change;
- Human population growth beyond the Earth's carrying capacity;
- National and global failures to understand and act preventatively on these risks;
- Nuclear weapons and other weapons of mass destruction;
- Pandemics of new and untreatable diseases;

- Rising food insecurity and failing nutritional quality;
- The advent of powerful, uncontrolled new technology.

Among these threats, "national and global failure to understand and act preventatively on these risks" is the most important. This is the incapability of the governments [55] and the public [56] to understand and take actions against threats that are most likely to, or definitely, lead to a catastrophe. This issue is the root cause of the failure of AI solutions—even if they target sustainability [57]—as they are mainly used to improve business efficiency and economic productivity in our cities [58] rather than actually tackling the aforementioned global threats that are mostly anthropogenic in origin [59].

The flourishing of humankind over the last 10,000 years in the Holocene is a consequence of the planet's beneficial conditions, that is, the perfect climate and ecosystem [60]. As such, investigating the ways in which AI can help establish conditions in which humans and the planet can thrive in the Anthropocene has been the subject of much recent scholarly work [61].

For example, Vinuesa et al. [62] explored the role of AI in achieving the UN's sustainable development goals (SDGs). Their study found that "AI may act as an enabler on 134 targets (79%) across all SDGs, generally through a technological improvement, which may allow to overcome certain present limitations. However, 59 targets (35%, also across all SDGs) may experience a negative impact from the development of AI". In another study, Gupta et al. [63] assessed whether AI is an enabler or an inhibitor of sustainability, measured via SDGs. Their study disclosed that "when SDGs related to Society, Economy, and Environment were analyzed, it was observed that the Environment category has the highest potential, with 93% of the targets being positively affected, whereas Society has the largest negative effect with 38% of the targets exhibiting a negative interaction with AI".

Likewise, Goranski and Tan [64] examined the role of AI in accelerating the progress of SDGs. Their investigation revealed that "AI can generate data for more intelligent targeting of intervention, reduce waste and losses in production and consumption, create new applications that will transform entire industries and professions, and provide the necessary improvements in connectivity and cost reductions that brings the benefits of the rapid pace of technological development to many people worldwide". While AI represents an opportunity for achieving the SDGs, as stated by Dwivedi et al. [65], an AI-supported delivery of SDGs will "require significant investment from governments and industry together with collaboration at an international level to effect governance, standards and security". Figure 1 shows the 17 SDGs.

Additionally, in recent years, we have witnessed an increase in academic literature reporting the outcomes of AI technology applications for social good, and in tackling social and environmental issues [66]. The environmental areas in which AI applications are utilized range from air pollution monitoring [67] to wastewater treatment [68], from endangered species protection [69,70] to climate change detection [71], from natural disaster prediction [27] to ecosystem service assessment [72], and other applications in environmental sciences [73].

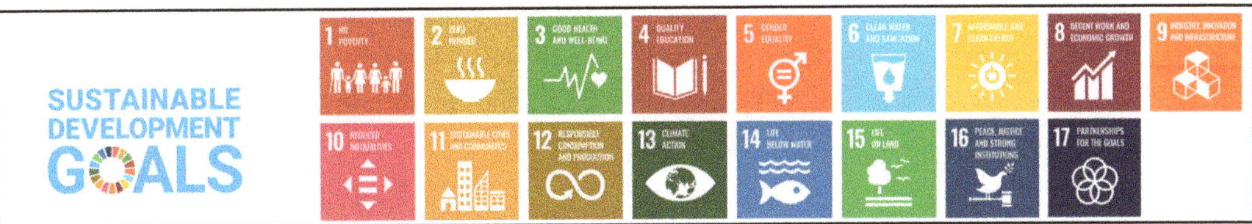

Figure 1. Sustainable development goals [74]. Source: https://www.un.org/sustainabledevelopment/news/communications-material/ (accessed on 9 August 2021).

While the existing and potential benefits of AI for the environment have been presented in the abovementioned studies, said studies also underlined the critical importance of addressing the risks involved. For instance, studies emphasized the critical importance of:

- Being supported by the necessary regulatory insight and oversight for AI-based technologies to enable sustainable development, and avoid gaps in transparency, safety and ethical standards [62];
- Going beyond the development of AI in sectorial areas, so as to understand the impacts AI might have across societal, environmental and economic outcomes [63];
- Offering a constructive, rather than optimistic or pessimistic, outlook on AI for promoting desired sustainable outcomes [75].

The most common negative effects of AI on the environment include increases in electricity usage (computation and transmission power consumption) and the resulting carbon emissions, along with errors in critical decisions due to user and data bias [76–78]. Given that global technology uptake is growing at an exponential rate, the impact of these externalities is expected to be immense [79]. Just to give an example, cryptocurrency mining in recent years has led to increased energy consumption globally.

As stated by Cuen [80], the bitcoin electricity consumption index of the University of Cambridge indicates that "bitcoin miners are expected to consume roughly 130 Terawatt-hours of energy (TWh), which is roughly 0.6% of global electricity consumption. This puts the bitcoin economy on par with the CO_2 emissions of a small developing nation like Sri Lanka or Jordan".

These undesired externalities call for a sustainable approach to AI that adopts a green-based technological perspective, including switching to a sustainable AI infrastructure [81–85]. In their study on the climate cost of global computation, Dobbe and Whittaker [86] made the following recommendations for tech-aware climate policy and climate-aware tech policy:

- Account for the entire tech ecosystem;
- Address AI's impact on climate refugees;
- Curb the use of AI to extract fossil fuels;
- Integrate tech and climate policy;
- Make non-energy policy a standard practice;
- Mandate transparency;
- Watch for rebound effects.

Making AI green and sustainable, i.e., the green AI approach, requires a bias-free (besides a reasonable environmental bias or positive discrimination), inclusive, trustworthy, explainable, ethical and responsible approach to technology that aims to alleviate the developmental challenges of the planet in a sustainable way [30,87]. This approach—using AI to solve sustainability challenges and using AI in a more sustainable way—will also serve as an enabler of smart city transformation [88,89].

4. Green Sensing, Communications and Computing

Now that we have discussed a number of policy and high-level issues related to green AI, we will define and discuss issues that relate to the development of digital infrastructure for green AI. Our intention is to discuss these infrastructural issues here, together with the high-level issues, and to provide a holistic overview such that different communities working in policy and infrastructure research can understand the cross-disciplinary issues, and collaboratively devise holistic and globally optimum solutions.

Sensing, communications, and computing have never been so interdependent as they are now due to the emergence and convergence of technologies, including the miniaturization of sensors, Internet-of-Things (IoT), data-driven methods, AI-driven optimizations, and cloud, fog and edge computing. The whole ecosystem of smart applications and systems is converging due to the need for these applications to be intrinsically collaborative and distributed.

These smart applications and systems require a certain level of smartness that enhances our ability to engage, sense and act on our environments, analyze them, and make timely, effective and sustainable decisions [90]. The trend is an increase in the number of IoT devices, expected to reach 25 billion by 2030 [91], which in turn increases the requirements for data pre-processing, storage, communications, and processing. AI provides the brain of the smartness, i.e., it undertakes the analyses and the decision-making processes. AI, while requiring significant computational resources (storage, communications and processing), has the capability to improve the efficiency of the whole infrastructural ecosystem via the holistic analysis and optimization of the system.

To aid in the development of green AI, both at the policy and the infrastructure level, we herein introduce and define the concept of "green sensing" as physical and virtual green sensing to enable triple bottom line (social, environmental and economic) sustainability. The definition proposes the development of methods and technologies to sense and measure social, environmental and economic sustainability. Sustainability is affected by challenges such as security, privacy, the safety of people, ethical standards and compliance, and so on, and therefore these are included in our definition of green sensing. These methods and technologies should be green in terms of their efficiency and energy usage.

What are the potential examples of green sensing methods and technologies in the broader sense of the term green sensing, as we have defined it above? A physical sensor to measure the sustainability of urban infrastructure and environmental pollution can be considered a green sensor [92,93]. A virtual sensor, such as that using big data or social media data to detect congestion on the road, can be considered a green sensor for environmental sustainability, in the sense that it detects the environmental pollution that may be caused by a high intensity of pollution in the geographical area where congestion is happening [94].

The same can also be considered a green sensor for social sustainability (or a social sustainability sensor) because it can detect people's anguish, and the potential health-related harms to the people living in or travelling through said geographical area [95]. AI-based virtual sensors can be developed by analyzing various literature, news media, or government regulations in order to understand their economic impacts and sustainability. The possibilities are almost endless. Sensors fit into two broad types: the ones that measure the impact of phenomena directly (e.g., the power or gas/petrol consumption of a community, or real-time or future crowd detection in public spaces) [96,97], and those that measure the impact indirectly (e.g., via social media) [98,99].

The concept of green sensing used here is different from and much broader than the earlier usage of this term, which refers to the methods used to save energy in the sensing process. The earlier use of the term mostly appeared in the context of wireless sensor networks (WSNs) [100]. A range of methods and technologies have been developed under the umbrella of wireless sensor networks to reduce the energy usage of sensors. These techniques have been naturally extended to the IoT paradigm. WSNs and IoTs have been used for many applications, including for environmental monitoring and protection, such as forest fire detection, ambience monitoring, and so on. The main motivation for developing these techniques has been to reduce the energy required by the sensing devices, which are typically wirelessly connected and battery powered.

The techniques to save energy include duty cycling (periodically turning the radio on/off to save energy), wake-up radio (on-demand radio on/off), sensor selection or scheduling (selection or scheduling of a subset of sensors instead of all the sensors), adaptive sampling (adapting the selection of sensors based on the context and application), and more. These green sensing techniques can not only reduce the energy required for sensing, but they also reduce the generated data, reducing the energy needed for data storage or pre-processing. Energy harvesting techniques by all, or a selection of, sensors using renewable energies or electric signals have also been proposed to make the sensing process greener [101].

The data sensed through IoT and other media are usually transferred to a central location, such as a master node or a cloud computing center, for their analysis. An astonishingly large amount of energy is required to transfer data across networks. Naturally, a range of techniques have been developed to reduce data communication energy and improve network efficiency. In addition to reducing data generation through the various green sensing techniques mentioned earlier, various data pruning methods have been developed, such as using data compression to reduce communication and bandwidth requirements.

Probably the most important development in this respect is fog and edge computing-based solutions, which reduce the data transfer and bandwidth requirements of smart applications by processing data at the edge or fog layers, near the sensors and devices, while offering other benefits such as data security and privacy. Several works have investigated the energy efficiency and other benefits of fog- and edge-based solutions [102].

For instance, Janbi et al. [103] developed a framework for the provision of distributed AI as a service (DAIaaS) in future environments. The framework divides "the actual training and inference computations of AI workflow into smaller computations that are executed in parallel according to the level and capacity of resources available with cloud, fog, and edge layers" [103]. They consider "multiple provisioning scenarios for DAIaaS in three case studies comprising nine applications and AI delivery models and 50 distinct sensor and software modules", and report the energy consumption and other benefits of edge- and fog-based AI delivery solutions.

Mohammed et al. [104] proposed the UbiPriSEQ framework, which aims to holistically and adaptively optimize energy efficiency, security, privacy, and quality of service (QoS). They reported an implementation of the proposed framework using deep reinforcement learning (DRL), which allows for devising policies and making decisions related to important parameters, such as transmit power, the specific fog nodes to be selected for offloading data and computation, and so on.

A number of key technological developments are shaping the development of high-speed and low-latency networks, such as fifth-generation networks (5G), which offer a powerful system for ubiquitous environments with advanced sensing and processing capabilities, along with IoT [104]. The requirements of next-generation societies are underpinned by the need to ubiquitously deliver smart services (AI and data-driven) and to increase the numbers of sensors, and these requirements are expected to be met by 6G, the next generation of cellular networks [103–106].

The 6G networks are expected to meet these requirements by use of higher-spectrum and multiple communication technologies [107], ultra-dense heterogeneous networking [106], terrestrial and non-terrestrial communications [108], and the use of AI to optimize service-oriented network operations [109,110]. An important feature of the 6G networks would be their support of ubiquitous AI services. More importantly, energy efficiency is considered the critical requirement of 6G, and this feature is expected to be ten to a hundred times better than that of 5G, achieved using novel antenna designs, zero-energy nodes, and other technologies related to low-rate sensing applications [105,111].

Energy efficiency is a grand challenge in the design of large-scale computing systems, such as supercomputers and computational clouds. For example, the system that is currently ranked number one on the list of the top 500 supercomputers, Fugaku, has over seven million cores and requires 28.3MW of power for its operation. While AI algorithms consume large amounts of power, they can be used to reduce the energy requirements of computations while optimizing performance [112], thus allowing for the concept of green (virtual) sensing and optimizations to be introduced into computing systems.

5. Final Remarks: Policy Directions for Making AI Greener and Cities Smarter

This perspective paper generally contributes to the growing AI literature by underlining the fundamental shortfalls in mainstream AI system conceptualizations and practices, and by advocating the need for a green AI approach to further support smart city transformation and SDGs.

The paper also provides a perspective on the green AI concept. It defines and elaborates on the concept, and discusses why a consolidated effort is needed in the area, including the benefits of a strengthened green AI approach. The elaborations are supported by the literature from diverse disciplines, including computer, environmental and social sciences, and urban studies. The paper also discusses issues that relate to the development of digital infrastructure for green AI. The intention is to discuss these infrastructural issues together with other high-level issues, and to provide a holistic overview such that different communities working in policy and infrastructure research can understand cross-disciplinary issues and collaboratively devise holistic and globally optimum solutions.

Moreover, in order to aid in the development of green AI, both at the policy and infrastructural levels, the paper introduces and defines the concept of "green sensing" as physical and virtual green sensing to enable triple bottom line (social, environmental and economic) sustainability. This paper also highlights the importance of, and advocates the need for, the development of methods and technologies to sense and measure social, environmental and economic sustainability.

This perspective piece makes an invaluable contribution to the emerging field of green AI, as there is no scholarly literature that discusses the policy and infrastructural issues of the given topic in an abstracted way. Our approach here allows readers to gain a holistic understanding of the issues related to green AI via a relatively succinct perspective piece, and presents prospective research and development directions.

We conclude the paper with the following remarks, as it is highly important to have timely, effective and efficient government policy in place for making AI greener and our cities smarter.

Firstly, we underline that the field of AI is growing rapidly; technological advancements are exponential, and applications are disruptive [113,114]. In such a situation, without appropriate government intervention, the business-as-usual scenario will create extended negative risks and consequences for our society and the planet [115]. Unfortunately, these risks and consequences have not yet been fully understood by governments, which means they do not act upon them or take preventative measures [116].

Secondly, as evidenced in the literature, there are colossal policy challenges in the way of making AI green [117]. Perhaps the most critical one is the need for governments to develop legal and ethical frameworks for AI and its use [118]. Expanding on this issue, Dwivedi et al. [65] listed fairness and equity, accountability and legal issues, ethics, misuse protection, transparency and auditing, and digital divides and data deficits as the fundamental public and environmental policy challenges of AI. Another study, by Jobin et al. [119], disclosed "the primary AI ethical principles as follows: transparency, justice and fairness, non-maleficence, responsibility, privacy, beneficence, freedom and autonomy, trust, sustainability, dignity, and solidarity". These principles are critical to AI projects' ability to deliver the desired outcomes to all.

Thirdly, up until now, no country has passed an AI law yet, and only a small number of countries have attempted to introduce AI ethical frameworks and regulation guidelines—such as the European Union's AI ethics guidelines, intended to inform future regulation, and other examples include AI ethical frameworks in Australia, Germany, Singapore and the UK [120]. The most popular existing practice for most governments seems to be adopting a "wait-and-see" approach to AI ethics and regulations [121]. Furthermore, as stated by Hagendorff [122], in most cases, the existing ethics frameworks fail to serve their purposes, as they lack any reinforcement mechanisms—in other words, there are no consequences if these ethical principles are not followed.

Furthermore, having no regulation at present does not mean that the AI domain will not be regulated in the near future—in this regard, we note the EU's recently released pioneering AI regulation [123]. It is very likely that, as has happened in cases of sharing economy applications, e.g., Uber and AirBnB [124,125]—governments will eventually regulate the AI practice to alleviate its undesired consequences [126]. As stated by Yara et al. [87], "with the development of technology, changes are needed in the legal regulation of AI so that the consequences of its use become useful for the whole society. Market forces with their own resources will not ensure successful development for the whole population. Thus, legal regulation of AI is inevitable."

Lastly, there is a critical role to be played by all stakeholders, e.g., public and private sectors, academia and the public, to make sure that forthcoming AI plans, ethics and regulations also bring efficiency, sustainability and equity perspectives to the technology domain, which will ultimately help in achieving SDGs [127,128]. This renewed green AI approach and capacity will also consolidate the efforts made to transform our cities into smart ones, and support the smart and sustainable development of our cities and communities [129]. In other words, as also indicated by Fisher [130], we need to put our best effort into making AI an efficient, sustainable and equitable technology for establishing smart cities and sustainable futures.

Author Contributions: Conceptualization, methodology, investigation, and writing—original draft preparation, T.Y.; investigation, and writing—review and editing, R.M. and J.M.C. All authors have read and agreed to the published version of the manuscript.

Funding: This research did not receive any specific grant from funding agencies in the public, commercial or not-for-profit sectors.

Institutional Review Board Statement: Not applicable as the study did not involve humans or animals.

Informed Consent Statement: Not applicable as the study did not involve humans or animals.

Data Availability Statement: The paper did not use any data beyond the literature, where cited and quoted literature pieces are listed at the Refences.

Acknowledgments: The authors thank the managing editor and anonymous referees for their constructive comments.

Conflicts of Interest: The authors declare no known competing financial interests or personal relationships that could have appeared to influence the study reported.

References

1. Morrar, R.; Arman, H.; Mousa, S. The fourth industrial revolution (Industry 4.0): A social innovation perspective. *Technol. Innov. Manag. Rev.* **2017**, *7*, 12–20. [CrossRef]
2. Lee, M.; Yun, J.; Pyka, A.; Won, D.; Kodama, F.; Schiuma, G.; Park, H.; Jeon, J.; Park, K.; Jung, K.; et al. How to respond to the fourth industrial revolution, or the second information technology revolution? Dynamic new combinations between technology, market, and society through open innovation. *J. Open Innov. Technol. Mark. Complex.* **2018**, *4*, 21. [CrossRef]
3. Kirwan, C.; Fu, Z. *Smart Cities and Artificial Intelligence: Convergent Systems for Planning, Design, and Operations*; Elsevier: Oxford, UK, 2020.
4. Shukla, A.; Janmaijaya, M.; Abraham, A.; Muhuri, P. Engineering applications of artificial intelligence: A bibliometric analysis of 30 years (1988–2018). *Eng. Appl. Artif. Intell.* **2019**, *85*, 517–532. [CrossRef]
5. Cugurullo, F. Urban artificial intelligence: From automation to autonomy in the smart city. *Front. Sustain. Cities* **2020**, *2*, 1–14. [CrossRef]
6. Singh, S.; Sharma, P.; Yoon, B.; Shojafar, M.; Cho, G.; Ra, I. Convergence of blockchain and artificial intelligence in IoT network for the sustainable smart city. *Sustain. Cities Soc.* **2020**, *63*, 102364. [CrossRef]
7. Mora, L.; Deakin, M.; Reid, A. Strategic principles for smart city development: A multiple case study analysis of European best practices. *Technol. Forecast. Soc. Chang.* **2019**, *142*, 70–97. [CrossRef]
8. Kassens-Noor, E.; Hintze, A. Cities of the future? The potential impact of artificial intelligence. *AI* **2020**, *1*, 192–197. [CrossRef]
9. Yigitcanlar, T.; Kankanamge, N.; Regona, M.; Maldonado, M.; Rowan, R.; Ryu, A.; Desouza, K.; Corchado, J.; Mehmood, R.; Li, R. Artificial intelligence technologies and related urban planning and development concepts: How are they perceived and utilized in Australia? *J. Open Innov. Technol. Mark. Complex.* **2020**, *6*, 187. [CrossRef]
10. Yu, Y.; Zhang, N. Does smart city policy improve energy efficiency? Evidence from a quasi-natural experiment in China. *J. Clean. Prod.* **2019**, *229*, 501–512. [CrossRef]
11. Corchado, J.M.; Chamoso, P.; Hernández, G.; Gutierrez, A.S.R.; Camacho, A.R.; González-Briones, A.; Pinto-Santos, F.; Goyenechea, E.; Garcia-Retuerta, D.; Alonso-Miguel, M.; et al. Deepint. net: A rapid deployment platform for smart territories. *Sensors* **2021**, *21*, 236. [CrossRef]
12. Wang, S.; Cao, J. AI and deep learning for urban computing. In *Urban Informatics*; Shi, W., Goodchild, M., Batty, M., Kwan, M., Zhang, A., Eds.; Springer Nature: Singapore, 2021.
13. Ortega-Fernández, A.; Martín-Rojas, R.; García-Morales, V. Artificial intelligence in the urban environment: Smart cities as models for developing innovation and sustainability. *Sustainability* **2020**, *12*, 7860. [CrossRef]

14. Gutberlet, T. Data-driven smart sustainable cities: Highly networked urban environments and automated algorithmic decision-making processes. *Geopolit. Hist. Int. Relat.* **2019**, *11*, 55–61.

15. Erkal, B.G.; Hajjar, J.F. Laser-based surface damage detection and quantification using predicted surface properties. *Autom. Constr.* **2017**, *83*, 285–302. [CrossRef]

16. Mohammadi, M.E.; Watson, D.P.; Wood, R.L. Deep learning-based damage detection from aerial SfM point clouds. *Drones* **2019**, *3*, 68. [CrossRef]

17. Nasimi, R.; Moreu, F. Development and implementation of a laser–camera–UAV System to measure total dynamic transverse displacement. *J. Eng. Mech.* **2021**, *147*, 4021045. [CrossRef]

18. Ghosh Mondal, T.; Jahanshahi, M.R.; Wu, R.T.; Wu, Z.Y. Deep learning-based multi-class damage detection for autonomous post-disaster reconnaissance. *Struct. Control. Health Monit.* **2020**, *27*, e2507. [CrossRef]

19. Faisal, A.; Kamruzzaman, M.; Yigitcanlar, T.; Currie, G. Understanding autonomous vehicles: A systematic literature review on capability, impact, planning and policy. *J. Transp. Land Use* **2019**, *12*, 45–72. [CrossRef]

20. Dennis, S.; Paz, A.; Yigitcanlar, T. Perceptions and attitudes towards the deployment of autonomous and connected vehicles: Insights from Las Vegas, Nevada. *J. Urban. Technol.* **2021**. [CrossRef]

21. Allam, Z.; Dey, G.; Jones, D. Artificial intelligence (AI) provided early detection of the coronavirus (COVID-19) in China and will influence future urban health policy internationally. *AI* **2020**, *1*, 156–165. [CrossRef]

22. Tiddi, I.; Bastianelli, E.; Daga, E.; d'Aquin, M.; Motta, E. Robot–city interaction: Mapping the research landscape—A survey of the interactions between robots and modern cities. *Int. J. Soc. Robot.* **2020**, *12*, 299–324. [CrossRef]

23. Ashfaq, M.; Yun, J.; Yu, S.; Loureiro, S. I, Chatbot: Modeling the determinants of users' satisfaction and continuance intention of AI-powered service agents. *Telemat. Inform.* **2020**, *54*, 101473. [CrossRef]

24. Engin, Z.; Treleaven, P. Algorithmic government: Automating public services and supporting civil servants in using data science technologies. *Comput. J.* **2019**, *62*, 448–460. [CrossRef]

25. Wirtz, B.; Weyerer, J.; Geyer, C. Artificial intelligence and the public sector: Applications and challenges. *Int. J. Public Adm.* **2019**, *42*, 596–615. [CrossRef]

26. Yigitcanlar, T.; Corchado, J.; Mehmood, R.; Li, R.; Mossberger, K.; Desouza, K. Responsible urban innovation with local government artificial intelligence: A conceptual framework and research agenda. *J. Open Innov. Technol. Mark. Complex.* **2021**, *7*, 71. [CrossRef]

27. Sun, W.; Bocchini, P.; Davison, B. Applications of artificial intelligence for disaster management. *Nat. Hazards* **2020**, *103*, 2631–2689. [CrossRef]

28. Desouza, K.; Dawson, G.; Chenok, D. Designing, developing, and deploying artificial intelligence systems: Lessons from and for the public sector. *Bus. Horiz.* **2020**, *63*, 205–213. [CrossRef]

29. Yigitcanlar, T.; Han, H.; Kamruzzaman, M.; Ioppolo, G.; Sabatini-Marques, J. The making of smart cities: Are Songdo, Masdar, Amsterdam, San Francisco and Brisbane the best we could build? *Land Use Policy* **2019**, *88*, 104187. [CrossRef]

30. Yigitcanlar, T.; Cugurullo, F. The sustainability of artificial intelligence: An urbanistic viewpoint from the lens of smart and sustainable cities. *Sustainability* **2020**, *12*, 8548. [CrossRef]

31. Valentine, S. Impoverished algorithms: Misguided governments, flawed technologies, and social control. *Urban. Law J.* **2019**, *46*, 364–427.

32. Nishant, R.; Kennedy, M.; Corbett, J. Artificial intelligence for sustainability: Challenges, opportunities, and a research agenda. *Int. J. Inf. Manag.* **2020**, *53*, 102104. [CrossRef]

33. Shin, D. Ubiquitous city: Urban technologies, urban infrastructure and urban informatics. *J. Inf. Sci.* **2009**, *35*, 515–526. [CrossRef]

34. Yigitcanlar, T.; Kamruzzaman, M. Smart cities and mobility: Does the smartness of Australian cities lead to sustainable commuting patterns? *J. Urban. Technol.* **2019**, *26*, 21–46. [CrossRef]

35. Batty, M. Artificial intelligence and smart cities. *Environ. Plan. B Urban. Anal. City Sci.* **2018**. [CrossRef]

36. Ullah, Z.; Al-Turjman, F.; Mostarda, L.; Gagliardi, R. Applications of artificial intelligence and machine learning in smart cities. *Comput. Commun.* **2020**, *154*, 313–323. [CrossRef]

37. Yigitcanlar, T.; Butler, L.; Windle, E.; Desouza, K.; Mehmood, R.; Corchado, J. Can building 'artificially intelligent cities' protect humanity from natural disasters, pandemics and other catastrophes? An urban scholar's perspective. *Sensors* **2020**, *20*, 2988. [CrossRef] [PubMed]

38. Zhang, Y.; Geng, P.; Sivaparthipan, C.; Muthu, B. Big data and artificial intelligence based early risk warning system of fire hazard for smart cities. *Sustain. Energy Technol. Assess.* **2021**, *45*, 100986.

39. Yigitcanlar, T.; Desouza, K.; Butler, L.; Roozkhosh, F. Contributions and risks of artificial intelligence (AI) in building smarter cities: Insights from a systematic review of the literature. *Energies* **2020**, *13*, 1473. [CrossRef]

40. Eubanks, V. A Child Abuse Prediction Model Fails Poor Families. 2018. Available online: https://www.wired.com/story/excerpt-from-automating-inequality (accessed on 27 April 2021).

41. Cooper, Y. Amazon Ditched AI Recruiting Tool that Favored Men for Technical Jobs. 2018. Available online: https://www.theguardian.com/technology/2018/oct/10/amazon-hiring-ai-gender-bias-recruiting-engine (accessed on 27 April 2021).

42. Ledford, H. Millions of Black People Affected by Racial Bias in Health-Care Algorithms. 2019. Available online: https://www.nature.com/articles/d41586-019-03228-6 (accessed on 27 April 2021).

43. Hill, K. The Secretive Company That Might End Privacy as We Know It. 2020. Available online: https://www.nytimes.com/2020/01/18/technology/clearview-privacy-facial-recognition.html (accessed on 27 April 2021).
44. Whiteford, P. Robodebt Was a Fiasco with a Cost We Have Yet to Fully Appreciate. 2020. Available online: https://theconversation.com/robodebt-was-a-fiasco-with-a-cost-we-have-yet-to-fully-appreciate-150169 (accessed on 27 April 2021).
45. Pasquinelli, M. How a machine learns and fails. *Spheres J. Digit. Cult.* **2019**, *5*, 1–17.
46. Kuziemski, M.; Misuraca, G. AI governance in the public sector: Three tales from the frontiers of automated decision-making in democratic settings. *Telecommun. Policy* **2020**, *44*, 101976. [CrossRef]
47. Allam, Z.; Dhunny, Z. On big data, artificial intelligence and smart cities. *Cities* **2019**, *89*, 80–91. [CrossRef]
48. Floridi, L.; Cowls, J.; Beltrametti, M.; Chatila, R.; Chazerand, P.; Dignum, V.; Luetge, C.; Madelin, R.; Pagallo, U.; Rossi, F.; et al. AI4People—An ethical framework for a good AI society: Opportunities, risks, principles, and recommendations. *Minds Mach.* **2018**, *28*, 689–707. [CrossRef]
49. Wearn, O.; Freeman, R.; Jacoby, D. Responsible AI for conservation. *Nat. Mach. Intell.* **2019**, *1*, 72–73. [CrossRef]
50. Falco, G. Participatory AI: Reducing AI bias and developing socially responsible AI in smart cities. In Proceedings of the 2019 IEEE International Conference on Computational Science and Engineering and IEEE International Conference on Embedded and Ubiquitous Computing, New York, NY, USA, 1–3 August 2019; pp. 154–158.
51. Metaxiotis, K.; Carrillo, F.J.; Yigitcanlar, T. *Knowledge-Based Development for Cities and Societies: Integrated Multi-Level Approaches: Integrated Multi-Level Approaches*; IGI Global: Hersey, PA, USA, 2010.
52. Yigitcanlar, T.; Kamruzzaman, M. Planning, development and management of sustainable cities: A commentary from the guest editors. *Sustainability* **2015**, *7*, 14677–14688. [CrossRef]
53. Dizdaroglu, D.; Yigitcanlar, T. Integrating urban ecosystem sustainability assessment into policy-making: Insights from the Gold Coast city. *J. Environ. Plan. Manag.* **2016**, *59*, 1982–2006. [CrossRef]
54. Hunter, A.; Hewson, J. There Are 10 Catastrophic Threats Facing Humans Right Now, and Coronavirus Is Only One of Them. 2020. Available online: https://theconversation.com/there-are-10-catastrophic-threats-facing-humans-right-now-and-coronavirus-is-only-one-of-them-136854 (accessed on 28 April 2021).
55. Copland, S. Anti-politics and global climate inaction: The case of the Australian carbon tax. *Crit. Sociol.* **2020**, *46*, 623–641. [CrossRef]
56. Hornsey, M.; Fielding, K. Understanding (and reducing) inaction on climate change. *Soc. Issues Policy Rev.* **2020**, *14*, 3–35. [CrossRef]
57. Yun, J.; Lee, D.; Ahn, H.; Park, K.; Lee, S.; Yigitcanlar, T. Not deep learning but autonomous learning of open innovation for sustainable artificial intelligence. *Sustainability* **2016**, *8*, 797. [CrossRef]
58. Sharma, G.; Yadav, A.; Chopra, R. Artificial intelligence and effective governance: A review, critique and research agenda. *Sustain. Futures* **2020**, *2*, 100004. [CrossRef]
59. Mortoja, M.; Yigitcanlar, T. Local drivers of anthropogenic climate change: Quantifying the impact through a remote sensing approach in Brisbane. *Remote. Sens.* **2020**, *12*, 2270. [CrossRef]
60. Shockley, K. The great decoupling: Why minimizing humanity's dependence on the environment may not be cause for celebration. *J. Agric. Environ. Ethics* **2018**, *31*, 429–442. [CrossRef]
61. Stahl, B.; Andreou, A.; Brey, P.; Hatzakis, T.; Kirichenko, A.; Macnish, K.; Laulhé Shaelou, S.; Patel, A.; Ryan, M.; Wright, D. Artificial intelligence for human flourishing: Beyond principles for machine learning. *J. Bus. Res.* **2021**, *124*, 374–388. [CrossRef]
62. Vinuesa, R.; Azizpour, H.; Leite, I.; Balaam, M.; Dignum, V.; Domisch, S.; Felländer, A.; Langhans, S.D.; Tegmark, M.; Nerini, F.F. The role of artificial intelligence in achieving the sustainable development goals. *Nat. Commun.* **2020**, *11*, 1–10. [CrossRef] [PubMed]
63. Gupta, S.; Langhans, S.; Domisch, S.; Fuso-Nerini, F.; Felländer, A.; Battaglini, M.; Tegmark, M.; Vinuesa, R. Assessing whether artificial intelligence is an enabler or an inhibitor of sustainability at indicator level. *Transp. Eng.* **2021**, *4*, 100064. [CrossRef]
64. Goralski, M.; Tan, T. Artificial intelligence and sustainable development. *Int. J. Manag. Educ.* **2020**, *18*, 100330. [CrossRef]
65. Dwivedi, Y.; Hughes, L.; Ismagilova, E.; Aarts, G.; Coombs, C.; Crick, T.; Duan, Y.; Dwivedi, R.; Edwards, J.; Eirug, A.; et al. Artificial intelligence (AI): Multidisciplinary perspectives on emerging challenges, opportunities, and agenda for research, practice and policy. *Int. J. Inf. Manag.* **2019**, *57*, 101994. [CrossRef]
66. Wamba, S.; Bawack, R.; Guthrie, C.; Queiroz, M.; Carillo, K. Are we preparing for a good AI society? A bibliometric review and research agenda. *Technol. Forecast. Soc. Chang.* **2021**, *164*, 120482. [CrossRef]
67. Ye, Z.; Yang, J.; Zhong, N.; Tu, X.; Jia, J.; Wang, J. Tackling environmental challenges in pollution controls using artificial intelligence: A review. *Sci. Total. Environ.* **2020**, *699*, 134279. [CrossRef]
68. Zhao, L.; Dai, T.; Qiao, Z.; Sun, P.; Hao, J.; Yang, Y. Application of artificial intelligence to wastewater treatment: A bibliometric analysis and systematic review of technology, economy, management, and wastewater reuse. *Process. Saf. Environ. Prot.* **2020**, *133*, 169–182. [CrossRef]
69. Jothiswaran, V.; Velumani, T. Jayaraman, R. Application of artificial intelligence in fisheries and aquaculture. *Biot. Res. Today* **2020**, *2*, 499–502.
70. Corcoran, E.; Denman, S.; Hamilton, G. Evaluating new technology for biodiversity monitoring: Are drone surveys biased? *Ecol. Evol.* **2021**. [CrossRef]

71. Huntingford, C.; Jeffers, E.; Bonsall, M.; Christensen, H.; Lees, T.; Yang, H. Machine learning and artificial intelligence to aid climate change research and preparedness. *Environ. Res. Lett.* **2019**, *14*, 124007. [CrossRef]
72. Pelorosso, R.; Gobattoni, F.; Geri, F.; Leone, A. PANDORA 3.0 plugin: A new biodiversity ecosystem service assessment tool for urban green infrastructure connectivity planning. *Ecosyst. Serv.* **2017**, *26*, 476–482. [CrossRef]
73. Haupt, S.; Pasini, A.; Marzban, C. *Artificial Intelligence Methods in the Environmental Sciences*; Springer Science & Business Media: Singapore, 2008.
74. United Nations. The 17 Goals, Sustainable Development. 2015. Available online: https://sdgs.un.org/goals (accessed on 24 May 2021).
75. Bjørlo, L.; Moen, Ø.; Pasquine, M. The role of consumer autonomy in developing sustainable AI: A conceptual framework. *Sustainability* **2021**, *13*, 2332. [CrossRef]
76. Jin, L.; Duan, K.; Tang, X. What is the relationship between technological innovation and energy consumption? Empirical analysis based on provincial panel data from China. *Sustainability* **2018**, *10*, 145. [CrossRef]
77. MacCarthy, M.; Kenneth, K. Machines Learn that Brussels Writes the Rules: The EU's New AI Regulation. 2021. Available online: https://www.brookings.edu/blog/techtank/2021/05/04/machines-learn-that-brussels-writes-the-rules-the-eus-new-ai-regulation (accessed on 24 May 2021).
78. Unwin, T. Digital Technologies are Part of the CLIMATE change Problem. 2020. Available online: https://www.ictworks.org/digital-technologies-climate-change-problem/#.YJDJ9S0RpYs (accessed on 5 May 2021).
79. Ritchie, H.; Roser, M. Technology Adoption (2017). Available online: https://ourworldindata.org/technology-adoption (accessed on 4 May 2021).
80. Cuen, L. The Debate about Cryptocurrency and Energy Consumption. 2021. Available online: https://techcrunch.com/2021/03/21/the-debate-about-cryptocurrency-and-energy-consumption (accessed on 4 May 2021).
81. Song, M.; Cao, S.; Wang, S. The impact of knowledge trade on sustainable development and environment-biased technical progress. *Technol. Forecast. Soc. Chang.* **2019**, *144*, 512–523. [CrossRef]
82. Dhar, P. The carbon impact of artificial intelligence. *Nat. Mach. Intell.* **2020**, *2*, 423–425. [CrossRef]
83. Schwartz, R.; Dodge, J.; Smith, N.; Etzioni, O. Green AI. *Commun. ACM* **2020**, *63*, 54–63. [CrossRef]
84. Yang, X.; Hua, S.; Shi, Y.; Wang, H.; Zhang, J.; Letaief, K. Sparse optimization for green edge AI inference. *J. Commun. Inf. Netw.* **2020**, *5*, 1–15.
85. Candelieri, A.; Perego, R.; Archetti, F. Green machine learning via augmented Gaussian processes and multi-information source optimization. *Soft Comput.* **2021**. [CrossRef]
86. Dobbe, R.; Whittaker, M. AI and Climate Change: How They're Connected, and What We Can Do about It. 2019. Available online: https://medium.com/@AINowInstitute/ai-and-climate-change-how-theyre-connected-and-what-we-can-do-about-it-6aa8d0f5b32c (accessed on 4 May 2021).
87. Yara, O.; Brazheyev, A.; Golovko, L.; Bashkatova, V. Legal regulation of the use of artificial intelligence: Problems and development prospects. *Eur. J. Sustain. Dev.* **2021**, *10*, 281. [CrossRef]
88. Gould, S. Green AI: How Can AI Solve Sustainability Challenges? 2020. Available online: https://www2.deloitte.com/uk/en/blog/experience-analytics/2020/green-ai-how-can-ai-solve-sustainability-challenges.html (accessed on 4 May 2021).
89. Yigitcanlar, T.; Foth, M.; Kamruzzaman, M. Towards post-anthropocentric cities: Reconceptualizing smart cities to evade urban ecocide. *J. Urban. Technol.* **2019**, *26*, 147–152. [CrossRef]
90. Alotaibi, S.; Mehmood, R.; Katib, I.; Rana, O.; Albeshri, A. Sehaa: A big data analytics tool for healthcare symptoms and diseases detection using Twitter, Apache Spark, and Machine Learning. *Appl. Sci.* **2020**, *10*, 1398. [CrossRef]
91. Holst, A. IoT Connected Devices Worldwide 2019–2030. 2021. Available online: https://www.statista.com/statistics/1183457/iot-connected-devices-worldwide (accessed on 24 May 2021).
92. Lee, S.H.; Yigitcanlar, T.; Han, J.H.; Leem, Y.T. Ubiquitous urban infrastructure: Infrastructure planning and development in Korea. *Innovation* **2008**, *10*, 282–292. [CrossRef]
93. Bibri, S.E. The IoT for smart sustainable cities of the future: An analytical framework for sensor-based big data applications for environmental sustainability. *Sustain. Cities Soc.* **2018**, *38*, 230–253. [CrossRef]
94. Alomari, E.; Katib, I.; Albeshri, A.; Yigitcanlar, T.; Mehmood, R. Iktishaf+: A big data tool with automatic labeling for road traffic social sensing and event detection using distributed machine learning. *Sensors* **2021**, *21*, 2993. [CrossRef] [PubMed]
95. Trettin, C.; Lăzăroiu, G.; Grecu, I.; Grecu, G. The social sustainability of citizen-centered urban governance networks: Sensor-based big data applications and real-time decision-making. *Geopolit. Hist. Int. Relat.* **2019**, *11*, 27–33.
96. Arshad, R.; Zahoor, S.; Shah, M.A.; Wahid, A.; Yu, H. Green IoT: An investigation on energy saving practices for 2020 and beyond. *IEEE Access* **2017**, *5*, 15667–15681. [CrossRef]
97. Garcia-Retuerta, D.; Chamoso, P.; Hernández, G.; Guzmán, A.S.R.; Yigitcanlar, T.; Corchado, J.M. An efficient management platform for developing smart cities: Solution for real-time and future crowd detection. *Electronics* **2021**, *10*, 765. [CrossRef]
98. Wang, S.; Paul, M.J.; Dredze, M. Social media as a sensor of air quality and public response in China. *J. Med. Internet Res.* **2015**, *17*, e22. [CrossRef]
99. Hayes, J.L.; Britt, B.C.; Evans, W.; Rush, S.W.; Towery, N.A.; Adamson, A.C. Can social media listening platforms' artificial intelligence be trusted? Examining the accuracy of Crimson Hexagon's (Now Brandwatch Consumer Research's) AI-driven analyses. *J. Advert.* **2021**, *50*, 81–91. [CrossRef]

100. Gupta, V.; Tripathi, S.; De, S. Green sensing and communication: A step towards sustainable IoT systems. *J. Indian Inst. Sci.* **2020**, *100*, 383–398. [CrossRef]
101. Gabrys, J. *Program Earth: Environmental Sensing Technology and the Making of a Computational Planet*; University of Minnesota Press: Minneapolis, MN, USA, 2016.
102. Kumar, N.; Rodrigues, J.J.; Guizani, M.; Choo, K.K.; Lu, R.; Verikoukis, C.; Zhong, Z. Achieving energy efficiency and sustainability in edge/fog deployment. *IEEE Commun. Mag.* **2018**, *56*, 20–21. [CrossRef]
103. Janbi, N.; Katib, I.; Albeshri, A.; Mehmood, R. Distributed artificial intelligence-as-a-service (DAIaaS) for smarter IoE and 6G environments. *Sensors* **2020**, *20*, 5796. [CrossRef]
104. Mohammed, T.; Albeshri, A.; Katib, I.; Mehmood, R. UbiPriSEQ: Deep reinforcement learning to manage privacy, security, energy, and QoS in 5G IoT hetnets. *Appl. Sci.* **2020**, *10*, 7120. [CrossRef]
105. Latva-aho, M.; Leppänen, K.; Clazzer, F.; Munari, A. Key Drivers and Research challenges for 6G Ubiquitous Wireless Intelligence (2020). Available online: https://elib.dlr.de/133477 (accessed on 24 May 2021).
106. Giordani, M.; Polese, M.; Mezzavilla, M.; Rangan, S.; Zorzi, M. Towards 6G networks: Use cases and technologies. *IEEE Commun. Mag.* **2020**, *58*, 55–61. [CrossRef]
107. Docomo, N. 5G Evolution and 6G. 2021. Available online: https://www.nttdocomo.co.jp/nglish/binary/pdf/corporate/technology/whitepaper_6g/DOCOMO_6G_White_PaperEN_v3.0.pdf (accessed on 24 May 2021).
108. Zhang, Z.; Xiao, Y.; Ma, Z.; Xiao, M.; Ding, Z.; Lei, X.; Karagiannidis, G.K.; Fan, P. 6G wireless networks: Vision, requirements, architecture, and key technologies. *IEEE Veh. Technol. Mag.* **2019**, *14*, 28–41. [CrossRef]
109. Letaief, K.B.; Chen, W.; Shi, Y.; Zhang, J.; Zhang, Y.J. The roadmap to 6G: AI empowered wireless networks. *IEEE Commun. Mag.* **2019**, *57*, 84–90. [CrossRef]
110. Gui, G.; Liu, M.; Tang, F.; Kato, N.; Adachi, F. 6G: Opening new horizons for integration of comfort, security, and intelligence. *IEEE Wirel. Commun.* **2020**, *27*, 126–132. [CrossRef]
111. Khan, L.U.; Yaqoob, I.; Imran, M.; Han, Z.; Hong, C.S. 6G wireless systems: A vision, architectural elements, and future directions. *IEEE Access* **2020**, *8*, 147029–147044. [CrossRef]
112. Usman, S.; Mehmood, R.; Katib, I.; Albeshri, A.; Altowaijri, S.M. ZAKI: A smart method and tool for automatic performance optimization of parallel SpMV computations on distributed memory machines. *Mob. Netw. Appl.* **2019**. [CrossRef]
113. Lauterbach, A. Artificial intelligence and policy: Quo vadis? *Digit. Policy Regul. Gov.* **2019**, *21*, 238–263. [CrossRef]
114. Camillus, J.; Baker, J.; Daunt, A.; Jang, J. Strategies for transcending the chaos of societal disruptions. *Vilakshan–XIMB J. Manag.* **2020**, *17*, 5–14. [CrossRef]
115. Girasa, R. *Artificial Intelligence as a Disruptive Technology: Economic Transformation and Government Regulation*; Springer Nature: Singapore, 2020.
116. Smuha, N. From a 'race to AI' to a 'race to AI regulation': Regulatory competition for artificial intelligence. *Law Innov. Technol.* **2021**, *13*, 57–84. [CrossRef]
117. Toll, D.; Lindgren, I.; Melin, U.; Madsen, C.Ø. Values, benefits, considerations and risks of AI in government: A study of AI policies in Sweden. *JeDEM-eJournal eDemocracy Open Gov.* **2020**, *12*, 40–60. [CrossRef]
118. Margetts, H.; Dorobantu, C. Rethink government with AI. *Nature* **2019**, *568*, 163–165. [CrossRef]
119. Jobin, A.; Ienca, M.; Vayena, E. The global landscape of AI ethics guidelines. *Nat. Mach. Intell.* **2019**, *1*, 389–399. [CrossRef]
120. Floridi, L. Establishing the rules for building trustworthy AI. *Nat. Mach. Intell.* **2019**, *1*, 261–262. [CrossRef]
121. Walch, K. AI Laws Are Coming. 2020. Available online: https://www.forbes.com/sites/cognitiveworld/2020/02/20/ai-laws-are-coming/?sh=2a12fe00a2b4 (accessed on 5 May 2021).
122. Hagendorff, T. The ethics of AI ethics: An evaluation of guidelines. *Minds Mach.* **2020**, *30*, 99–120. [CrossRef]
123. McNally, A. Creating Trustworthy AI for the Environment: Transparency, Bias, and Beneficial Use. 2020. Available online: https://www.scu.edu/environmental-ethics/resources/creating-trustworthy-ai-for-the-environment-transparency-bias-and-beneficial-use (accessed on 4 May 2021).
124. Leshinsky, R.; Schatz, L. I don't think my landlord will find out: Airbnb and the challenges of enforcement. *Urban. Policy Res.* **2018**, *36*, 417–428. [CrossRef]
125. Stein, E.; Head, B. Uber in Queensland: From policy fortress to policy change. *Aust. J. Public Adm.* **2020**, *79*, 462–479. [CrossRef]
126. De Almeida, P.; dos Santos, C.; Farias, J. Artificial intelligence regulation: A framework for governance. *Ethics Inf. Technol.* **2021**. [CrossRef]
127. Di Vaio, A.; Palladino, R.; Hassan, R.; Escobar, O. Artificial intelligence and business models in the sustainable development goals perspective: A systematic literature review. *J. Bus. Res.* **2020**, *121*, 283–314. [CrossRef]
128. Fatima, S.; Desouza, K.; Dawson, G.; Denford, J. Analyzing Artificial Intelligence Plans in 34 Countries. 2021. Available online: https://www.brookings.edu/blog/techtank/2021/05/13/analyzing-artificial-intelligence-plans-in-34-countries (accessed on 24 May 2021).
129. Zhou, Y.; Kankanhalli, A. AI regulation for smart cities: Challenges and principles. In *Smart Cities and Smart Governance: Towards the 22nd Century Sustainable City*; Estevez, E., Pardo, T., Scholl, H., Eds.; Springer: Singapore, 2021; pp. 101–118.
130. Fisher, D.H. Computing and AI for a Sustainable Future. *IEEE Intell. Syst.* **2011**, *26*, 14–18. [CrossRef]

 *sustainability* 

Perspective

Towards Australian Regional Turnaround: Insights into Sustainably Accommodating Post-Pandemic Urban Growth in Regional Towns and Cities

Mirko Guaralda [1], Greg Hearn [1], Marcus Foth [1,*], Tan Yigitcanlar [2], Severine Mayere [2] and Lisa Law [3]

[1] QUT Design Lab, Queensland University of Technology, 2 George Street, Brisbane 4000, Australia; m.guaralda@qut.edu.au (M.G.); g.hearn@qut.edu.au (G.H.)

[2] School of Built Environment, Queensland University of Technology, 2 George Street, Brisbane 4000, Australia; tan.yigitcanlar@qut.edu.au (T.Y.); severine.mayere@qut.edu.au (S.M.)

[3] Division of Tropical Environment and Societies, James Cook University, PO Box 6811, Cairns 4870, Australia; lisa.law@jcu.edu.au

* Correspondence: m.foth@qut.edu.au; Tel.: +61-7-3138-8772

Received: 26 November 2020; Accepted: 12 December 2020; Published: 15 December 2020

Abstract: The COVID-19 pandemic has made many urban policymakers, planners, and scholars, all around the globe, rethink conventional, neoliberal growth strategies of cities. The trend of rapid urbanization, particularly around capital cities, has been questioned, and alternative growth models and locations have been the subjects of countless discussions. This is particularly the case for the Australian context: The COVID-19 pandemic heightened the debates in urban circles on post-pandemic urban growth strategies and boosting the growth of towns and cities across regional Australia is a popular alternative strategy. While some scholars argue that regional Australia poses an invaluable opportunity for post-pandemic growth by 'taking off the pressure from the capital cities'; others warn us about the risks of growing regional towns and cities without carefully designed national, regional, and local planning, design, and development strategies. Superimposing planning and development policies meant for metropolitan cities could simply result in transferring the ills of capital cities to regions and exacerbate unsustainable development and heightened socioeconomic inequalities. This opinion piece, by keeping both of these perspectives in mind, explores approaches to regional community and economic development of Australia's towns and cities, along with identifying sustainable urban growth locations in the post-pandemic era. It also offers new insights that could help re-shape the policy debate on regional growth and development.

Keywords: regional towns; regional cities; regional Australia; regional lifestyle location; regional innovation system; regional turnaround; post-pandemic urban growth; COVID-19 impact; regional planning; sustainable urban development

---

1. Introduction

The ongoing COVID-19 pandemic has affected many countries and millions of people all over the globe. This most severe global health crisis impacted public health, disrupted consolidated neoliberal economies, labor markets, and other facets of social and individual life. Subsequently, the global economy has hit a recession, where it is now gradually turning into a global financial crisis [1]. Today, we are witnessing substantial changes in various social activities and behaviors of individuals, such as reducing economic and social engagements, social distancing, delaying school, staying more at home, working and studying remotely, and moving to safer and less denser locations [2–4].

COVID-19 is a reminder of the vulnerability of major cities as dominant centers of economic and social life; the pandemic has seriously challenged our economic, social, and urban systems and it is still too early to assess its long-term impacts. As Sir Norman Foster has stated [5], our cities will survive, but we need to rethink our model of growth and economic structure. Some cities, so far, have done better than others [6]. However, in Australia, the factors affecting the spread of the virus such as international access points and difficulties of confining spread in geographical clusters, have meant that large cities have become the epicenter [7]. Regional towns and cities may not have had as dramatic health impacts, but certainly their economies have been severely affected in concert with the national economy as a whole. In parallel, the unprecedented move to remote work, nomadic/mobile work, and home-based work has accentuated the use of digital technologies, which obviate the need for geographic co-location to conduct business. These trends have been further accelerated by the pandemic [8–10]. This potentially opens up opportunities for Australia's regional towns and cities, especially for realizing a regional turnaround. Nevertheless, if ever there was a time for innovative approaches to decentralizing Australia's population and economic resources, it is now.

This opinion piece explores and discusses current opportunities for Australian regional towns and cities. The Australian borough of statistics defines towns as settlements with a population between 10,000 and 100,000 [11], but for our purposes we find it helpful to specify that regional towns and cities are those that lie beyond the major capital cities of Sydney, Melbourne, Brisbane, Perth, Adelaide, and Canberra. We follow the Regional Australia Institute's [12] nuanced definitions where regional cities are more than 50,000 people, and towns can include connected lifestyle regions, industry and service hubs, and heartland regions. In other words, these areas are variegated and a one size fits all is not possible. Following this introduction, Section 2 provides background to the regionalism discourses in Australia. Sections 3 and 4 present growth challenges and growth opportunities of Australian regional towns and cities, respectively. Lastly, Section 5 concludes the opinion piece with some remarks on the insights into sustainably accommodating post-pandemic urban growth in regional Australia.

2. The Growth of Regional Towns and Cities

In Australia, most of the population is clustered around a few big urban centers (Figure 1). In a country that is almost 90% urban, more than two-thirds of the population are housed by the capital cities and their metropolitan regions. Although relatively few people might live in regional towns, the Regional Australia Institute [12] argues that regional towns and cities collectively house 9.45 million Australians (versus the 15.9 million in the metropolitan capital cities). Moreover, they argue that regional cities and connected lifestyle regions are growing at significant rates (5.4 and 6.8%, respectively) which is comparable to the growth of metropolitan capitals at 8%. In recent decades, like many cities across the world, numerous Australian major cities have been struggling with unprecedented growth, rising costs, and the changing nature of employment [13]. Urban population growth puts pressures on infrastructure, housing, environment, social, and community resilience. The greenfield model of development adopted so far as the prominent strategy to manage urban growth erodes fertile agricultural lands and has a consistent financial impact due to the need of new infrastructures to move people and goods [14].

At the same time, some regional and rural communities fight to keep a sustainable population [15] and experience skills shortages [16], leaving them with the challenge of attracting skilled and motivated workers [17]. Although COVID is projected to slow Australia's population growth [18] and presumably alter these dynamics, the deep contradictions of our cities are the economic and geographic structures that promote 'a winner takes all' urbanism of the talented and advantaged clusters, which succeed in leaving everybody else behind [19]. Population shifts and their urban dynamics already unfold differently in regional towns and cities, where urban landscapes reflect and help reproduce contradictions and inequalities in distinct ways [20]. Nonetheless, even in a century where more than half the world's population is now urban (and rapidly increasing), 'in every place and in every century, there have been alternatives' [21].

A number of planning and development policies are relevant to this discussion. The Regional Australia Institute argues that further agglomeration in capital cities does not improve national economic performance and will impair livability particularly in terms of housing affordability and congestion [22]. They advocate the development of regional settlement strategies, analysis of infrastructure investment options, optimizing land use policies, and better strategies for migration of skilled workers to regional cities. The 'Planning for Australia's Future Population' report also suggests taking the pressure off capital cities, and argues for similar policy interventions, along with an emphasis on well-functioning communities and community services [23]. The Australian Productivity Commission's report on 'Transitioning Australia's Regions' [24] cautions against ad hoc interventions, but also argues for optimizing land use and planning along with rigorous and transparent evaluation of regional development strategies. Their methodology is particularly relevant to our discussion. They developed an index of Regional Adaptive Capacity that emphasizes human, social, natural, financial, and physical capital as well as other economic indicators such as regional economic diversity [24].

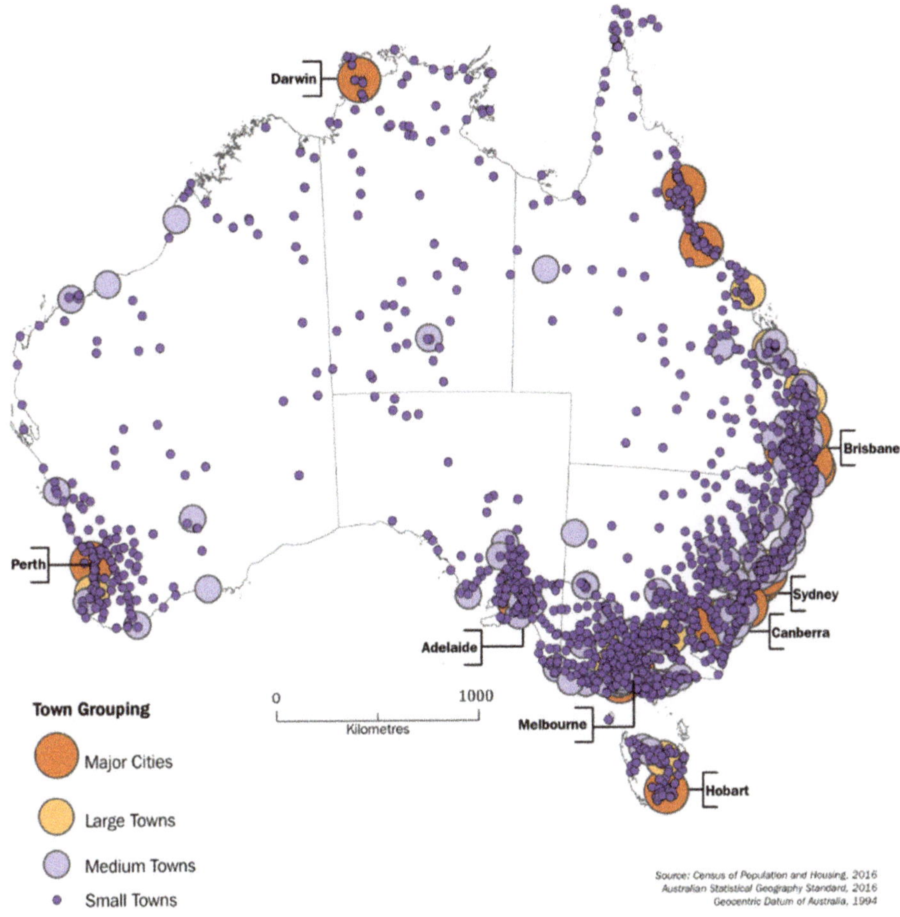

Figure 1. Australian settlements by population size groupings [11].

In addition to government-led interventions through policy, we also argue that for these policies to work, regional industries and businesses must have the capacity to innovate and adapt to make use of the opportunities afforded, e.g., better land use and more availability of a high-skilled workforce. This means continuing transforming industries such as agriculture, manufacturing, and tourism through an emphasis on developing high value-added products and services, as well as accelerating the adoption and tailoring of advanced technology to regional contexts. The emerging technological reality of the modern professional is that an increasingly large number of knowledge workers can choose their desired location and lifestyle while seamlessly collaborating with customers, clients, and stakeholders online [25–27]. These technological forces are doing more than merely shifting economic activities from one area to another. In addition to this geographic movement, these innovations have also been a catalyst for new economic growth and renewal.

Current urban development paradigms replicate at a smaller scale the same pattern used in major cities. The central business district (CBD)-Suburbia model is often implemented with a cookie-cutter approach that does not take into consideration the peculiarities and lifestyle advantages of regional centers, their socioeconomic structure, and the synergies between different types of urban forms [28,29]. Promoting growth in regional towns—a.k.a., medium (1000–50,000 inhabitants) and large (50,000–100,000 inhabitants) towns as per ABS definition [1], as well as regional cities of 50,000 or more—has to be supported by an investigation into suitable urban designs and urban structures to maximize the use of existing infrastructures and capitalize on the local sense of place and community identity [30–33].

Some scholars argue that Australian regional towns and cities pose an invaluable opportunity for post-pandemic growth by 'taking off the pressure from the capital cities' [22,34]. On the other hand, some scholars warn us on the risks of growing regional towns and cities without carefully designed national, regional, and local planning and development strategies, and transferring the ills of capital cities to regions [35,36]—such as unsustainable development and heightened socioeconomic inequalities. As Archer et al. [37] stress, "it is crucial that environmental sustainability, social inclusion, and livability considerations are also included in the deliberations to develop population-based initiatives … not only 'How many Australians?' but 'Where will they live?'" (p. 124). Indeed, there is an urgent need for a national discussion about the Australia-wide distribution of urban areas as well as the long-term planning for different settlement patterns [38].

Against this backdrop, the discussion in this opinion piece sets out to explore approaches to regional community and economic development for recovery and growth of Australia's regional towns and cities, along with identifying sustainable growth locations in the post-pandemic era for Australia.

The discussion is guided by three themes:

i. Attracting population to regional cities;
ii. Increasing employment in regional cities with a focus on technology and innovation;
iii. Support urban growth in regional cities exploring suitable sustainable urban development models for third-tier centers.

As an opinion piece, we do not aim to provide empirically grounded answers to these themes. Rather, the purpose of this article is to offer a discussion informed by a critical review of the relevant literature with a view to stimulate additional voices to chime in and further the discourse in the field.

3. The Challenges

While, arguably, regional Australia faces many challenges, within the scope of this article we have selected three interrelated issues to focus on. They are: (Section 3.1) The population counter-trend of an undercurrent of people leaving metropolitan cities and migrating and settling in regional areas—a.k.a. counter-urbanization—and the challenge of retaining them and providing career pathways for them; (Section 3.2) the role and impact of digital technology for a sustainable approach to population shifts, settlement strategies, and regionalization; and (Section 3.3) the challenge of reconciling population growth aspirations with protecting lifestyle benefits and a sustainable natural environment.

3.1. Offering Sea-Change, e-Change or Flee-Change Locations

The phenomenon of urbanism increasingly impacts on Australian society through population growth in urban areas or decline in rural and regional areas [39]. Population growth puts pressures on infrastructure, housing [40], environment, social networks, and communities. Like cities all over the world, Australian cities struggle with unprecedented growth, rising costs, the changing nature of work—particularly Sydney, Melbourne, and Brisbane [13], and uncertainty in a complex political arena [19]. Economic geographies and associated policies tend to focus on urban concentrations. Yet, there is a lack of attention on realizing significant opportunities for the development of robust and sustainable knowledge and innovation economies in regional areas [41,42]. We note that not all regional cities have the same characteristics; for instance, Gold Coast and Sunshine Coast have been experiencing rapid growth. Yet, the growth faced by metropolitan areas, the shrinking population of the regional centers, which often try to compete in an increasingly globalized economy, and the impact of lifestyle shifters can disrupt the delicate balance of sensitive coastlines and other key environmental features [43]. The impacts are global game changes that threaten Australia's productivity, social cohesion, and livability as we know it. Nonetheless, urbanization has never been either inevitable or without countervailing tendencies—in other words, path dependency matters. However, in every place and in every century, there have been alternatives [21].

Traditionally, coastal cities of Australia have been the locations for 'sea-change' [44]. However, Gurran and Blakely [45] argue coastal migration is incidental to urban dominance, and unlikely to be a factor in the prosperity of Australia's regional cities located outside metropolitan areas. In recent years, as part of a counter-urbanization trend, 'e-change' is underway in Australia—migrating from the capital cities to nearby lifestyle towns [46] and working remotely [4,47]. The latest trend is 'flee-change'. People particularly from VIC and NSW have started to move to coastal cities/towns to escape COVID-19 risks and strict pandemic restrictions [48]. Nevertheless, regional cities outside metropolitan areas, away from the coast and metropolitan CBDs, have been less likely to attract large migrant populations. It is unclear if COVID-19 and the wider embracement of new technology will alter this.

3.2. Offering Broadband Access, Innovation, and the Digital Economy Conditions

There is an acknowledged overlap between creative industries, other knowledge work, and remote work, and an established trend for them to seek out lifestyle opportunities [4,49–51]. The connectivity promised by Australia's National Broadband Network (NBN) has also opened up possibilities for regional centers to market themselves as lifestyle towns for telework—also linked with the above-discussed e-change [46,52,53]. Yet, there are benefits and challenges to location choices [54]. There are pressures on some of these lifestyle choices such as the rapid population increases—e.g., Gold Coast, QLD [21]. Nevertheless, the contrasting and highly criticized dystopian effect of high population growth on the Gold Coast raises issues of urban and transport planning and design policy questions for the promotion and growth of sensitive and sustainable regional centers [21,55].

Within the digital economy, knowledge and creative occupations—and digital nomadism—are important [56] and have been growing with each census since 2001 [57,58]. Lobo et al. [59] argue that higher densities of educated creative and knowledge workers spread innovations throughout the economy and impact metropolitan productivity broadly. A number of studies have documented the dynamics of knowledge and creative work outside CBD areas where they are strongest [60–63]. In addition to the knowledge and creative industries, there may be a general 'employment dividend' by embedding knowledge and creative workers [64] in other larger sectors such as manufacturing [57,58,65,66].

Regional employment opportunities through innovation are critical. Some of the success stories include the Ideas Lab in Cairns (formerly the Cairns Innovation Centre)—a self-sustaining start-up and innovation ecosystem in Far North Queensland [67]—and Australian Tropical Sciences and Innovation Precinct—a world class tropical research hub located in Townsville [68]. Nonetheless, both projects are public/academic sector investments, and there is only a negligible level of private sector driven large investments in regional Australia—excluding mining and agricultural operations. The lack of private sector led innovation activities in regional cities and the lack of effective partnerships pose a risk for the adoption of knowledge and innovation economy [69,70].

3.3. Offering Regional Liveability and Sustainability

The concepts of livability and sustainability are increasingly used to define our cities [71]. Pressures on metropolitan urban areas are creating vast disparities between inner suburbanites and the outer regions of a city. Consequently, while this crisis is often reported as a statistical phenomenon, it is an urban reality for millions of individuals, their lives, their work, and the livability of their neighborhood. As housing prices soar in metropolitan cities of Australia, there is a generation of youth questioning if they will ever be able to enter the housing market and an older generation trying to maintain home ownership under increasing economic pressure and uncertainty [72]. Besides, commuting robs families and communities of time and social input, undermining the social structures of our society—especially in the case of Sydney and Melbourne. At the same time, many regional and rural communities fight to keep a sustainable population [15] and experience skills shortages [16], leaving them with the challenge of attracting skilled and motivated workers.

This issue brings the importance of developing and delivering new urban planning, place-based design, and placemaking strategies tailored to regional Australia [73–76] that avoid regional cities and towns from being propelled to join their larger counterparts in the new urban crisis. This also assists in addressing one of the practical research challenges identified by the Commonwealth government of developing more resilient urban, rural, and regional infrastructure. Without a desired level of livability and sustainability, regional urban offerings are unlikely to become a growth alternative to the primate—i.e., Sydney—and second-tier cities of Australia—other state capitals and some cities within their metropolitan area [77].

4. The Opportunities

In response to the triad of challenges we identified and presented above, we now turn our attention to a similarly interrelated set of three areas of opportunities for regional Australia. They are: (Section 4.1) The policy response; (Section 4.2) the planning and design response, and (Section 4.3) the governance response.

4.1. The Need for a Distinctive Policy for Regional Knowledge and Innovation Industry Dynamics

In line with the international focus on the knowledge and innovation industries as a source of potential cultural and economic innovation and growth, numerous regional areas have identified the stimulation of knowledge and innovation industries as a key element for sustainable growth [78]. Further, knowledge and innovation industries and knowledge workers are not restricted to urban areas: These industries exist as a site of strong potential innovation and development beyond major cities [79]. In fact, many knowledge workers opt to move themselves and their work away from urban centers. Evidence signals that many knowledge workers are attracted to specific regional areas as part of a wider 'e-change' ex-urban migration trend [54]. However, there is a gap in research on knowledge/creative industries in regional locations [80].

This issue brings the following questions to mind: (a) How do knowledge and innovation industries work in regional locations? (b) What can knowledge and innovation industries and workers be attracted to regional locations? (c) In what types of places do regional knowledge and innovation industries thrive best? (d) How is technology—specifically the shift to telecommuting—playing a role in shaping regional innovation cultures and practices? There is limited research conducted on these questions, which signals that knowledge and innovation industries in regional locations do not operate in the same ways as they do in the CBDs of first- and second-tier cities [81,82].

Understanding how Australian regional knowledge industries operate on the ground could provide a foundation for regional-place specific innovation policies and initiatives in Australia. This could also inform the development of policies and programs that avoid replicating metropolitan urban strategies, which may well miss the mark. Hence, uncovering the economic and cultural geographies of knowledge industries innovation in regional Australia is critical to provide a strong basis for evidence-based policymaking.

4.2. The Policy, Planning, and Design Response

The literature highlights the impact of placemaking and the built environment on the livability of regional centers and their ability to become globally competitive as places for people to live and work [83]. It is, hence, imperative to explore the factors that make the biggest and most pertinent impacts, supporting knowledge industries and workers and their spatial needs, and develop urban planning and design policy recommendations tailored to a digital economy in regional Australia. Urban planning and design strategies adopted regionally are often simply pale imitations of metropolitan approaches that do not resonate with local communities [84]. Given e-changers and flee-changers are moving to regional centers (mostly coastal) to avoid overly crowded and congested urban milieus—along with others with different relocation reasons—it is important to work closely with local communities to deliver novel urban planning and design strategies that are engineered to maximize and protect the local identity and capitalize on local opportunities. In that perspective, it is useful to adopt innovative participatory action research and participatory design methodologies to address this component [85,86].

4.3. The Governance Response

Enabling local governments in regional Australia to navigate the challenges outlined here will require a balance of impact, lifestyle, and shifting economies. Achieving this balance will in turn require a concerted and collective effort to foster innovation, strong policy direction, and an understanding of the key drivers and influences. Issues of pollution, inequalities, rising housing costs and affordability, cost of living, transport networks and capacities, community and social connection place pressures on lifestyle and health and, in doing so, compromise the livability for the masses [87–89]. Florida [19] suggests the 'new urban crisis' is not just a crisis of our cities but of our age, of a highly urbanized knowledge-based capitalism. Cities have increasingly become a patchwork of concentrated advantage and much larger swathes of disadvantage. Yet, this patchwork model is also a faithful reflection of the methodological approaches that have emerged in the literature for addressing this suite of challenges. These topics have been separately examined through the lenses of pollution [90], economics [91,92], and social science [93], but prior investigations in this area have so far failed to provide any kind of integrative analysis that draws together these disparate bits into a more cohesive, unified perspective that can guide local governance. In sum, strategic governance highly matters for further development of Australian regional cities [94], and effective urban planning, design, policy, and monitoring—along with the planner as "an orchestrator and enabler of planning regional futures" [95]—are the key vehicles to achieve the desired outcomes and regional futures.

5. Concluding Remarks

The COVID-19 pandemic has forced us to rethink our cities from the prism of health, livability, and sustainability [96,97]. Referring to climate change and COVID-19, Bauman [98] argues that, "the existential emergencies we face require a wholesale reimagining of how we live, work and play in urban spaces". Specifically in the context of the post-pandemic recovery driven by a renewed attention to regional Australia, we argue that this will require a continuing focus on food and agriculture [99,100], as well as radically reframing our unhealthy relationship with nature and the planet towards a post-anthropocentric design agenda [72,101] that prioritizes revegetation, rewilding, and more-than-human perspectives [102–104].

The pandemic has also created an opportunity to revisit our options on human settlement, urban habitat, and where and how to locate the future population (and also economic) growth in Australia [105]. We have a chance to rethink the structure of our urban systems and pursue alternative models to the consolidated paradigms of Euclidean planning. This opinion piece underlined the following regional challenges to attract and retain migrants to regional cities (particularly coastal ones): (a) Offering sea-change, e-change, and flee-change locations; (b) offering broadband access, innovation, and the digital economy conditions; and (c) offering regional livability and sustainability. It also highlighted the crucial importance of the following mechanisms for sustainably accommodating the post-pandemic urban growth in regional Australia: (a) A distinctive policy for regional knowledge and innovation industry dynamics; (b) the planning and design response; and (c) the governance response.

Our enquiry has revealed that the three original themes used to structure our discussion can be articulated in a series of diverse and detailed questions, to further the understanding of opportunities and dynamics in regional towns:

- Attracting population to regional cities: (a) What factors affect the choice of location in moving to regional Australia? (b) What placemaking strategies and urban design features are necessary and desired to enable knowledge workers to work effectively in Australian regional lifestyle locations? (c) How can the unique local qualities and the 'sense of place' of regional Australia be maintained, while catering to the demands and needs of new workers' families? (d) How can we attract skilled migrants to regional Australia through cultural interventions? (e) What are the new models of affordable housing for regional Australia? (f) How can the pandemic migrants be retained in regional Australia post pandemic?

- Increasing employment in regional cities with a focus on technology and innovation: (a) How can existing businesses grow their revenues and jobs? (b) What specific skills are needed in regional Australia? (c) How can creative and knowledge services grow revenues of existing businesses in regional Australia? (d) What models of start-up and innovation hubs are best suited to regional Australia? (e) How can the community, industry, and local government sectors come together to spur innovation in regional Australia? (f) What are the profiles of jobs, professions, and education that are most likely to choose and succeed in relocating to regional Australia? (g) How can the economic development strategies (including land use policies) in regional Australia best respond to these insights? (h) What is the prospect of new service providers and business models emerging to support innovation across regional Australia? (i) How can we create new and sustainable industries in regional Australia?

- Support urban growth in regional cities exploring suitable sustainable urban development models for third-tier centers: (a) What models of cities can optimize the sustainable growth in regional Australia? (b) What is the role of land use policy and developmental regulations in accelerating the creation of innovation precincts/districts in regional Australia? (c) How can innovation/coworking hubs be designed to enable affordable workspace and access to equipment, and knowledge networks in regional Australia? (d) How can more inclusive and diverse innovation ecosystems be developed in regional Australia? (e) How can urban redevelopment be rethought to better suit the context and opportunities of regional Australia? (f) How can these changes impact the economic prospects and prosperity for regional Australia?

We conclude this opinion piece—that explores approaches to regional community and economic development of Australia's regional towns and cities (Australian regional turnaround), along with providing sustainable urban growth locations in the post-pandemic era—by restating the perspective by Archer et al. [5]: We need "through better evidence, to ensure that distributing growth to regional areas achieves its proper place in the policy debate and that government agencies invest the time and effort required to make good decisions on these opportunities" (p. 35). The themes highlighted in this piece can contribute to re-shaping the policy debate on regional growth and development and identify new avenues for research and practice.

Author Contributions: Conceptualization, formal analysis, and original draft preparation; G.H. and M.F.; formal analysis, and writing—original draft preparation, T.Y.; formal analysis, and writing—review and editing, M.G., S.M., and L.L. All authors have read and agreed to the published version of the manuscript.

Funding: This research received no external funding.

Acknowledgments: This research did not receive any specific grant from funding agencies in the public, commercial or not-for-profit sectors. The authors thank Nick Osbaldiston of James Cook University, and three anonymous referees for their invaluable comments on an earlier version of the manuscript.

Conflicts of Interest: The authors declare no conflict of interest.

References

1. Peiró, T.; Lorente, L.; Vera, M. The COVID-19 Crisis: Skills That Are Paramount to Build into Nursing Programs for Future Global Health Crisis. *Int. J. Environ. Res. Public Heal.* **2020**, *17*, 6532. [CrossRef]

2. Donthu, N.; Gustafsson, A. Effects of COVID-19 on business and research. *J. Bus. Res.* **2020**, *117*, 284–289. [CrossRef]

3. Ling, G.H.; Chyong Ho, C.M. Effects of the Coronavirus (COVID-19) pandemic on social behaviors: From a social dilemma perspective. *Tech. Soc. Sci. J.* **2020**, *7*, 312–320.

4. Zenkteler, M.; Foth, M.; Hearn, G. The role of residential suburbs in the knowledge economy: Insights from a design charrette into nomadic and remote work practices. *J. Urban Design.* **2021**. [CrossRef]

5. Foster, N. The Pandemic Will Accelerate the Evolution of Our Cities. 2020. Available online: https://www.theguardian.com/commentisfree/2020/sep/24/pandemic-accelerate-evolution-cities-covid-19-norman-foster?CMP=share_btn_link (accessed on 25 September 2020).

6. Sparke, M.; Anguelov, D. Contextualising coronavirus geographically. *Trans. Inst. Br. Geogr.* **2020**, *45*, 498–508. [CrossRef]

7. Yigitcanlar, T.; Kankanamge, N.; Preston, A.; Gill, P.S.; Rezayee, M.; Ostadnia, M.; Xia, B.; Ioppolo, G. How can social media analytics assist authorities in pandemic-related policy decisions? Insights from Australian states and territories. *Health Inf. Sci. Syst.* **2020**, *8*, 37. [CrossRef]

8. Kramer, A.; Kramer, K.Z. The potential impact of the Covid-19 pandemic on occupational status, work from home, and occupational mobility. *J. Vocat. Behav.* **2020**, *119*, 103442. [CrossRef] [PubMed]

9. Manzini Ceinar, I.; Mariotti, I. The effects of Covid-19 on coworking spaces: Patterns and future trends. In *Shared Workplaces in the Knowledge Economy*; Springer: Berlin/Heidelberg, Germany, 2020.

10. Rogers, Y. Is Remote the New Normal? Reflections on Covid-19, Technology, and Humankind. *Interactions* **2020**, *27*, 42–46. [CrossRef]

11. ABS. 2071.0—Census of Population and Housing: Reflecting Australia—Stories from the Census. 2016. Available online: https://www.abs.gov.au/ausstats/abs@.nsf/Lookup/2071.0main+features1132016 (accessed on 23 November 2020).

12. Regional Australia Institute. What is Regional Australia? 2019. Available online: http://www.regionalaustralia.org.au/home/what-is-regional-australia (accessed on 9 December 2020).

13. FYA. *The New Work Order*; Foundation for Young Australians (FYA): Melbourne, Australia, 2015.

14. Mortoja, G.; Yigitcanlar, T. Local Drivers of Anthropogenic Climate Change: Quantifying the Impact through a Remote Sensing Approach in Brisbane. *Remote. Sens.* **2020**, *12*, 2270. [CrossRef]

15. Forth, G.; Howell, K. Don't cry for me Upper Wombat: The realities of regional/small town decline in non-coastal Australia. *Sustain. Reg.* **2002**, *2*, 4–11.

16. Sharma, K.; Oczkowski, E.; Hicks, J. Skill shortages in regional Australia: A local perspective from the Riverina. *Econ. Anal. Policy* **2016**, *52*, 34–44. [CrossRef]

17. Foth, M.; McQueenie, J. Creatives in the Country? Blockchain and AgTech Can Create Unexpected Jobs in Regional Australia. 2019. Available online: https://theconversation.com/creatives-in-the-country-blockchain-and-agtech-can-create-unexpected-jobs-in-regional-australia-117017 (accessed on 28 October 2020).

18. Charles-Edwards, E.; Wilson, T.; Bernard, A.; Wohland, P. *How Will COVID-19 Impact Australia's Future Population? A Scenario Approach*; Working Paper 2020/03; Queensland Centre for Population Research, The University of Queensland: St Lucia, Australia, 2020. [CrossRef]

19. Florida, R. *The New Urban Crisis: How Our Cities Are Increasing Inequality, Deepening Segregation and Failing the Middle Class*; Basic Books: New York, NY, USA, 2017.

20. Luckman, S.; Gibson, C.; Lea, T. Mosquitoes in the mix: How transferable is creative city thinking? *Singap. J. Trop. Geogr.* **2009**, *30*, 70–85. [CrossRef]

21. Connell, J.; McManus, P. *Rural Revival? Place Marketing, Tree Change and Regional Migration in Australia*; Routledge: New York, NY, USA, 2016.

22. Archer, J.; Houghton, K.; Vonthethoff, B. *Regional Population Growth: Are We Ready? The Economics of Alternative Australian Settlement Patterns*; Regional Australia Institute: Canberra, Australia, 2019.

23. Australian Government. Planning for Australia's Future Population. 2019. Available online: https://www.pmc. gov.au/resource-centre/domestic-policy/planning-australias-future-population (accessed on 11 October 2020).

24. Australian Productivity Commission. Transitioning Regional Economies. 2017. Available online: https: //www.pc.gov.au/inquiries/completed/transitioning-regions/report (accessed on 11 October 2020).

25. Alizadeh, T.; Sipe, N.G.; Dodson, J. Spatial Planning and High-Speed Broadband: Australia's National Broadband Network and Metropolitan Planning. *Int. Plan. Stud.* **2014**, *19*, 359–378. [CrossRef]

26. Benson, M.; O'Reilly, K. From lifestyle migration to lifestyle in migration. *Migr. Stud.* **2016**, *4*, 20–37. [CrossRef]

27. Yigitcanlar, T.; Kamruzzaman, M. Smart Cities and Mobility: Does the Smartness of Australian Cities Lead to Sustainable Commuting Patterns? *J. Urban Technol.* **2019**, *26*, 21–46. [CrossRef]

28. Mitchell, W.; Carlson, E. Exploring employment growth disparities across metropolitan and regional Australia. *Australas. J. Reg. Stud.* **2015**, *11*, 25–40.

29. Pugalis, L.; Tan, S. Metropolitan and Regional Economic Development: Competing and Contested Local Government Roles in Australia in the 21st Century. 2017. Available online: https://opus.lib.uts.edu.au/ handle/10453/81975 (accessed on 11 October 2020).

30. Brabazon, T. *Unique Urbanity? Rethinking Third Tier Cities, Degeneration, Regeneration and Mobility*; Springer: London, UK, 2014.

31. Gray, I.; Lawrence, G. *A Future for Regional Australia: Escaping Global Misfortune*; Cambridge University Press: London, UK, 2001.

32. Law, L.; Safarova, S.; Campbell, A.; Halawa, E. Design for liveability in tropical Australia. In *Leading from the North: Rethinking Northern Australia Development*; Wallace, R., Harwood, S., Gerritsen, R., Prideaux, B., Brewer, T., Rosenman, L., Dale, A., Eds.; ANU Press: Canberra, Australia, 2021; in press.

33. Zhang, H.; Li, L.; Hui, E.C.-M.; Li, V. Comparisons of the relations between housing prices and the macroeconomy in China's first-, second- and third-tier cities. *Habitat Int.* **2016**, *57*, 24–42. [CrossRef]

34. Rolfe, J.; Kinnear, S. Populating regional Australia: What are the impacts of non-resident labour force practices on demographic growth in resource regions? *Rural. Soc.* **2013**, *22*, 125–137. [CrossRef]

35. Gurran, N.; Ruming, K. Less planning, more development? Housing and urban reform discourses in Australia. *J. Econ. Policy Reform* **2015**, *19*, 262–280. [CrossRef]

36. Gurran, N.; Norman, B.; Hamin, E. Population growth and change in non-metropolitan coastal Australia. In *Parallel Patterns of Shrinking Cities and Urban Growth: Spatial Planning for Sustainable Development of City Regions and Rural Areas*; Routledge: London, UK, 2016; pp. 165–184.

37. Archer, J.; McCluskey, S.; Ford, B.; Swan, C.; Dwyer, J.; Ashton-Wyatt, A.; Nguyen, T.; Rennie, M. *Population Dynamics in Regional Australia*; Regional Australia Institute: Canberra, Australia, 2015.

38. Weller, R.; Bolleter, J.A. *Made in Australia: The Future of Australian Cities*; UWA Publishing: Perth, Australia, 2013.

39. Mc.Guirk, P.; Argent, N. Population growth and change: Implications for Australia's cities and regions. *Geogr. Res.* **2011**, *49*, 317–335. [CrossRef]

40. Dufty-Jones, R. A historical geography of housing crisis in Australia. *Aust. Geogr.* **2017**, *49*, 5–23. [CrossRef]

41. Millar, C.C.J.M.; Choi, C.J. Development and knowledge resources: A conceptual analysis. *J. Knowl. Manag.* **2010**, *14*, 759–776. [CrossRef]

42. Yigitcanlar, T.; Edvardsson, I.R.; Johannesson, H.; Kamruzzaman, M.; Ioppolo, G.; Pancholi, S. Knowledge-based development dynamics in less favoured regions: Insights from Australian and Icelandic university towns. *Eur. Plan. Stud.* **2017**, *25*, 2272–2292. [CrossRef]

43. Caton, B.; Harvey, N. *Coastal Management in Australia*; University of Adelaide Press: Adelaide, Australia, 2010.

44. Burnley, I.H.; Murphy, P. *Sea Change: Movement from Metropolitan to Arcadian Australia*; UNSW Press: Sydney, Australia, 2004.

45. Gurran, N.; Blakely, E. Suffer a Sea Change? Contrasting perspectives towards urban policy and migration in coastal Australia. *Aust. Geogr.* **2007**, *38*, 113–131. [CrossRef]

46. Salt, B. Super Connected Lifestyle Locations: The Rise of the "E-change" Movement. 2016. Available online: https://www.nbnco.com.au/corporate-information/media-centre/media-statements/e-change-is-the-new-sea-change-in-2016.html (accessed on 28 October 2020).

47. Glover, A.; Lewis, T. Fancy an e-Change? How People Are Escaping City Congestion and Living Costs by Working Remotely. 2019. Available online: https://theconversation.com/fancy-an-e-change-how-people-are-escaping-city-congestion-and-living-costs-by-working-remotely-123165 (accessed on 14 October 2020).

48. Jacques, O. Queensland Schools in Demand with Families Fleeing Coronavirus-hit States. 2020. Available online: https://www.abc.net.au/news/2020-10-08/qld-schools-in-demand-for-families-fed-up-with-covid-19-lockdown/12726284 (accessed on 14 October 2020).

49. Baum, S.; Connor, K.E.O.; Yigitcanlar, T. The implications of creative industries for regional outcomes. *Int. J. Foresight Innov. Policy* **2009**, *5*, 44. [CrossRef]

50. Flew, T.; Gibson, M.A.; Collis, C.; Felton, E. Creative suburbia: Cultural research and suburban geographies. *Int. J. Cult. Stud.* **2012**, *15*, 199–203. [CrossRef]

51. Gibson, C.; Brennan-Horley, C. Goodbye Pram City: Beyond Inner/Outer Zone Binaries in Creative City Research. *Urban Policy Res.* **2006**, *24*, 455–471. [CrossRef]

52. DBCDE. National Digital Economy Strategy. 2011. Available online: https://www.communications.gov.au/departmental-news/teleworkers-are-vital-australias-workforce-challenges (accessed on 13 July 2018).

53. Infrastructure Australia. A Report to the Council of Australian Governments. 2018. Available online: http://infrastructureaustralia.gov.au/policy-publications/publications/Report-to-COAG-2008.aspx (accessed on 11 October 2020).

54. Argent, N.; Tonts, M.; Jones, R.; Holmes, J. The Amenity Principle, Internal Migration, and Rural Development in Australia. *Ann. Assoc. Am. Geogr.* **2014**, *104*, 305–318. [CrossRef]

55. Alizadeh, T.; Irajifar, L. Gold Coast smart city strategy: Informed by local planning priorities and international smart city best practices. *Int. J. Knowl.-Based Dev.* **2018**, *9*, 153–173. [CrossRef]

56. Australian Bureau of Statistics. Australian National Accounts: Cultural and Creative Activity Satellite Accounts (No5271.0). 2014. Available online: http://www.abs.gov.au/ausstats/abs@.nsf/Latestproducts/5271.0Main%20Features32008-09 (accessed on 11 October 2020).

57. Hearn, G.; Bridgstock, R.; Bilton, C.; Cummings, S. The curious case of the embedded creative: Creative cultural occupations outside the creative industries. In *Handbook of Management and Creativity*; Edward Elgar Publishing: Cheltenham, UK, 2014; pp. 39–56.

58. Hearn, G.; Bridgstock, R.; Goldsmith, B.; Rodgers, J. (Eds.) *Creative Work Beyond the Creative Industries: Innovation, Employment and Education*; Edward Elgar: Cheltenham, UK, 2014.

59. Lobo, J.; Mellander, C.; Stolarick, K.; Strumsky, D. The Inventive, the Educated and the Creative: How Do They Affect Metropolitan Productivity? *Ind. Innov.* **2014**, *21*, 155–177. [CrossRef]

60. Bilandzic, A.; Foth, M.; Hearn, G. The role of Fab Labs and Living Labs for economic development of regional Australia. In *Regional Cultures, Economies, and Creativity: Innovating through Place in Australia and Beyond*; Routledge: Abingdon, UK, 2019; pp. 174–197. [CrossRef]

61. Flew, T. Creative suburbia: Rethinking urban cultural policy? the Australian case. *Int. J. Cult. Stud.* **2012**, *15*, 231–246. [CrossRef]

62. Gibson, C. *Creativity in Peripheral Places: Redefining the Creative Industries*; Routledge: London, UK, 2014.

63. McIntyre, P.; Kerrigan, S. Pursuing extreme romance: Change and continuity in the creative screen industries in the Hunter Valley. *Stud. Australas. Ciné.* **2014**, *8*, 133–149. [CrossRef]

64. Goldsmith, B. Embedded digital creatives. In *Creative Work Beyond the Creative Industries: Innovation, Employment and Education*; Hearn, G., Bridgstock, R., Goldsmith, B., Rodgers, J., Eds.; Edward Elgar: Cheltenham, UK, 2014; pp. 128–144.

65. Hearn, G. Creative occupations as knowledge practices: Innovation and precarity in the creative economy. *Cult. Sci. J.* **2014**, *7*, 83. [CrossRef]

66. Rodgers, J. Embedded creatives in the Australian manufacturing industry. In *Creative Work Beyond the Creative Industries: Innovation Education and Employment*; Hearn, G., Bridgstock, R., Goldsmith, B., Rodgers, J., Eds.; Edward Elgar: Cheltenham, UK, 2014; pp. 111–127.

67. Haines, T. Developing a startup and innovation ecosystem in regional Australia. *Technol. Innov. Manag. Rev.* **2016**, *6*, 24–32. [CrossRef]

68. CSIRO. Townsville, Australian Tropical Sciences and Innovation Precinct. 2020. Available online: https://www.csiro.au/en/Locations/Qld/Townsville-ATSIP (accessed on 14 October 2020).

69. Foth, M.; Adkins, B. A Research Design to Build Effective Partnerships between City Planners, Developers, Government and Urban Neighbourhood Communities. *J. Community Inform.* **2006**, *2*, 116–133. [CrossRef]

70. Yigitcanlar, T.; Adu-McVie, R.; Erol, I. How can contemporary innovation districts be classified? A systematic review of the literature. *Land Use Policy* **2020**, *95*, 104595. [CrossRef]

71. Yigitcanlar, T.; Kamruzzaman, M. Planning, Development and Management of Sustainable Cities: A Commentary from the Guest Editors. *Sustainability* **2015**, *7*, 14677–14688. [CrossRef]

72. Saberi, M.; Wu, H.; Amoh-Gyimah, R.; Smith, J.F.; Arunachalam, D. Measuring housing and transportation affordability: A case study of Melbourne, Australia. *J. Transp. Geogr.* **2017**, *65*, 134–146. [CrossRef]

73. Dufty-Jones, R.; Wray, F. Planning regional development in Australia: Questions of mobility and borders. *Aust. Plan.* **2013**, *50*, 109–116. [CrossRef]

74. Morrison, T.H.; Lane, M.B.; Hibbard, M. Planning, governance and rural futures in Australia and the USA: Revisiting the case for rural regional planning. *J. Environ. Plan. Manag.* **2014**, *58*, 1601–1616. [CrossRef]

75. Thompson, S.; Maginn, P. *Planning Australia: An Overview of Urban and Regional Planning*; Cambridge University Press: London, UK, 2012.

76. Van Staden, J.-W.; McKenzie, F.H. Comparing contemporary regional development in Western Australia with international trends. *Reg. Stud.* **2019**, *53*, 1470–1482. [CrossRef]

77. Tapsuwan, S.; Mathot, C.; Walker, I.; Barnett, G. Preferences for sustainable, liveable and resilient neighbourhoods and homes: A case of Canberra, Australia. *Sustain. Cities Soc.* **2018**, *37*, 133–145. [CrossRef]

78. Yigitcanlar, T.; Dur, F. Making space and place for knowledge communities: Lessons for Australian practice. *Australas. J. Reg. Stud.* **2013**, *19*, 36–63.

79. Pancholi, S.; Yigitcanlar, T.; Guaralda, M.; Mayere, S.; Caldwell, G.A.; Medland, R. University and innovation district symbiosis in the context of placemaking: Insights from Australian cities. *Land Use Policy* **2020**, *99*, 105109. [CrossRef]

80. Daniel, R.J. Creative industries in the tropics: Reflections on creativity and northeastern Australia as place. *Singap. J. Trop. Geogr.* **2015**, *36*, 215–230. [CrossRef]

81. Beer, A.; Clower, T. Specialisation and Growth: Evidence from Australia's Regional Cities. *Urban Stud.* **2009**, *46*, 369–389. [CrossRef]

82. Plummer, P.; Tonts, M.; Martinus, K. Endogenous growth, local competitiveness and regional development: Western Australia's regional cities, 2001–2011. *J. Econ. Soc. Policy* **2014**, *16*, 146.

83. Foth, M. Participatory urban informatics: Towards citizen-ability. *Smart Sustain. Built Environ.* **2018**, *7*, 4–19. [CrossRef]

84. Daniels, T.L.; Keller, J.W.; Lapping, M.; Daniels, K.; Segedy, J. *The Small Town Planning Handbook*; American Planning Association: Chicago, CA, USA, 2007.

85. Foth, M.; Brynskov, M. Participatory Action Research for Civic Engagement. In *Civic Media: Technology, Design, Practice*; MIT Press: Cambridge, MA, USA, 2016; pp. 563–580.

86. Gün, A.; Pak, B.; Demir, Y. Responding to the urban transformation challenges in Turkey: A participatory design model for Istanbul. *Int. J. Urban Sustain. Dev.* **2020**, 1–24. [CrossRef]

87. Badland, H.; Pearce, J. Liveable for whom? Prospects of urban liveability to address health inequities. *Soc. Sci. Med.* **2019**, *232*, 94–105. [CrossRef]

88. Gudes, O.; Kendall, E.; Yigitcanlar, T.; Pathak, V.; Baum, S. Rethinking Health Planning: A Framework for Organising Information to Underpin Collaborative Health Planning. *Heal. Inf. Manag. J.* **2010**, *39*, 18–29. [CrossRef] [PubMed]

89. Power, A. Social inequality, disadvantaged neighbourhoods and transport deprivation: An assessment of the historical influence of housing policies. *J. Transp. Geogr.* **2012**, *21*, 39–48. [CrossRef]

90. Brack, C. Pollution mitigation and carbon sequestration by an urban forest. *Environ. Pollut.* **2002**, *116*, S195–S200. [CrossRef]

91. Bell, S.; Keating, M. *Fair Share: Competing Claims and Australia's Economic Future*; Melbourne University Publishing: Melbourne, Australia, 2018.

92. Stilwell, F.; Jordan, K. *Who Gets What? Analysing Economic Inequality in Australia*; Cambridge University Press: London, UK, 2007.

Sustainability **2020**, *12*, 10492

93. Badland, H.; Whitzman, C.; Lowe, M.; Davern, M.; Aye, L.; Butterworth, I.; Hes, D.; Giles-Corti, B. Urban liveability: Emerging lessons from Australia for exploring the potential for indicators to measure the social determinants of health. *Soc. Sci. Med.* **2014**, *111*, 64–73. [CrossRef]

94. Eversole, R.; Walo, M.T. Leading and following in Australian regional development: Why governance matters. *Reg. Sci. Policy Pr.* **2020**, *12*, 291–302. [CrossRef]

95. Harrison, J.; Galland, D.; Tewdwr-Jones, M. Regional planning is dead: Long live planning regional futures. *Reg. Stud.* **2020**, 1–13. [CrossRef]

96. Batty, M. The Coronavirus crisis: What will the post-pandemic city look like? *Environ. Plan. B Urban Anal. City Sci.* **2020**, *47*, 547–552. [CrossRef]

97. Kareem, B. Do Pandemics Disrupt or Seed Transformations in Cities? A Systematic Review of Evidence. *SSRN Electron. J.* **2020**. [CrossRef]

98. Bauman, I. How the Twin Disasters of Climate Change and Covid-19 Could Transform Our Cities. 2020. Available online: http://www.theguardian.com/commentisfree/2020/aug/18/climate-change-covid-19 (accessed on 20 October 2020).

99. Blay-Palmer, A.; Carey, R.; Valette, E.; Sanderson, M.R. Post COVID 19 and food pathways to sustainable transformation. *Agric. Hum. Values* **2020**, *37*, 517–519. [CrossRef] [PubMed]

100. Loker, A.; Francis, C. Urban food sovereignty: Urgent need for agroecology and systems thinking in a post-COVID-19 future. *Agroecol. Sustain. Food Syst.* **2020**, *44*, 1118–1123. [CrossRef]

101. Yigitcanlar, T.; Foth, M.; Kamruzzaman, M. towards Post-Anthropocentric Cities: Reconceptualizing Smart Cities to Evade Urban Ecocide. *J. Urban Technol.* **2019**, *26*, 147–152. [CrossRef]

102. Clarke, R.; Heitlinger, S.; Light, A.; Forlano, L.; Foth, M.; DiSalvo, C. More-than-human participation: Design for sustainable smart city futures. *Interactions* **2019**, *26*, 60–63. [CrossRef]

103. Gibson-Graham, J.K.; Hill, A.; Law, L. Re-embedding economies in ecologies: Resilience building in more than human communities. *Build. Res. Inf.* **2016**, *44*, 703–716. [CrossRef]

104. Loh, S.; Foth, M.; Caldwell, G.A.; Garcia-Hansen, V.; Thomson, M. A more-than-human perspective on understanding the performance of the built environment. *Arch. Sci. Rev.* **2020**, *63*, 372–383. [CrossRef]

105. Liaros, S. Implementing a new human settlement theory: Strategic planning for a network of regenerative villages. *Smart Sustain. Built Environ.* **2019**, *9*, 258–271. [CrossRef]

MDPI

St. Alban-Anlage 66

4052 Basel

Switzerland

Tel. +41 61 683 77 34

Fax +41 61 302 89 18

www.mdpi.com

Sustainability Editorial Office

E-mail: sustainability@mdpi.com

www.mdpi.com/journal/sustainability